Infusing the Teaching of Critical and Creative Thinking into Secondary Science

A Lesson Design Handbook

Robert J. Swartz, Stephen David Fischer & Sandra Parks

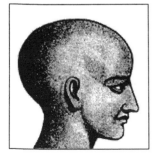

© 1998

CRITICAL THINKING BOOKS & SOFTWARE

www.criticalthinking.com

P.O. Box 448 • Pacific Grove • CA 93950-0448

Phone 800-458-4849 • FAX 408-393-3277

ISBN 0-89455-703-3

Printed in the United States of America

These handbooks are dedicated to all of the teachers with whom we have worked, whose commitment to helping their students achieve their full potential as clear and careful thinkers has stimulated many of the ideas we have included in these handbooks.

Robert Swartz received his B.A. and Ph.D. in philosophy from Harvard and studied as part of a Fulbright Scholarship at universities in Oxford and Cambridge in England. He is a faculty member at the University of Massachusetts at Boston. He is director of the National Center for Teaching Thinking. Through the Center, he provides staff development to educators across the country on restructuring curriculum by infusing critical and creative thinking into content area instruction. He has authored numerous articles and books on critical thinking and has acted as a testing consultant for the National Assessment of Educational Progress.

Stephen Fischer received his B.A. in Biology at the University of California at San Diego and his M.S.T. in teaching science from Rowan University in Glassboro, New Jersey. He presently teaches Biology and Physics at Atlantic City High School in Atlantic City, New Jersey, and also is an instructor in bioethics and anatomy and physiology at Richard Stockton College of New Jersey. Mr. Fischer is a staff-development consultant through the National Center for Teaching Thinking, working with secondary schools on infusing critical and creative thinking into science instruction.

Sandra Parks received an M.A. in Education, specializing in Curriculum Development, from the University of South Florida and pursued doctoral studies at Indiana State University. She also received a certificate of advanced study from Harvard University. She has conducted National Curriculum Studies Institute workshops annually for the Association for Supervision and Curriculum Development and provides staff development to schools across the country. Sandra served as co-director of the National Center for Teaching Thinking.

PREFACE

This book explores what can be accomplished when we combine effective classroom techniques for teaching students to become good thinkers with effective strategies to engage students in thoughtful learning of the regular secondary school science curriculum. The technique of lesson design and instruction that results is called *infusing critical and creative thinking into content instruction*. The infusion lesson design framework and the tools that we introduce in this handbook to facilitate designing and teaching infusion lessons are powerful devices to accomplish the basic objectives of education.

This lesson design handbook follows an earlier handbook, published in 1994, on elementary school instruction, and is one of a cluster of secondary school handbooks that will eventually include secondary history and social studies and secondary English and language arts.

Infusion has not only grown out of our own ideas and experience, but also builds on previous work in the field of teaching thinking. We expand on ideas developed by Art Costa and Ron Brandt in their important distinction between teaching for thinking and the teaching of thinking. We draw on the insights of Costa, Robert Sternberg, and others on the importance of metacognition in instruction, and on those of David Perkins, Gavi Solomon, and others on ways of facilitating the transfer of thinking skills into new contexts.

The deepest roots for the concepts we employ in the framework for infusion lessons lie in two previous collaborations: (1) the book *Teaching Thinking: Issues and Approaches* and the article "The Nine Basics of Teaching Thinking," by Robert Swartz and David Perkins and (2) the two lesson-book series *Building Thinking Skills* and *Organizing Thinking* by Sandra Parks and Howard Black. A number of the illustrative graphics we use in this handbook are modifications of diagrams used in *Teaching Thinking: Issues and Approaches*, and the articulation of the "ladder" of metacognition first occurred there. The idea of thinking defaults was first devel-

oped in "The Nine Basics of Teaching Thinking." Similarly, some of the graphic organizers used in infusion lessons are modifications of graphic organizers appearing in the *Organizing Thinking* series.

As important as these sources have been for the work we have done on infusion, the most significant influence on the ideas in this book has occurred in secondary school classrooms, especially those of Stephen Fischer at Atlantic City High School, where many of the lessons were taught. To assure effectiveness, the model lessons included in this handbook were assessed and revised based on their implementation with such secondary school students. We thank the multitude of additional classroom teachers with whom we have worked for their involvement in this process. This includes teachers from a variety of other schools who worked with Robert Swartz on science lessons originally published in *Biology: The Study of Life*, and *Chemistry: The Study of Matter*, both published by Prentice Hall, Inc. These lessons were the prototypes for some of the science lessons in this handbook.

In particular, we wish to give special thanks to Joseph Carroll, of Atlantic City High School, for developing the lesson "Sound Waves," and Mary Ann Brearton and Sarah C. Duff of the Baltimore City Schools and William W. Duff of the MITRE Corporation, whose lessons and lesson ideas, originally published in *Chemistry: The Study of Matter* by Prentice-Hall, Inc., were the basis for two of the lessons that now appear in this handbook, though much revised ("Sulfuric Acid" and "The Best Way to See It").

We are also indebted to the keen insights of the editorial staff at Critical Thinking Books and Software, Inc., whose understanding of thinking instruction and the practicalities of classroom procedures helped us express our ideas clearly.

This secondary school science handbook draws on the framework for teaching thinking and the techniques of infusion included in the earlier volume, *Infusing the Teaching of Critical and Creative Thinking into Content Instruction: A Lesson*

Design Handbook for the Elementary Grades, co-authored by Robert Swartz and Sandra Parks. Indeed, Chapter 1, the commentaries on the different types of thinking that begin each chapter, and the concluding chapters of this new book are based on the text in this earlier volume, suitably revised to focus on teaching science in middle and high school. Some of the science lessons that appear in that earlier book—those that were taught at the 6th grade level (e.g., "The Bottom of the Ocean" and "The Kestrel")—have also been included in this volume, while upgraded versions of others from the elementary school handbook, as these upgraded versions have been taught in middle and/or high school, appear in this volume as well (e.g., "The Extinction of the Dinosaurs" and "Alternative Energy Sources"). The remainder of the lessons—about fifteen—and the ideas for other lessons on the same thinking skills that are included at the end of each chapter are new to this volume. The revisions of the material from the elementary school handbook, as well as all of these new lessons and lesson ideas, were composed by Robert Swartz and Stephen Fischer. (The infusion lessons new to this handbook were written to reflect their use in the actual classroom settings in which they have been taught.) Sandra Parks, whose name also appears on the title page of this book, is listed as one of the authors of this book because of the valuable contributions she made as co-author of the material from the elementary grades design handbook included in this volume.

Infusion, as we describe it in this book, combines techniques that many teachers who will use this book have been employing in their classrooms for many years. If you already guide your students' thinking by prompting questions, you will not be surprised to find that asking a series of prompting questions is a key technique in guiding organized thinking in infusion. If you regularly ask your students to plan the way they think through problems and issues, you will not be surprised to find that this is a key metacognitive technique that dramatically enhances learning in infusion lessons. If you already teach students strategies for thinking in a direct and explicit way, you will not be surprised to find that this is also a key technique in infusion lessons.

However, these individual practices do not in themselves constitute infusion. The infusion framework builds on what many teachers, to their credit, already do in their classrooms to enhance thinking. What is different about infusion is the way that these and other techniques are combined to complement each other to produce a complete and effective framework for lesson design and instruction. Any teacher at any grade level and in any content area can create infusion lessons to enhance his or her instruction using this framework.

While we are indebted to many of our colleagues who have done significant research and who have developed ways of clarifying the domain of thinking and instruction in thinking, we wish to dedicate this book to the many reflective teachers with whom we have worked over the past fifteen years and who are dedicated to helping their students realize their full potential as thoughtful learners and good thinkers. What you find in these pages may come to you from both of us, but their voices are represented in it as well.

Robert Swartz & Stephen Fischer,
Newton Center, Massachusetts, and Atlantic City, New Jersey
August, 1997

TABLE OF CONTENTS

HOW TO USE THIS HANDBOOK

FAMILIARIZE YOURSELF WITH THE DESIGN OF INFUSION LESSONS

Acquaint yourself with the structure of infusion lessons by reading Chapter 1: What Is Infusion?

SELECT A THINKING SKILL THAT YOU WOULD LIKE TO TEACH.

If you learn better by seeing examples, read the four sections below beginning here.

If you learn better by understanding the whole structure, read the four sections below beginning here.

EXAMINE A MODEL LESSON

Review a model lesson based on the thinking skill you chose. Relate the thinking map and student responses to the questions in the lesson and to the graphic organizer. Teachers' statements are shown in bold type. Students' answers are italicized, and directions for teachers are in plain type.

CLARIFY WHAT MAKES THE THINKING SKILLFUL

Clarify the thinking skill that you chose by reading the explanation which introduces the chapter on that skill. Examine the thinking map and the graphic organizer.

CLARIFY METACOGNITION

Read Chapter 19: Metacognition. Familiarize yourself with the types of questions that can prompt students to think about their thinking.

CLARIFY METHODS FOR TEACHING INFUSION LESSONS

Read Chapter 18: Instructional Methods. Clarify the types of methods that are used in all infusion lessons and those that are helpful in teaching the content.

AFTER YOU HAVE READ THE SECTIONS ABOVE, BEGIN PREPARING YOUR OWN INFUSION LESSON!

FIND AN INFUSION LESSON CONTEXT IN THE CURRICULUM YOU TEACH

Examine the lesson menus that follow the model lessons. They offer lesson ideas; also reading them may suggest other ideas. Read Chapter 20: Selecting Contexts for alternative methods of selecting infusion lesson contexts. Check textbooks or curriculum guides for good contexts. Identify a context in which you could teach the thinking skill or process.

DESIGN YOUR INFUSION LESSON

Use the lesson plan form (pp. 26–28) to jot down what you will ask students in each part of the lesson. Be brief. The detailed comments in the model lesson plans are intended to help readers; your plan may not be as lengthy. Fill in the graphic organizer to predict students' responses. If possible, work with a colleague.

TEACH YOUR LESSON

Try out your lesson. Save any transparencies or chart paper for review and possible revision. Plan how you will reinforce this skill with other activities as the school year progresses.

WORK ON ANOTHER THINKING SKILL

Select another thinking skill for implementation and repeat the same process of lesson design.

PART 1

THE DESIGN OF INFUSION LESSONS

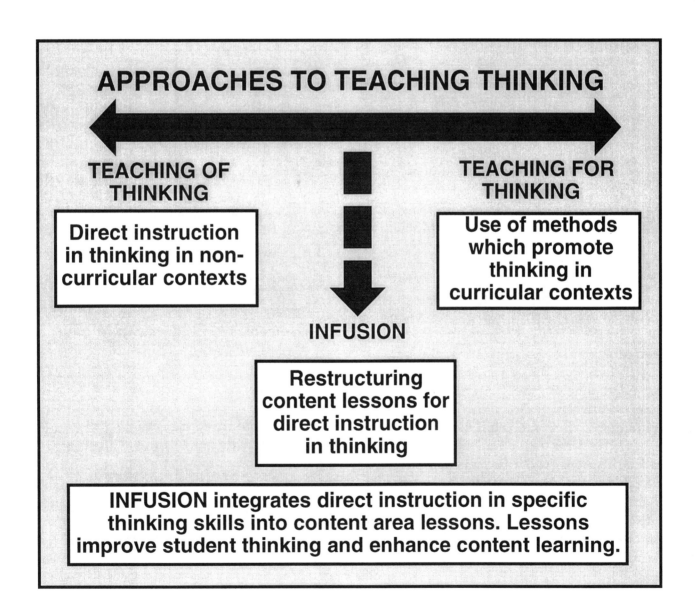

APPROACHES TO TEACHING THINKING

TEACHING OF THINKING

Direct instruction in thinking in non-curricular contexts

TEACHING FOR THINKING

Use of methods which promote thinking in curricular contexts

INFUSION

Restructuring content lessons for direct instruction in thinking

INFUSION integrates direct instruction in specific thinking skills into content area lessons. Lessons improve student thinking and enhance content learning.

CHAPTER 1
WHAT IS INFUSION?

Helping Our Students Become Better Thinkers

Improving the quality of student thinking is an explicit priority of current educational reform efforts. Recommendations from groups ranging from education commissions to the nation's governors support this priority and affirm that good thinking is essential in meeting the challenge of living in a technologically oriented, multicultural world.

Although these recommendations have been advanced primarily because of the projected demands of the work force in the 21st century, they also reflect an awareness that knowledgeable thinkers have a better chance of taking charge of their lives and achieving personal advancement and fulfillment. Our students must be prepared to exercise critical judgment and creative thinking to gather, evaluate, and use information for effective problem solving and decision making in their jobs, in their professions, and in their lives.

Making good thinking an educational goal affirms that growth in thinking is obtainable by *all students*. This goal also reflects confidence that *all teachers* can help students to become better thinkers whatever the learning level, socioeconomic background, and culture of the students.

Although textbooks and tests are changing to reflect this aim, it is the classroom teacher who, through day-to-day instruction, must assume the main responsibility for helping our students become better thinkers. The effort that is required to meet this goal must, therefore, be directed at effective classroom implementation. This handbook presents a teacher-oriented approach to improving student thinking that blends sound theory and effective classroom practice and can be used by every teacher.

What does it mean to emphasize good thinking as a major educational goal? Students already use a variety of types of thinking in their personal lives. They compare and contrast when choosing friends. They predict that they will soon eat when they stand in line in the cafeteria. They make numerous decisions in and out of school every day. They do not have to be taught to do thinking.

Performing such thinking tasks, however, does not necessarily mean performing them *skillfully*. For example, sometimes a person feels inclined to do something and may not think much about it before doing it. A person may purchase an automobile just because he likes the way it looks. Such hasty and ill-considered decisions may lead to disappointing and unexpected surprises, such as high repair bills. In contrast, if we think about many options, search for new alternatives, think about the significant factors in making the choice, consider the consequences of our actions, and plan how to carry out our choice, our decision may be a more effective one. It is *ordinary thinking done well* that is our goal when we "teach thinking."

How can we teach students to improve the quality of their thinking? The thinking skills movement of the '80s produced special programs and emphasized instructional methods to foster thinking. Three principles emerged from these efforts:

- The more explicit the teaching of thinking, the greater impact it will have on students.
- The more classroom instruction incorporates an atmosphere of thoughtfulness, the more open students will be to valuing good thinking.
- The more the teaching of thinking is integrated into content instruction, the more students will think about what they are learning.

These principles provide the basic rationale for infusing critical and creative thinking into content instruction.

Infusion is a natural way to structure lessons. The curriculum is not a collection of isolated bits of information; rather, it is the material that informed, literate people use to make judgments. We expect that information about nutrition should influence students' dietary habits. We expect that an understanding of American political history should affect how citizens vote. We expect that a deep understanding of a character's motivation and actions in a story should inform a discerning reader about his or her conduct and responsibility.

It is, therefore, essential that we teach students how to use information and concepts that they learn in school to make decisions and solve problems effectively. Infusion, as an approach to teaching thinking, is based on the natural fusion of information that is taught in the content areas with forms of skillful thinking that we should use every day to live productively.

Improving Student Thinking in Content Area Instruction

This handbook spells out how we can perform ordinary thinking activities skillfully. Key questions that effective thinkers raise and answer when making sound judgments are organized into thinking plans that can be used to guide good thinking.

The curriculum contains a multitude of natural contexts to teach skillful thinking. Utilizing such contexts, any teacher can design well-crafted infusion lessons that dramatically enhance student content learning.

Kevin O'Reilly, a teacher from the Hamilton-Wenham School District in Massachusetts, introduces a lesson on determining the reliability of sources of information in history by staging a scuffle in the corridor outside his classroom. He then asks student witnesses to describe what happened. He draws an analogy between his students' differing accounts and the variety of accounts regarding who fired the first shot at the Battle of Lexington, the first battle of the Revolutionary War, in 1775. As O'Reilly's students attempt to determine which of the eyewitnesses gave accurate accounts, they reflect on why some historical accounts may be more reliable than others. This reflection arms them with critical thinking skills that they draw on again and again in O'Reilly's classroom. These skills relate to assessing the reliability and accuracy of eyewitnesses, of observation, or of other sources of information—skills of great importance in our lives outside of the classroom.

In the immediate context of studying the Revolutionary War, O'Reilly's students use the skills of assessing the reliability of sources to examine the context of the battle and the biases that people might have had in describing it. They then make informed critical judgments about the accuracy of various textbook accounts of the Lexington incident. Students who are simply directed to read to "get the facts" typically do not make such judgments about material in their texts. O'Reilly's students gain a deep critical perspective on the role of firsthand reports in constructing a history and learn that histories can be written from different points of view.

Infusion lessons are similarly effective in the primary grades. Cathy Skowron, a first grade teacher at the Provincetown (Massachusetts) Elementary School, uses the same technique to teach the tale of Henny Penny. Many first grade teachers use this story to help students develop listening skills and vocabulary. While fulfilling these language-arts goals, Skowron also uses the story to teach students to think skillfully about the reliability of sources of information. Prompted by her questions, students discuss whether the other animals should have accepted what Henny Penny told them. How can they determine whether Henny Penny is a reliable source of information?

Skowron restructures her lesson by including questions that students might ask about *any* source of information. Raising questions about Henny Penny as a source of reliable information helps them understand the story at a deeper level. They then grasp the "moral" of the story: *hasty, unquestioning thinking can be dangerous.*

Skowron's lesson differs from O'Reilly's in the sophistication of the content, the level of vocabulary, and students' background knowledge. However, both groups of students consider factors that are often overlooked in thinking about sources. They develop strategies for asking and researching the relevant questions about reliability. Between Skowron's first grade and O'Reilly's ninth grade, Skowron's students have plenty of time to develop more and more sophisticated standards for the reliability of sources. When they get to O'Reilly's classroom, in fact, they may already have considerable sophistication in judging the reliability of both primary and secondary sources, and in applying these skills to a variety of content areas across the curriculum. O'Reilly's job will be to build on and reinforce this prior learning. This will take less class time than would be needed if he were introducing students to this skill for the first time in their educational careers, and may allow for much more sophistication of his approach to the study of American History than is usually possible with ninth-grade students.

If students have not been exposed to this kind of instruction before the secondary grades, however, it is incumbent upon secondary school teachers to introduce students to these skills as soon as possible. Skillful evaluation of sources can be taught, reinforced, and elaborated in many contexts, subjects, and secondary grade levels besides O'Reilly's ninth-grade classroom. When students are asked to do library research and then write on a topic, for example, teachers can ask them to compare a variety of books and articles on the topic in the library, and then to develop a list of questions they would need to answer to decide which sources are likely to give them accurate information on that topic. The students would consider relevant factors, such as the date of the publication, the expertise of the authors, whether the account is primary or secondary, whether the account is fictional, where the author got his or her information, etc. These questions can then be put in a more organized way and written down so that they serve as a guide to the students' thinking.

Based on the information they gather to answer their questions, students then make critical judgments about which of the book(s) being considered are likely to provide the most accurate information. Usually, when students make such judgments, their interest in the topic is enhanced, and better research projects result. Equally important, however, is that, as in O'Reilly's classroom, such students develop a strategy that they can use again and again to make informed judgments about the reliability of other sources of information.

The same content material can be used to teach other critical thinking skills. For example, Skowron introduces causal reasoning by prompting her students to think about whether Henny Penny had good evidence that the sky was falling. Could something other than falling

sky have caused the bump on Henny Penny's head? How could we find out? In general, what do we ask in order to find out what caused something to happen?

Causal reasoning, a fundamental skill of inference, involves considering a cluster of questions different from those involved in thinking about reliable sources. These questions prompt consideration of which possible causes are reasonable in light of the evidence. Asking and carefully answering these questions contrasts with hasty and ill-informed judgments that people often make about what caused something.

Skowron's students engage in causal reasoning by thinking about what evidence they would need to tell what really hit Henny Penny. The students then look at the pictures in the book for clues to determine what the cause might be. They contrast *careful* thinking about causes with Henny Penny's quick conclusion that the sky is falling, identifying her thinking as hasty thinking. They use the term "Henny Penny thinking" to describe someone who jumps to a conclusion about causes. This reminds them not to do the same thing but, instead, to look for evidence. Helping students think skillfully about causes in the primary grades can make this kind of thinking second nature as they progress through upper elementary and secondary school.

Secondary school teachers will find ample opportunities to introduce causal reasoning in what they teach. The causes of the Civil War, the events leading up to the stock market crash in 1929, the extinction of the dinosaurs, the ability of jet aircraft to take off and fly, and the poor (or good) performance of the school's football team are topics that can all generate lessons in which students not only gain a deep understanding of these events, but also learn strategies (that they can use repeatedly) for making well-informed judgments about causes. The secondary curriculum is replete with such contexts.

Causal explanation lessons can also be used in secondary school to introduce some of the more challenging themes that students study. For example, causal reasoning also clarifies human motivation and action. Questions like "What was Huckleberry Finn's motivation in not turning in Jim?" can generate lessons designed to help students learn how to use causal reasoning to answer them. Building on such activities, causal reasoning also helps us to determine the responsibility for things people do. Cathy Peabody, a high school English teacher in Groton, Massachusetts, asks her students causal questions (similar to the type asked by Cathy Skowron) as they study *Romeo and Juliet.* She recognizes in this play that chance, emotion, misunderstanding, and deliberate intent weave a tragic causal web that raises important questions about responsibility.

Specifically, Peabody helps her students spell out the causal chain that led to the deaths of Romeo and Juliet by helping students to identify possible causes of the tragedy and then to select the best explanations based on evidence in the text. Recognizing that various people played a role in this causal chain, Peabody poses the question, "Who, if anyone, is responsible for the deaths of Romeo and Juliet? The feuding parents? The Prince? Friar Lawrence? The lovers themselves? On what basis do we hold people responsible for things that happen?"

Through a detailed examination of the play, informed by their conclusions about the causes of the tragedy, Peabody and her students raise, and try to answer, such deep questions about responsibility. They develop an explicit set of standards that enables them to make a well-supported judgment about who should be held responsible. Some students, for example, think that Friar Lawrence should be held responsible. They then stage a "trial" of Friar Lawrence to determine whether he should really be held responsible for this tragedy. When they reach their conclusion, Peabody helps them extract the standards they used to judge the friar's responsibility. There is no substitute for careful thinking to answer questions like these.

When such activities are completed, Peabody usually helps her students see analogies between issues in the play and issues in their own experience in which questions of blame and responsibility have arisen. Peabody helps her students test their ideas about responsibility by applying them in these personal cases. Her intention is to expose them to this kind of thinking and to help them to transfer and use it reflectively in a variety of appropriate contexts.

Thinking carefully about causes is crucial in almost every profession. Effective work in science, engineering, accounting, journalism, nursing, and law enforcement, for example, involves the need for well-founded judgments about causes. This kind of thinking is also crucially important in our daily lives. We make judgments about causes in getting to work on time, preventing or treating illness, preparing a tasty meal, and minimizing stress in our lives. Helping students transfer the use of skillful causal explanation to these contexts enriches any infusion lesson on causal explanation.

These examples demonstrate how the infusion of key critical thinking skills into content learning adds richness and depth to the curriculum. These examples are representative of a multitude of other lessons that are designed to help students develop a wide range of additional thinking skills and processes. This handbook provides the basic tools for such lessons.

Thinking Skills and Processes Featured in this Handbook

The types of skillful thinking we discuss in this handbook form a core of important thinking skills that cut across the various content areas. They fall into the three main categories: skills at generating ideas, skills at clarifying ideas, and skills at assessing the reasonableness of ideas. Generative skills are creative thinking skills: they stretch our thinking and develop our imaginations. Skills of clarification involve analysis: they enhance our understanding and the ability to use information. Skills at assessing the reasonableness of ideas are critical thinking skills: they lead to good judgment. Both examples discussed so far fall into the category of skills at assessing the reasonableness of ideas.

When we engage in natural thinking tasks, these skills of good thinking are rarely used in isolation. Many thinking tasks that we face in our lives or professional work involve decision making and/or problem solving. Thinking skills from each of the three categories blend together for thoughtful decision making and problem solving. We should try to generate original solutions to problems; we should base our decisions on relevant information; and we should assess the reasonableness of each option to select the best one.

These broader thinking processes are also discussed in this handbook. The strategies we present for skillful decision making and problem solving provide the link between the more circumscribed thinking skills that appear in each of the three categories and the authentic thinking tasks students must engage in both in and out of school.

The outline in figure 1.1 shows the thinking skills and processes featured in this book.

In figure 1.2 (page 7), these thinking skills and processes are shown within the more comprehensive context of the thinking domain.

Figure 1.3 (page 8) shows how various thinking skills from each of these categories are combined in decision making.

Teaching the thinking skills of clarification, creative thinking, and critical thinking without helping students learn how to use them in decision making and problem solving accomplishes only part of the task of teaching thinking. Teaching strategies for problem solving and decision making, without teaching students the skills needed to use these strategies effectively, is similarly limited. If we teach lessons on individual thinking skills *and* lessons on decision making and problem solving, we can show how these thinking skills are connected with good decision making and problem solving. Students will then have the thinking tools they need to face their most challenging tasks in using information and ideas.

The Structure of Infusion Lessons

Infusing critical and creative thinking into content instruction blends features of two contrasting instructional approaches to teaching thinking that educators have taken: (1) direct instruction of thinking in noncurricular contexts and (2) the use of methods that promote thinking in content lessons.

Infusion lessons are similar to, but contrast with, both of these types of instruction. The diagram in figure 1.4 (page 8) represents this triad.

The *teaching of thinking* by direct instruction means that, in a time period designated for thinking instruction, students learn how to use explicit thinking strategies, commonly guided by the teacher. Such lessons employ the language of the thinking task and procedures for doing it skillfully. Usually the *teaching of thinking* occurs in separate, self-contained courses or programs with specially designed materials and is taught outside the standard curriculum. For example, students are guided in using the terms and procedures of classification to classify but-

Thinking Skills and Processes Featured in Infusion Lessons in This Handbook

THINKING SKILLS

I. **Skills at Generating Ideas**
    1. **Generating Possibilities**
        *   Multiplicity of Ideas
        *   Varied Ideas
        *   New Ideas
        *   Detailed Ideas
    2. **Creating Metaphors**
        A. Analogy/Metaphor

II. **Skills at Clarifying Ideas**
    1. **Analyzing Ideas**
        A. Compare/Contrast
        B. Classification/Definition
        C. Parts/Whole
        D. Sequencing
    2. **Analyzing Arguments**
        A. Finding Reasons/Conclusions
        B. Uncovering Assumptions

III. **Skills at Assessing the Reasonableness of Ideas**
    1. **Assessing Basic Information**
        A. Accuracy of Observation
        B. Reliability of Sources
    2. **Inference**
        A. Use of Evidence
            a. Causal Explanation
            b. Prediction
            c. Generalization
            d. Reasoning by Analogy
        B. Deduction
            a. Conditional Reasoning (if ... then)

THINKING PROCESSES

I. **Goal Oriented Processes**

    A. Decision Makin
    B. Problem Solving

Figure 1.1

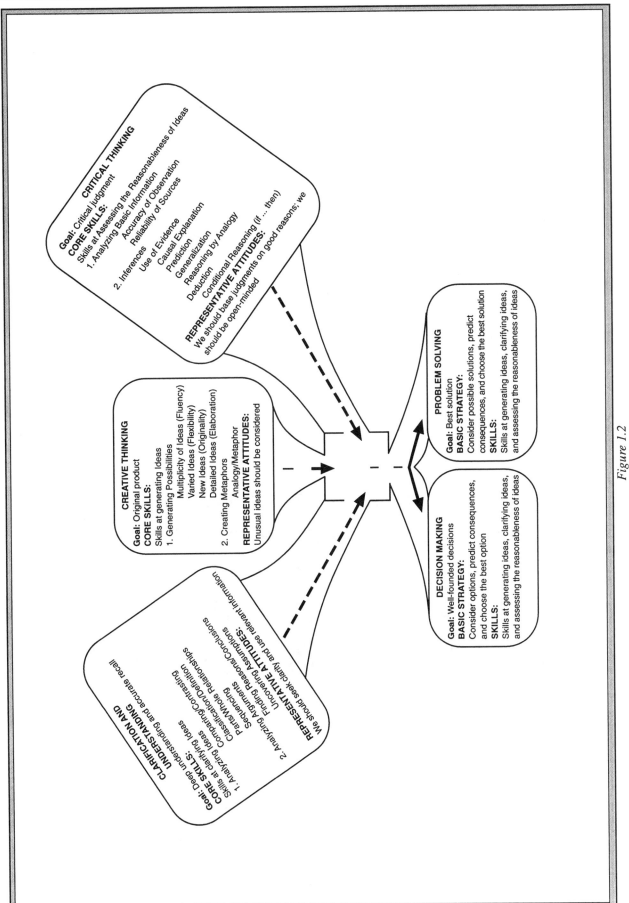

Figure 1.2

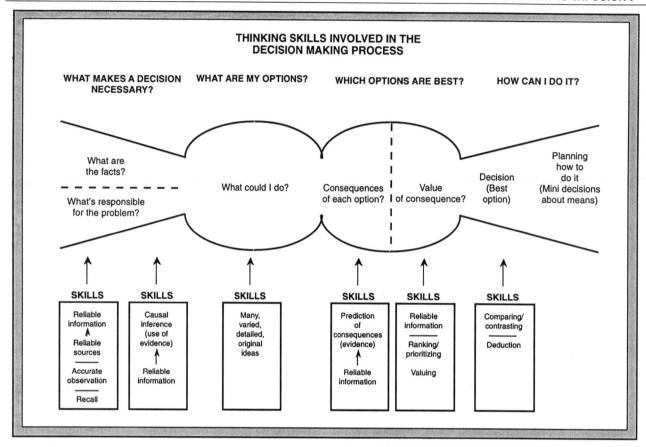

Figure 1.3

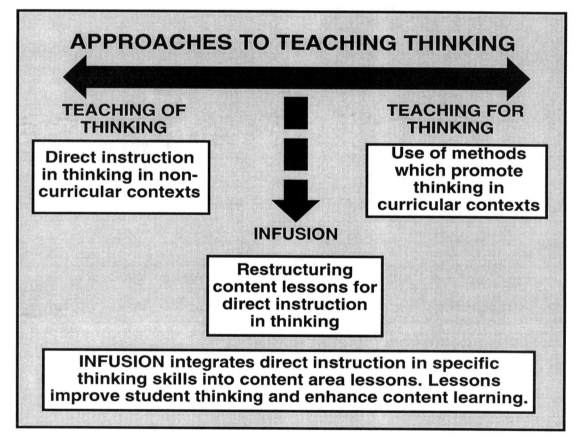

Figure 1.4

tons, to demonstrate and practice the thinking skill, or they are asked to assess arguments from text books on critical thinking, to practice skills in logic. Since the skills are taught using examples that are not curriculum-related, they must then be bridged into the curriculum if students are to apply them to content learning.

In contrast to this approach, infusion lessons are not taught in separate courses or programs outside the regular curriculum. They do, however, employ direct instruction in the thinking skills and processes that they are designed to improve. In infusion lessons, direct instruction in thinking is blended into content lessons.

Teaching for thinking involves employing methods to promote students' deep understanding of the content. Such methods include using cooperative learning, graphic organizers, higher order questioning, Socratic dialog, manipulatives, and inquiry learning. While students may respond thoughtfully to the content, no thinking strategy is taught explicitly. In contrast, although infusion lessons also feature such methods, infusion lessons are characterized by direct instruction in thinking skills and processes.

Educators often confuse infusion with using methods that promote thinking. For example, many teachers employ "higher order questioning" or "Socratic dialogue" to stimulate more thinking about the content than asking standard recall-oriented questions. They typically ask "Why," "What if," and "How" questions. For example, a question like "Why did the plague spread so rapidly in medieval Europe?" is a challenging question and unlike the question "What were the dates of the plague in medieval Europe?" provides an opportunity for higher order thinking.

This kind of questioning, however, remains content-oriented. Its goal is usually to yield a deeper understanding of what is being taught. When students respond by mentioning factors like lack of sanitation or lack of medical knowledge, teachers usually ask students to elaborate so that the class can develop a rich understanding of conditions that could cause such an epidemic. The product (student answers), rather than the process (student thinking), is the focus in these lessons.

Typically, when using such methods as higher order questioning, teachers spend little or no classroom time discussing the thinking students engage in when they respond to such questions. *How* students arrive at their responses remains implicit. Some students may respond thoughtfully; others may respond hastily and unsystematically. Some students may not respond at all. In order to yield more thoughtful responses from more students, teachers must take time to clarify the skillful thinking needed to develop thoughtful responses to the questions asked.

Infusion lessons are crafted to bring into content instruction an explicit emphasis on skillful thinking so that students can improve the way they think. Classroom time is spent on the thinking skill or process, as well as on the content. Infusion lessons feature a variety of effective teaching practices that characterize the way thinking is explicitly emphasized in these lessons:

- The teacher introduces students to the thinking skill or process by demonstrating the importance of doing such thinking *well*.

- The teacher uses explicit prompts to guide students through the skillful practice of the thinking as they learn concepts, facts, and skills in the content areas.

- The teacher asks reflective questions that help students distance themselves from what they are thinking about, so they can become aware of how they are thinking and develop a plan for doing it skillfully.

- The teacher reinforces the thinking strategies by providing additional opportunities for students to engage in the same kind of thinking independently.

Conducting a lesson using this four-step strategy to teach thinking is time well spent and will maximize our chances for real improvement in student thinking.

To summarize: Although some people use the word "infusion" only to describe the techniques used to promote higher-order thinking about the content material, what we have been calling "infusion" lessons are also crafted to bring into content instruction *an explicit emphasis on skillful thinking*, together with the use of such thought-provoking methods, so that such lessons maximize the impact we have on helping all students improve how they think.

Classroom time is spent on the thinking skill or process, as well as on the content. Infusion lessons, therefore, have two sets of objectives: learning to do a type of thinking skillfully and achieving a deep understanding of the content being taught.

We will now analyze an infusion lesson to show how the four basic components designed to meet these two objectives can weave together into a well-crafted infusion lesson.

A Secondary School Science Lesson That Infuses Critical Thinking

The following lesson serves as a model for designing infusion lessons to teach any thinking skill or process featured in this handbook. It is a lesson in high school biology but has all the elements of infusion lessons in any content area. The lesson uses Rachel Carson's description of the impact of the overuse of pesticides on an American town from *Silent Spring* as a context for teaching skillful causal explanation.

The thinking skill objective of this lesson is to teach students to avoid making hasty and unfounded judgments about the causes of events. Rather, they are first to consider possible causes that they accept or reject as likely based on the evidence they determine to be relevant to the issue. This is a key critical thinking skill in the sciences and in many other fields of study.

The content objectives of this lesson include
1) understanding the background and historical significance of books like *Silent Spring;*
2) understanding the effects of pesticides such as DDT on animals and people. As you review this lesson, notice how the causal explanation activity leads students naturally to gather information that provides a deeper understanding of chemical pollution rather than to read passages from their textbook about such pollution.

The lesson introduction. The lesson introduction should give students a sense of the importance of the thinking process, explain what can be done to engage in this process skillfully, and link the thinking process to the content they are studying. The teacher starts this way:

• When things happen that we don't understand or don't like, we often try to find out what is causing them. For example, when my car doesn't start on cold mornings I want to find out why so that I can do something to prevent this from happening again. Or when my tomato plants produce really large tomatoes I want to find out if anything I did caused this (like how long I waited to plant the seedlings in my gardening, or my watering cycle) so that I can do it again. Can you think of times when you wanted to find out what caused something? Write down one or two in your notebooks.

• Finding a cause is often a simple process; turn on a light and the monsters of your imagination disappear. But sometimes, when things are a little more complicated or unfamiliar, we think we know the cause but are really jumping to a conclusion without any evidence. Are any of your examples ones in which you thought you knew the cause but were mistaken? If not, can you think of any such additional examples? Write one or two in your notebooks.

• Let's think about these examples now. Can you think of anything you might have considered before you decided on a cause that might have led to a better sense of what *really* happened—something that might have helped you avoid your mistake? How might you have done some more careful critical thinking in these cases? Write one or two ideas down in your notebooks.

The key in the lesson introduction is to relate skillful causal explanation to the students' own experiences and to demonstrate what they already know about making such judgments. This connection increases student acceptance of the lesson and makes it more meaningful.

The teacher begins the lesson introduction by being explicit about its thinking skill objective, in this case causal explanation. She then shifts the focus of the lesson immediately to the students: They will learn about causal explanation by reflecting on situations in their own lives in which they had to determine what caused something. By examining causal explanations that were mistaken and suggesting questions they might have asked in order to come up with a better explanation, the students will generate the key factors for skillful causal explanation. The teacher will formulate these responses into a strategy that can be used to guide causal explanation in the future by focusing attention, for example, on possible causes, possible evidence, actual evidence, and what the evidence shows about the possible causes, as students raise these questions. Drawing out students' own insights about good causal explanation contrasts sharply with instruction in which the teacher simply explains skillful causal explanation and asks students to practice it.

This activity also sets the tone for the whole lesson. It is designed to help these students manage and change the way they think about their own causal explanations.

After involving the students in some brief collaborative work, the teacher asks them to share their thoughts with the class. More sophisticated forms of collaborative learning appear later in the lesson.

• **Now that you have had a chance to discuss your examples of causal explanations with a partner, will someone tell the class about your example?**

As students report, the teacher uses the language of the thinking skill to help them identify what they could have thought about to make a better decision. For example, some students realize that some causes they might have considered would have been better choices. The teacher identifies these as *possible causes,* and writes the question, "What are some possible causes?" on the chalkboard. Other students report that they might have gathered more evidence than they did before they decided what the cause was. The teacher writes this on the board: "How can you prove which cause is the most likely?"

The teacher's comments emphasize the vocabulary of causal explanation and are a direct and explicit way of introducing students to what is involved in skillful causal explanation. These terms serve as verbal cues as the lesson unfolds. If students don't mention other important points

that should be considered in skillful causal explanation, the teacher may mention them directly. She may say, "Did anyone think about what evidence you might find showing one possible cause as the best explanation before you started looking for evidence?" If no one responds, she may then ask whether anyone can think of an example like that and add this to the list of important questions to ask in determining the best causal explanation.

• **Here's a way to put your ideas together into a series of important questions to ask and answer before you decide what caused something.**

The teacher and the class now discuss whether, for example, thinking about possible causes should come before thinking about evidence. They usually agree that it should, realizing that, if other possible causes are suggested later, they should be added.

The teacher then writes the series of questions they choose on the chalkboard and titles it "Skillful Causal Explanation." This "verbal map" of skillful causal explanation may look like the following:

SKILLFUL CAUSAL EXPLANATION

1. What are possible causes of the event in question?

2. What is possible evidence that would show that these causes are likely?

3. What actual evidence is there that is relevant to what caused the event in question?

4. Based on the evidence, what is the likely cause of the event?

Figure 1.5

Although there are variations in the way a strategy for a thinking skill or process can be introduced, it should always be made explicit. In this infusion lesson, the students develop this plan themselves. In other lessons, the teacher may present the students with the causal explanation plan more directly and then ask them to connect it with their own experience.

The teacher then relates skillful causal explanation to an example in science:

• Finding the cause of complex or unfamiliar phenomena in science also can require ingenious critical thinking. In medicine, for example, in order to find a cure for a disease, the first step is to isolate its cause. We're going to look at a classic example of a medical mystery and how one physician, Ignaz Semmelweis, solved it, to see if our ideas about skillful causal explanation coincide with what this physician did. The action takes place in Vienna, Austria, during the middle years of the 19th century. The place, the Vienna Hospital, is where Semmelweis worked as an assistant. The mystery to be solved is that at that hospital within a week or two after childbirth many mothers were dying from a disease called "childbed" or puerperal fever. Childbed fever was particularly common after hospital deliveries and was often referred to as "the terrible evil." It often produced fatal symptoms including high temperature, pain, abscesses, peritonitis, septicaemia, delirium, and heart failure. Read the passage in your source material handout titled *Saving Mothers*, which describes the conditions and attitudes of the time. In your groups, reconstruct a list of the possible causes of childbed fever considered at that time.

These comments create a transition from an emphasis on skillful causal explanation to its use in science: Ignaz Semmelweis' 19th century determination that microscopic organisms must be the cause of childbed fever. The importance of engaging in skillful decision about such a weighty matter is then stressed by the teacher.

This new example serves two purposes. It points out how important careful thinking about causes is in science. It also provides students with an example of the kind of thinking they have just developed as a model of skillful causal explanation so that they will become more familiar with how it is used.

To accomplish this second purpose, the teacher asks the students to analyze how Semmelweis thought through the issue about what was causing childbed fever. After they read the source material, they easily find that

Semmelweis used the same strategy that is in the thinking map of skillful causal explanation they developed. In this discussion the teacher keeps using the language of causal explanation, "possible causes," "possible evidence," "actual evidence," "most likely cause," etc. This sets an example that the students follow.

The teacher then makes the transition into the main activity of the lesson:

• Bacteria, like mycoplasma haminis, are not the only causes of human suffering. Sometimes we are our own worst enemy. We're now going to use causal explanation to solve another scientific "whatdunit": *The Mystery of Silent Spring*.

• Rachel Carson was an aquatic biologist with the United States Fish and Wildlife Service. She had changed her college major from English to Biology after a biology professor rekindled her childhood fascination with nature. Her book, *Silent Spring*, written in 1958, became hugely successful and equally controversial. Moreover, Carson's book was to influence a president (John F. Kennedy), anger an industry, and act as a catalyst for worldwide change. Today you are going to read a selection from *Silent Spring* and figure out why.

• Follow along with me as I read to you part of the first chapter of *Silent Spring*, "A Fable for Tomorrow."

"There was once a town in the heart of America where all life seemed to live in harmony with its surroundings. The town lay in the midst of a checkerboard of prosperous farms, with fields of grain and hillsides of orchards where, in spring, white clouds of bloom drifted above green fields. In autumn, oak and maple and birch set up a blaze of color that flamed and flickered across a backdrop of pines. Then foxes barked in the hills and deer silently crossed the fields, half hidden in the mists of fall mornings...

"Then a strange light crept over the area and everything began to change. Some evil spell had settled on the community: mysterious maladies swept the flocks of chickens; the cattle and sheep sickened and died. Everywhere was a shadow of death. The farmers spoke of much illness among their families. In the town the doctors had become more and more puzzled by new kinds of sickness appearing among their patients. There had been several sudden and unexplained deaths, not only among adults but even among children, who would be stricken suddenly while at play and die within a few hours..."

In order to help students assimilate the information in this paragraph, the teacher asks the students to use a special content-oriented graphic organizer called "About? Point?"

- **Something bad happened here. Let's work toward solving the mystery. We are going to begin our investigation by carefully thinking about what we've read and asking two questions about each paragraph: what is it about and what is the point? Meet in your groups, read the first paragraph, and then write a group statement for each paragraph in the quote.**

After students have done this, the class is ready for the thinking activity in the lesson.

The thinking activity. In order to have a real impact on the way that the students think about causal issues, it is important to guide them in skillful causal explanation. Simply discussing causal explanation is not enough. Let's see how guided practice is woven into the lesson:

- **Now that we have carefully read this quote from the first chapter of Rachel Carson's book, what question immediately comes to your mind?**

Students usually say that they want to know what happened to this town.

- **Why does this come up immediately?**

Typically, students say that the town was once prosperous and there has obviously been a change that has destroyed this.

The teacher then engages the students in a brief metacognitive excursion:

- **What kind of thinking is important to do well to answer this question satisfactorily?**

Students respond that it is causal explanation.

- **If we go back to the thinking map we used in thinking about Semmelweis and his attempt to explain causes, what should we avoid and what should we think about first in trying to uncover the causes of this change?**

Typically, students answer that we should avoid making a hasty judgment about the cause without evidence; rather, we should think about possible causes and then try to get evidence that will show which of these possibilities is the most likely explanation.

With the stage set, the main activity in the lesson now begins:

- **In your groups, read the passage again and make a list of all the possible causes you can think of to account for the strange and mysterious events described in *Silent Spring*. Include a variety of possibilities and some unusual possibilities on your list. What might cause such a community to suffer such a catastrophe?**

This is a brainstorming activity. Students are asked to write down the possible causes they have brainstormed. Simply asking the basic question, "What are some possible causes of the silent spring?" adapted from the causal explanation strategy, prompts this brainstorming. Referring students to the causal explanation map reinforces where this question fits in the overall thinking strategy.

The brainstorming activity can be conducted in a variety of ways:

- The teacher may ask each individual student to brainstorm his or her own list of possible causes;
- The class may brainstorm as a group;
- The class may work in small collaborative learning groups, each of which generates its own list of possible causes.

Notice that the teacher sets standards for the brainstorming. She requests many possible causes, asks for varied possibilities, and encourages the students to suggest some original ones. If the students in a group are having trouble generating possible causes, she may ask: "For example, could some poison have gotten into their food supply? Think of a number of different possibilities like this."

Like brainstorming practiced in noncurricular contexts (e.g., "Think of as many different uses of this paper clip as you can"), all student suggestions are acceptable at this point. Note, however, that this brainstorming is done in the broader setting of causal explanation, not as an end in itself. The students will shortly think about these possible causes critically in order to sort out which are viable.

A teacher may be more directive in setting this task. She may ask students to generate at least five possible causes or remind them to include at least one unusual possibility. Such prompting depends on whether students in this particular class typically come up with only a few or only routine possibilities when they brainstorm.

Then the teacher asks the students to report to the whole class from their groups:

- **Tell us one possible cause from your list.**

By asking for one possible cause from each group, the teacher can create a class list involving as many of the students as possible. This composite confirms the value of collaborative work in the class and gives the students an important message: *Thinking together can help us think of ideas that we might not ordinarily bring to mind.* The teacher may wish to stress to students that it is perfectly all right for them to work with the ideas of others.

As the students report, the teacher again uses an explicit prompt in the language of causal explanation. She writes the phrase "possible causes" on the board and lists students' suggested possibilities under it. The more the teacher uses this term, the more likely her students are to remember to think about possible causes in trying to determine what caused something.

The teacher also asks some of the students to elaborate on the possibilities they report. Students should be able to explain ideas they generate. This dialogue makes each student's knowledge available to the rest of the class.

Students generally suggest a variety of possible causes for the silent spring: radiation poisoning, an epidemic, fright, they all ate poisonous mushrooms, pollution by poisonous chemicals, etc. Most possibilities like these are derived from the common background knowledge of the class. The ones that only a few class members

know about become common knowledge when they are elaborated by those suggesting them.

The teacher is now ready for the next stage. She shifts her students' attention to evidence. She does this in a straightforward way:

- **As you have said, once we have a list of possible causes, we have to determine the best explanation from our list. How can we do that in an efficient way?**

This shifts the focus to the students to indicate where to go from a list of possible causes to thinking through what the real cause is. Students by this time respond fairly quickly that it is important to think about what evidence we might get that would tend to show one or another of these possibilities to be likely. Then they can look for this evidence to see how likely the possibility is. After this strategy has been reviewed, the teacher suggests that they try it out:

- **Let's try this out by taking a closer look at one of the possible causes, "Radiation from a nuclear disaster." In order to help us organize our thoughts about evidence, we are going to use a graphic organizer that highlights evidence. We will determine whether radiation from a nuclear disaster is a likely, unlikely, or uncertain cause of the silent spring, based on the evidence, by working through the graphic organizer together.**

The teacher introduces a graphic organizer for her students to use in their thinking and asks them to write, "Radiation from a nuclear explosion or nuclear accident." She has chosen a graphic organizer for causal explanation:

Graphic organizers like the diagram in figure 1.6 (page 15) enhance thinking dramatically. They are particularly helpful for visual learners but have benefits for all students. A well-constructed graphic organizer

- shows important relationships in the thinking process,
- guides the students through the thinking process,
- makes it possible for users to "download" information otherwise difficult to hold in memory,
- shows important relationships between pieces of information in a clear way.

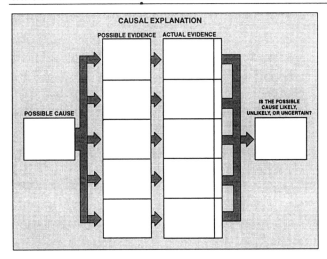

Figure 1.6

Not all graphic organizers serve these multiple functions. Some, such as appointment books or Venn diagrams, help only to organize information. Graphic organizers that also guide thinking processes, such as the specialized diagrams featured in this handbook, are especially useful in infusion lessons.

In this lesson, a teacher may draw the graphic organizer on the board, use a transparency of it, or create a large poster. Then, when she asks the class for ideas about evidence, she can write their responses on the diagram. As an alternative, she may ask the students to work again in collaborative learning groups, write their ideas on a "group" diagram, and report back to the class. Whichever technique she uses, she should write suggested evidence on the class graphic so that all students can see the results. Note that in this part of the activity, the whole class begins to think about one of the possible causes. The teacher's plan is to guide students' examination of one possibility and then ask them later in the lesson to examine another possibility in their collaborative learning groups.

- **Work in your groups again and think about the kind of evidence you would like to see that would point to radiation as being a likely cause of the silent spring. If radiation were the actual cause, what would you expect to find if you visited the town?**

In guiding her students to focus their attention on evidence, the teacher can set some standards. She can ask them to think about evidence that is direct, as well as circumstantial, so that her students consider a comprehensive set of evidence. She phrases these requests in directions to the students: "Make sure you think about evidence that is direct as well as consequential...." She may ask questions instead: "What are some pieces of direct evidence that might be found that would show pretty clearly that this possibility is a likely one?"

If, as the students report, the teacher notices that the students are listing evidence of the same kind, she may say, for example: "These are all pieces of evidence that could be found in the town. Is there any evidence that might be obtained from outside the town?"

The lesson continues by the teacher guiding the students to use their list of possible evidence to search for actual evidence.

- **Let us now use your list of possible evidence to research for actual evidence. If what we find includes what you have listed, what would that show about radiation as the basic cause of the devastating change described in *Silent Spring*?**

Students usually respond that if some of this evidence is found, it would increase the likelihood that radiation was the cause. If a lot of what we have listed is found, it would make radiation pretty likely. The teacher then writes: "What does the evidence show about the likelihood of the possible cause being the real cause?" This is to become the guiding question to be answered in the remainder of the activity. It expresses the sense in which this activity involves the students in critical thinking. Recognizing the need for evidence or support for this kind of judgment, and developing standards for the use of such evidence in determining whether a possible cause is, in all likelihood, the real cause, or whether this is unlikely, or uncertain, is the primary critical thinking objective of this lesson.

- **Let's first reread the selection from "A small town in America" and see if we can find any actual evidence cited that matches the possible evidence we listed on the graphic organizer. We will start with "Symptoms of radiation sickness in people and animals." Read each paragraph carefully and see if you can discover any clues** which would count for or against this possible piece of evidence. As you write this evidence in the actual evidence box next to "Symptoms of radiation sickness..." put a plus (+) next to it if it counts in favor of radiation as a cause, and a minus (-) if it counts against.

When the students have completed this task, the teacher asks each group to report what they have found. They discuss each item of evidence and reach a class consensus on whether it counts for or against radiation as the cause, and how strong it is. In considering the latter the students consider whether the information is circumstantial or definitively shows that radiation was the cause of the silent spring. In this case the students find the information all circumstantial—it is what you might expect if radiation were the cause, but its presence alone does not establish that it was radiation. It still could have been something else.

- **Now that the we have evaluated one piece of evidence and have decided on whether it counts in favor of or against the possible cause, it is time to work in groups and complete the "actual evidence" boxes for each of the remaining pieces of "possible evidence."**

Since this is the first time students are engaging with an activity of this sort in which they are evaluating evidence, the teacher is careful to move slowly and to be as explicit as possible about each piece of evidence. This is what she wants her students to learn to do with ease and facility. Later, when they show that they treat evidence with the same critical thoughtfulness, she need not spend as much time raising these questions about specific evidence.

Notice how helpful using the causal explanation graphic organizer becomes in carrying out this thinking process skillfully. Students need not keep all of this information in memory, where some of their ideas may slip away. The information is all recorded on the diagrams for them to take into account in order to make a well-considered critical judgment.

After student groups have completed the task and reported their findings and the teacher records them on the diagram, the teacher says:

- **The evidence we are looking at is limited to Rachel Carson's prose in the reading selection. Has she planted enough evidence in her story so that we can accept or reject radiation as an explanation for the afflictions that have fallen on this town? Is the possible cause likely, unlikely, or uncertain?**

Almost at a glance the class can see that the evidence is quite inconclusive about radiation as the cause of the silent spring. The group diagram may look like the sample in figure 1.7.

Acknowledging that the evidence that the students have used is extremely limited (it comes from what is included in *Silent Spring* only), the teacher prepares the students for a more significant search:

- **When you investigate a possible cause like we just did and you find inconclusive evidence in one source, what might you do to resolve the uncertainty?**

Students typically respond that they should look for more evidence elsewhere.

- **Where might you get more evidence?**

The teacher wants her students to develop plans for a broader evidence search before they undertake an assessment of another possible cause on their own. Their responses, which she writes on the board, usually include: looking in newspapers from the time, consulting the relevant governmental agencies (e.g., in the case of radiation, the Atomic Energy Commission), etc. Once again, the teacher has structured the activity so that it culminates in students' sharing their ideas with the whole class. In the course of doing this, she may ask probing questions about how to get information from such sources. Making this discussion public in her classroom further serves to speak to the needs of students whose background and experience is more limited than others. They learn from each other and develop a classwide approach with which all students identify.

The next task is then set by the teacher:

- **I'm going to ask you to investigate, in the same way, the other possible causes you've listed, and to expand your investigation to include such additional sources. The period in question is the late 1950s. Each** group should select another possible cause for the silent spring and use the graphic organizer for causal explanation to record your results. Begin by making a list of possible evidence you might find that would convince you that the possible cause in question was cause for the silent spring and use the graphic organizer for causal explanation to record your results. Begin by making a list of possible evidence you might find that would convince you that the possible cause in question was likely. Then reread the selection from *Silent Spring*. Look for actual evidence in the reading selection that matches the possible evidence you list, fill in the appropriate columns, and mark whether each item counts in favor of (+) or against (-) the suggested causal explanation. Then determine if the possible cause is likely, unlikely or uncertain, based on the evidence. If additional evidence is needed, use the packet of resource material which includes newspapers, magazines, and various reports to try to find other actual evidence that counts in favor or against the possible cause. Record that evidence on the graphic organizer also, and rate it as well.

An even more ambitious way to provide such resources is to insert these articles in facsimiles of newspapers, magazines, etc. so that the students have to search for them in a way that more closely represents what they would have to do if they were engaged in an authentic causal investigation. An alternative to the latter is to use a real case—e.g., the case of Love Canal—as the context for expanding the investigation of the causes of the silent spring in this town. Then real newspapers, etc., can provide authentic source material in which to search for real evidence to solve the mystery.

Reading reports for relevant information is a key content objective in the lesson. This ob-jective is well-served in this phase of the activity. Similarly, students' substantive learning about pesticides, for example, is also well served.

After their research, students share their results with the rest of the class. The teacher can either have students from each group write their results on the chalkboard or photocopy their completed graphic organizers and distribute them to everyone in the class. Then there is a basis for a discussion of which possible cause provides the *best* explanation for what happened to this town. Class consensus is usually that the most likely cause of the silent spring is pesticide contamination, although they also realize that multiple factors could have caused it.

The activity then concludes with a brief excursion into problem-solving, looking ahead to a time when that can be the focus:

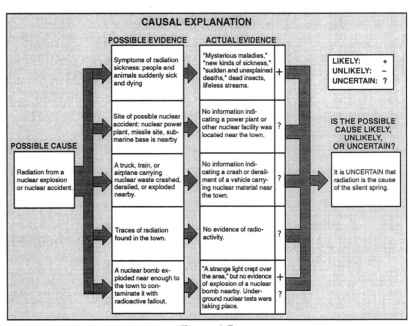

Figure 1.7

• **Now that you've made judgments about the likely causes of the silent spring in this American town, what are some possible things we might do to prevent such tragedies from happening again?**

To summarize: The thinking activity in this lesson is crafted to blend instruction in content and in thinking. When the students put themselves in the place of an investigator trying to discover why an environmental disaster has occurred, they get the kind of practice needed to develop better thinking habits. Using the causal explanation strategy prompts students to search for evidence that they might otherwise overlook. Searching for such evidence, and determining what causal explanations they support, produces a deep understanding of the science of such events.

Two key strategies are used to guide students through the lesson thinking activity:

- Using sequenced explicit verbal prompts (e.g., questions) in the language of the thinking skill or process;

- Using a graphic organizer that visually represents the process.

Using explicit prompts is one of the basic ways that infusion lessons differ from merely employing higher-order or Socratic questioning about the content.

Guiding students to do skillful thinking by using an organized sequence of explicit questions is a basic strategy common to all lessons in this book. The teacher typically reinforces the thinking strategy by writing focus words like "possible causes" on the board, emphasizing the vocabulary of the thinking process.

Throughout this book, "thinking maps," like the one for skillful causal explanation (figure 1.5) used in this lesson, verbalize what makes the thinking skillful. Thinking maps employ the vocabulary of the thinking skill or process and provide the key organizing questions that should be incorporated into the lesson.

Prompting students to reflect on how they are thinking. Thinking about what caused the silent spring is the central thinking activity in this sample lesson. The goal of this thinking activity is to involve students in deliberately making a considered judgment that they support with reasons. The teacher now helps the students understand how they carried out skillful causal explanation so that they can guide themselves in the future:

- **After such complex thinking, it's always a good idea to stop and think about your thinking, especially when you are trying to follow a plan for your thinking. Let's review what plan we followed and how well we were able to do it.**

- **What kind of thinking did you engage in and what was the sequence of questions you followed in doing so?**

This line of questioning distances students from the mystery of the silent spring so that they can reflect on their own thinking. This type of reflection is called "metacognition." Students are *thinking about their thinking*. Thinking about their thinking has dramatic effects on students' learning and is usually not a difficult or complicated task if prompted by the type of questions this teacher asks.

At first, many students may find it disorienting to shift from thinking about the mystery to thinking about their own thinking. Teachers may have to ask questions like, "Did you think about possible causes first?" After such a direct question, most students catch on.

Notice that the students are asked simply to *name* and *describe* how they thought. They are learning how to *monitor* their thinking, one of the key functions of metacognition. If students know *how* they think about their causal explanations, they can choose whether they want to continue to think about causal explanations in the same way or to choose some different way. Here, as elsewhere in the lesson, the language of thinking is important. The teacher expects students to describe their thinking using key terms: "possible causes," "possible evidence," etc.

The next series of questions takes students to the next step in metacognition: *evaluating* their thinking. The teacher is direct in the way she prompts students to think about this.

- **In this a good way to determine what caused something?** The teacher could ask this in more detail:

- **What do you think about this way of making decisions?**

- **How does this compare to the way you ordinarily make decisions?**

- **Is this a good way to do it? What are some of its pros and cons?**

- **Were there any aspects of this activity that you found particularly hard? Why? What might make it easier for you next time?**

Teachers who have conducted this metacognitive activity usually ask students to discuss their responses to these questions in the classroom. Students may also be asked to write about their thinking in a short essay or in their journals.

Students often say that they like this way of deciding what caused something because they think their explanations will more likely be correct. Some students say, however, that they find this way of thinking about causes difficult. This is usually because it takes time and effort to track down evidence. This is a legitimate reaction. If a student makes this comment, it is appropriate to ask the class how this can be done more easily next time.

Students often respond to this question thoughtfully and offer good suggestions. Asking metacognitive questions helps students to take ownership of a thinking strategy that will work for them.

Finally, the students are asked to *develop a plan* for the next time they are faced with a situation in which it is important to determine why something happened. In so doing, they are employing another type of metacognition that builds on the first two: *directing* their thinking.

- **How could you be sure to follow this plan in the future if I am not around?**

- **It's a good idea for you to put this plan for causal explanation in your own words so that you develop your own plan for thinking through a complex problem like the mystery of the silent spring. Please do this in your notebooks.**

Writing a plan for thinking clarifies what we need to think about and the order in which we may want to think about it. A thinking plan serves as a cue to remind us of the strategy that we think should be employed the next time we need to do this type of thinking.

Writing down a plan is, of course, a familiar technique for taking better control of *any* task that we undertake, like baking a cake, repairing a car, and Christmas shopping. The great insight about metacognition as a tool for good thinking is that our thinking itself is subject to our deliberate control, just as we can control our overt behavior. By taking charge of our own thinking, we can reorganize it to counter shortfalls that we detect in the way we ordinarily think.

To summarize: The metacognitive part of this lesson serves two functions:

- It brings to students' consciousness the structure of the thinking that they just did, using the language of the thinking skill. This is a retrospective function.

- It allows students to develop an explicit written plan for thinking that can guide their thinking in the future when a need for careful causal explanation arises. This helps prepare students for future thinking.

Helping students transfer the use of this skill to other examples. To finish, the teacher asks the students to apply their causal explanation strategy to three other examples.

This transfer activity should occur as soon as possible after students' metacognitive reflection. When the thinking activity and the students' reflection on their own thinking take a full class period, students should be asked to apply their thinking to transfer examples in the next available class period.

The teacher continues the lesson by focusing on another environmental issue:

- **During the past 25 years, the names of a number of animals have appeared on the list of endangered species that the government keeps. Using your plan for causal explanation, pick one of these animals and determine what caused them to become endangered .**

She also asks them to think about two other circumstances in which they can use their plan for skillful causal explanation.

- **Bacteria have been living on the earth for close to four billion years. What is the best causal explanation for their survival?**

- **Causal explanation is used not only in science, but also in history. Pick one of the following historical events that you have studied and determine the best causal explanation of the event: the industrial revolution, the exploration of space, the production and use of metal tools, the "age of exploration."**

This part of the lesson provides students additional practice in using the same thinking process. It is included to teach for *transfer*. Notice that it builds on the previous activities, referring back to the thinking plan students developed in the lesson.

Notice, also, that the students are asked to guide their own thinking. This shift is deliberate and important. If they need help, the teacher can provide it. However, the more students guide themselves, the more they will develop the habit of directing their own thinking according to the plan they developed. That is the ultimate goal of this lesson.

The first two transfer examples differ from the third. The first two are similar to the original lesson activity about silent spring. They are "near-transfer" examples. The third example is quite different. Although each makes use of the same causal explanation strategy, the third example is a "far-transfer" example, applying skillful causal explanation to a different curricular or noncurricular situation, in this case, historical events. Both types of transfer are important and should be incorporated into the initial lessons we do with students on new thinking skills, like the mystery of the silent spring.

The transfer component of the lesson can be conducted in a number of different ways. The teacher can vary the examples or ask students to suggest some themselves. However the application is selected, teaching for transfer is an essential part of the lesson, helping students develop the habit of asking important questions about causes and evidence to guide their judgments about what the best explanation is for specific events. This type of transfer activity goes hand-in-hand with the metacognitive activity prompted in the lesson, and both combine to yield a powerful internalization and learning strategy. Metacognition and transfer are the hallmarks of infusion lessons and occur in this combination in every infusion lesson in this book.

Thinking skillfully about causes should be reinforced regularly throughout the school year. Suggestions for reinforcement are included at the end of every lesson. For example, here is another near-transfer challenge, and two far transfer challenges, that the teacher might raise a little later in the school year for the same students to work on:

- **What are the main causes of acid rain? Tell why your explanation is best.**

- **Select a major character in one of the novels you are reading in English, and select an important turning point in the novel. Explain why the events that constitute this turning point occurred. Explain what evidence you have for your explanation.**

- **Select one of the following and explain why it occurred: an upswing in your performance in school, a downswing in your performance at school, your steady performance at school.**

When these topics arise in class, the teacher will prompt students to use their causal explanation plan to think these questions through. She may also ask students to apply skillful causal explanation to current problems, such as litter in the school, as a prelude to coming up with a plan to control or eliminate it. In each case, she expects her students to follow their plan for skillful causal explanation.

Transfer activities need not take nearly as much time in class as the initial thinking activity in the lesson. They can be assigned as homework, independent group projects, etc. Of course, asking students to report in class on the results and to discuss the way they went about making their decisions can be very helpful in the overall process of the refinement and reinforcement of skillful causal explanation.

To summarize: In the transfer portion of the lesson, students are given additional practice to help them develop habits of skillful causal explanation. This involves

- Additional examples for causal explanation similar to the one students considered in the main lesson activity (*near* transfer)

- Additional examples quite different from the one considered in the main activity in the lesson (*far* transfer)

- Reinforcement of the same thinking throughout the school year

In each instance, students are challenged to guide themselves in their thinking, usually by referring to the thinking plans that they themselves developed for skillful causal explanation.

Overview of Infusion Lessons

The lesson as a whole. The lesson *The Mystery of Silent Spring* is an initial lesson on causal explanation for high school science students. Students not only engage thoughtfully with what they are reading, but also their attention is focused on the thinking process that they are learning. Such an interrelationship distinguishes an infusion lesson from other practices.

The lesson introduction. Focusing students' attention on the thinking that they are learning is done differently in each of the four sections of the lesson. Students are introduced to the thinking skill goal of the lesson along with the content material. This is achieved by a discussion or activity designed to

- demonstrate to the students themselves what they already know about the thinking skill being taught;
- show students why this type of thinking is important;
- help them to relate its importance to their own experience;
- introduce them to the process of engaging in the thinking skillfully;
- introduce them to the significance of engaging in this kind of skillful thinking as they reflect on the content they are learning.

In the sample lesson plans in this handbook, the preceding activities listed form the *Introduction to the Content and to the Thinking Skill/Process*. Figure 1.8 summarizes the main emphasis in this part of the lesson.

INTRODUCING THE THINKING

1. Importance of the thinking.

2. Method of the thinking.

3. Importance of the content.

Figure 1.8

Thinking actively. Next, students engage in an activity in which they are guided through a skillful performance of the kind of thinking being taught. In this part of the lesson, teaching

the content and teaching the thinking skill are combined. Two explicit thinking prompts guide the thinking activity:

- Verbal prompts (usually questions)
- Graphic organizers

These activities are labeled *Thinking Actively* in each of the lessons in this book.

Thinking maps are furnished for each of the skills and processes of critical and creative thinking featured in the lessons in this handbook. These thinking maps provide the organizing questions that teachers should ask to guide students through the main thinking activity in each lesson. Graphic organizers to reinforce these questions are also included.

Thinking about thinking. In the next part of the lesson, the teacher engages the students in a reflective activity in which they distance themselves from the lesson's content so as to consider the thinking they did. Students map out their own thinking process explicitly, commenting on how easy or hard it was, how they might improve it, whether this was a productive way to think about such issues, and planning how they will do the same kind of thinking in the future.

We call this metacognitive section of infusion lessons *Thinking About Thinking*.

The thinking map in figure 1.9 provides guidance in designing the metacognitive section in infusion lessons by providing basic question types that can be adapted to individual lessons.

METACOGNITION MAP

1. What kind of thinking did you engage in?

2. How did you carry out this kind of thinking?

3. Is there an effective way to engage in this kind of thinking?

Figure 1.9

Using this map should not keep the teacher from asking students other important questions about the thinking that they have been doing. For example, one might also ask students how the thinking strategy in the lesson compares to their usual way of thinking. Furthermore, in-

---

cluding a special metacognitive section in these lessons should not preclude asking metacognitive questions in other sections of the lesson. **Applying the thinking.** Finally, the teacher helps the students apply the thinking skill or process taught in the lesson to other situations. These transfer activities should occur soon after the other three parts of the lesson have been completed and should be reinforced in other activities throughout the school year. Important additional practice is offered by both "near-transfer" examples and "far- transfer" examples from personal experience).

This section of the lesson is called *Applying Thinking*. In lessons throughout the handbook, the categories in figure 1.10 are used to remind readers of the importance of using different types of examples that demonstrate the versatility of the thinking skill or process.

TRANSFER MAP

1. Immediate transfer: practicing the skill again right after the lesson
   a. Near transfer: the example is similar
   b. Far transfer: the example is different

2. Reinforcement later: more practice later

Figure 1.10

Infusion lessons designed to teach any thinking skill or process and any content objective all have the same structure.

Figure 1.11 summarizes the components that will be used throughout this handbook.

INFUSION LESSON

Introduction

Thinking Actively

Thinking about your Thinking

Applying your Thinking

Figure 1.11

The flow chart in figure 1.12 details this structure. A blank reproducible version of the lesson plan form for use in designing infusion lessons appears at the end of the chapter.

Infusion as an Approach to Teaching Thinking and Enhancing Learning

In this chapter, we have provided a rationale for infusing the teaching of critical and creative thinking into content instruction, illustrated the breadth of contexts for designing infusion lessons throughout the curriculum, and commented on the variety of instructional strategies that can be combined to give infusion lessons a powerful structure. We believe that infusion lessons bring out our students' capabilities for quality thinking and learning. These lessons are not difficult to design and teach. The learning they engender will prepare students to enter an increasingly complex and technological world with skills that they will need to use information meaningfully, to make sound judgments, and to have confidence in themselves as thoughtful people.

COMPONENTS OF INFUSION LESSONS

INTRODUCTION TO CONTENT AND PROCESS

Teacher's comments to introduce the content objectives
The lesson introduction should activate students' prior knowledge of the content and establish its relevance and importance.

Teacher's comments to introduce the thinking process and its significance
The lesson introduction should activate students' prior experience with the thinking skill/process, preview the thinking skill/process, and demonstrate the value and usefulness of performing the thinking skillfully. The introduction serves as an anticipatory set for the thinking process and should confirm the benefits of its skillful use.

THINKING ACTIVELY

Active thinking prompted by teacher questioning and graphic maps
The main activity in the lesson interweaves the explicit thinking skill/process with the content. This is what makes the content lesson an infused lesson. Teachers guide students through the thinking activity by using questions phrased in the language of the thinking skill/process and by using graphic organizers.

THINKING ABOUT THINKING

Distancing activities that help students think about the thinking process
Students are asked direct questions about their thinking that prompt them to reflect about what kind of thinking they did, how they did it, and how effective it was.

APPLYING THE THINKING

Transfer activities that involve student-prompted use of the skill in other examples
There are two broad categories of transfer activities: (1) near or far activities that immediately follow the substance of the lesson and (2) reinforcement of the thinking later in the school year. Both types of transfer involve less teacher prompting of the thinking process than in the Thinking Actively component of the lesson.

Immediate transfer
Near transfer
Application of the thinking process within the same class session, or soon thereafter, to content similar to that of the initial activity in the lesson. Decreased teacher prompting of the thinking is involved.
Far transfer
Application of the thinking process within the same class session, or shortly thereafter, to content different from that of the initial activity in the lesson. Decreased teacher prompting of the thinking is involved.

Reinforcement later
Application of the thinking process later in the school year to a variety of both near and far transfer contexts. Teacher prompting of the thinking is at a minimum.

Figure 1.12

INFUSION LESSON PLAN EXPLANATION

Subject: **Grades:**

OBJECTIVES

CONTENT

Statement of content objectives from curriculum guide or text outline

THINKING SKILL OR PROCESS

Description of the thinking skill/process the students will learn

METHODS AND MATERIALS

CONTENT

Use of instructional methods to teach the content effectively

Expository methods
Inquiry methods
Co-operative learning
Graphic organizers
Advance organizers
Higher order questions

Using manipulatives
Discourse/Socratic dialog
Integrated arts
Directed observation
Specialized software

THINKING SKILL OR PROCESS

Use of instructional methods to teach the thinking process effectively

Structured questioning strategies
Specialized graphic organizers
Collaborative learning, including
    Think/Pair/Share
Direct or inductive explanation of thinking
    processes
Learner-generated cognitive maps
    (diagrams and pictures)

LESSON

INTRODUCTION TO CONTENT AND THINKING SKILL/PROCESS

Teacher's comments to introduce the content objectives

The lesson introduction should activate students' prior knowledge of the content and establish its relevance and importance.

Teacher's comments to introduce the thinking process and its significance

The lesson introduction should activate students' prior experience with the thinking skill/process, preview the thinking skill/process, and demonstrate the value and usefulness of performing the thinking skillfully. The introduction serves as an anticipatory set for the thinking process and should confirm the benefits of its skillful use.

THINKING ACTIVELY

Active thinking involving verbal prompts and graphic maps

The main activity of the lesson interweaves the explicit thinking skill/process with the content. This is what makes the content lesson an infused lesson. Students are guided through the thinking activity by verbal prompts (e.g., questions) in the language of the thinking skill/process and by graphic organizers.

Figure 1.13

THINKING ABOUT THINKING

Distancing activities that help students think about the thinking process
Students are asked direct questions about their thinking. The metacognition map guides the composition of these questions. Students reflect about what kind of thinking they did, how they did it, and how effective it was.

APPLYING THINKING

Transfer activities that involve student-prompted use of the skill in other examples
The two broad categories of transfer activities are (1) near or far activities that immediately follow the substance of the lesson and (2) reinforcement later in the school year. Both types of transfer involve less teacher prompting of the thinking process than in the Thinking   Actively component of the lesson.

Immediate Transfer

Near transfer: Application of the process within the same class session or soon afterward to content similar to that of the initial infusion lesson. Decrease teacher prompting of the thinking.

Far transfer: Application of the process within the same class session or soon afterward to content different from that of the initial infusion lesson. Decrease teacher prompting of the thinking.

Reinforcement Later

Application of the process later in the school year to content different from that of the infusion lesson. Decrease teacher prompting of the thinking.

OPTIONAL EXTENSION ACTIVITIES
(Can occur at any time during the lesson)

REINFORCING OTHER THINKING SKILLS AND PROCESSES: Working on additional thinking skills/processes from previous infusion lessons that can play a role in this lesson.

RESEARCH EXTENSION: Gathering additional information that may be useful in reaching a conclusion or an interpretation in this lesson.

USE OF SPECIALIZED ASSIGNMENTS TO REINFORCE THE THINKING: Assigning written or oral tasks or projects that may further illustrate students' thinking about the content in this lesson.

ASSESSING STUDENT THINKING

Extended written or oral assignments or performance assessments of the effective use of the thinking skill or process.

Figure 1.13

INFUSION LESSON PLAN

INFUSION LESSONS
Introduction
Thinking Actively
Thinking about Thinking
Applying Your Thinking

TITLE:

SUBJECT: **GRADE:**

OBJECTIVES
CONTENT   THINKING SKILL/PROCESS

METHODS AND MATERIALS
CONTENT   THINKING SKILL/PROCESS

LESSON
INTRODUCTION TO CONTENT AND THINKING SKILL/PROCESS

INTRODUCING THE THINKING
1. Importance of the thinking.
2. How do you do the thinking?
3. Importance of the content.

THINKING ACTIVELY

THINKING ABOUT THINKING

THINKING ABOUT THINKING
1. Kind of thinking?
2. How did you do it?
3. Is it effective?

APPLYING THINKING

Immediate Transfer

APPLYING YOUR THINKING

1. Immediate transfer
   a. Near transfer
   b. Far transfer

2. Reinforcement later

Reinforcement Later

EXTENSION

ASSESSING STUDENT THINKING

CHAPTER 2
CRITICAL AND CREATIVE THINKING IN SCIENCE ———

Science is not just a collection of information about the natural world. Rather, science is a living enterprise involving a range of human activities all focused on finding out how the world we live in works and on applying this knowledge to serve our purposes in this world. The information that often goes under the name "science" is the *product* of these activities. Thinking is involved in all of these activities, and to do them well, the thinking we do must, correspondingly, be done skillfully and well.

How does the framework for thinking we have introduced into this work play itself out in the enterprise of science? Is there any sort of thinking that is distinctively "Scientific Thinking?" If so, what are its characteristics?

"The Scientific Method"

The most common answer to the second and third of these questions is often a resounding "Yes" and an account of what has been called "Scientific Method." Scientific method, in its most general form, is a special methodology for gathering data to test hypotheses, usually about the causes of things. This methodology involves gathering observational data that is public and reproducible, and provides relevant support or counter-support to possible explanations. This data is often gathered through experimentation in which other possibly explanatory factors are controlled so the data provided can be used to genuinely rule in out specific possible causes.

For example, in trying to determine whether worms of a certain sort found in the diet of game hens that had mysteriously died at a game-hen farm had caused their death, an experiment was performed in which similar game-hens were fed the same food with the worms, and other hens were given the same diet without the worms. Here *the variables were controlled*—that is, the only thing that varied from one group to the other, as far as the experimenters knew, were the worms. When the hens in the group that ate the diet with the worms died, and the others didn't, the experimenter had observational data

that supported the hypothesis that the worms were poisonous to these hens.

We should, of course, be cautious about thinking that these data show more than that the worms are poisonous to these game hens. If someone said that these data showed that these worms were poisonous to *all* game hens, this person would be mistaken according to the principles of scientific investigation. This is because, for all these investigators know, these worms, together with certain microorganisms that are present in the digestive system of the type of hens tested only, are *together* causing the deaths. To rule this out, another experiment of a different sort would have to be conducted. In fact, to be in a position to generalize from the data, we need to know certain things about the sample we are relying on, as articulated in skillful generalization.

The thinking that goes into judging whether it is reasonable to think that these worms killed the hens based on the data in this experiment is, of course, *critical thinking*, and using the standards in science that relate to the need to make sure that all other variables are controlled is what makes this kind of *judgment of causal explanation* skillful. But another important critical thinking skill must be involved as well: we must make sure that the observational reports we base these judgments on are accurate and reliable. Are the investigators in this experiment reliable judges of exactly what point in time a hen has died? Maybe it hasn't died but has lapsed into unconsciousness. To be sure that the data recorded is accurate, a good scientific investigator must skillfully determine whether his or her observation is a reliable source of this information. Here, the standards that we use to *determine the accuracy of an observation* are obviously important to use in making this judgment of critical thinking as well.

Moreover, consider the other kinds of skillful thinking that go into the gathering of data from such an experiment. Skillfully *comparing and contrasting* the experimental and control groups is quite essential, *classifying* the data that results, making sure of the *sequence of events* in the experiment, etc. are all involved to some degree.

When we back up a bit and think about what kind of thinking goes into the process of *designing an experiment of this sort*, we realize that prior knowledge of what sorts of results would show or count against the hypothesis under consideration is quite essential. When we design such experiments, we are, in fact, thinking about what *evidence we expect to find* that could rule in or rule out the particular hypothesis and how we *could* gather that evidence. This involves us in more skillful thinking if we are to design an experiment well. For example, we have to *predict* how certain things will react in the experiment. And we have to do this in the context of making skillful decisions about the best way to run the experiment. In short, we have to make use of—or gather—a wide range of scientific and practical information using the strategies of *skillful decision making* to design such experiments well.

Do we always have to conduct experiments like the game hen experiment described above in order to make truly scientific judgments about what is causing the deaths of the game hens? Certainly if no one knows why, or has relevant data about why, the hens died, such an experiment may well be called for. But science is a social and public enterprise. In order to find out whether certain microorganisms cause a certain disease, we may not have to conduct any experiment ourselves. The experiments may already have been carried out to establish conclusively the existence of a cause/effect relationship. Then all we have to find out is that these experimental results have already been obtained by someone else. Of course, since we are relying on information from another source, we should make skillful judgments about the *credibility of this source*. But we may, indeed, find that the source is impeccable and that the experiments have been repeated and the results confirmed many times. Then it may be silly to do this all over again. Likewise, we may identify the worms in the feed and find out, by reading a relevant book or article, or asking someone who knows about such worms, that work has already been done that establishes that these worms are poisonous to these game hens.

All of this is proper science just as much as laboratory work is. In fact, some of the greatest scientific discoveries of the century have involved creatively putting together data derived from the work of others to *generate new ideas* that furnish us with powerful and well-supported explanatory hypotheses that are real breakthroughs in science. Fundamentally, the work that Crick and Watson did in *developing* a double-helix *model* for the structure of the DNA molecule, work for which they were awarded a Nobel Prize, involved drawing on the results of others, especially the crystallography work of Rosalind Franklin and the work of Linus Pauling, and was not derived from new experimentation of their own.

Applying Science to Solve Problems

Is this what science is all about, and does this portrayal of scientific thinking capture the essence of this great enterprise? Many who discuss science say that it is. In science, we are told, we hypothesize, collect data, analyze the data, and summarize our results. But this is too narrow a view of science to capture its richness and to explain its achievements. If we think for a moment about many of the great scientific advances of this century like the development of manned flight, of vaccination against diseases, of computers, and of the space shuttle we realize that science, and scientific thinking, is a broader enterprise than just developing an understanding of how the world we live in works.

Without diminishing the importance of this kind of thinking, let's set the example of trying to ascertain what is killing the game hens in a broader context. This context is one in which, obviously, the hen farmers want to do things that will *prevent* their hens from dying—indeed, that will keep their hens healthy and flourishing. Their perception of this as a problem is the overall context in which the experiment described above becomes relevant. Information about what is causing the hens to die, once confirmed by the experiment, can then be used to make judgments about how to solve the problem of the dying hens. In fact, overall, most contexts in which such experiments in "pure" science are conducted are contexts like these that call for careful *problem solving and decision making*—so-called "applied" science. Here, the whole array of important types of thinking that are involved in problem solving and decision making are essential to do well if the problem is to be solved effectively. Problems that call for our action must be *diagnosed and defined*, *possible solutions should be generated*, their *consequences assessed*, and the *best solution chosen* based on this careful assessment.

One should not take too lightly the kind of thinking that is needed to solve such problems well, even after we know what is causing them. For example, knowing that it is the worms that are killing the hens may seem to lead to a simple solution—get rid of the worms. But that may not be so easy and may not be the best solution. It may cost a lot of money to ferret out the worms, and they may keep coming back.

Less routine and more *creative solutions* may be necessary to solve this problem. Maybe moving the hens, or the whole farm, to a non-infested area is a better solution, or maybe there is something that can be put in their feed that will counteract and neutralize the effects of the worms. All of these possibilities have to be explored with the same high standards as experiments in the laboratory in order to determine the best thing to do, while fully respecting the need for credible observational data to support judgments of cause and effect when they have to be made.

Let's consider another better-known example. When it was determined that the distortion in the images sent back to Earth by the Hubble telescope was due to the misgrinding of the reflecting lens in the telescope, one obvious solution to the problem was to replace the lens with one that was ground properly. But that was not the best solution. Rather, the team of scientists working on the project opted for a correcting mechanism that saved millions of dollars and worked just as well, rather like eyeglasses work for those of us with eyes that distort the images we see. The thinking that goes into this kind of problem solving is as much the enterprise of science as the work in the laboratory that

yields some of the basic knowledge needed to solve such problems.

A Scientific Frame of Mind

It is easy to focus attention just on processes of thinking and how to do them well when discussing scientific thinking. These are so important and seem to define science to many. But there is more to being a scientific thinker than just knowing how to engage in these processes well and practicing good scientific thinking.

Good persistent scientific practice, for example, often reveals something important about the overall *attitudes and mental dispositions* of the practitioner towards gaining knowledge about the world. Good scientific practice is certainly not a "quick fix" and usually takes time and effort. People willing to expend this time and effort show their *respect for and commitment to the procedures of science* as a route to knowledge and understanding. If someone painstakingly engages in an investigation to determine the cause of a plane crash and does not accept any of the hypotheses until there is incontrovertible evidence that shows that one of them, and that one only, could have caused the crash, that person is displaying a disposition not to embrace a theory until it has been scientifically proved, and hence, a commitment to respect the standards of good scientific procedure in gaining knowledge, and to keep at it until all leads have been exhausted.

In the case of the game hens, at least as we have contrived it, economics is the obvious motivator for finding out both *why* the hens died and *how best* to resolve this problem. Less evident, but just as pertinent to understanding what inspires good science, may be another driving force: a circumspect and probing frame of mind. Something unusual, or even not so unusual, but not well understood, happens, and we don't know why. Why, then, did it happen? What caused it? This is not simply curiosity, but a deeper bent of mind motivated by the desire to understand how things around us work and why things are the way they are. It is manifested by a person raising seemingly simple questions like, "Why did these hens die," "Why did that plane explode," "Why and how does DNA replicate itself," "Why is the universe expanding?" We know that the answers to such questions are often not at all simple or easy to get. A person who raises such questions and is motivated to seek answers to them using the best scientific techniques is a person who has the kind of scientific frame of mind that has guided, and at times, inspired the great discoveries in science. These attitudes and commitments have always been as much a part of scientific thinking as the actual procedures used by scientifically minded people to gain understanding and solve problems using what we know in science. Both a scientific frame of mind and sound scientific practice are essential to expanding the body of scientific knowledge.

Science Literacy and Scientific Thinking

Do only professional scientists use scientific thinking? Suppose that when we inquire about how we might deal with the poisonous worms we find out that there is a well-known and totally effective low-cost remedy for such poisoning that completely counteracts the effects of these worms. Then we may find that our problem can be solved quite simply. Is the thinking we are doing still respectable scientific thinking? We believe that even such apparently "lowest-common-denominator" thinking is still scientific if practiced with the same standards as any other type of thinking we do in science. The scientifically literate person—a person who has a basic understanding of the concepts and ideas of modern science and knows how to get scientific information, how to certify that it is accurate and reliable, and then how to use it to solve problems well—can be engaging in quite respectable scientific problem solving even though he or she may not have and practice the technical and experimental skills of the research scientist.

Likewise, it doesn't take a research chemist to recognize that the laundry detergent "experiments" featured in the well-known "ring-around-the-collar" television commercial are suspect. Two dirty shirts with heavily soiled collars are placed in separate washing machines with the sponsor's product in one and some other brand in the other. After a little time lapse, the dirt around the collar of the shirt that seemed to be the one washed in the sponsor's detergent was gone, while the collar of the shirt washed in the other brand looked the same as it had looked before—with a "ring-around-the-collar." When a person resists such an advertisement because they realize that we don't know that variables like the water temperature, amount of detergent used, etc. have been controlled, this person is using good scientific thinking to avoid making an ill-founded inference, even though this person may not be a research scientist.

Teaching Science is Teaching Basic Scientific Knowledge and Scientific Thinking to All Students

It is scientific thinking, in this broad sense, that we believe science education ought to aim at helping students respect and do well and not just limit itself to basic scientific information and the more narrow skills of laboratory and experimental procedure. Hence, the lessons we include speak to how we can do our best to infuse instruction in critical and creative thinking into science instruction in secondary school to maximize our impact in teaching *all* students to be careful, skillful, scientific thinkers in ways that they can use in their lives, even if they are not going to become laboratory technicians or research scientists.

PART 2

SKILLFULLY ENGAGING IN COMPLEX THINKING TASKS

Chapter 3: Decision Making

Chapter 4: Problem Solving

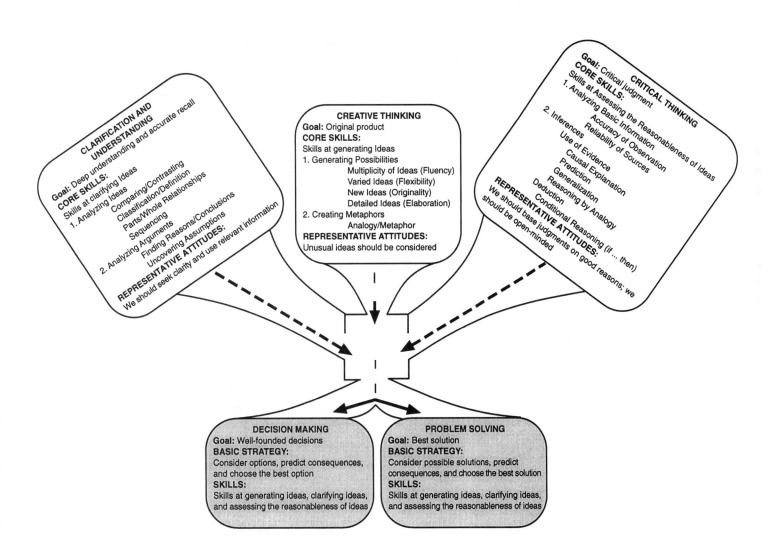

CLARIFICATION AND UNDERSTANDING
Goal: Deep understanding and accurate recall
CORE SKILLS:
Skills at clarifying Ideas
1. Analyzing Ideas
    Comparing/Contrasting
    Classification/Definition
    Parts/Whole Relationships
    Sequencing
2. Analyzing Arguments
    Finding Reasons/Conclusions
    Uncovering Assumptions
REPRESENTATIVE ATTITUDES:
We should seek clarity and use relevant information

CREATIVE THINKING
Goal: Original product
CORE SKILLS:
Skills at generating Ideas
1. Generating Possibilities
    Multiplicity of Ideas (Fluency)
    Varied Ideas (Flexibility)
    New Ideas (Originality)
    Detailed Ideas (Elaboration)
2. Creating Metaphors
    Analogy/Metaphor
REPRESENTATIVE ATTITUDES:
Unusual ideas should be considered

CRITICAL THINKING
Goal: Critical judgment
CORE SKILLS:
Skills at Assessing the Reasonableness of Ideas
1. Analyzing Basic Information
    Accuracy of Observation
    Reliability of Sources
2. Inferences
    Use of Evidence
    Causal Explanation
    Prediction
    Generalization
    Reasoning by Analogy
    Deduction
    Conditional Reasoning (if ... then)
REPRESENTATIVE ATTITUDES:
We should base judgments on good reasons; we should be open-minded

DECISION MAKING
Goal: Well-founded decisions
BASIC STRATEGY:
Consider options, predict consequences, and choose the best option
SKILLS:
Skills at generating ideas, clarifying ideas, and assessing the reasonableness of ideas

PROBLEM SOLVING
Goal: Best solution
BASIC STRATEGY:
Consider possible solutions, predict consequences, and choose the best solution
SKILLS:
Skills at generating ideas, clarifying ideas, and assessing the reasonableness of ideas

PART 2
SKILLFULLY ENGAGING IN COMPLEX THINKING TASKS: DECISION MAKING AND PROBLEM SOLVING

Often in our personal and professional lives it is important to think carefully about what we are going to do before we do it. For example, suppose there is a school holiday this week and students need time for collaborative work and a set of field trips in order to complete projects for an exhibit next week. Making arrangements so that students can complete these projects effectively requires careful thought and planning.

We often think through complicated arrangements like these without mishap: arranging transportation, helping students plan for the kinds of equipment they will need (notebooks, cameras, etc.), making accurate predictions about the time students will need at the field trip site, and obtaining reliable information about when facilities allow student visits. Moreover, we need to keep all of this information in mind when scheduling the field trip.

The daily thinking tasks that govern our actions are not all as complex as the one just described. Most of us easily arrange to go out for an evening's entertainment, purchase items for personal or home use, and have friends over. However, even these tasks require careful thought.

In this book, we emphasize two overarching types of thinking that are involved in almost all our personal and professional actions. Each is crucially important to engage in skillfully:

- Deciding what to do when a choice is needed (decision making); and

- Solving the problems that arise when circumstances make it difficult to do what we want or need to do (problem solving).

Skillful decision making and problem solving require that we blend *a range of different types of thinking skills,* including skills at generating ideas, clarifying ideas, and assessing the reasonableness of ideas. In addition, decision making and problem solving require the selection and use of *available information* with a keen sense of its relevance.

Good decision makers and problem solvers also manifest key *attitudes* and *dispositions.* For example, good decision makers reserve judgment until they have appropriately explored the decision or problem. They are persistent in their thinking. They monitor and guide themselves so that their thinking is as thorough as the circumstances permit.

Since decision making and problem solving blend thinking skills from each of the other three categories (generating ideas, clarifying ideas, and assessing the reasonableness of ideas), we have a choice of the order in which we teach students these thinking skills and processes. We can work with students on the specific thinking skills in each of the three categories and then help them apply these to skillful decision making and problem solving. Or we can begin by teaching students decision making and problem solving, and later we can refine students' use of the thinking from the three basic categories by teaching those skills directly.

The order in which the thinking skills and processes are presented in this handbook favors the second of these approaches; decision making and problem solving are introduced before the specific skills of generating, clarifying, and assessing the reasonableness of ideas. However, where you start is not as important as making sure that you work with students at both levels — the complex tasks involved in decision making and problem solving and the individual thinking skills that we engage in to perform decision making and problem solving well.

In the following section, we stress the importance of managing our thinking well when we face decisions and problems. By using a plan for thinking and then following it, we can avoid the disappointments of poorly thought-out actions and bring about results that we can deservedly feel proud to have organized.

CHAPTER 3
DECISION MAKING —————————————————————————

Why is Skillful Decision Making Needed?

When we think about decision making, we often think only of "big" choices like buying a car, choosing a college, or deciding for whom to vote. Opportunities for decision making are plentiful, though. Our decisions may also be about "little" things, like which route to take on a trip or what clothing to wear to work. Whenever we want to do something and believe we have a choice, we make a decision. Recognizing the multitude of circumstances in which we make decisions is the first step to making them better.

Causes and effects of poor decisions. Whatever the circumstance, good decision making is important. Purchasing a car that breaks down frequently can cost more than the original purchase price. Choosing the wrong street on the way to work can cause a delay that can be embarrassing and can put one at risk of job loss.

Sometimes, things beyond our control cause our chosen plans to work out poorly. A water main may break on the street we decide to take to work. A car purchase may be a lemon.

The fault is sometimes ours, however. We may not consider everything that we might in making a decision. Suppose that there is readily available information showing that a car I am considering for purchase has a very bad repair record but I don't think to seek that information. Then, if the car needs frequent repairs and I regret having purchased it, the fault is mine. I could have taken the time to get sufficient information in order to make a better decision. We can all recall situations in which we might have thought about a choice more carefully.

Problems with the way we think about decisions. A number of common shortcomings limit our decision making. One is that we often make snap decisions—we decide to do the first thing that comes to mind. We may have more options than we realize, however. Some of these options may work out better than others. For example, someone may say, "I stained my blouse; I'll put water on it to get the stain out." The water may set the stain; cleaning fluid may do a better job. If we don't stop to think about other options, we'll never find out whether there are any alternatives that might be better.

Sometimes we fail to consider alternatives, not because we are hasty in our decision making, but because we think that our decisions are "black or white." We sometimes hear, "I can either do x or not do x." Choices are seldom so simple. I may believe that I must either pay the asking price for a car or not buy it. Usually, "not doing x" masks a variety of other options. I might negotiate with the seller for a better price, trade in my old car, or wait for a sale. Thinking about the different ways of *not* doing something may reveal choices we didn't realize we had.

A second common problem in decision making arises when we don't take the time to think about all the important consequences of our choices before we make them. Considering outcomes can show us whether a decision is a wise one. Often, however, we consider only the most immediate and obvious consequences. I may just think about the initial cost of the car I am interested in and not consider other important factors like gas mileage, reliability, availability of service, etc. If I find out after I purchase the car that it needs constant repair, it is too late for this consequence to influence my decision. Knowing about this consequence beforehand might have deterred me from buying the car.

These problems with decision making are instances of two common defaults (failures to think effectively) in thinking: *hastiness* and *narrowness*. There are other problems as well.

Sometimes I may miss important options or consequences because my thinking is *scattered* and *sprawling*. I may jump from one idea to another without exploring any fully. I may think about going to the mall, and that may generate images of some of the stores at the mall. That, in turn, may make me think about how much I like to find bargains. It may not enter my mind to think about whether I have the time to make this shopping trip, whether I have other more important things to do, or whether I really need anything at the mall. I may simply not be aware that I am missing important factors.

Finally, I may make poor decisions because my thinking is *fuzzy*. I may blur together a number of quite distinct options, not being aware of differences between them. I may think that travelling to Florida means a trip to Miami because people I know always go to Miami. Thus, fuzziness in my thinking is another habit that can limit an appreciation of my range of choices and the different implications of each.

In summary, four habits of thought limit our decision making. They are listed in figure 3.1.

COMMON PROBLEMS WITH OUR THINKING ABOUT DECISIONS

1. We make quick decisions without thinking much about them. (Our decision making is **hasty**.)

2. We make decisions based on very restricted information. (Our decision making is **narrow**.)

3. Our thoughts about our decisions are disconnected and disorganized. (Our decision making is **scattered**.)

4. Lack of clarity about important aspects of a decision causes us to overlook important considerations. (Our decision making is **fuzzy**.)

Figure 3.1

These problems can affect all of the thinking we do, not just decision making. In the case of decision making, however, such defaults can lead us to be unaware of our options or miss important consequences that make a difference. Being aware of these defaults can motivate us to develop better decision-making habits.

How To Make Decisions Skillfully?

Good decision makers. We can all think of times when we have made poor decisions. If we ask how good decision makers might have made decisions in those circumstances, various remedies come to mind. First, good decision makers understand why a decision is needed. Understanding what creates the need for a decision can help us set our standards for a good choice—

one that resolves the problem. For example, if I am regularly late for work, I may think that I have to choose a better route to get to work in the morning. When I realize getting to work on time may depend also on how early I leave I expand my options to include leaving earlier.

Second, it is always important *to consider as many options as possible*. I might ask myself, "What are some different ways to get to work?" or "At what different times can I leave home?" An effective way to answer these questions is to brainstorm. We can generate many ideas by brainstorming. When we brainstorm in the context of decision making, the ideas we generate increase our chances of finding a really good option.

Good decision makers attend to three other matters before they make a decision: *They consider a range of consequences of their options; they consider the likelihood of those consequences; and they consider their significance.*

Considering a range of consequences prevents narrowness in our thinking. For example, it is often important to note short- and long-term consequences for ourselves and for others who might be affected by our actions. If leaving for work earlier in the morning will create a conflict because my children and I will have to use the same bathroom before they catch an early school bus, I should certainly take this into account.

It is also important to consider *how likely a consequence may be*. If I don't, I may exaggerate the significance of consequences that are unlikely, if not far-fetched. I may initially think that there will be a conflict with my children over using the bathroom if I leave for work earlier, but I remember that they usually get up early to review their homework. They are often out of the bathroom before I would want to use it. When in doubt, we should judge consequences by looking at evidence that supports or counts against their likelihood. Careful consideration of the likelihood of consequences is often missed in strategies for decision making that emphasize listing only pros and cons of our options.

We should also think about *how important the consequences are*. Are they a cost or a benefit? How serious is that cost or benefit? In leaving earlier for work, I may have to sacrifice watching the morning news. Is that serious? Should I weigh it heavily in my decision making? In thinking about this, I may realize that although I like watching the morning news, it may not be as significant as the stress of rushing to get to work on time. In addition, if continuing to be late is likely to result in losing my job, I should take that outcome even more seriously. Thinking about the significance of possible consequences is important in good decision making.

Of course, *all of these critical judgments should be based on accurate information from reliable sources.* For example, we usually trust our watches. Most of the time they are accurate, but sometimes they aren't. Checking them periodically is a good idea, especially if it is important to be some place on time. Good decision makers use information from only trustworthy sources.

When a decision is being made in a context in which other people are considering the same question, it is important to state the reasons for our choices. If other people disagree with us, making our reasons explicit will locate specific points of disagreement. This, in turn, facilitates dialogue. We should always allow for the possibility of changing our minds in the light of such dialogue. This kind of *open-mindedness* is also an important mark of a good decision maker.

Tips for good decision making. How do people who are skillful decision makers make sure they attend to all of these issues? One way is to prompt thinking explicitly before or during the making of a decision. We can do this by reminding ourselves of key questions and the order in which we should answer them. Here is an organized list of questions that reflect what good thinkers ask when they make decisions.

- What makes this decision necessary? What is creating the need for a decision?
- What are my options? Are there unusual ones that I should consider?
- What consequences would result if I took these options? Are there long-term consequences, consequences for others, or consequences that I may not ordinarily consider?
- How likely are these consequences? Why? Why should I think they are likely?
- Do these consequences count in favor of or against the options being considered?

- How important are these consequences — not just for me, but for all those affected by them? Are there some consequences that are so important that they should count more in my thinking than others?
- When I compare and contrast the options in light of consequences, which option is best?
- How can I carry out this decision?

Thinking about our decisions in this way counters each of the thinking defaults. Organizing our thinking by following this sequence of questions prevents our decision making from being sprawling. Considering a wide range of options and consequences prevents it from being either narrow or hasty. Systematically thinking about my options and consequences prevents my thinking from being fuzzy.

Tools for good decision making. Using a thinking map (a series of guiding questions) for skillful decision making is an easy way to remember these questions. The thinking map in figure 3.2 contains a simpler list of core questions to guide decision making. This thinking map can be elaborated, if needed, for more complex decision issues.

SKILLFUL DECISION MAKING

1. What makes a decision necessary?

2. What are my options?

3. What are the likely consequences of each option?

4. How important are the consequences?

5. Which option is best in light of the consequences?

Figure 3.2

Supplementing these questions with a graphic organizer can be very helpful in organizing thinking and managing information in skillful decision making. The diagram reproduced in figure 3.3 (next page) serves this purpose.

The graphic organizer is designed to allow consideration of one option from the many options generated. Notice that this graphic focuses on questions 2–4 from the thinking map (figure 3.2). These questions are the heart of good decision making. Question 1 on the thinking map

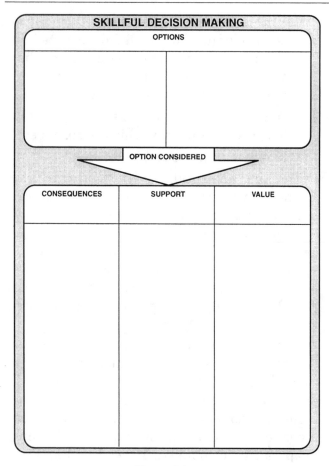

Figure 3.3

should be discussed before using this diagram. Reviewing the completed graphic organizers for many different options should serve as a basis for answering question 5.

Using the graphic organizer for skillful decision making. The graphic organizer in figure 3.3 provides both structure to guide our thinking as we make decisions and space to record our thoughts. It is designed to prompt the organized use of critical and creative thinking. The following explanation describes how it should be used:

Options. As an important first step in most decision-making situations, brainstorming techniques should be employed to generate a list of options. Record the options in the options box before giving any further consideration to any of them. Make sure to include unusual as well as ordinary options in the list. (Adding options later is perfectly acceptable.)

Consequences. Select one option to consider. The diagram guides us to project as many consequences as we believe might occur as a result

of this option, again using standard brainstorming techniques. List these proposed consequences in the first column. We should make sure to include consequences for others as well as for ourselves and also long-term as well as short-term consequences. If we list primarily negative consequences, we should strive to include a balance of positive ones as well.

Support for the consequences. The second column prompts a search for evidence to evaluate how realistic the consequences are. We can use either of two strategies to do this.

The first strategy is to consider each consequence, asking what evidence we have, or can find, to support its likelihood. When we find good evidence, we should write it next to the consequence. Then we should put a check mark next to the consequence as an indication that it is a reasonable one. On the other hand, if we don't find evidence in support of the consequence or if we find evidence against it, we can cross it out.

For example, suppose one of the cars you are considering for purchase is a foreign import. You have listed "Higher cost for repairs and service than American-made cars" as a consequence. After you acquire opinions from your friends who own foreign imports and from those who own American-made cars, you find that the labor and parts costs on the foreign imports have been consistently higher than the comparable work on U. S. made cars. This information would go in the "Support" column next to this consequence, and you would check the consequence as reasonable.

To follow this strategy through, we should continue to work horizontally after we have assessed the likelihood of a projected consequence. We should move to the last column and decide the importance of that consequence.

The second way to evaluate the likelihood of the consequences is to consider specific pieces of relevant information (e.g., an article from a consumer magazine about a product we are thinking about purchasing) and ask what this information shows about each of the projected consequences. If we find that a projected consequence is well supported by the information, it should be checked. We should then note the information we base these judgments on—e.g., the magazine article—in the middle column as a reference, though not necessarily next to the consequence(s) it supports.

In employing either of these strategies, we should, of course, make sure that the information we use to determine the consequences' likelihood is reliable and accurate.

The value of the consequences. The third column prompts us to think about the force and importance of the consequences. This, too, requires careful critical judgment.

We have a number of choices in the way we use this graphic organizer to indicate our judgments about the value of the consequences. The first question to ask is whether the consequence counts in favor of or against the option. A simple way to mark our judgment is to put a "plus" for the pros and a "minus" for the cons to the left of the consequences on the diagram. That will make it easy to see at a glance the pros and cons.

Our judgments about the relative importance of certain consequences are usually based on our values. We can express these judgments of relative importance in a number of ways. We may indicate, simply, that some consequences are important and others are not. We can, however, express our judgments of priority more finely. For example, we might want to indicate whether a consequence is *very important, moderately important,* or *not too important.* Comments such as these should be placed in the "Value" column beside the appropriate consequence.

In whatever way we identify importance, the graphic organizer provides space for us to accompany our judgment with written comments about *why* we think the consequence is important. Then, when we compare options, we can make sure we have assigned values consistently.

Making a choice. Our final choice should be made after comparing and contrasting a number of these diagrams, appropriately filled in with relevant information. The balance of pros and cons—now easily discerned—should be compared from option to option. Usually, the best choice is clear. The diagram can be used to show our reasons for this choice.

Using a data matrix to manage skillful decision making. Some decision issues require a considerable amount of information that is not initially available. As we research an issue, we do not want to become so overwhelmed that we forget important information. In addition, we want to make sure to consider everything available that is relevant. We can follow the same strategy for decision making, represented on the thinking map (figure 3.2), but use a decision-making matrix, which is more suitable for complex decisions. This appears in figure 3.4.

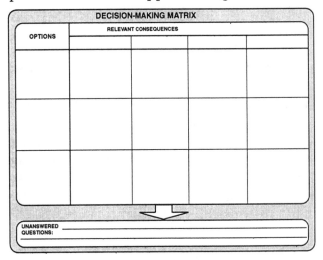

Figure 3.4

Before using this matrix, we should brainstorm a list of options. These options—or a smaller group that we consider the most promising options—can be recorded in the left-hand column on the matrix. Then we should ask what we need to find out about the options to make a well-thought-out decision. This will generate a list of factors to consider. These factors should be written at the top of each column. The list of factors can guide our search for data relevant to the decision we are trying to make. This data can then be recorded in the cells on the matrix.

For example, if you are carefully researching an expensive purchase, you can list the factors to consider (cost, durability, ease of use, safety, etc.) across the top of the matrix. Then you can use this list to guide your search for information. When you get the information, you can record it in an orderly and systematic fashion in the cells.

The box below the matrix provides space for writing important unanswered questions that arise about the information we are recording in the cells. For example, an advertisement for one of the items you are researching may include the comment that it is safe to use. You may write this information in the appropriate cell under "Safety." However, you may also realize that this information appears in an advertisement prepared by the manufacturer of the item and that it, therefore, may not be reliable. In this case you may record the question, "Is there a safety record for this item from independent sources that don't have a vested interest in selling this product?" Questions like this can direct further inquiry in order to make sure that the recorded information is accurate and complete.

As we continue the decision-making process, we should ask the same questions discussed above about the force and importance of the data recorded. We can add a "plus" or a "minus" to the information to indicate whether it counts for or against the decision. Similarly, we can circle the information that is especially significant. Comments about the basis for these judgments, however, should be written out separately; this graphic organizer does not provide a place to record such reflection.

The decision-making matrix provides a compact way of organizing information relevant to a decision so that we can easily compare and contrast options to make an informed choice. One addition to it that can be helpful in cases of great complexity is a summary column placed after each row for our overall appraisal of the strengths and weaknesses of each of the options. With or without such a summary column, the decision-making matrix, suitably filled in, can be used to provide the documentation we may need to explain the choices we make.

Comparing the two graphic organizers for decision making. The data matrix in figure 3.4 is primarily an information-organizing graphic. Using it in the process of decision making requires a heavier reliance on verbal prompts than is needed in using the previous graphic organizer in figure 3.3. If you are using the data matrix in an instructional context, for example, you will have to guide students through the decision-making process by asking more prompting questions than are necessary when you use the basic decision-making graphic organizer, where these questions explicitly appear. The virtue of using the data matrix, however, lies in its usefulness in organizing a large

amount of information. The basic graphic organizer for decision making depicts less complex data but guides us through the process of decision making more explicitly than the matrix.

How Can We Teach Students to Make Decisions Skillfully?

Teaching students to internalize the organized series of questions that guide skillful decision making can help them improve the way they think through their own decisions. Simply asking students what they think someone (e.g., a character in a story they are reading or a historical figure) should decide will not accomplish this. Such a request gives students an *opportunity* for decision making but does not *teach them how to do it skillfully*. Students who do not usually make well-thought-out decisions will continue to practice poor decision-making.

Teaching skillful decision making explicitly. A better approach is to teach skillful decision making explicitly. You can explain the strategy directly and ask your students to practice it on an example from the regular curriculum, or you can demonstrate it using an activity in which you guide students through such a decision. When you do the latter, you can use the thinking map to generate the structured questions or directions that guide thinking. In both cases, supplementing your verbal prompts with a graphic organizer is important.

Reinforcing the process. Your goal in teaching these lessons is to help your students develop, remember, and internalize strategies for skillful decision making so that they can guide their own thinking when they have to make a decision. This requires continued reinforcement. Make the graphic organizers for skillful decision making available in your classroom and encourage your students to use them on their own. Help them to practice skillful decision making deliberately in curricular contexts other than the one in which you introduced it ("far transfer") as well as others which are similar to the topic of your original lesson ("near transfer"). School-related decisions, including behavioral issues or interpersonal difficulties, are also natural contexts in which you can help students transfer what they learn about thoughtful decision making in your infusion lessons. You will find that as the process becomes familiar, students will use it without your guidance.

Contexts in the Curriculum for Decision-making Lessons

The curriculum offers numerous opportunities to teach skillful decision making. Contexts in the curriculum in which decision is a *natural* response provide such opportunities. Decisions made by characters in novels and plays (e.g., Huck Finn's not turning Jim in), by historical figures (e.g., Lincoln's issuing of the Emancipation Proclamation), as well as controversial issues (What should our energy policy be?) provide rich contexts for decision making lessons that will enhance content learning greatly.

In science, a great many decisions relate to the use of scientific knowledge and technology in the world outside the laboratory or science classroom. President Truman's decision to drop the atomic bomb on Hiroshima, genetic engineering, and the use of computer technology for communication (e.g., through the Internet) are good examples of these. Typically, such decision-making contexts for infusion lessons appear in science curricula that include a science, technology, and society (STS) focus. Other decision making contexts in straightforward science curricula reflect decisions that are frequently made within science. Decisions about the design of experiments, the use of resources and equipment, and even subjects of investigation all provide contexts for infusion lessons on skillful decision making. In general, there are two types of decision-making contexts in the secondary science curriculum materials:

- *Important decisions have been made in pure or applied science.* Students may re-enact a historical decision (e.g., the decision to eradicate yellow fever, to put a man on the moon, to explore the chemistry of plastics) using the decision-making strategy and then evaluate the actual decision made by the historical figure. This gives students opportunities to exercise decision making, to understand the values of those who actually made these decisions more deeply, and to think critically about these values.

- *What students are learning has application to or generates decision-making issues.* Students can be asked to design experiments or make decisions about equipment that would be most useful in conducting specific research projects (e.g., on crystal growth). The study of the uses of technology in STS programs also gives us many opportunities to infuse this strategy into science instruction. For example, students can be asked to think through issues about air pollution, nuclear power, or damming rivers.

A menu of suggested contexts for infusion lessons on decision making in secondary science can be found at the end of the chapter.

Model Lesson On Decision Making

The model lesson on decision making included in this chapter is a middle school science lesson on alternative energy sources that can accompany most sixth-grade textbooks. The lesson uses the decision-making matrix.

As you review these lessons, think about the following key questions:

- How does the thinking skill instruction interweave with the content in these lessons?

- Can you distinguish the four components of infusion lessons in these examples?

- Can you identify additional transfer examples to add to these lessons?

Tools for Decision-making Lessons

The thinking map for skillful decision making provides focus questions to guide students' decision making in infusion lessons. The questions may be stated as shown or expressed in students' own words. The two graphic organizers described earlier follow the thinking map. They supplement and reinforce the guidance in decision making provided by the thinking map.

The thinking maps and graphic organizers can guide you in designing the critical thinking activity in the lesson and can also serve as photocopy or transparency masters, or as models that can be used as posters. Reproduction rights are granted for use in single classrooms only.

SKILLFUL DECISION MAKING

1. What makes a decision necessary?

2. What are my options?

3. What are the likely consequences of each option?

4. How important are the consequences?

5 Which option is best in light of the consequences?

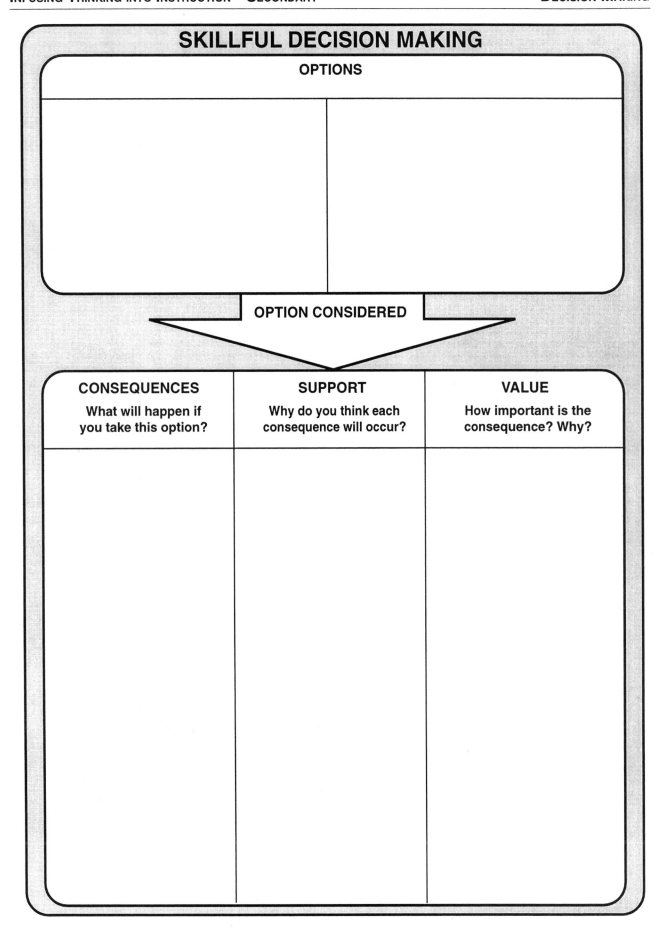

SKILLFUL DECISION MAKING

OPTIONS

OPTION CONSIDERED

CONSEQUENCES	SUPPORT	VALUE
What will happen if you take this option?	Why do you think each consequence will occur?	How important is the consequence? Why?

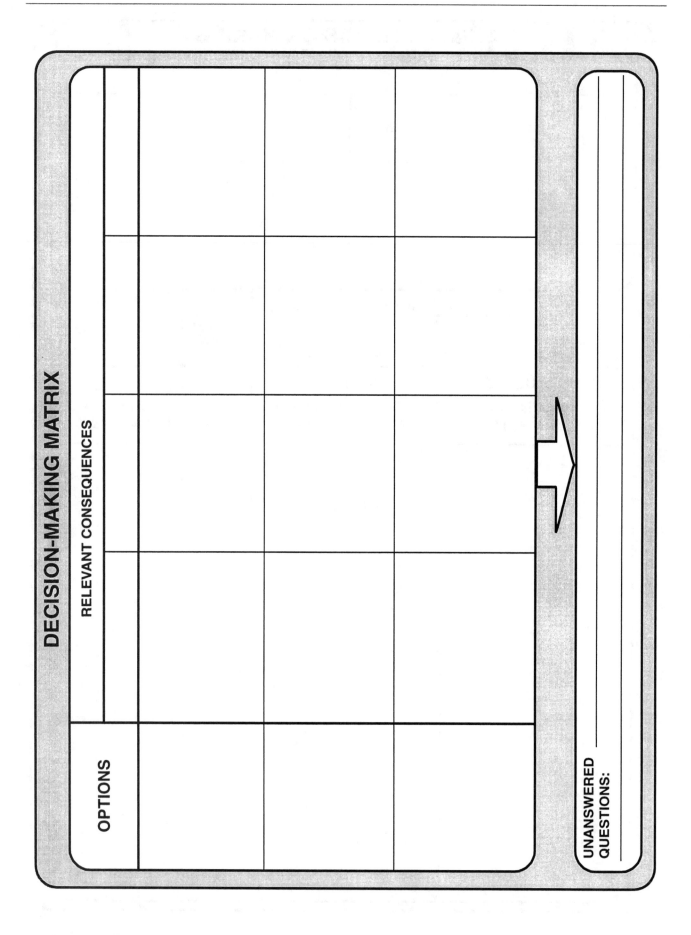

DECISION-MAKING MATRIX

RELEVANT CONSEQUENCES

OPTIONS

UNANSWERED QUESTIONS:

ALTERNATIVE ENERGY SOURCES

General Science, Biology, Environmental Science **Grades 6–12**

OBJECTIVES

CONTENT

Students will learn how energy can be derived from the major alternative energy sources; they will learn of the availability, renewability, and environmental impact of these energy sources.

THINKING SKILL/PROCESS

Students will learn to consider options and their consequences and the importance of these consequences in making decisions. Students will also recognize the need for reliable information in making decisions.

METHODS AND MATERIALS

CONTENT

Students will use textbook material on the subject of alternative energy. This lesson includes a research option in which students gather reliable information about alternative energy from a variety of sources.

THINKING SKILL/PROCESS

Structured questioning about options and consequences and the use of a data matrix guide students through the decision-making process. Collaborative learning groups brainstorm options and consequences.

LESSON

INTRODUCTION TO CONTENT AND THINKING SKILL/PROCESS

- We've been studying energy as it is used to meet human needs, and, in particular, different sources of energy. In this lesson, we're going to focus our attention on these different sources of energy and make some judgments about the scope and limits of their use in meeting our needs in this country as we approach the 21st Century. As we do this, we will also learn some ways to think about our decisions that can help us make good ones more frequently. Let's spend a little time on decision making first. **How many of you have made decisions recently that you think are good ones?** Ask for a show of hands and a few examples.

- **How many of you have made other decisions that you weren't too happy with?** Ask for a show of hands. **Write down some details about a decision that you've made recently that didn't work out too well.** Allow a few minutes.

- **Now tell your partner about your decision. Discuss two things: Why wasn't this such a good decision? What might you have thought about before you made it that could have resulted in a decision you'd feel better about now? Write these ideas down as they come up in your discussion.** Give the groups a few minutes to work together.

- **Now that you have had a chance to discuss your decision with a partner, will someone tell the class about your decision?** Ask for three or four examples. Write the decisions on the chalkboard, along with students' suggestions about how they could have made better decisions. Responses tend to include examples in which students didn't realize some of the consequences of their decisions. They usually indicate that, if they had thought about what might happen, or had gotten information about the consequences, they would have made a different decision. Other responses tend to include examples in which students didn't think about other alternatives that were better. Write the options examples in one column and the consequence examples in another. Then ask students to label what is in each column. ANSWERS INCLUDE: Alternatives, possibilities, choices, things to do, options (column 1); results, consequences, what will happen (column 2). Write these words on the board at the top of the appropriate column. Categorize students' responses using the words "Options" and "Consequences" for these two columns. Label the columns "Think about Options" and "Think about Consequences."

- Now, you have some ideas about what to think about when you're making another decision. Let's put these ideas together with some others that have been suggested about making decisions and develop an organized plan for good decision making. Write the thinking map of skillful decision making on the board:

SKILLFUL DECISION MAKING

1. What makes a decision necessary?

2. What are my options?

3. What are the likely consequences of each option?

4. How important are the consequences?

5 Which option is best in light of the consequences?

- Good decision making is especially important when it comes to decisions our country has to make about issues that affect all of us. As you know, one issue that this country has faced is what energy source should be the one we rely on as our dominant source of energy to meet our needs. When our country makes such a decision and shapes its energy policy on it, the decision should be based on scientific facts. Suppose that you were an advisor to the government on this matter and wanted to make a careful and thorough recommendation. You would certainly want to think about this issue according to our plan for making good decisions. Let's start that process by thinking about what we already know about energy and energy sources. Ask the students to review what they have studied about energy, in general, and about alternative sources of energy, in particular.

THINKING ACTIVELY

- Now let's follow our plan for skillful decision making. Why are people today concerned about energy? Can you remember a time when you heard someone discussing this question? What made them concerned about it? POSSIBLE ANSWERS: *We use a lot of energy. We rely on oil for our main energy needs, and the supply of oil may run out sometime within the next fifty years. We depend on oil from other countries, and some of those countries have closed off our oil supply in the past. We'll be needing more energy in the future. Pollution affects our health and the environment. Oil can be used in more valuable ways. The price of oil continues to go up.*

- What are some of our options regarding energy sources? Which should be our major source? We now rely on oil. Maybe oil should still be our dominant source. However, let's think about other possibilities and try to decide which is best. Work together in groups of four and list as many energy sources as possible. Try to include some that are unusual. Think about how energy is produced from each source. Ask the groups to report at random but to mention only one energy source from their lists. Then ask for sources that haven't been mentioned, creating a list from as many students as possible. Write the energy sources on the chalkboard or on a transparency under the heading "Options." When uncommon sources are mentioned, ask if anyone knows how energy is produced from those sources. This taps students' prior knowledge and contributes to the collaborative nature of the activity. POSSIBLE ANSWERS: *Nuclear power, wind, water power from dams, solar power, burning garbage, ethanol from grain, the tides, heat from the earth (geothermal), magnetism, lightning, animal power, human power, wood, oil, coal, methane gas, natural gas, steam, gravity, and chemical reactions (batteries).*

- When you are trying to make a complicated decision like this one, it's a good idea to think about what information you might need in order to decide. What would you want to know about the consequences of relying on each of these energy sources? Make a list of the things you need to know about a type of energy in order to decide whether it is a good source for our country to rely on. Your list might include, for example, how easy it is to produce the energy. What else would you add? The students should work in groups again and then report. Write student responses on the chalkboard or a transparency with the heading, "Factors to Take into Account." POSSIBLE ANSWERS: *Costs (production, transportation, costs to the consumer, storage, distribution, finding the source, and research), abundance, safety, environmental impact, ease and cost of converting to a new energy source, technology needed to produce and transmit the energy, whether the source is renewable, jobs lost or created, ease of use, consumer comfort, ease of production, public acceptability.*

- Each group should now pick three sources from its list of options. Gather information about the consequences of relying on these different forms of energy with regard to the factors you have listed. You'll use a diagram called a "data matrix," which is a chart having columns and

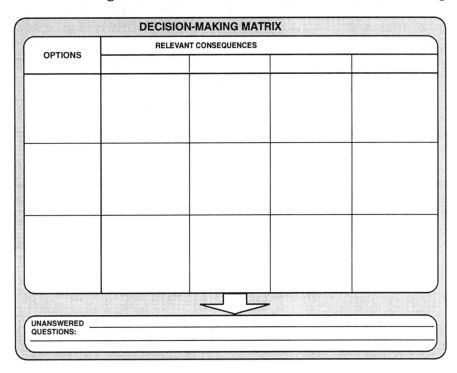

rows. Write the energy forms down the side under "options," and write the factors to consider across the top. Get the appropriate information from your textbooks or other sources to fill in the boxes. If you don't have information, if you have only partial information, or if you're not sure about its reliability, put a question mark in the cell. Write your unanswered questions on the lines below the diagram. For example, suppose solar was a source that you were working on. Your textbook provides you with a certain amount of information that you can put in your matrix. But it may not tell you how much a solar panel costs, and you may want to get that information in order to determine how costly converting to solar power may be for consumers. You would then write "How much does a solar panel cost?" in the "Unanswered Questions" box.

Allow sufficient time for your students to complete this activity. A good plan for this part of the lesson is to have students do it in two stages. Initially, and in class, they can use their textbooks, perhaps supplemented by some additional resource material you provide on alternative energy sources. Then you can ask them to do some outside research to get answers to their unanswered questions (e.g., using resources in the school library). Help them to develop a research plan by identifying possible sources for the information they want. Sources may include libraries, governmental agencies that deal with energy, TV documentaries, etc. Give students a few days to a week to get this information and then to report back to their groups. (If you want to conduct this lesson in less time, you can collect more information sources on alternative energy yourself and bring these into the class for students to work with there.)

- **What does this information about the consequences of using these energy sources show about the options you are considering? Put a plus next to the factors that you think count in favor of choosing the energy source and a minus next to those that count against choosing it. Put an asterisk in the boxes you think contain information that is very important. For each entry, explain why you put a plus, a minus, or an asterisk.** ANSWERS WILL VARY. Ask each group to report its findings to the rest of the class and explain why they rated the consequences as they did. Make a "class matrix" on the chalkboard or on a large piece of poster board. Fill in the matrix based on the reports of each group. Encourage students to add information from the group matrix to their individual ones. More than one copy of the matrix can be used if students need more space. (See examples at the end of this lesson.)

- **Work together in your groups and decide what you would pick as the best source of energy for this country to rely on, given the information on the chart. Discuss why it is the best energy source.**

Ask each group to prepare an oral or written recommendation explaining the reasons for their choices. To make this lesson an ongoing research project, students should continue to gather information to answer any remaining unanswered questions and periodically report their findings. When they get additional information, students may add it to their diagrams and report it to the class. Encourage them to reconsider their recommendations in the light of new information. A diagram may also be kept on a bulletin board for students to fill in.

THINKING ABOUT THINKING

- **How did you go about thinking through your decision? Describe what you did first, next, etc. Draw a diagram that represents a flow chart of your thinking.** Ask students to display their diagrams. Student descriptions will vary, but their flow charts should contain the five key questions on the thinking map for decision making.

- **Did you find any aspects of this activity particularly difficult? Why? How might you do this more easily next time?** ANSWERS WILL VARY.

- **What do you think about this way of making a decision? Is this a good way to do it? What are some of its pros and cons?** Discuss this strategy with the class and ask students whether the pros outweigh the cons or whether they have a better way to make a decision. POSSIBLE ANSWERS:

    Pros: *Helps us think about wider energy options, reduces narrow thinking, presents an organized way to think about a complex subject, improves confidence in choices, encourages active involvement in the topic, makes us aware of the important information in determining desirability of relying on the source, makes it possible to record relevant information in an organized way so that we won't forget it and can compare different options easily.*

Cons: *Takes longer, requires that we write on the diagram the information we find, depends on the reliability of the information we are using.*

Most students favor this approach for important decisions, though many say that it is too much trouble for decisions that are not too important. They still favor a shorter version of this strategy for even these choices—one that involves taking time to think about options and consequences before the decision.

- **What plan for careful decision making works best for you?** Allow students to map out their own plan for skillful decision making. If they omit one of the five attention points, suggest that they include something comparable. Students may include any points that they think will help them avoid difficulties they may have encountered with the energy activity.

- **How does this compare to the way you ordinarily make decisions? What's a good way to make sure that you follow this new plan instead?** Students sometimes suggest writing down their plan in their notebooks or thinking portfolios or posting it on the wall of their classroom.

APPLYING THINKING

Immediate Transfer

- Select a decision that you are trying to make right now or that you will have to make soon. Think it through, following the plan for decision making you just developed. List any unanswered questions that you may have so that you can continue to think about your decision making and research it after you leave class. After you reach a decision, indicate how confident you feel about it, based on the questions you've been able to answer through your research.

- Imagine that your parents are considering whether they should use alternative energy sources in their apartment or home. Help them decide by using your decision-making strategy.

Reinforcement Later

Later in the school year, introduce these reinforcing activities:

- **As we study the atmosphere of the earth, we will find out that it is polluted with many different particles and chemicals. Decide what to do about some types of this pollution.**

- **As we study endangered species, select one endangered animal and recommend what to do to prevent its extinction.**

- **We will be studying the way many immigrants came to this country from Europe in the early years of this century. Suppose you were a senator living at the time. What immigration policy would you support in Congress. Why? Use your decision-making strategy to decide what you think is the best thing to do. Would you support the same policy with regard to the Asians and Latin Americans who wish to immigrate to this country now? Explain why or why not.**

THINKING SKILL EXTENSION

Read an advertisement about nuclear power. What does it tell us about nuclear power? What would you want to find out about the advertisement and its source to help you determine whether the information it includes is reliable? Make a checklist of things that you might find out that would help you decide whether the information given in the advertisement is reliable. How could you get this information? Try to get this information so that you can judge whether the ad information is likely to be reliable.

Now think about whether the ad leaves unanswered any questions you have about nuclear power. If so, pick two of these questions. How could you get additional, reliable information to answer your unanswered questions? List some sources you think will be reliable and some that you are not sure about. Explain why you think they are or are not reliable.

ASSESSING STUDENT THINKING ABOUT DECISIONS

To assess this skill, ask students to write about any of the application questions. For example, ask them to think through a personal decision or to write about a controversial social issue, such as disposing of hazardous waste or environmental pollution. You may ask a similar question about a major historical decision, like Lincoln's decision to issue the Emancipation Proclamation (making sure that students have enough background knowledge to answer it). Their writing can be in the form of a standard essay, a recommendation to someone, or a letter to the editor of a local newspaper. Ask students to make their thinking explicit in their writing. Determine whether they are attending to each of the steps in the thinking map for decision making.

If you choose the extended form of this lesson, in which students gather additional information and periodically return to the question, you can use portfolio assessment techniques. Ask your students to state their learning and thinking goals in this activity and to include their comments in a special portfolio for this project. They should include products in their portfolios that indicate how well they are meeting these goals (filled-in matrices, notes on energy sources from their research, their written recommendations, etc.). They should comment on how well they believe they are meeting their goals, both with regard to gathering information about energy sources and in their decision making. The students should include these comments in their portfolios.

WRITING EXTENSION

Write a recommendation to one of your state representatives explaining why it is necessary to make a decision about which energy source this country should rely on as its dominant source, what you recommend, and why you recommend it. Make sure you explain the options you've considered, their main pros and cons, and why you choose the option you do despite its cons.

Sample Student Responses • Alternative Energy Sources

DECISION-MAKING MATRIX

RELEVANT CONSEQUENCES

OPTIONS	ABUNDANCE/ RENEWABILITY	ACCESSIBILITY	COST OF PRODUCTION	COST TO CONSUMERS	SAFETY	ENVIRONMENTAL IMPACT	EASE OF CONVERSION
OIL	U.S. has less than a decade of oil reserves left. Oil is nonrenewable; worldwide resources are also being depleted. Reserves can run out. ✳	Oil is underground. More than 60,000 oil wells are drilled in U.S. alone. New oil reserves are very difficult to discover; locating oil will get tougher. –	In 1980, $20 billion spent on oil search. Refining, storing, transporting and delivery to consumers will continue to be expensive. –	Petroleum products (primarily oil and gasoline) are usually readily available worldwide at mostly affordable prices. + ✳	Petrochemicals are flammable, toxic, and can pose extreme production and transport risks. Increased demand may make safety risks higher. –	Combustion of fossil fuels produce airborne pollutants; oil seepage and spills into ground and oceans can threaten life. Increased use will increase damage to the environment. ✳	Minimal: most vehicles have internal combustion engines; homes have furnaces easily converted to oil. Distribution of fuel oil and gasoline is already at the neighborhood level and will remain so. +
NATURAL GAS	U.S. has substantial reserves of natural gas, but much in places that are uneconomical to develop. LNG available from Arab nations; possible to gassify coal. Gas may be available for the long haul. + ✳	Only one-fourth of all wells drilled are for gas. Unconventional sources like "tough sands" need to be fractured with high pressure liquids. These less accessible sources will have to be tapped. –	Natural gas does not have to be refined therefore not much at present but if demand increases costly sources such as "tight sand" and shale deposits will have to be used. –	In the U.S., natural gas readily available to consumers in developed areas is generally low in cost. Technology to get to consumers in use for years. + ✳	Gas is explosively combustible and poses extraordinary hazards to handlers; extraction, production, transport, delivery and consumer use of gas is dangerous. Increased use will increase risks. –	Natural gas burns very cleanly. Leaks into the atmosphere are relatively rare. Increased use should not harm the environment beyond the addition of more drill rigs throughout U.S. + ✳	Expensive: furnaces might readily be converted to gas, but vehicles require new technology. It would require lots of money to build natural gas filling stations. –
COAL	Coal is the most plentiful fossil fuel; U.S. has about 25% of world's supply. The world has enough to last for at least 200 years. There is enough for the long haul. + ✳	Forty percent of coal is dug out of deep mines; the rest comes from surface strip mines. Much is too deep for economical extraction. Increased demand could require developing new mining technology. –	Most coal not refined, delivered directly to electric utilities. Only other major expense is transportation. But increased use for other purposes might require costly processing. –	Though no longer widely used in U.S., coal is available and very low in cost. It is simple to transport and has been in use for many years. + ✳	In typical year, 100 miners die, 1000s are injured. Coal, though combustible, is not explosively flammable or dangerous to transport or deliver. If demand rises, so will casualities. –	Smokestack emissons from coal-burning industries and increased carbon dioxide from combustion may cause acid rain and global warming. Strip mine sites need to be cleaned up. Increased demand means more pollution. ✳	Expensive: few homes or industries and no vehicles are equipped with coal-burning furnaces. Coal is delivered by rail. Internal combustion engines & home heating systems must be changed and the means of delivery provided. –
GEO-THERMAL	Geothermal energy is plentiful. 2,300,000 exploitable acres being explored. If new technology is developed to get it, it may be enough for the long run. + ✳	Little geothermal energy can be utilized. In most of the U.S., a significant temperature gradient lies too deep for practical use. However, the West and the eastern seaboard are ideal areas for development. –	Cost will be high to develop new technologies to reach and use deep lying geothermal energy sources (hot rock, steam, magma). If demand increases, cost will skyrocket. –	Geothermal energy sources are scattered throughout the U.S.; though the energy is low-cost once produced, the technology to make it widely used is not yet available, hence potentially very costly. – ✳	No substantial dangers attributed to geothermal production of energy. Steam and hot water are already safely utilized. Geothermal energy is not flammable and poses little risk. +	Geothermal energy is used to heat water. The heated water, if hot enough, generates electricity. Warm water recycled into the ground is not especially harmful to the environment. + ✳	Easy to heat with warm water but impractical to replace the gasoline engine with portable steam. Would have to build new power plants, produce electricity from steam and replace the engine with batteries. –

KEY   ✳ important   + pro   — con

Sample Student Responses • Alternative Energy Sources

DECISION-MAKING MATRIX

RELEVANT CONSEQUENCES

OPTIONS	ABUNDANCE/ RENEWABILITY	ACCESSIBILITY	COST OF PRODUCTION	COST TO CONSUMERS	SAFETY	ENVIRONMENTAL IMPACT	EASE OF CONVERSION
NUCLEAR	(+) ✳ Uranium in U.S. reserves will power existing reactors only thirty years; breeder reactors produce more fuel than they use and can meet increased future demands.	(+) One-fourth of world's uranium is in U.S. in 300 mines. If demand increases, mining would remain a practical process.	(+) A pound of uranium fuel has 3 million times the energy of a pound of coal; refining uranium is very expensive, but smoothly running plants produce cheap energy. Future plants likely to be more efficient.	(+) ✳ Energy produced from existing nuclear power plants readily available, abundant, and affordable. Nuclear power plants can be built almost anywhere.	(−) Radioactive material is extremely dangerous. At Three Mile Island and at Chernobyl, serios nuclear accidents have occurred. More reactors mean more risk.	(−) ✳ Safe and long-term disposal of used reactor fuel is a big problem. Leaked radioactivity can sicken and kill people and cause long-term damage to ecosystem. More reactors means more risk.	(+) Easy, although it is not feasible to power vehicles directly with nuclear energy. nuclear power plants produce electricity, which is used along the existing power distribution network.
SOLAR	(+) ✳ The sun potentially supplies 500 times more energy than we consume each year, more than we will likely ever need. Solar energy is renewable resource.	(+) Usable radiant energy also diffuses through clouds. The sun is the most accessible of all energy sources and will remain available regardless of future demand.	(−) Sunlight is expensive to harness. Home solar collectors can cost $5000. Photo-voltaic cells generate electricity only in small amounts. Increased demand on solar energy would be expensive.	(−) ✳ Although solar panels are costly, once in place, the energy produced is virtually free. For those who live in regions that get little sun, transportation costs for the energy make it more expensive.	(+) Sunlight is not ordinarily dangerous. It is not flammable, does not explode, does not leak, does not create pollutants. Harnessing more solar energy poses no unusual risks or danger.	(+) ✳ The sun is not only a part of nature, it is a requirement for the survival of life on Earth. Without the energy of the sun, the planet's temperature would plummet to 450°F below zero.	(−) Using solar energy to heat bath water is one thing; using it to power industry and vehicles is another. Would be extremely difficult covering major power utilities to solar energy.
WIND	(+) ✳ Areas of strong, prevailing continuous wind are not common-place in the world. Wind is seasonal; in most places the amount varies from night to day, season to season.	(−) Where wind blows continuously, it is often usable for producing electricity by windmills. But most areas having prevailing winds, like open oceans or mountain ranges, are impractical to exploit.	(−) Windmill turbines expensive to build and maintain, takes hudreds to generate a small amount of electricity. Increased supply of windmill energy would be very costly.	(+/−) ✳ In regions in which there is regular wind and windmills are in place, consumer costs are very low. Costs are determined by maintenance and transportation. In low-wind areas, costs would be higher.	(+) Modern windmills are simple machines that stay anchored in the ground. Neither the wind nor the windmill poses any extraordinary danger to those maintaining them or those using them.	(+) ✳ Windmills little threat to environment. They don't produce toxic chemicals or endanger wildlife. Other then the property cleared for a windmill "farm," they are environment-friendly.	(−) Very difficult: It takes many windmills to generate a limited amount of power; it is unlikely that those areas with adequate wind would host the thousands of windmills necessary to produce significant power.
HYDRO-ELECTRIC	(−) ✳ Water is a renewable resource. However, availability of new construction sites for dams and hydroelectric plants are limited by environmental concerns.	(−) You need a fast-flowing river, a dam site, and room for a plant. Many end users of electricity are too remote from dammable rivers to benefit from them.	(−) Enormous initial investment to build the dam and power plant. However, the water is free. But if demand increased, new dams would be built at great expense.	(+) ✳ Energy from hydroelectric plants is low cost, once the dams and other technology are in place. However, because sites for dams are limited, the cost to transport the energy.	(+) Modern dams only rarely breach. The power is produced cleanly, and maintenance of water turbines is routine. There is little danger to operators.	(−) ✳ Hydroelectric plants produce "clean" energy and emit no pollutants into the air or water. Interrupt the natural flow of rivers, which has threatened the habitat of some organisms.	(−) Very difficult: a substantial increase in hydroelectrict capacity would involve building hundreds of new dams, which would take years; the problem of auto and truck pollution would remain.

KEY
✳ important + pro − con

DECISION-MAKING LESSON CONTEXTS

The following examples have been suggested by classroom teachers as contexts to develop infused lessons. If a skill or process has been introduced in a previous infused lesson, these contexts may be used to reinforce it.

GRADE	SUBJECT	TOPIC	THINKING ISSUE
6–8	Science	Animal rights	What should be considered in deciding whether or not to experiment with animals?
6–8	Science	Waste management	What should be taken into account in deciding whether recycling should be a common practice, and how it should be practiced, or whether more standard methods of waste management should be retained?
6–8	Science	Global warming	Which causes of global warming would be most worthwhile spending money and resources to control?
6–8	Science	Space exploration	Should we spend large amounts of money on pushing further into outer space? Are the benefits likely to justify the expense? Or should we adopt a more limited type of space program? Explain.
6–8	Science	Endangered species	What can we do to protect old-growth forest given that the lumber industry depends on wood from these forests?
6–8	Science	Simple machines	How would you decide which simple machine is best to lift a heavy bundle in a given situation?
6–8	Science	Experiment design	Design an experiment so that you will be able to gather data that will confirm a specific hypothesis about metals, heat, and conductivity.
9–12	Biology	AIDS	What should be considered in deciding whether public health or individual privacy is most important in making mandatory AIDS testing the law? Does a better alternative exist?
9–12	Biology	Genetics	What should a couple who know they carry a defective gene consider when deciding whether to have a child?
9–12	Biology	Bioethics	What should be taken into account in forming a national policy on organ transplants?
9–12	Biology	The environment	What should be taken into account when deciding whether and where to build and locate a new nuclear power plant?
9–12	Biology	The environment	What should be taken into account when deciding which method to use to get rid of pests—pesticides, sterile organisms, natural predators, destruction of habitat, etc.?
9–12	Biology	Genetics	What characteristics should be taken into account when deciding which of a number of plants would work best in inheritance experiments?

DECISION-MAKING LESSON CONTEXTS

GRADE	SUBJECT	TOPIC	THINKING ISSUE
9–12	Biology	Fossil dating	What should be taken into account when deciding which method —radiocarbon, potassium/argon/tree rings—to use in dating a fossil?
9–12	Biology	Immune system	What would you take into account in deciding the use or availability of cancer-fighting drugs that act on the immune system?
9–12	Biology	Environment	What factors should be taken into account when deciding whether or not, or to what extent, land development will be allowed in wetlands?
9–12	Biology/ Social studies	Bioethics	What should be taken into account in forming a national policy on organ transplants, genetic engineering, or financial responsibility for children with physical or mental handicaps?
9–12	Biology/ Social studies	Bioethics	What should be taken into account to decide whether DDT should be banned?
9–12	Chemistry	Chemical indicators	What characteristics—pH, functional groups, shape of molecule, etc.—should be taken into account when deciding which chemical indicator to choose?
9–12	Physics	Simple machines	What factors should be considered when deciding which simple machine would be best to perform a specific task?
9–12	Physics/ Social studies	Atomic energy	What should be taken into account in deciding how and where nuclear waste is stored?
9–12	Science	Laboratory work	How would you best arrange a science laboratory for accessibility of instruments, safety, and ease in conducting experiments?

CHAPTER 4
PROBLEM SOLVING

Why is Skillful Problem Solving Necessary?

Solving problems well is very important. Good problem solution often improves the quality of the lives it affects. In fact, many of the benefits we enjoy today would not be available to us if, along the way, there had not been problem solvers who tackled and mastered difficulties they perceived. The cures of diseases, advances in technology, and our present standard of living are all the results of careful and successful problem solving.

Realizing that we face a problem and trying to do something about it is the first step in problem solving. If we fail to deal with problems, difficulties usually do not go away. In most cases, the intervention by a problem solver is the only way that change takes place.

Success in problem solving means that we have eliminated a difficulty for ourselves or others. Problem solving has not occurred until we have succeeded in implementing the solution. This makes problem solving a somewhat more complex process than decision making. In problem solving, as in decision making, we think through and choose the best solution to a problem. However, carrying out a solution may pose a new set of problems that are as important to solve well as is choosing the best solution.

Sometimes the term "problem solving" refers to using mathematical operations. In trying to figure out the best way to get to work, I may have to calculate how long it will take me to get to the bus in order to decide whether taking the bus will get me to work on time. I arrive at an answer to this mathematical "problem" based on the application of certain mathematical operations using given facts (how far away the bus stop is, how fast I can walk). In typical mathematics problems of this sort, "problem solving" refers to selecting the appropriate mathematical operations and getting the right answer by doing them accurately.

Many science "problems" involve the application of science principles in ways analogous to solving mathematics problems. Learning to solve such mathematical and scientific problems well can help us to learn the mathematical and scientific principles involved but will not teach us the thinking strategies we need to carry out everyday problem solving skillfully.

Common Difficulties about the Way We Choose Solutions to Problems

Problem solving, like decision making, occurs much more frequently than we think it does. It is not just big problems, like trying to find a cure for AIDS, that engage us in problem solving. Hardly a day goes by that we do not tackle a multitude of "little" problems, often with some degree of success. Suppose, as I leave for work and get into my car, I realize that I do not have my car keys. That's a problem because I can't start my car, and I need it to get to work. This problem may be easily solved, however. I may remember that I left the keys in the house. I can return to get them.

This is not to say that we tackle every problem that we face, that finding good solutions is easy, or that we find solutions as effectively as we could. When I get to work, I may notice that it's awfully hot in my office. Unless I recognize this as a problem, I might remain uncomfortable. Often, situations present difficulties, but we do not recognize a problem until it's too late to do anything about it.

Furthermore, to do effective problem solving, I should be clear about exactly what the problem is. I may realize that the heat is a problem because it makes me uncomfortable and think that to solve this problem I have to make myself comfortable. So I may change my clothes, putting on shorts and a light shirt. But that may cause more problems because informal dress is frowned on where I work. Moreover, it's only a short-term solution. The heat may be a continual problem. Even if it were permitted, it may not be practical to wear shorts every day.

On the other hand, if I identify the problem as getting the temperature lower in my office, there are a whole range of other solutions that may be more viable than the shorts solution. I may open the window, turn on a fan, install air conditioning, turn the heater down, have the heating system repaired, etc.

Identifying the problem in a comprehensive way opens the door to a variety of possible solutions, but it does not guarantee that we will consider them. There is a tendency in problem solving, as in decision making, to adopt the first solution that comes to mind without asking whether there are any better or more creative solutions. A quick solution like opening a window may not be the best way to solve the problem. It may reduce the temperature but may be costly since the lower temperature will surely keep the heater running (if the thermostat is causing the problem). As in decision making, many people fail to generate alternative solutions and consider consequences. This often leads to adopting solutions that have many unwanted side effects.

Choosing solutions hastily often means not considering the feasibility of carrying out the solution. I may think that installing air-conditioning in my office is the best solution because it is a long-term solution and will lead to better comfort and productivity. But in order to have air-conditioning installed, my school may have to bring in air-conditioning specialists from 250 miles away. That may be time-consuming and so costly that the school's resources for programs like staff development would have to be cut. This solution may not, therefore, be a feasible one. Although this is not the direct consequence of installing air-conditioning, it is a consequence of my school's adopting this solution to the problem and should be considered along with all the other relevant consequences.

These five basic shortcomings with the way we choose the best solutions to problems are listed in figure 4.1.

What Does Skillful Problem Solving Involve?

Countering tendencies for choosing hasty or poorly considered solutions involves strategies and skills that most of us can learn easily. But, of course, we also need to develop the disposition

---

COMMON PROBLEMS WITH THE WAY WE SOLVE PROBLEMS

1. We often do not recognize problem situations as problems; hence, we do nothing about them.

2. We often conceive of the problem to be solved in very narrow ways.

3. We make a hasty choice of the first solution we think of; we fail to consider a variety of possible problem solutions.

4. We do not consider many, if any, consequences of adopting the solutions we are considering.

5. We do not consider how feasible it would be to undertake this solution.

Figure 4.1

---

to use these strategies and skills when faced with a problem.

Identifying a situation as problematic. When we become aware of circumstances that create *problems*, we become aware that these conditions conflict with our purposes, interests, or needs. Not having my car keys is a problem because I need them to start my car and get to work on time, and I need to be on time because I have planned to take my students on a field trip.

Although this example is a problem for me as an individual, we can identify other situations that are problems for groups of people. Unemployment in the United States is a problem for the whole country. The depletion of the ozone layer is a problem for people in general. These conditions are problems because they frustrate, or otherwise conflict with, our basic needs or interests. Raising the question of whether any such conflict exists is the first step in identifying a problem. Once we have determined that there is a conflict, we recognize that the existing situation or condition creates a problem. Identifying a situation as problematic, in turn, prompts us to recognize the need for change.

A special case of this occurs when we are thinking about man-made objects and their functioning. Machines that we create, for example, are designed to serve certain purposes. Unless we recognize that they are not serving these purposes as well as they could, we will accept the machines as they are and not try to improve them.

For example, less than 20 years ago, computers tended to be large and cumbersome, even though their capability for processing information was greater than hitherto known. Typically, for individuals to gain the benefits of using computers, they had to link up with large mainframe computers. Recognizing that such large computers, while serving their purpose, were by no means perfect led to creative problem solving that has yielded small personal computers that have tremendous capability. For many of us, the availability of such devices has revolutionized our lives.

Skillful problem solvers do not ask just how things function. They also consider the purposes that things serve. Then they are in a better position to devise improvements.

Defining the problem. Before we can actually engage in trying to solve a problem, it must be defined. If I do not have my car keys and define my problem as finding my car keys, I may be defining it too narrowly. I may have lost my keys yesterday and not have known it. Taking the time now to try to find them may be nonproductive. Maybe I can get my car started without the keys by bypassing the ignition. Then I am defining the problem as getting my car started. I could also try to get a spare key.

How I might get my car started, however, may also be too narrow a definition of the problem. When I realize that getting the car started and driving it to my destination serves the purpose of getting me to work, I may define my problem as getting to work. Since I usually use my car, but now I can't start it, perhaps I can get other transportation. Maybe I can catch a bus or ask my neighbor to drive me there.

These possible solutions will, of course, have to be thought through carefully. For example, I will have to find out whether I can still catch the bus. But catching the bus and having my neighbor drive me to work will occur to me as possible solutions only if I define the problem broadly, not simply as a problem of finding my keys.

Effective problem definition can be facilitated by taking the time to think about the kinds of problems generated by the unsatisfactory situation. These resulting problems can be listed, then organized into types of problems that include problems about *means*, like finding the keys, as well as broader problems about *goals*, like getting to work. Problems related to goals should be given priority over problems about the means to accomplish those goals.

Generating and Assessing Solutions

Generating solutions. Some specialists in problem solving have described the process of skillful problem solving as accordion-like: We engage in divergent, open, thinking and then converge by analyzing and evaluating; then we engage in divergent thinking again, etc. One insight this image gives us into the process of skillful problem solving is the interplay of critical and creative thinking in the problem-solving process. Once we have defined a problem, we should not, of course, jump at the first solution that occurs to us. As in skillful decision making, it is well worth the time to generate a number of different possible solutions. Expanding the range of possible solutions "moves us out" from the convergent thinking that is involved in defining the problem into open, divergent thinking. As in skillful decision making, brainstorming is a technique that works well.

In situations in which the problem is an important one and solution finding is complex, a variety of solutions, not limited to those that initially seem the most promising, should be considered. Even alternatives that seem "far out" may be worth considering in order to make sure that we conduct a thorough search for the best solution.

In many approaches to problem solving, unusual solutions, however unworkable, are valued because they are "creative." Just as in skillful decision making, it is important to generate original and unusual solutions. However, we should not prefer these solutions unless our investigation of their consequences shows them to be preferable to other, more ordinary remedies. If a routine solution to a problem works better than an unusual one, it should be chosen. It might be interesting to take the space shuttle to work, but more routine ways of getting there will, in most circumstances, be preferable.

At the same time, we often face situations in which old solutions to problems do not work too well. Then, more creative solutions are in order. One effective strategy is combining possible solutions to overcome the shortcomings of each. For example, if I decide to take the bus to work, I may miss the bus that will get me there on time and have to wait for a later one that will make me arrive late. If I ride with my neighbor, she can leave right away, but she can take me only part of the way. Walking the rest of the way may make me late. Perhaps, however, she can drop me off at the bus stop, which is on her way, and I can catch the early bus, which will get me to work on time. It is well worth asking whether any of the proposed solutions can, either in whole or in part, be combined for a more creative solution than either would offer.

Assessing possible solutions. Like skillful decision making, skillful problem solving requires careful consideration of the consequences before picking the best solution. The strategy for a thorough investigation of consequences that is used in skillful decision making can also be used in problem solving. Think about consequences for others, as well as for yourself; think about both short- and long-term consequences. Weigh the important consequences more heavily than the less significant ones. Asking my neighbor to go out of her way to drive me to work may get me there on time but may strain our friendship. Our relationship may be more important to me than getting to work late.

To be more rigorous in our consideration of the consequences, we may wish to adopt a set of criteria for good solutions beforehand, setting out the factors we should consider and weighing how important they are. For example, I might note that to find a good solution to the problem with my car keys, I should consider whether a proposed solution to the problem will get me to work, whether I can arrive on time, how costly the transportation is, whether the way I get to work offends anyone, etc. I might determine that getting to work on time is the most important consideration, indicating a willingness to pay more if necessary. These criteria can then guide me in getting information to assess possible solutions. As always, we should make sure that the information we use is accurate and reliable.

A key consideration in deciding which solution is best is whether or not a particular solution is feasible. Solutions that solve problems but are not practical are too idealistic. For example, giving everyone free medical care will solve the problem of inadequate health care for those who can't afford it. But is this solution practical? This is not a good option if implementing it will bankrupt the health-care system.

Implementing problem solutions. The strategies we have suggested for problem solving are strategies for choosing the best solution to the problem at hand. However, problem solving involves more than deciding what to do. Many problem-solving experts have stressed two additional considerations: developing an implementation plan and convincing others that the solution is a good one (sometimes called "acceptance finding"). You do not really solve problems unless you actually put into practice the solution you think is best and it proves to be effective. Implementation often requires that the solution is acceptable to and endorsed by others whom it involves.

Developing an implementation plan and convincing others that the solution is best become new situations for problem solving. In the first instance, the problem is how to carry out the solution we have chosen. In the second, the problem is convincing others. Each of these calls for careful thinking. The thinking should be guided by the same strategy used to develop the original problem solution while considering two additional questions:

- What resources are needed to put this solution into effect?
- What obstacles are likely to be encountered in trying to implement this solution?

These strategies for skillful problem solving are useful for any problem we face, including everyday problems, like the problem about the lost keys, and problems faced in more technical contexts, as in engineering, business, and communications. There are some problems, of course, that we cannot now solve. All of the possible solutions we come up with fail, and we do not have enough information to generate other workable solutions, as in the present case of the AIDS epidemic. We won't ever find a solution in the future unless we keep trying, however. Using strategies for skillful problem solving can guide us even in the most difficult cases.

Tools for Finding the Best Problem Solutions

The thinking map in figure 4.2 provides questions to ask as we go through the process of skillful problem solving.

SKILLFUL PROBLEM SOLVING

1. Why is there a problem?

   a. What is the present situation?

   b. What purpose, interest, or need makes it desirable to improve on the present situation?

2. What is the problem?

3. What are possible solutions to the problem?

4. What would the consequences be if these solutions were adopted?

   a. What types of consequences are important to consider?

   b. What are these consequences?

   c. How important is each consequence?

5. What is the best solution to the problem based on this information?

Figure 4.2

Two graphic organizers are useful for skillful problem solving: one for problem definition (figure 4.3) and one for finding the best solution (figure 4.4). Figure 4.3 shows the graphic organizer for problem definition.

To use this graphic organizer, write the facts of the problem situation in the first box. Write the purposes, interests, or needs that are frustrated by these facts in the second box, and write the problems generated in the spaces to the right. Then the best problem statement can be selected and written in the bottom box. We should make sure that the problem we select is not too narrow and that it does not presuppose that one of the other problems identified is solved.

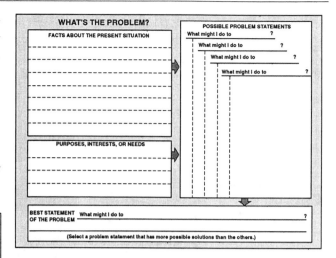

Figure 4.3

For example, in the first box I would list the fact that I can't find my car keys and that my job is a distance from my home. In the second box, I would note that one of the purposes of my car is to transport me to my job; I need to be at work shortly, and I need my keys to start the car. Then, on the right, I might write the following problems: "How might I find my keys?" "How might I get my car started?" and "How might I get to work?" I would then write the most comprehensive problem statement in the box at the bottom. The overriding problem to be solved would then read, "What might I do to get to work?"

The graphic organizer for finding the best solution (figure 4.4) should be used after we have defined the problem. We brainstorm possible solutions and, as in skillful decision making, use the graphic organizer to explore one of the possibilities. We do this by first listing the different kinds of factors we should take into account in order to choose a well-thought-out solution. We might, for example, be interested in how quickly an alternative means of transportation will get us to work, how much it will cost, whether it has an adverse effect on others, whether it is legal, etc. Working collaboratively with others to define our criteria for good solutions can help us avoid bias in our selection of criteria.

Once we have thoughtfully selected a comprehensive set of criteria, they can be used to guide us in determining what to expect if we adopt a particular solution. If the projected result serves our purposes, interests, or needs, it

---

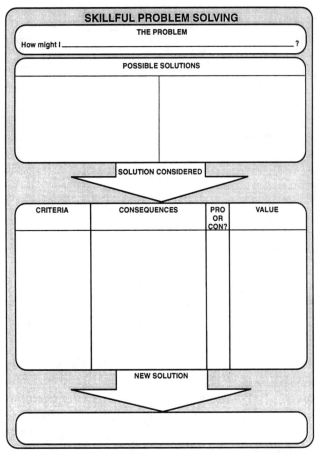

Figure 4.4

should be counted as a "pro." If it conflicts with or frustrates our purposes, interests, or needs, it should be counted as a "con." For example, if taking a taxi will get me to work on time, that is a "pro." If it will cost a large amount of money, that will be a "con." Stealing someone else's car may get me to work on time; that's a "pro." However, stealing is illegal and puts me in jeopardy of punishment. That's a "con."

Finally, I can modify the solution I am considering to make it work better. Combining parts of two possible solutions is often an effective strategy for making such modifications.

As in decision making, if you compare different solutions that have been thought through by using these diagrams, it can be very easy to determine which solution is preferable.

How Can We Teach Students to Solve Problems Skillfully?

To teach skillful problem solving, it is not enough to give students problems to solve. Skillful problem solving involves asking and an-

swering various questions that guide us through the problem-solving process in an organized and reflective way. Students usually do not attend to such questions as they solve problems. To teach students skillful problem solving, we should teach them what these guiding questions are and help them develop the habit of asking and answering them when they are faced with problems.

When teaching students this approach to skillful problem solving, it is, likewise, not enough to ask students to brainstorm possible solutions to problems and then select the best one. More articulation of the strategy for skillful problem solving is needed. Problem definition should be done thoughtfully. Careful judgment about the likelihood and weight of the consequences of the different solutions is also necessary. These considerations are sometimes overlooked in approaches to problem solving. Using the thinking map in figure 4.2 to guide this type of thinking, supplemented by the two graphic organizers, provides this needed articulation.

A Simplified Strategy for Teaching Students Skillful Problem Solving

A simplified approach to problem solving that preserves the fundamental outline of the problem-solving strategy is appropriate for students who are being introduced to thinking skills for the first time through problem solving lessons. Students consider what a problem is, consider why it is a problem, generate many possible solutions, and assess each solution by weighing its pros and cons. Wherever possible,

---

SKILLFUL PROBLEM SOLVING

1. Why is there a problem?

2. What is the problem?

3. What are possible solutions to the problem?

4. What would happen if you solved the problem in each of these ways?

5. What is the best solution to the problem?

Figure 4.5

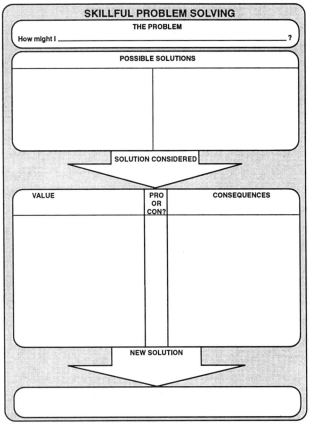

Figure 4.6

students should be encouraged to combine solutions to come up with ones that are better than those they initially generate.

The thinking map for skillful problem solving in figure 4.5 (previous page) captures this simplified strategy.

The graphic organizer in figure 4.6 is also a modification of the more complex diagram. It can be used easily to supplement the simplified strategy for problem solving.

This graphic organizer omits the step in the problem-solving strategy in which students introduce the criteria for their choice of the best solution. They simply list consequences that count in favor of or against the solution. The teacher must prompt students orally to take a range of different considerations into account.

After students are introduced to problem solving via the simplified strategy, you can help them learn the more advanced strategy. The more advanced strategy is, in a way, merely a fine-tuning of the simplified strategy. However, if your students have already been introduced to lessons in which they learn strategies for skillful thinking, you can skip the simplified

strategy and use the more advanced one from the outset.

Skillful problem solving will become second nature to our students if they start it early and it is reinforced throughout their schooling.

Contexts in the Curriculum for Lessons on Problem Solving

The curriculum already contains many contexts in which students study how various real-world problems have been solved. They are sometimes directed to solve problems themselves. Authentic problems in any curriculum area may provide excellent contexts for teaching students skillful problem solving. Moreover, characters in short stories, novels, and plays that they read grapple with problems, as do historical figures that they study. Students can be asked to "become" these characters or historical figures and to use the resources available to them in these contexts to solve the problems they face.

Specialized problem-solving examples often found in mathematics and science texts may not provide adequate contexts to teach skillful problem solving, however. The range of thinking skills that students use in such problem solving is quite restricted. Such activities are intended to give students practice in applying mathematical and scientific principles, rather than grappling with problems that require definition, the selection of the best solution, and implementation.

Contexts in science. On the other hand, broader and more ill-defined problems play a central role in the work done in the sciences. Hence, they also have a place in the science curriculum. In fact, what has been called "The Scientific Method" is a cluster of related procedures designed to solve such problems. What is causing AIDS? What can cure AIDS? How can we prevent an AIDS epidemic? These are typical—and extremely challenging—problems of the sort that science turns its methodology of empirical investigation on to solve.

Many such problems in science are intellectual problems the resolution of which is achieved by greater understanding. Therefore, skillful problem solving, which shares structural features with skillful decision making, has a much broader scope in science than does decision making. The goal of skillful decision making, as we treat it in this book, is *action* only. Hence, there is an overlap between the two (those decisions that are problematic) while problems that are intellectual in character fall outside of decision making, and decision making that is not generated by a problem—including some of the more routine decisions that we make—fall into the category of decision making but not problem solving.

There are two basic contexts for problem-solving lessons in the science curriculum:

- *Students study how important problems have been defined and the solutions that people have tried.* Insofar as students study the problem solving scientists have engaged in as science has advanced, they can be put in the position of someone who faces the same problem and asked to think through how they might solve it. They can then compare their solution to what the historical scientist did. Why did the Tacoma Narrows bridge collapse in a windstorm? What could be done to correct its construction so that it would not happen again? Why are birds that inhabit areas sprayed extensively with DDT not producing as many offspring as they used to? What can be done to correct this?

- *Students have to solve problems themselves as part of their academic work.* For example, in laboratory work in biology, chemistry, and physics, students are asked to conduct experimentation that will yield certain results (e.g., the purification of certain chemicals found in impure forms in nature). When the students have to figure out the best way to do such things in the laboratory, they are solving problems that could provide rich contexts for infusion lessons on problem solving. Similarly, when they work on such problems in environmental science as the protection of certain specific endangered species, they are working on problems that could likewise serve as rich contexts for infusion lessons on problem solving.

A menu of suggested contexts for infusion lessons on problem solving can be found at the end of the chapter.

Problem-based learning activities as contexts for lessons on skillful problem solving. *Problem-based learning* is an approach to learning that reshapes the way a standard curriculum is delivered to students. Instead of following the pattern of learning provided in most standard textbooks, teachers challenges students with authentic problems whose solutions require specific content knowledge and conceptual understanding. As the students identify things that they have to find out in order to solve the problem effectively, the teacher helps the students to locate resources through which they can find these things out. Hence, the content knowledge that students are exposed to in standard textbooks, piecemeal and out of the contexts of application that such problems provide, is acquired through the students' active research in trying to solve the problem they face.

Usually, problem-based learning is almost exclusively content-focused. It does not typically involve direct instruction in strategies for solving problems, and it is not usually accompanied by the metacognitive opportunities structured into infusion lessons. To teach skillful problem solving in problem-based learning activities, teachers must infuse instruction in the problem-solving process into these activities.

Introducing problem-based learning is one way various subjects in the curriculum can be integrated. For example, in middle school, students learn in some depth about the 19th century westward movement. Material in standard textbooks gives them information about gold in California, wagon trains, American Indian tribes, the railroad, etc. When teachers teach this material through problem-based instruction, they do not follow a standard textbook approach. Rather, they may ask students to simulate being pioneers who were making the long trip from east of the Mississippi to the West by planning out a trip and constructing and outfitting a wagon that might be used for the trip. When students work on this problem, their learning often cuts across standard subjects like geography, social studies, and mathematics, and the teacher helps them locate resources across these content areas in which they can learn what they need to solve the problem effectively.

Problem-based learning can also be designed within specific disciplines. For example, students might engage in a measurement project to determine how much carpeting is needed for the school. A science project can involve students in determining the best way to keep plants and animals alive in their classroom, to design various rides at an amusement park, or to redesign the school cafeteria so that the excessive noise there during lunch period can be diminished. In each of these cases, the activities are designed specifically so they require students to learn and use concepts and information that is part of the regular curriculum.

Practical problems like these examples provide excellent contexts for designing problem-solving lessons. The teacher creates the problem-based learning activity and then designs an infusion lesson in skillful problem solving in that context.

Model Lessons on Skillful Problem Solving

One model lesson on skillful problem solving is in this handbook. It is a lesson in environmental sciences in which students grapple with how to solve some of the major problems caused by chemical pollution of the soil, air, and drinking water due to the use of pesticides. It is based on Rachel Carson's book *Silent Spring* and is best used after students have worked on the lesson on the same topic, "The Mystery of Silent Spring," in which they use causal explanation to diagnose the problems in the "small town in America" that Rachel Carson describes. This lesson can also serve as the cornerstone of a problem-based learning unit.

As you read this lesson, ask yourself the following questions:

- What are alternative ways of introducing this thinking process to students before the main activity in the lesson?

- How can the lesson be transformed into a hands-on lesson?

- What other contexts or activities reinforce this thinking process in other content areas in the elementary school curriculum?

Tools for Designing Problem-solving Lessons

Two thinking maps for skillful problem solving are used in this handbook: the standard problem-solving strategy, and the simplified strategy. Similarly, you will find two graphic organizers for skillful problem solving. The one on page 65 is for use with the simplified strategy, and the other (p. 64) is for use with the more advanced strategy. Finally, a basic diagram is used for problem defining at all grade levels.

The thinking maps and graphic organizers can guide you in designing the critical thinking activity in the lesson and can also serve as photocopy masters, transparency masters, or as models that can be enlarged and used as posters in the classroom. Reproduction rights are granted for use in single classrooms only.

SKILLFUL PROBLEM SOLVING

1. **Why is there a problem?**

   a. **What is the present situation?**

   b. **What purpose, interest, or need makes it desirable to improve on the present situation?**

2. **What is the problem?**

3. **What are possible solutions to the problem?**

4. **What would the consequences be if these solutions were adopted?**

   a. **What types of consequences are important to consider?**

   b. **What are these consequences?**

   c. **How important is each consequence?**

5. **What is the best solution to the problem based on this information?**

SKILLFUL PROBLEM SOLVING

1. **Why is there a problem?**

2. **What is the problem?**

3. **What are possible solutions to the problem?**

4. **What would happen if you solved the problem in each of these ways?**

5. **What is the best solution to the problem?**

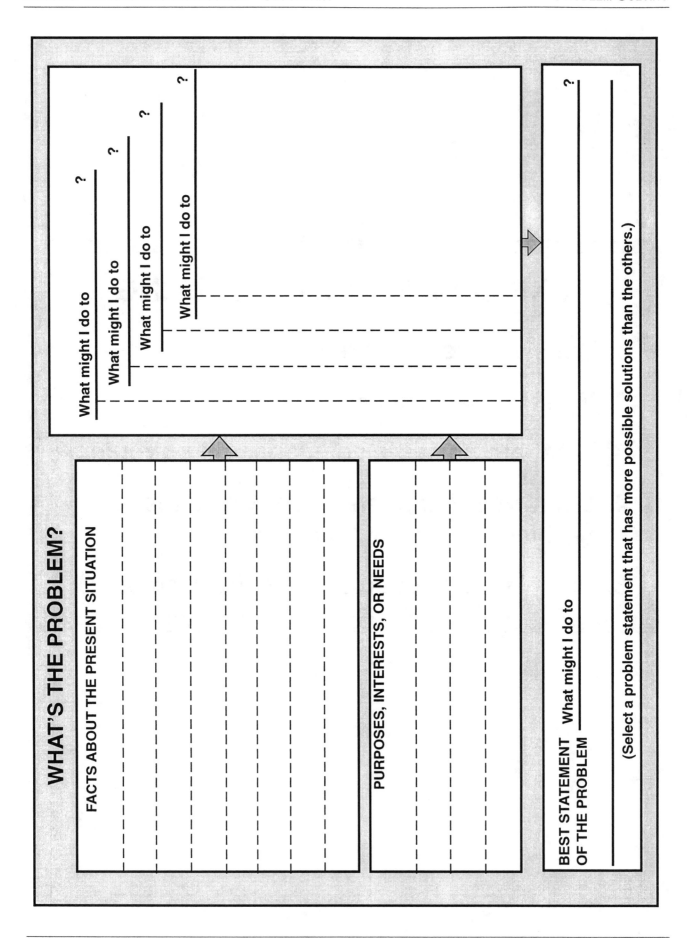

WHAT'S THE PROBLEM?

FACTS ABOUT THE PRESENT SITUATION

PURPOSES, INTERESTS, OR NEEDS

What might I do to ?

What might I do to ?

What might I do to ?

What might I do to ?

BEST STATEMENT OF THE PROBLEM   What might I do to ?

(Select a problem statement that has more possible solutions than the others.)

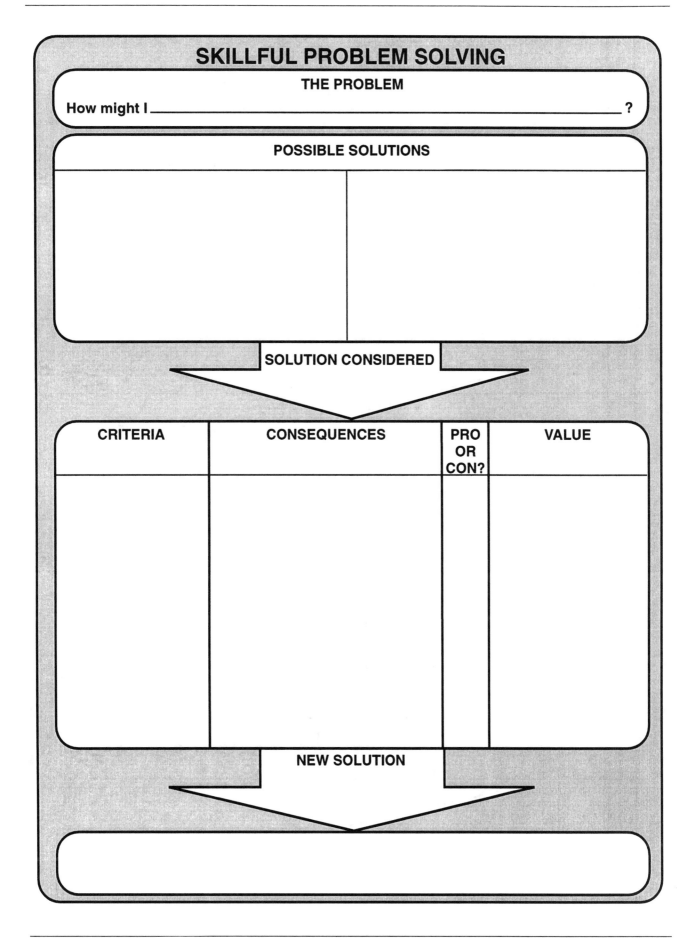

SKILLFUL PROBLEM SOLVING

THE PROBLEM

How might I _____ ?

POSSIBLE SOLUTIONS

SOLUTION CONSIDERED

CRITERIA	CONSEQUENCES	PRO OR CON?	VALUE

NEW SOLUTION

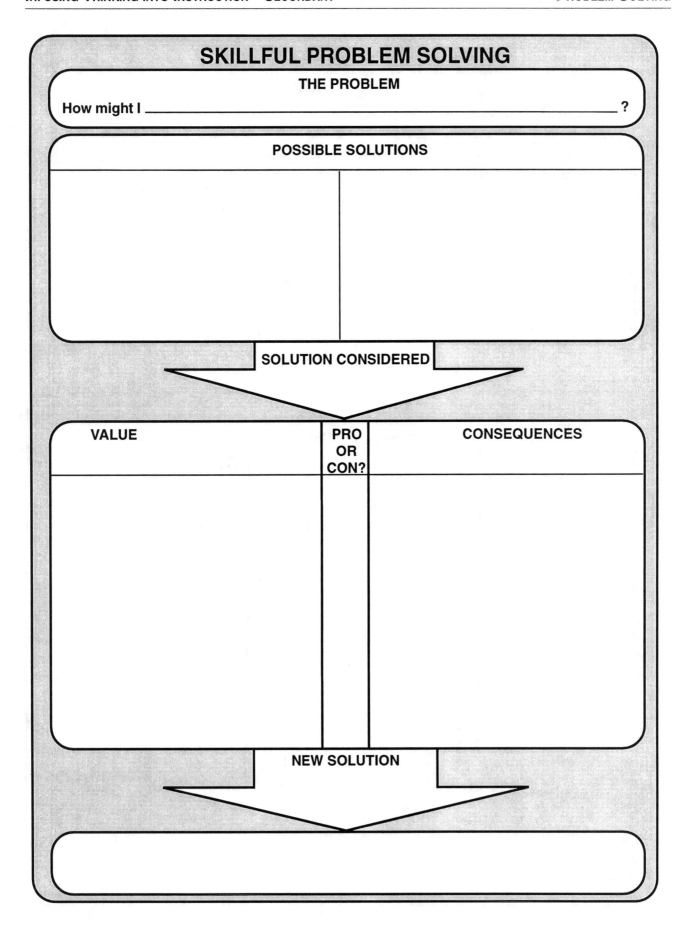

SKILLFUL PROBLEM SOLVING

THE PROBLEM

How might I _____ **?**

POSSIBLE SOLUTIONS

SOLUTION CONSIDERED

VALUE	PRO OR CON?	CONSEQUENCES

NEW SOLUTION

THE PROBLEM OF SILENT SPRING

Biology, Environmental Science   **Grades 9–12**

OBJECTIVES

CONTENT

Students will learn about major environmental disasters and how they were handled. Students will learn about the characteristics of pesticides, and in particular DDT, and the relationship between ecological pyramids, bioaccumulation of toxins, and human health.

THINKING SKILL/PROCESS

Students will learn to solve problems skillfully by determining what the problem is, by generating possible solutions, and by selecting the best solution on the basis of consequences.

METHODS AND MATERIALS

CONTENT

Students will read source material describing environmental disasters at Love Canal, Chernobyl, and Bhopal. Students will refer to relevant text material describing the nature and source of environmental pollutants.

THINKING SKILL/PROCESS

An explicit thinking map, graphic organizers, and structured questioning emphasize a thinking strategy for problem solving. Problem solving selections from the movie *Apollo 13* demonstrate the thinking skill. Collaborative learning enhances the thinking. Questioning strategies for metacognition are also employed.

LESSON

INTRODUCTION TO CONTENT AND THINKING SKILL/PROCESS

- Sometimes finding a solution to a problem is urgent. Not only do we have to come up with a feasible solution—one which is both practical and effective—but we also have to come up with the solution quickly. Sometimes there is real danger involved. Usually there is a clock ticking. Often, it takes the best solution to save the day; a poor solution usually leads to injury and death.

- The events which transpired on board Apollo 13, during four days in April 1970, presented just such a problem. Two days after Apollo's launch, while outbound to the moon, an oxygen tank exploded and catastrophically damaged life support systems on board. The supply of oxygen fell below that needed to successfully complete the mission. The problem quickly defined itself as "How might we safely return our astronauts to earth in a damaged spaceship that does not have enough oxygen to make the return trip?" Mission control, scientists, and other astronauts worked feverishly to solve the problem. In the end they used materials available on Apollo 13 to recycle oxygen purified by removing excess carbon dioxide. The astronauts were instructed in how to adapt lithium hydroxide canisters found in one part of the ship to the situation in which they were needed. Using an exact replica of the spacecraft, technicians explored ways to conserve power. Aquarius, the lunar descent module, was used as a lifeboat, its engines providing enough thrust to speed up the trip to the moon, whose gravity was used to whip the spacecraft back toward earth. The crew transferred to the Command Module when they neared the earth and safely landed. Within four days, a spacecraft was reconfigured,

> **SKILLFUL PROBLEM SOLVING**
>
> 1. Why is there a problem?
> 2. What is the problem?
> 3. What are possible solutions to the problem?
> 4. What would happen if you solved the problem in each of these ways?
> 5. What is the best solution to the problem?

a flight plan was rewritten and a challenging problem solved, resulting in the saving of three astronauts from certain death.

- Let's take a closer look at how this problem might have been solved. Here is a plan for skillful problem solving. Show students the Skillful Problem Solving verbal map. Use a transparency or put it on the chalkboard. **What do you notice about this plan?** STUDENT RESPONSES: *It is easy to follow; it proceeds one step at a time; it requires you to think about many solutions, not just one; it is easy to understand; it would be easy to follow.*

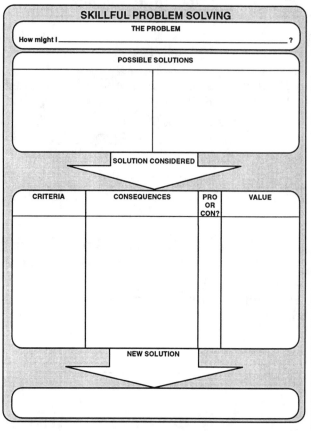

- Let's apply this plan to the Apollo 13 emergency so that we can see if it is an effective design for solving problems. If we can show that the thinking that so effectively solved that problem might have followed from a plan like this, then perhaps we can use this plan to solve other problems effectively. In your groups, pretend that you were on hand at Mission Control during the Apollo 13 mission. You have just heard Navy Captain Jim Lovell's famous understatement "Houston, we've had a problem." You are meeting with your team to think about solving the problem. In your groups, brainstorm answers to the first two questions on the skillful problem solving thinking map. **Be prepared to report back to the class.** Students readily explain that there is a problem because there was an explosion on board the Apollo 13 spacecraft. Students also have little difficulty defining the problem as *getting the men back to Earth* or *doing something about the oxygen shortage.* Have each group contribute and fill in responses on a class graphic organizer similar to the one at right.

- Let's think about Step 3. Take a few minutes to think about possible solutions to the problem. What might we do to "get the men home safely?" Allow adequate time for students to consider the problem. Then use guided questioning to elicit responses. Encourage "far out" and unusual solutions. Then add responses to the transparency or on the board. TYPICAL STUDENT RESPONSES: *Send up a manned rescue mission in another Apollo spacecraft; see if the Russians can help; send up an unmanned rocket with extra oxygen, which the astronauts can transfer to the Aquarius module after they enter into an earth orbit; put all but one of the astronauts into deep sleep to conserve oxygen; kill one of the astronauts so that there will be one-third more oxygen left.*

- Let's examine one of these possibilities, *Send up a manned rescue mission in another Apollo spacecraft.* **What would happen if you tried to solve the problem in this way; what would be the consequences?** STUDENT RESPONSES: *It might take too much time to prepare, check out, and launch an Apollo rescue mission. It would cost millions of dollars to use an Apollo spacecraft to rescue astronauts. There might not be enough room in the rescue command module for both crews. One of the future Apollo missions would have to be scrubbed because its spacecraft had been used in the rescue mission. If the rescue mission failed, we might lose two Apollo crews.* Students usually determine that any solution that might take a long time or one that would put others at risk would not be worthwhile. Students usually decide that the problem's solution must be implemented on board by the astronauts.

- **Do you think that the scientists involved in successfully bringing the Apollo 13 astronauts home might have followed a plan like this?** Most students point out that they certainly must have assessed the consequences of a number of possible solutions to arrive at the solution they did. Many also remark that they obviously didn't have the thinking map for causal explanation there before them, so they must be so used to these kinds of problems and to having to ask these questions that they've learned to ask them naturally. **Do you think that this plan is an effective one for determining the best solution to a problem?** Students agree that this kind of plan would work well to solve a problem.

- **Usually, there is the pressure of time associated with problem solving; however, dangerous situations—problems requiring fast and effective solutions—are often resolved over a much longer span of time than the few days of frenzied activity during the Apollo 13 mission. For example, global problems like overpopulation and epidemics are treated with a similar urgency; if solutions are not found, people may die. In your groups, take a few minutes and think about the world around you; then make a list of some of its urgent problems.** Discuss each response and use guided questioning to expand the list. STUDENT RESPONSES: *The hole in the ozone layer; war; diseases like AIDS; toxic rain, global warming and the greenhouse effect; air pollution from fossil fuels; toxic waste, water pollution; chemical pollution from pesticides and other chemicals; oil spills; poverty and malnutrition; radiation leaks from reactors like Chernobyl and Three Mile Island; nuclear stockpiles; biological and chemical weapons storage.*

THINKING ACTIVELY

- **Let's examine one of these problems more carefully; let's look at pollution from pesticides. In another lesson, we have already discussed Rachel Carson's book *Silent Spring*. What did we determine was the main cause of the problem?** Students usually recall that they concluded that the "silent spring" probably resulted from the toxic effects of pesticides like DDT. **Given this scenario—a small town in America and the moribund countryside and farmland surrounding it are dying because of pesticide contamination—let's use the skillful problem-solving plan we just evaluated to find a feasible solution to the town's problem. In order to help us organize our thoughts, we will be using a skillful problem-solving graphic organizer.** Show students the graphic organizer on the right. Use a transparency or draw it on the chalkboard. Be sure to point out that this graphic has been designed to examine, in depth, one possible solution at a time.

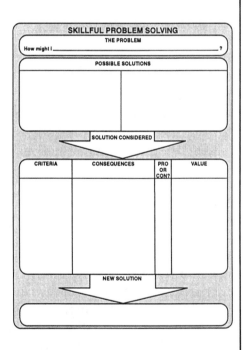

- **In your groups, write a statement that begins "How might I best" that clearly states the problem. For example in the Apollo 13 dilemma, the problem statement might have been written "How might I best return the astronauts to Earth safely?"** Students usually have little trouble in defining the problem. TYPICAL STUDENT RESPONSES: How might I *best protect the townspeople from the harmful effects of pesticides* or *best reverse environmental damage from pesticide contamination.* Sometimes students view the problem as one of *cleaning up and disposing of the pesticides* or less frequently one of *preventing future use of dangerous pesticides.* Use higher order questioning to generate a final class problem statement that the class accepts. Explain that the problem statement shouldn't be too narrow, (like "How might I rid my land of contamination?" or too remote (like "How might I prevent future uses of dangerous pesticides?"). Rather, the problem statement should focus on the immediate problem for the town and not presuppose that other problems have been solved. Write this in "The Problem" box on the graphic.

- **Let's think about possible solutions to this problem. In your groups, make a list of solutions you feel might solve this problem: "How might I best protect human health and limit environmental damage resulting from pesticide contamination of the town and its surroundings?"** STUDENT RESPONSES: *Permanently evacuate the town. Relocate the human population to an uncontaminated area. Temporarily evacuate the area and then defoliate the countryside. Demolish the town. Drain the pond. Destroy all wildlife; strip the soil. Transport all contaminated material to a toxic waste dump and then rebuild the town and resettle it. Quarantine the area by keeping it off limits and then monitor until pesticide levels come down to an acceptable level, then resettle the town. Relocate the population and then use the area as a new toxic waste dump site. Develop a chemical to neutralize the pesticide contamination and treat the countryside and environment with it. Develop an antidote to pesticide poisoning and treat people in the town with it. Make it illegal to use dangerous pesticides.*

- **Let's look at one of these possibilities,** *Permanently evacuate the area.* **Discuss in your group what might be the likely results of evacuating the town and relocating the human population. Make a list of three or four consequences and be prepared to report back to the class.** Allow adequate time for each group to brainstorm its list, and then have each group contribute one or two consequences. Record these on the skill-problem-solving transparency or use the chalkboard. STUDENT RESPONSES: *People would be completely out of danger. The ecosystem would be left alone and not destroyed, compared with toxic waste cleanup. Evacuation would cost much less money. Pesticides would remain in the countryside and pose a continued threat to nature. People would be immediately removed from the danger—no time wasted. The town would be "out of business." Livelihoods would be destroyed. The economy of the region might be crippled.*

- **Now that you have listed the consequences of permanently evacuating the area, let's consider whether each consequence argues for or against adopting this solution. Let's think about each of the consequences together and determine whether to mark "PRO" or "CON" next to it. Does the consequence that people would be completely out of harm's way count for or against the feasibility of this solution?** Students readily judge that the safety of people would count for adopting this solution. Brainstorm each of the remaining consequences and complete the PRO or CON column on the graphic organizer. Students usually find that most of the consequences count in favor of adopting this solution.

- **Sometimes, a consequence may count against a solution but it isn't considered very important when deciding whether a possible solution is feasible; for example, one consequence of having worked around the clock to reconfigure energy consumption on board the damaged Apollo 13 was the huge amount of overtime paid to technicians. Although this consequence would count against working all night, it was not considered very important in the context of saving the astronauts' lives. Hence, it was given much less weight than saving their lives. It is, therefore, necessary to evaluate just how important each consequence is before deciding whether a solution is worthwhile. In your groups, evaluate one consequence of permanently evacuating the area and write in the "Value" column whether that consequence is Very Important, Important, Not Very Important, or Unimportant. Be sure to state your reasons for making each judgment.** Assign a different consequence to each group. Allow adequate time for students to discuss and record their judgment. Then have them report back. Enter their responses on the class graphic organizer. Students usually decide that getting people out of harm's way and getting them away from dangerous pesticides immediately are very important considerations. *The population would no longer be exposed to pesticides.* They find that having to leave pesticides in the environment is also important but perhaps not as important because contamination will diminish over time. However, the future side effects on other living things of leaving the environment contaminated often weighs heavily with some students. The value of other consequences vary, usually between important and very important.

- **Do you think that this would be a viable solution, one which would protect human health and limit environmental damage?** Students immediately reach consensus that this would be a good solution for the human population but usually feel uncomfortable about deserting wildlife living in a countryside laden with toxic pesticides. Students also feel that because all the town's businesses and farms would be out of business, there might be economic hardship in the region that other solutions might remedy.

- **How might we make this solution an even better one? Let's make a list on the chalkboard of modifications to the original solution,** *permanently evacuate the area,* **which might make it even more feasible.** STUDENT RESPONSES: *You could make a maximum effort to collect mammals, birds, fish, and other animals and evacuate them to a new home also. You could set up a job-retraining program to help ease evacuees into new types of work. You could lobby the government to compensate farmers and businessmen for their losses and help them with start-up money at their new location.*

- **Let's take this information and use it to rewrite the original solution statement so that there will be fewer consequences counting against using it.** Work with students to generate a new solution statement. For example, *After permanently evacuating all people to a safe location, you could transfer as much wildlife as feasible to a pesticide-free environment and make available, through government funding, jobs retraining, and lost business compensation.*

- **As a final part of this activity, I'd like you to do some homework tonight. Members of each group will think through a skillful problem-solving graphic organizer for one of the remaining solutions. Tomorrow, during class, each group will present their findings, and then we will discuss which solution we believe is the best one.** Distribute blank photocopies of the graphic organizer and assign a different solution to each group. When students report back the next day, try to reach consensus on which solution might be best. Then ask the students to read the source material on Love Canal and compare it to their silent spring solution.

THINKING ABOUT THINKING

- **When you did the lesson on "The Mystery of Silent Spring," you took some time to think about your thinking after you tried to solve the mystery. Let's do the same thing now that you've worked on the problem caused by contamination with pesticides in Rachel Carson's book. How important do you think it was to try to solve the mystery before you tackled solving the problem faced by the town?** STUDENT ANSWERS INCLUDE: *Without knowing what caused the problem, it's pretty difficult to either figure out what the problem is or to try to solve it; it's like shooting in the dark. We need to do causal explanation first; it's the first question on the thinking map of problem solving. If we don't try to figure out what caused the problem, we won't really be able to solve it easily.*

- **What else besides finding the cause of the problem is it important to think about in problem solving?** Students quickly identify stating the problem, finding possible solutions, figuring out the consequences of the possible solutions, and then deciding which is the best solution by deciding which consequences are the focal points for skillful problem solving.

- **How does this way of problem solving compare to other ways that you've thought about solving problems in the past?** Students sometimes say that they think about possible solutions sometimes, or that they do think about the consequences of solutions to problems sometimes.

Most students recognize, however, that this strategy helps them to focus careful attention on each of these questions, as well as others, when they are problem solving. This, they say, makes their thinking more organized, and makes them feel that they've done a more careful job and that their solution is probably a good one.

- **Work in your groups now and draw a "picture" of the thinking process of skillful problem solving, incorporating these important focal points.** STUDENT DRAWINGS VARY.

APPLYING YOUR THINKING

Immediate Transfer

Soon after your students have completed the lesson on the problem of the silent spring, ask them to work on the following:

- **A species has become endangered. If nothing is done, the species will become extinct. Use the problem-solving strategy to determine what is the best remedy for saving this species from extinction.**

- **For 30 years, plutonium used in nuclear weapons was manufactured at a site in Hanford, Washington. A number of huge deteriorating tanks filled with high-level radioactive waste are close to leaking their contents into the atmosphere. What problems can you identify? Use the strategy for problem solving to decide what to do about them.**

Reinforcement Later

- **What is the best solution to the problem of acid rain that you diagnosed earlier?**

- **Landfills, as a way of disposing of waste, are becoming more and more scarce. What are viable alternate ways of waste management in this country?**

- **Identify a major problem faced by your school. Use the problem-solving strategy to think through what might be done to solve this problem.**

THINKING SKILL EXTENSION

Instead of the informal problem definition in this lesson, use the "What's the Problem" graphic organizer to prompt a more organized attempt at problem definition. Student results are included at the end of this lesson.

WRITING EXTENSION

Write a report recommending the courses of action that you think provide the best solution t the problem of the silent spring. Include in your report an explanation of why these are the best solutions to these problems. Reference any relevant supporting information.

CONTENT EXTENSION

What do we know about the effects of massive uses of DDT on the environment? Is this information well-supported?

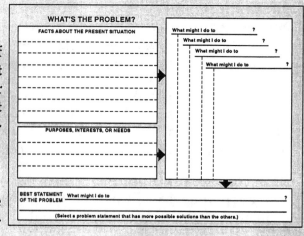

ASSESSING STUDENT PROBLEM SOLVING

Any of the transfer examples can be used as prompts to provide information on the basis of which your students' problem-solving skills can be assessed. Ask students to write out why they think the solutions they choose are best. Be sure that they are following the steps in the thinking map for problem solving.

THE PROBLEM OF SILENT SPRING

SOURCE MATERIAL ON CONTAMINATION AT LOVE CANAL, CHERNOBYL, AND BHOPAL

Love Canal excerpts from Love Canal, My Story, Louis Marie Gibbs, ISBN0-87395-587-0

"If you drove down my street *before Love Canal* (that's what I call what happened to us), you might have thought it looked like a typical American small town that you would see in a TV movie — neat bungalows, many painted white, with neatly clipped hedges or freshly painted fences. The houses are generally small but comfortable; at that time ("before Love Canal" in 1978) they sold for about $30,000. If you came in the summertime, you would have seen men painting their houses or adding an extra room, women taking care of gardens, and children riding bicycles and tricycles on the sidewalks or playing in the back yards.

"You would see something quite different today. Since Love Canal, the houses nearest the canal area have been boarded up and abandoned. Many have homemade signs and graffiti, vividly telling what happened to make this a ghost town. The once-neat gardens are overgrown, the lawns uncut. A high chain-link fence surrounds the houses nearest the canal. The area is deserted. The fence is a reminder of the 22,000 tons of poisons buried there, poisons that can cause cancer, that can cause mothers to miscarry or give birth to deformed children, poisons that can make children and adults sick, many of them in ways doctors only dimly understand ... there was a school within walking distance. The school's playground was part of a big, open field with houses all around.

"It is really something, if you stop and think of it, that underneath that field were poisons, and on top of it was a grade school and a playground. We later found out that the Niagara Falls School Board knew the filled-in canal was a toxic dump site. We also know that they knew it was dangerous because, when the Hooker Chemical Corporation sold it to them for one dollar, Hooker put a clause in the deed declaring that the corporation would not be responsible for any harm that came to anyone from chemicals buried there. That one-dollar school site turned out to be some bargain!

What is the Love Canal?

"The Love Canal is a hazardous waste dumpsite located in the center of a middle class community in Niagara Falls, New York; we are the first dump-site to be recognized, of the thousands of dumps across the nation. The EPA has recently estimated there are about 30,000 to 50,000 toxic waste dumps across the United States.

The History of Love Canal

"In 1892 William T. Love (where the name Love Canal originated), proposed connecting the upper and lower Niagara River, by digging a canal 6 to 7 miles long. However, the country fell into an economic depression and financial backing for the project slipped away. Love then abandoned the project, leaving behind a partially dug section of the canal. In 1920, the land was sold at public auction and became a municipal and chemical disposal site until 1953. The principal company that dumped their waste in the canal was Hooker Chemical Corporation, a subsidiary of Occidental Petroleum. The City of Niagara Falls and the United States Army used the site as well, with the city dumping garbage and the Army dumping possible chemical warfare material and parts of the Manhattan project.

"In 1953, Hooker, after filling the canal and covering it with dirt, sold the land to the Board of Education for $1.00. The deed contained a stipulation which said if anyone incurred physical harm or death because of their buried wastes, Hooker would not be responsible. Hooker continuously tells us they properly warned the city and the board; we wonder.

"Soon after the land changed owners, home building began adjacent to the 16-acre rectangle which was once the canal. The families were unaware of Love Canal when purchasing their homes. In 1955, an elementary school was opened; it had been erected near the corner of the canal.

"Residents began to complain about children being burnt, nauseous odors, and black sludge in the later 1950s, but nothing was done. It was not until the later 1970s that the government finally decided to investigate the complaints.

What Does the Canal Contain and How Far Has it Gone?

"There are 200 different compounds that have been identified so far, in and around the Canal. There are at least 12 known carcinogens, some human, other animal. Benzene is one which is well known for causing leukemia in people. Dioxin, the most deadly of all chemicals, has also been found in and around the Love Canal ...

"Hooker admits to burying about 21,800 tons of various chemicals in the Canal, but this is all they will admit to. The Army denies burying wastes, yet there are residents who testified to seeing Army personnel and trucks on site.

"The extent of chemical migration is still in question. Many of the air and soil-and-water tests have found chemicals throughout a 10-block residential area, in our creeks and the Niagara River ...

What are the Health Effects?

"Our Association, with the help of other scientists, conducted a health survey of our community. We were forced to do our own study because the governmental agencies would not conduct a good objective scientific study. We looked at the area as a whole, and found families who lived on underground streambeds had the highest incidence of disease, but also many families in the area were affected with an abnormally high rate of illnesses.

"The results of our studies showed above-normal amounts of miscarriages, 50-75 percent, while living in Love Canal. A birth defect rate in the past five years was 56 percent.

"Also, an increase in Central Nervous System disease including epilepsy, nervous breakdowns, suicide attempts, and hyperactivity in children, a greater chance of contracting urinary disorder, including kidney and bladder problems, an increase of asthma and other respiratory problems.

"Our most recent survey showed, out of the last 15 pregnancies in Love Canal women, we have had only two normal births. The rest resulted in a miscarriage, stillborn, or birth-defected babies.

The solution to the toxic waste problem implemented at Love Canal involved evacuating people from the immediate area of contamination, erecting fences to keep people out and later installing a drainage system consisting of perforated clay pipes to intercept contaminated water before it can leave the site. The polluted water is then treated by an activated charcoal system. The filtration system was then buried under topsoil and clay.

The Love Canal Homeowners Association is unhappy and have found "many problems with this system:

(1) At best it will contain wastes in the canal; it will not remove them.

(2) It will not remove or address any chemical wastes that have moved out into the ten-block area or through the storm sewer system.

(3) There are no monitoring wells placed; therefore, no one can tell us if it is even working.

(4) Will the clay crack, as it has done in the past, only to cause further contamination of our environment and threaten public health?"

Chernobyl: reference "Reassessing Nuclear Power: The Fallout from Chernobyl" *Worldwatch Pater #75*, Washington, D.C.: Worldwatch Institute

There was nothing particularly foreboding or unusual about the evening of April 25th, 1986 in the small Russian village of Pripyat in the remote Russian Ukraine. The villagers, as was their custom, retired early in the evening. But, as they slept, events were unfolding at the nearby Chernobyl Nuclear Power Plant which would lead to the worst disaster in the history of nuclear energy. Many villagers would not survive the night; even more would be made horribly sick as a result of their exposure to radioactivity. Ultimately, many more would die.

Like so many disasters in the history of technology, this one was also caused by human error. Chernobyl had been designed with many safeguards to defend against errors in human judgment. The design included elaborate safety systems with redundant backups. However, during the early morning hours of April 26th, as the village of Pripyat slept, engineers at the plant *intentionally* disconnected a number of safeguards including emergency core-cooling pumps.

The engineers chose to disregard standard operating procedures and to bypass safeguards so that they could perform an experiment; they wanted to answer the question: If there was an emergency shutdown of the reactor, would there be enough residual energy left in the plant's spinning turbines to operate the plant for a while. To answer this question, they planned on 1) slowing down one of the reactors to a small percentage of its normal operating power in order to mimic a shutdown and 2) turning off all electrical appliances and equipment not absolutely necessary to operate the plant in order to minimize emergency power consumption. The emergency core-cooling pumps were shut down for this reason.

Disaster struck when the graphite core of reactor number four became unstable as its operating power was cut back. Russian nuclear reactor technology depends on the use of graphite components. Graphite, used in reactor cores, tends to become unstable at low operating outputs. As reactor output sunk to about 6 percent of its maximum, the graphite core became unstable and the temperature within the core unexpectedly rose quickly. Ordinarily, core temperature is moderated with control rods; when they are fully extended into the core, the reactor shuts down; as they are withdrawn, the reactor pile undergoes increased nuclear fission and heats up. Ordinarily, should the core begin to overheat, additional cooling water can be brought on line to cool it down; or in the event the temperature rises very rapidly, the control rods can be slammed back into the core ("scrambled") and the reactor shut down. Because the emergency cooling pumps were disconnected, additional cooling water could not be brought on line; the only way to control the rising temperature in the core was to scramble the control rods. However, excessive heat warped the channels through which the rods needed to pass and they could not be extended deep enough to shut the core down.

The temperature within the reactor's core rose out of control until it reached "meltdown." Resulting explosions blew a hole in the concrete roof of the reactor; radioactive molten fuel, radioactive steam from vaporized cooling water, and tons of soot and ash became airborne.

Radioactive fallout in appreciable amounts reached France, Scandinavia, Germany, and Italy. Trace levels of radiation were detected in North America. As many as 300 people died as a result of their exposure to radioactivity. There are estimates that as many as 100,000 people will get cancer during the next 50 years because of this disaster. A significant rise in leukemia and birth defects has already been discovered in areas close to the accident.

Bhopal selection from *Environmental Science*, Jane L. Person, 1990, J.M. LeBel Enterprises, Inc.

The site of the world's largest industrial accident was the Union Carbide Pesticide Plant in Bhopal, India. On December 3, 1984, water accidentally entered a tank containing methyl isocyanate (MIC), a chemical used in making pesticides. The chemical reaction caused the pressure in the underground storage tank to increase. The safety system failed, and the very toxic MIC gas filled the air.

The Human Cost

Some of the MIC broke down into hydrogen cyanide, a poison used in the Nazi gas chambers in World War II. Unfortunately, no one realized this and a cyanide antidote was not used. Government figures show that 2,500 people died and another 200,000 people were affected. People are still dying because of lung or heart problems caused by the gas. Some complain of chest pains, breathlessness, and pain in their muscles. Those with lung damage cannot get enough oxygen for hard physical labor.

Some are improving and will be healthy again. Others will require medical care for the rest of their lives. It depends on how much gas they breathed. More than 2,000 of the women in neighborhoods affected by the leak were pregnant. The number of these women that miscarried was more than twice the usual number.

Sample Student Responses • The Problem of Silent Spring

SKILLFUL PROBLEM SOLVING

THE PROBLEM
How might I best protect human health and limit environmental damage from pesticide contamination in the town that experienced "the silent spring"?

POSSIBLE SOLUTIONS

• Permanently evacuate the town; relocate human population to an uncontaminated area.

• Temporarily evacuate the area; defoliate countryside, demolish town, drain pond, destroy all wildlife, strip soil, remove contaminated material to toxic waste dump and then rebuild and resettle.

• Quarantine area by keeping it off limits and then monitor until pesticide levels come down to an acceptable level; then resettle.

• Relocate population and then use the area as a new toxic waste dump site.

• Develop a chemical to neutralize the pesticides.

SOLUTION CONSIDERED
Permanently evacuate
the town

CONSEQUENCES	PRO OR CON?	VALUE
People would be completely out of harm's way.	PRO	**Very important**: The population would no longer be exposed to pesticides.
The ecosystem would be left alone.	PRO	**Important**: Without humans, no more pesiticides would be used. Nature could recover over time.
Compared with a toxic waste cleanup, evacuation would cost much less money.	PRO	**Important**: Money is hard to raise and would take time. Moving a population is cheaper than rebuilding a town.
Pesticides would remain in the countryside and pose a continued threat to nature.	CON	**Not important**: Pesticide contamination would diminish over time. Damage is already done.
People would be immediately removed from the dangerous pesticides. No time wasted.	PRO	**Very Important**: Pesticides are bioaccumulative. Urgent to get out of area as soon as possible.
The town would be "out of business." Livelihoods would be destroyed. The economy of the region might be crippled.	CON	**Important**: Although making a living is important, "living" is much more important.

NEW SOLUTION

After permanently evacuating all people to a safe location, you could transfer as much wildlife as feasible to a pesticide-free environment. Through government funding, you could make job retraining and lost business compensation available to the townspeople, merchants, and farmers.

Sample Student Responses • The Problem of Silent Spring

SKILLFUL PROBLEM SOLVING

THE PROBLEM
How might I best protect human health and limit environmental damage from pesticide contamination in the town that experienced "the silent spring"?

POSSIBLE SOLUTIONS

• Permanently evacuate the town; relocate human population to an uncontaminated area.

• Temporarily evacuate the area; defoliate countryside, demolish town, drain pond, destroy all wildlife, strip soil, remove contaminated material to toxic waste dump and then rebuild and resettle.

• Quarantine area by keeping it off limits and then monitor until pesticide levels come down to an acceptable level; then resettle.

• Relocate population and then use the area as a new toxic waste dump site.

• Develop a chemical to neutralize the pesticides.

SOLUTION CONSIDERED
Temporarily evacuate, defoliate, demolish town ... rebuild and resettle

CONSEQUENCES	PRO OR CON?	VALUE
The countryside would have to be replanted.	CON	**Very important**: Wildlife and plants would die, an ecosystem would be destroyed.
The countryside would be completely defoliated.	CON	**Important:** Could not replace old-growth forest and biodiversity.
The town would be demolished.	CON	**Very important**: Established businesses would be destroyed.
The town would have to be rebuilt.	CON	**Important:** Town landmarks would be destroyed.
Most of the pesticide contamination would be removed.	PRO	**Very important:** People could return to a safe environment.
It would cost a tremendous amount of money to relocate, house, and feed townspeople; defoliate; demolish; transport toxic waste.	CON	**Very important:** Extraordinary cost would likely run into tens of millions of dollars, well beyond the likelihood of government funding.

NEW SOLUTION

You could evacuate the townspeople but defoliate only those areas of the countryside with the worst pesticide contamination. You could preserve those buildings in town that have historical value. You might raise private funds by holding a telethon or selling T-shirts nationally.

WHAT'S THE PROBLEM?

FACTS ABOUT THE PRESENT SITUATION

In a once beautiful town in the countryside, animals, plants, and people have recently become very sick. A town that was once healthy and prosperous is now dying. It appears that environmental pollution caused by pesticides is the most likely cause of the problem. There is evidence that pesticides such as DDT are present in large amounts in the city and countryside. The water supply and food crops have been compromised and animals such as chickens, other birds, and pigs are not producing enough or healthy eggs. Humans have gotten sick, and many have abandoned the town.

PURPOSES, INTERESTS, OR NEEDS

The townspeople need a safe place to live and uncontaminated food to eat. They need relief from the toxic effects of pesticide poisoning.

POSSIBLE PROBLEM STATEMENTS

How might I:

- clean up pesticide contamination in the city and countryside?

- prevent the disaster from happening again?

- cure and treat sick animals and people?

- best protect human health and limit environmental damage from pesticide contamination?

- identify sources of pollution?

- restore the ecosystem?

- purify the water supply?

- develop treatments?

- raise money?

BEST STATEMENT OF THE PROBLEM

How might I best protect human health and limit environmental damage from pesticide contamination?

PROBLEM-SOLVING LESSON CONTEXTS

The following examples have been suggested by classroom teachers as contexts to develop infused lessons. If a skill or process has been introduced in a previous infused lesson, these contexts may be used to reinforce it.

GRADE	SUBJECT	TOPIC	THINKING ISSUE
6–8	Science	Deforestation	Old-growth forests in the Pacific Northwest are used as an important source of wood for the lumber industry. But old-growth forests are also home to many species, including the threatened spotted owl. What problems does this situation create? What might be done and why? Which is the best solution?
6–8	Science	Automobile emissions	Nearly all trucks and cars are powered by the internal combustion engine; we depend on it for cheap transportation. But car engines are powered by combustion that turns fossil fuel (gasoline) into carbon dioxide, water, heat, and air pollutants such as nitrous oxides and sulfur dioxides. What problems will the continuing use of the gasoline engine give us, and what might be done to solve these problems? Which solution might be the best one?
6–8	Science	Food production	Pretend that you are members of a "think tank" that is planning a trip to a distant planet—a trip that will take 5 years out and 5 years back. You have solved other problems and now you must think about the most difficult problem: How can the astronauts get enough food to survive the journey? What are possible solutions to this problem? What might be the best solution?
6–8	Science	Nuclear power plants	Fossil fuel reserves are becoming quickly depleted. The use of nuclear fission to produce heat to make steam to turn turbines to make electricity to power cities is considered by many to be an efficient and unavoidable alternative source of power for the future. However, the safe operation of nuclear power plants has long been a source of controversy. What problems will the continued and expanded use of nuclear energy confront us with, and what might be done to solve these problems? What might the best solution be?
9–12	Biology	CO_2/O_2, nitrogen and water cycles	You would like to set up a terrarium in your classroom. Starting from scratch, what will you need to build it, and what problems are you likely to be faced with when maintaining it? What are possible solutions? Which are best and why?
9–12	Biology	Hunting and the environment	Hunting in America has evolved from being a primary means of getting food to a sport. Some feel that hunting is wrong and should stop; others feel that responsible hunting is a tradition worth continuing but also necessary to keep populations of animals under control. Identify the environmental problems inherent in each point of view and their consequences and come up with a solution that might satisfy both sides of the controversy. Explore the best solution, in depth, and present your recommendations to the class.

PROBLEM-SOLVING LESSON CONTEXTS

GRADE	SUBJECT	TOPIC	THINKING ISSUE
9–12	Biology	Nuclear waste and the environment	For 30 years, plutonium used in nuclear weapons was manufactured at a 1500 sq. km. site at Hanford, Washington. This "nightmarish agglomeration of decaying, contaminated facilities" is leaking into the nearby Columbia River. Moreover, 177 huge deteriorating tanks filled with high-level radioactive waste are threatening to vent into the atmosphere. Identify potential clean-up problems. Consider possible solutions and then choose the best one. (ref: "Hanford's Nuclear Wasteland," *Sci Amer*, May, 1996)
9–12	Biology	Taxonomy, the phylogenic tree, and observation	You are walking in the woods and discover an animal you have never seen before. You would like to study it without disturbing it. What might you do so that you could learn enough about it to classify it and also describe its behavior without disturbing it very much? What are possible solutions to this problem? Which might be best? Why?
9–12	Biology	The Miller-Urey experiment	You are asked to recreate, in your school's science lab, the environment that existed on earth about the time that the first molecules of life evolved. What problems might you have in designing this experiment and completing the exercise? What would be the best solution? (ref: "Change Over Time," FAST, University of Hawaii, Page 225-6)
9–12	Chemistry	Radiation and its effects	We microwave, can, freeze-dry and freeze some of the foods we eat. These processes have proven to be safe. There is a new treatment that greatly increases the shelf life of food—irradiation. Irradiated foods don't sprout, and radiation kills microorganisms that cause food to decay. But irradiation is not being used widely. Identify possible problems resulting from this process. What might be a good solution? Why? (ref: *Chemistry*, Addison-Wesley, 1997, p. 762)
9–12	Chemistry	The combined gas law	If you live in remote areas, chances are your cooking is done by liquid propane (LP), which is stored in refillable pressurized tanks outside your home. Liquid natural gas (LNG) is another product. What problems might you face if you wanted to store and refill LNG the same way as LP? What would you have to know in order to determine the feasibility of doing so? What might be the best solution? Why?
9–12	Physics	Motion of falling bodies	You are asked to devise an experiment that would show that the law followed by falling bodies is not "the heavier the object, the faster it falls." What problems might you have in designing this experiment? What are possible experimental designs you might try? Which one is best? Why?

PART 3

SKILLS AT CLARIFYING IDEAS: THINKING FOR UNDERSTANDING

ANALYZING IDEAS

ANALYZING ARGUMENTS

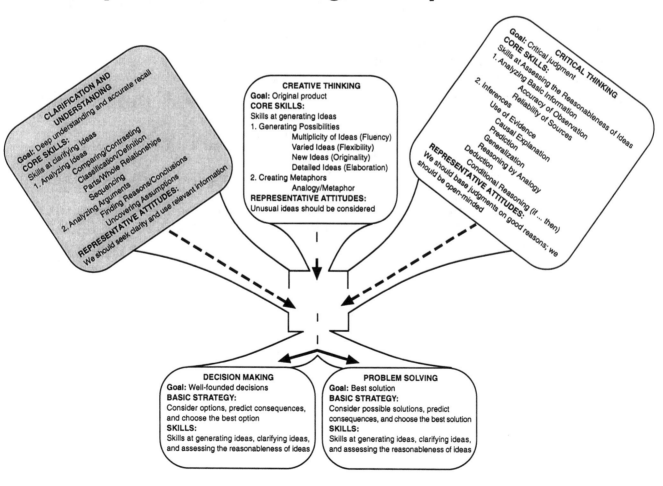

PART 3
SKILLS AT CLARIFYING IDEAS: THINKING FOR UNDERSTANDING

We seek deeper understanding of ideas for many different reasons. For example, I may read an ad for a low-priced vacation at a seaside resort. Before I decide to go there, I want to clarify what vacationing at this site will involve. Do the benefits pointed out in the advertisement give me good reasons for choosing this resort, or do I need more information? For example, how does this resort compare to the one I enjoyed visiting last year? What kind of resort is it: a health spa, a gambling casino, a sports facility? How is the restaurant rated? What will a vacation there be like; do they have any organized activities for guests, like sight-seeing or fishing trips? In deciding to go there, am I taking anything for granted, such as what the weather will be like?

Answering these questions skillfully will give me a deeper understanding of what vacationing at this resort is like and will enable me to make an informed choice. Not answering these questions may leave me with a superficial understanding that could result in an unwise decision. The difference between these two situations lies in the degree to which I attempt to *clarify* what I am considering. To clarify ideas, I use *analytical* thinking skills to provide me with insight and understanding.

Clarification involves basic analysis skills that fall into two categories. The first category involves analyzing ideas or things by:

- Determining their key properties and what having these properties implies (classification/definition);
- Breaking them into their parts and determining how these parts function in relation to the whole (determining parts-whole relationships);
- Comparing them to other things in order to bring out similarities and differences (comparing and contrasting); or
- Locating them in a sequence relative to other things that have different degrees of the same properties (sequencing/ranking).

The second type of analysis involves identifying the ingredients in the thinking that people engage in when they communicate ideas:

- Finding the reasons people (including ourselves) offer to support the ideas they try to convince others to adopt (finding reasons and conclusions); and
- Detecting what we take for granted, often without knowing it, in our actions and beliefs (uncovering assumptions).

Clarification yields deeper understanding. While understanding, alone, is not all there is to good thinking, we cannot do without it if we are to think about the world around us in an informed way. Teaching students the skills of clarification covered in this handbook will improve their abilities to achieve this understanding when it is needed.

CHAPTER 5
COMPARING AND CONTRASTING

Why is Skillful Comparing and Contrasting Important?

Comparing and contrasting involves detecting a variety of similarities and differences between two or more objects, events, organisms, institutions, or ideas in order to achieve certain specific purposes. Comparing and contrasting always involves analyzing features that match and features that do not and drawing out the implications of this analysis. Comparing and contrasting is helpful to gain a deeper understanding of the things compared in order to make well-considered decisions or to clear up confusion. Comparing and contrasting is also involved in more complex thinking tasks such as classification, definition, and reasoning by analogy.

Purposes of comparing and contrasting. We compare and contrast for a variety of purposes. Many everyday decisions, like shopping or choosing a route to work, involve comparing and contrasting. A manufacturer might compare and contrast his company with more successful companies to get ideas about improving productivity.

Our goal in comparing and contrasting may be to gain insight and understanding. We may compare and contrast a friend's new job to her old one because we notice that she is more relaxed than she was. We wonder whether her new job is contributing to this change.

Problems with the way we compare and contrast things. We compare and contrast with varying degrees of thoroughness. Sometimes we attend only to surface characteristics, like how things look, when other factors are more relevant.

Suppose I want to buy a car and am trying to decide between two models. If I consider only how they look, I may miss important differences in their gas mileage, durability, and overall performance. Making sure we consider all the relevant points of comparing and contrasting is very important.

While we often make only rough comparisons and contrasts, precision may be particularly important in certain contexts. I may note that both cars I'm considering cost under $10,000, yet determining the price precisely may reveal an important difference.

In determining equality, such as the congruence of two mathematical shapes or equality in job opportunities, we sometimes make rough comparisons and contrasts. This imprecision may lead us to think that the figures are equal or that any qualified individual's chances for being hired for a specific job are comparable when, in fact, subtle differences may result in significant inequalities.

Another problem arises when we do not realize the implications of the similarities and differences we note. I may realize that a car I am considering gets 10 miles per gallon less than another, but I may not consider what effect the better gas mileage will have on the long-term operating costs of the vehicle. In statistical analysis, difference in raw scores may seem important until we perform other procedures and find out whether the difference in scores is significant. Understanding the implications of similarities and differences is very important for successful comparisons and contrasts.

The four basic problems with comparing and contrasting are summarized in figure 5.1. The first default relates to quantity, the second to depth, the third to precision, and the fourth to completeness.

COMMON PROBLEMS WITH THE WAY WE COMPARE AND CONTRAST

1. We identify only a few similarities and differences.

2. We identify only superficial similarities and differences.

3. We make rough and imprecise judgments of similarity or difference.

4. We don't draw out the implications of the similarities and differences we have identified.

Figure 5.1

How Do We Compare and Contrast Skillfully?

Comparing and contrasting can lead to more effective choices and deeper insights if we focus our attention on a variety of important similarities and differences and take the time to think about what the similarities and differences show. There are two basic ways to compare and contrast skillfully:

- Open comparison and contrast;
- Focused comparison and contrast.

Open comparison and contrast. One way to counter making too narrow a comparison and/or contrast is to use brainstorming to identify as many different similarities and differences as we can. We then select those similarities and differences that are significant or relevant to our goals, explicitly drawing out their implications. We call this broad consideration of similarities and differences *open compare and contrast.*

Suppose I am comparing two cars to decide which one I want to buy and I prefer the color of one of them. Instead of deciding just on the basis of color, however, I then list other similarities and differences in order to make sure that I consider as many factors as possible. I then sort out which of the similarities and differences are significant, look for patterns, and ask what these significant similarities and differences show about the cars.

This broad comparison and contrast may reveal a variety of features relevant to my purchase. For example, I may discover certain hidden costs for one car that are not present for the other. This discovery may make a difference in my choice.

Open comparing and contrasting generates a broad consideration of what is being compared. It brings to light similarities and differences that we ordinarily might not take into account. The thinking map of open comparing and contrasting in figure 5.2 can guide us along the way.

Using a graphic organizer for open compare and contrast. The graphic organizer for open comparing and contrasting in figure 5.3 reinforces the process highlighted on the thinking map and provides spaces in which to record information.

OPEN COMPARE AND CONTRAST

1. How are they similar?

2. How are they different?

3. What similarities and differences seem significant?

4. What categories or patterns do you see in the significant similarities and differences?

5. What interpretation or conclusion is suggested by the significant similarities and differences?

Figure 5.2

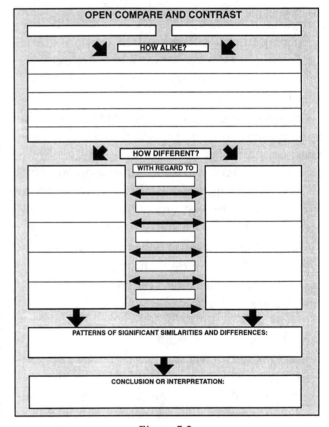

Figure 5.3

The use of this graphic organizer is very straightforward. First note similarities and write them in the "How Alike?" box. Then note differences and write them in the two outer columns under "How Different?" For each difference recorded, clarify the kind of difference you are recording by filling in the category word in the middle column.

For example, in comparing two cars, I might first notice that one costs $7,000 and the other $9,000. I would list this difference in the two columns. Then I should ask, "What kind of difference is this?" It is, of course, a difference in cost. "Cost" should be written in the corresponding space in the middle column.

We shouldn't stop with just a few similarities and differences or those of the same kind. We should search for as many varied similarities and differences as we can find. Taking time for this investigation is important, and working with someone else can be helpful. Brainstorming what to look for is a good way to assure the openness and breadth of this search.

When we categorize differences, we determine what these differences describe. This is very important in helping us articulate what the comparison and contrast shows when we later draw a conclusion. Categorizing the differences can also broaden our thinking by prompting us to consider other related differences that we may have overlooked, such as operating expenses.

The *patterns* step is important in considering complex examples. Asking whether there are important patterns revealed in the comparison and contrast prompts us to scrutinize the range of similarities and differences we've found to determine whether common themes emerge. I might notice, for example, that there is a pattern of "extras" on one car and not on the other. Since this is worth taking into account in making my purchase, I would write down "More extras on car A" in the *patterns* box. When I formulate a conclusion about the two cars, more "extras" may be a factor that I would want to mention.

Focused compare and contrast. There is a second way to compare and contrast that makes our search for similarities and differences more organized from the outset. We *first* determine the types of similarities and differences that we should consider in order to achieve our goal. These factors then guide our search for specific similarities and differences. When we have located information about these factors, we then draw conclusions about the two things compared. This is called *focused compare and contrast.*

In the case of the two cars, instead of noting similarities and differences as I found them, I would first ask, "What factors should I consider if I need a reliable car that I can afford and that will serve me well in the hot summer and cold winter?" I might then list factors like cost, performance in cold weather, etc. Noting these factors in advance will then guide my search for relevant similarities and differences. When these similarities and differences have been taken into account, I will be in a position to make a well-considered judgment about which car will serve my purpose better.

Notice that the thinking map for focused compare and contrast (figure 5.4) organizes the questions in a different sequence from the open compare and contrast thinking map (figure 5.2).

FOCUSED COMPARE AND CONTRAST

1. What is the purpose of the comparison and contrast?

2. What kinds of similarities and differences are significant to the purpose of the comparison and contrast?

3. What similarities fall into these categories?

4. What differences fall into these categories?

5. What pattern of similarities and differences are revealed?

6. What conclusion or interpretation is suggested by the comparison and contrast that is significant to its purpose?

Figure 5.4

Using a graphic organizer for focused compare and contrast. Notice that the graphic organizer in figure 5.5 reflects the different sequence of questions for focused comparing and contrasting. In using this graphic organizer, first state the purpose of the comparison and contrast in the purpose box, and then in the next box list the relevant factors to consider to accomplish this purpose.

For example, in answering the question "What factors should I consider in making my car purchase?" I would write "cost" and "performance in cold weather" in the "Factors to Consider" box, along with any other relevant factors that I should take into account. I would use this as a checklist to

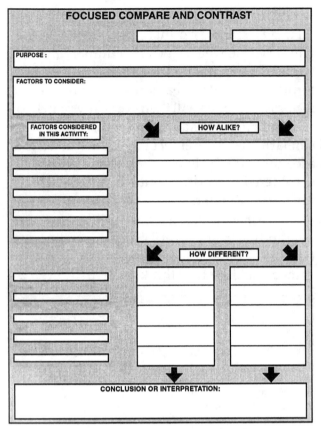

Figure 5.5

guide my search for relevant similarities and differences in the cars I am considering.

After gathering information, I write it in either the similarities or differences boxes, depending on what I find. If both cars cost $7,000, for example, then I would record this information in the similarities box and "Cost" in the corresponding "Factors Considered" box. If, on the other hand, one costs $6,000 and the other costs $9,000, I would write this information in the difference columns and write "Cost" in the corresponding box next to these columns.

Open vs. focused compare and contrast. Both the open and focused forms of comparing and contrasting go beyond merely listing similarities and differences. Both bring more organization and depth to comparing and contrasting than we ordinarily find when we just list similarities and differences. Both generate a conclusion or interpretation suggested by the comparisons and differences. Your choice of one or the other of these two strategies depends on whether you have a specific goal in comparing and contrasting or whether you are exploring the things compared to see what you can learn about them.

Other thinking skills that supplement comparing and contrasting. It is worth noting the interrelationship between comparing and contrasting and two other thinking skills: assessing the reliability of sources and drawing well-founded inferences.

Both open and focused comparing and contrasting presume that the information used is reliable and accurate. If we use misinformation, the interpretation or conclusion that is suggested by comparing and contrasting may be distorted or erroneous. When there is a question about the quality of the information, examining the reliability of the sources should be considered to supplement comparing and contrasting.

In both open and focused comparing and contrasting, when we draw out the implications of the similarities and differences, we do not necessarily follow guidelines for inference to determine whether or not our conclusion is well-supported. In some cases, conclusions may reflect faulty reasoning. Any conclusions or interpretations drawn from the similarities and differences should, therefore, be treated as suggestions only. We underscore this tentativeness by using the phrase "Conclusion Suggested" on both graphic organizers. If you are unsure whether your conclusion is well-supported, or if someone challenges it, you may check it out by using appropriate critical thinking skills of inference, described in the section on assessing the reasonableness of ideas (Chapters 13–18).

How Can We Teach Students to Compare and Contrast Skillfully?

Tips for teaching skillful comparing and contrasting. To teach comparing and contrasting well, it is not enough to ask students simply to list similarities and differences between two things. We must teach students explicitly what questions to ask to generate many kinds of similarities and differences, to select relevant similarities and differences for certain purposes, and to draw out the implications of the similarities and differences they've found. The goal of this instruction is to help students internalize these questions so they guide themselves in thinking skillfully when doing compare and contrast activities.

The thinking maps for comparing and contrasting (figures 5.2 and 5.4) contain questions that can be adapted to guide students through a compare and contrast activity in any instructional area. These guiding questions can be supplemented by using one of the appropriate graphic organizers. Guided activities in which students compare and contrast what they are studying will deepen their understanding of what they are comparing and contrasting. This active-thinking component of the lesson is an important step toward students' learning how to compare and contrast skillfully.

Prompting metacognition is another important step in teaching this skill. In order to use comparing and contrasting in other contexts, students must distinguish the thinking strategy from the content. To focus students' attention on the guiding questions, you can use several techniques: prompt them to recall the process by discussing it or writing about it, show them the thinking map, or help them construct a similar thinking map.

Make sure you give your students an opportunity to reflect on whether asking and answering these questions is a useful way to compare and contrast. Asking students for examples of things that might be compared and contrasted can enhance this kind of reflection.

Then, as in all infusion lessons, provide additional opportunities for students to practice skillful comparing and contrasting deliberately. Always strive to let your students guide themselves in this kind of thinking and to check themselves by referring to the thinking map and the graphic organizer. Repeated, increasingly self-directed practice in comparing and contrasting can lead your students to use this thinking skill as a matter of course.

Contexts in the content areas for compare and contrast lessons

Students have many opportunities to compare and contrast things. Repeatedly asking for comparisons and contrasts, however, can lead to ineffective overuse. To avoid this, employ compare and contrast lessons only to serve important content objectives in ways that reflect when we find it useful to compare and contrast in situations outside school, either in our professional work or in our everyday lives.

For example, we often compare and contrast unfamiliar things with things we know to get a better sense of what the new thing is like. This can be a great learning tool throughout the curriculum. Students who are familiar with the Revolutionary War in American history can compare and contrast it to the Civil War. When they realize that the latter war pitted Americans against Americans, students will understand how the Civil War had the greatest American casualties of any war that America has fought. Similarly, students can compare characters or works of fiction they are just starting to study with characters or works of fiction they are already familiar with to gain deeper insight into the new work they are studying.

In science, comparing and contrasting is an invaluable thinking activity. In the active work done in science, we often compare and contrast molecules, types of energy, chemical compounds, bodily functions, biological needs, habitats, and a myriad of other things. We do this type of comparison and contrast in science in order to understand the things we are comparing better and often to choose the best thing to serve a specific purpose. Besides having students compare and contrast the unfamiliar with the familiar, it is particularly helpful to infuse instruction in comparison and contrast when

- *Certain concepts are likely to be confused*, such as work and force, RNA and DNA, or endothermic and exothermic reactions;

- *Concepts are complex or abstract and important to distinguish from other concepts*, such as work and force, kinetic and potential energy, covalent and ionic bonds, and vector and scalar quantities;

- *Students will benefit from applying, clarifying, extending, or refining other key concepts* such as photosynthesis, genetic engineering, radioactivity, and infinity. Typically, compare and contrast lessons in this category work best when specific things are compared, one of which manifests the concept and the other doesn't (e.g., lead and uranium).

Reinforcing the skill. Once you have taught comparing and contrasting, you can use a variety of examples to reinforce it. These, too, should serve your content objectives. You can, for example, make comparing and contrasting a regular habit when you introduce a new concept in science (e.g., a new type of chemical reaction). You can ask your students to compare its properties to other types of reactions that they already know.

A menu of suggested contexts for infusion lessons on comparing and contrasting is provided at the end of the chapter.

Model Lessons on Comparing and Contrasting

In this chapter, we include two model lessons on comparing and contrasting: a middle school science lesson on plant cells and animal cells and a high school biology lesson on RNA and DNA. Both utilize open compare and contrast, the first to clarify two types of cells and their relationship to the organisms that they compose and the second to clarify molecules often confused by students. A focused compare and contrast version of the graphic organizer for compare and contrast is also included in the RNA/DNA lesson. As you review the lessons, try to answer these questions:

- What content objectives does comparing and contrasting activities in the lessons enhance? How?

- What verbal prompts guide students through the process of comparison and contrast so that they attend to a range of similarities and differences and draw out their implications?

- How does the graphic organizer reinforce this process?

- What methods are used in these lessons to enrich students' thinking?

- Can you think of other ways to reinforce skillful comparison and contrast after these lessons have been completed?

Tools for Designing Comparing and Contrasting Lessons

The thinking maps included in this chapter provide questions to guide students' thinking in open or focused comparing and contrasting lessons. Each of these thinking maps can also be used to help students engage in metacognitive reflection on how to engage in comparing and contrasting skillfully.

The graphic organizers for open or focused comparing and contrasting lessons are at the end of the chapter. Each of these graphic organizers supplement and reinforce the sequenced questions on the thinking maps. The thinking maps and graphic organizers can serve as photocopy masters, transparency masters, or as models that can be enlarged and used as posters in the classroom. Reproduction rights are granted for use in single classrooms only.

OPEN COMPARE AND CONTRAST

1. **How are they similar?**

2. **How are they different?**

3. **What similarities and differences seem significant?**

4. **What categories or patterns do you see in the significant similarities and differences?**

5. **What interpretation or conclusion is suggested by the significant similarities and differences?**

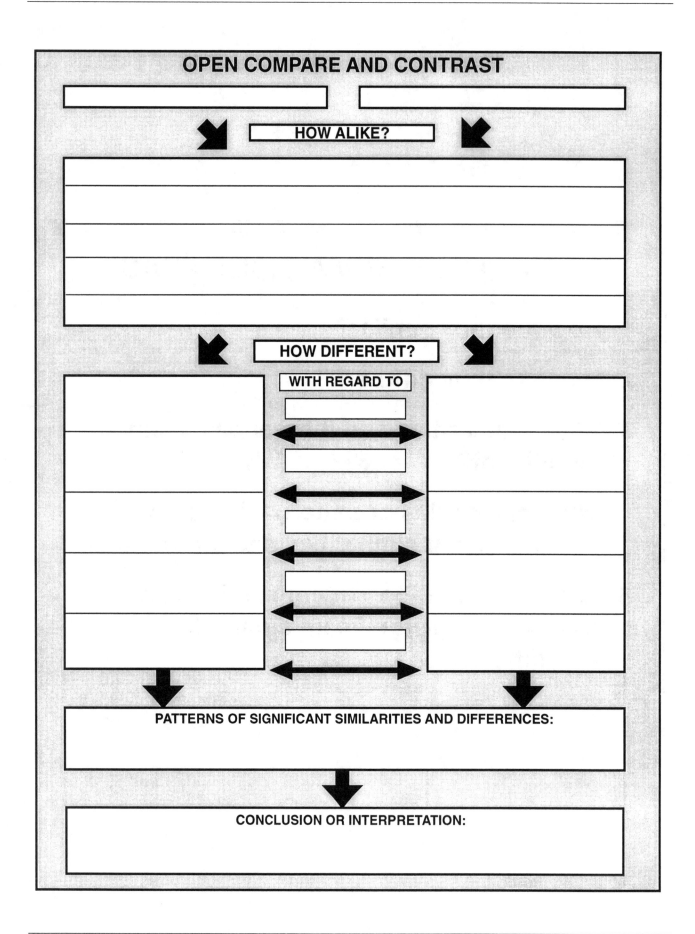

OPEN COMPARE AND CONTRAST

HOW ALIKE?

HOW DIFFERENT?

WITH REGARD TO

PATTERNS OF SIGNIFICANT SIMILARITIES AND DIFFERENCES:

CONCLUSION OR INTERPRETATION:

FOCUSED COMPARE AND CONTRAST

1. What is the purpose of the comparison and contrast?

2. What kinds of similarities and differences are significant to the purpose of the comparison and contrast?

3. What similarities fall into these categories?

4. What differences fall into these categories?

5. What patterns of similarities and differences are revealed?

6. What conclusion or interpretation is suggested by the comparison and contrast that is significant to its purpose?

FOCUSED COMPARE AND CONTRAST

PURPOSE :

FACTORS TO CONSIDER:

FACTORS CONSIDERED IN THIS ACTIVITY:

HOW ALIKE?

HOW DIFFERENT?

CONCLUSION OR INTERPRETATION:

PLANT CELLS AND ANIMAL CELLS

General Science, Biology, Anatomy, Physiology   **Grades 6–12**

OBJECTIVES

CONTENT

Students will learn the role of cells as building blocks of all living structures. They will differentiate between plant and animal cells with regard to their structure, function, and components.

THINKING SKILL/PROCESS

Students will learn to compare and contrast skillfully by finding significant similarities and differences, determining patterns in the similarities and differences, and by reaching a conclusion based on the comparison and contrast.

METHODS AND MATERIALS

CONTENT

Students use diagrams of a typical plant and animal cell and relevant textbook material to learn about their structure and function.

THINKING SKILL/PROCESS

This lesson features structured questioning, a compare and contrast graphic organizer, brain storming and metacognitive reflection to develop a plan for comparing and contrasting skillfully.

LESSON

INTRODUCTION TO CONTENT AND THINKING SKILL/PROCESS

- Ever notice how two things can do the same job but look different.? For example, let's consider bricks and cinderblocks. Both are basic building materials used in the construction of buildings. **Can you think of similar ways that these materials are used?** Ask for responses from the class. TYPICAL STUDENT RESPONSES INCLUDE: *When piled on top of each other they make up foundations which support an entire building. They are used to make walls that support floors and the roof of a building. Both are used to make fireplaces inside buildings.* **Can you think of other uses of these materials that are similar?** TYPICAL STUDENT RESPONSES INCLUDE: They *can also be used for building roads and for creating pathways. Both can be used to make retaining walls by landscapers. Both are found in the construction of dams.* **Things used to make buildings are called "materials of construction." Can you name other materials of construction? Think about building a house.** Have students brainstorm a list of common building materials. STUDENT RESPONSES INCLUDE: *lumber, nails, siding, cement, mortar, plywood, roof shingles, insulation, pipes, girders, rebars, sheetrock, screws, molding, doors, hinges, etc.*

- Although similar building materials are often used to accomplish the same building purpose, there are times when the differences between them make one a much better choice than the other. Architects earn their living by knowing all about building materials. They have learned to tell the differences between similar materials of construction and then choose which is best for a given job. For example, brick walls on a house look nice, but cinderblocks, if not plastered over with stucco or some other covering, would often look unattractive. Usually an architect will choose brick instead of cinderblocks for the exterior walls of a fine home because the characteristic of attractiveness is very important. Although cinderblocks could do the same job, they are rarely used for house exterior walls. Where do you usually find cinderblocks used as a material of construction for walls? Have students brainstorm a list in the class and report the items on their list. Ask for one item only from each student who responds. TYPICAL RESPONSES INCLUDE: *supermarkets, warehouses, lumber yards, gas stations, factories, underground foundations, basement walls, the walls of a school building.* Discuss with students the fact that in cases like these function is more important than good looks.

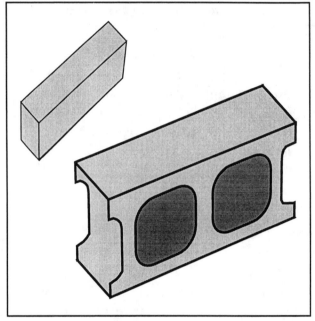

- Meet in your groups and make a list of differences between bricks and cinderblocks that might determine different uses, just as the look of these two types of building materials is what determines their use in the situations we just discussed. Pass out a diagram of a typical brick and cinderblock. Also, if available, bring a brick and a cinderblock to class so that students can examine them. TYPICAL RESPONSES INCLUDE: *Bricks are solid, cinderblocks are hollow. Bricks are smooth, cinderblocks are rough. Bricks are small; cinderblocks are much larger. Bricks come in colors like red, tan, and orange; cinderblocks are usually gray. Bricks are solid; cinderblocks have passageways through which pipes can travel. Bricks cannot be reinforced; cinderblocks are sometimes filled with concrete and iron rods.*

- Can you think of other building materials which can do the same job but look different? ANSWERS VARY BUT SOMETIMES INCLUDE: *Lumber and steel, vinyl siding and stucco, plastic (PVC) and lead pipes, tile and wood flooring, paneling and paint, screws and nails, flathead and roundhead screws, brads and nails.*

- What other characteristics do you think building materials might have that would determine how and when they are used? TYPICAL STUDENT RESPONSES INCLUDE: *how easily they bend (flexibility), how brittle they are, if they are waterproof, whether they are coarse or smooth, how much they weigh, their size, their shape, how long they last, their availability, their cost.*

- Another name for the process of looking at similarities and differences in order to reveal important characteristics of things is called comparing and contrasting. In the example of bricks and cinderblocks, we compared the similarities they both shared and then we contrasted their differences. Once we know these distinctions we are in a good position to draw an intelligent conclusion about which material would work best for a particular job. This is a thinking map of the questions that can guide us in engaging in this kind of comparing and contrasting. Notice how these questions include, but go beyond, asking only about similarities and differences. This is a typical way of comparing and contrasting. We're going to use this map to guide us through an activity in science in which we will explore the basic building blocks of living material.

> **OPEN COMPARE AND CONTRAST**
>
> 1. How are they similar?
> 2. How are they different?
> 3. What similarities and differences seem significant?
> 4. What categories or patterns do you see in the significant similarities and differences?
> 5. What interpretation or conclusion is suggested by the significant similarities and differences?

THINKING ACTIVELY

• Just as a house or a factory are made up of lots of basic building units like bricks and cinderblocks, the structures of nature—trees and fish and people—are also made up of basic materials of construction. The primary building blocks of living things are called cells. Cells, in fact, are the smallest units of construction which are able to carry on all the activities of living things. In this lesson, we are going to compare and contrast animal and plant cells to see what we can learn about these building blocks of living things.

• First, let's read some material about the structure and function of plant and animal cells. Distribute copies of the Animal Cell/Plant Cell Source Material to each student. **As you read about plant and animal cells, be sure to look closely at the diagrams of each so that you can identify their structures.** Allow students 6 or 7 minutes to become familiar with the material.

• Now that each of you has had time to become familiar with the layout of plant and animal cells, **organize yourselves into groups of four or five students each and examine the reading material and the diagrams more closely. Reread the material on animal and plant cells and look for similarities and differences between them. Use the compare and contrast diagram to record how they are alike and different.** Distribute a copy of the Open Compare and Contrast graphic organizer to each group. **When you discover similarities, list them in the box headed "How Alike?" The differences between the animal cell and the plant cell are less obvious. Be sure to read the material very carefully and examine the cell diagrams closely. When you find a difference, put it in the boxes under "How Different?" and think about what kind of difference it is, recording this in the box under "With Regard To." For example, you will notice immediately that plant cells are green and animal cells are colorless. That's a difference with regard to what?** The cell's color. **Record the word "color" in the box under "With Regard To."** If another example is needed, refer back to the earlier discussion of the differences between bricks and cinderblocks. Recall that during your discussion of bricks and cinderblocks they noted that bricks are small and cinderblocks are large. Ask what characteristic is being referred to. They should readily answer "size." Allow 9 or 10 minutes for group work on the compare and contrast graphic organizer.

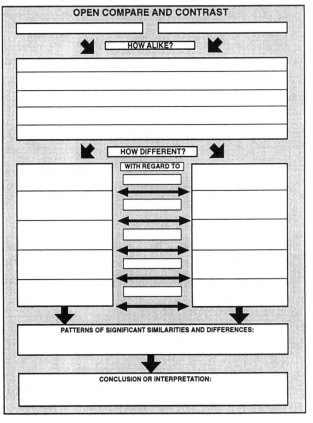

• **How are animal and plant cells alike?** After students complete significant work on their group graphic organizers, have one member of each team report back to the class by describing one similarity. Record responses on the chalkboard or a transparency made from the blackline master in this chapter. As students report on the similarities their group found, ask them questions of clarification that help them to elaborate the similarity and/or note its consequences or what it reveals about these cells. Encourage students to make use of information from the source material they have just read and anything they already know about these cells. TYPICAL

STUDENT RESPONSES: *They are both basic building blocks of living material. They both have a nucleus. They both have a cell membrane. They both have cytoplasm. They both are usually very small and require a microscope to see. They both reproduce by cell division. They both contain chromosomes. They both come in many sizes and colors. They both carry genetic material called DNA. They both can grow. They both can repair themselves. Plants and animals have many cells. Many different kinds of cells are found in plants and animals. The proper operation of the cells in both plants and animals keeps the plants and animals alive.*

- **How are animal and plant cells different?** As the reporter from each group contributes a difference, record the difference and ask him or her what kind of difference this is. Note the type of difference in the boxes beneath the "With Regard To" heading on the transparency or diagram you have written on the chalkboard. Ask the student who mentions the difference what it reveals about these cells. STUDENT RESPONSES INCLUDE: *Plant cells have a cell wall; animal cells have membranes only; Plant cells have one huge central vacuole; animal cells have several small vacuoles, if any. Plant cells have chloroplasts and mitochondria for energy production; animal cells have only mitochondria for energy production. Plant cells undergo photosynthesis; animal cells undergo cellular respiration. Plant cells are rigid; animal cells are flexible. Plant cells are usually green; animal cells are usually colorless. Plant cells have chlorophyll; animal cells don't. Plant cells need light to function properly; animal cells can function in darkness. Many kinds of animal cells have flagella for motion; only some plant sperm cells have flagella. Plant cells take in carbon dioxide from the atmosphere and give off oxygen; animal cells take in oxygen from the atmosphere and give off carbon dioxide.*

- **Now that you have stated how animal and plant cells are similar and different, we will use this information to give us insight into how and why these cells could be so much the same, yet in some ways very different. To do this, we will follow the thinking map for comparing and contrasting by asking the remaining three questions about the similarities and differences. Meet in your groups and first determine which of the similarities and differences you've come up with are significant. If a similarity or difference seems not very significant, draw a line through it. Then, see if you can discover some patterns of similarities and differences or major themes in what remains in the "How Alike" and "How Different" boxes. Write these, using only a word or short phrase, in the "Patterns" box on your diagram.** It may be useful here to have students remember the earlier example of the brick and the cinderblock. Ask them to brainstorm an answer to this question: **"Can you come up with a pattern of similarities and differences that might explain why bricks are used sometimes and cinderblocks preferred at other times?"** Students easily recognize that there is a pattern relating to appearance, cost, strength, and versatility in both bricks and cinder blocks, with bricks exemplifying a higher degree of these qualities. After giving students a few minutes for reflection, discussion, and recording their responses on the graphic organizer, ask for reports from a few groups using the same technique as used in getting reports about the similarities and differences they found. STUDENT RESPONSES INCLUDE: *Structure and components relate to what the cells do. Many cells make up both plants and animals. Both kinds of cells are basic building blocks. Internal mechanisms lead to cell reproduction and the growth of the living thing the cell is part of (variety of functions, variety of components). What happens inside these cells remains constant while changes are produced (energy users and producers).*

- **In your groups, think about and discuss these patterns of significant similarities and differences. Then express one or more of these patterns in some important insight or conclusion you come to about these two types of cells. Formulate your conclusion in one sentence only and write it in the "Conclusion or Interpretation" box on your team's graphic organizer.** If students aren't sure what you mean when you ask them to draw a conclusion, explain that a conclusion is not a summary of the similarities and differences but rather a statement that goes beyond what is in the list of similarities and differences yet is supported by them. Go back to the example of bricks and cinder blocks and give an example like "Cinder blocks are important to consider for buildings in which strength is needed and cost is a factor; bricks are important when these building materials show and looks matter." Explain to students how this goes beyond the similarities and differences stated earlier, yet is supported by them. You may also wish to tell students that you want them to come up with more substantive conclusions than just that both kinds of cells have some similarities and some differences. STUDENT RESPONSES INCLUDE: *The structure of typical plant cells reflects the basic features of plants, for example rigid walls for stem and trunk strength and chlorophyll for photosynthesis; animal cells similarly reflect basic features of animal life, for example flexible membranes and no rigid walls for mobility. Animal and plant cells both use raw materials to make the products and energy essential to sustain the activity and life of the things that they are part of. Cells are like factories using complicated activities for producing the ingredients that keep living things alive; but typical plant cells do this through mechanisms that involve photosynthesis triggered by light, and typical animal cells do this by using mechanisms that involve the breakdown of foods like glucose.*

• **Each member of your group should now pair up with a member of some other group for an activity called "Think-Pair-Share." In this activity, I would like one member of the pair to read his or her conclusion to the other, and the other member of the pair to help that person clarify and extend their thinking about their conclusion. The way the second member will do this is only by asking questions, not by making statements. There are three types of questions that can be asked:**

> <u>Questions of clarification</u>: **If you don't understand what a word or a statement means, you may ask questions to help you understand what is being said. For example, you may ask "What do you mean when you say_____?"**

> <u>Questions that extend the idea</u>: **If you think your partner is saying something interesting, but it is too brief, you can ask for more details about your partner's idea. You might say something like "What more can you tell me about_____?"**

> <u>Questions to challenge what is said</u>: **If you think the speaker is misled or confused, you may ask questions you think may prompt your partner to rethink or restate some part of his or her statement like "Why do you think_____?" Maybe the speaker will explain why and you won't think the statement is confused anymore, or maybe the speaker will reconsider aspects of the statement.**

After two minutes of reflection, signal students to change roles. After both partners have served as speaker and listener, allow students an opportunity to rewrite their statement in any way they see fit. Then ask if anyone would like to read the sentence to the whole class. Accept two or three students reading their sentences. Ask these students to read their sentences twice and ask the other students to listen to each statement, once for content and once for the kind of statement that is being read (comparison, contrast, comparison and contrast, generalization, etc.). Then, ask the class to suggest what type of information from the similarities and differences noted could be offered to support the statement if the statement was the main idea for an essay assignment. Create a composite bulletin board of students' conclusions about the two types of cells.

THINKING ABOUT THINKING

• **Let's stop thinking about plant cells and animal cells and focus our attention on what we did to think about these two types of cells. What kind of thinking did we do?** Students rapidly identify the type of thinking as comparing and contrasting. **What did we do to compare and contrast the two types of cells? What, for example, did you think about first? Next?** Prompt students to recall the steps in the process. Record their strategy on the board or use a transparency of the thinking map that was identified in the introduction, uncovering each step as students identify it. Review the discussion for each step of the thinking map of open compare and contrast.

- **How was the compare-and-contrast process different from just identifying similarities and differences? Is this way of comparing and contrasting more or less helpful in thinking about things? How?** Student answers usually focus on how this way to compare and contrast helps them to think about what they are comparing and contrasting more carefully than they would if they just listed similarities and differences, and to understand what they are comparing and contrasting much better. They comment that having to draw a conclusion, especially, gives them a chance that they rarely have to formulate ideas of their own about what they are comparing and contrasting.

- **How did the way that you compared and contrasted the two types of cells differ from the way you usually study important concepts in science?** Students say that comparing and contrasting helps them look for important information as they read different passages, instead of trying to learn everything in case it will be on a test.

- **Was using the graphic organizer helpful to you? How?** Students comment that using the diagram assists them in recording details that they notice and might otherwise forget. They also say that the graphic organizer helps to lead them to draw a conclusion from the similarities and differences they have listed.

- **In the Think-Pair-Share activity, was writing out your statement beforehand important?** Students recognize that, for clarity and ownership, having their thoughts written down before discussion frees them to examine the meaning and implications of their conclusions.

APPLYING THINKING

Immediate Transfer

- Compare and contrast two parallel processes in plant and animal cells—for example, meiosis and mitosis, or photosynthesis and cellular respiration.

- Compare and contrast two types of plant or animal cells: for example, cells from leaves and cells from roots, or epidermis cells and brain cells.

- Compare and contrast two breakfast cereals in order to decide which is a better buy and which is more nutritious.

- Use the compare and contrast strategy in a subject other than science to compare and contrast two ideas, stories, characters, historical figures, or countries that you have been studying in order to learn something important about them.

Reinforcement Later

- Compare and contrast two different pieces of music.

- Compare and contrast two animals or two plants that you are studying in science.

- Use the compare and contrast strategy to help you decide how you will spend some block of free time that you have on the next weekend.

WRITING EXTENSION

Write a compare and contrast essay about plant and animal cells, using your conclusion as the topic sentence or main idea. Use the first paragraph to explain your conclusion, the second to state important similarities between plant and animal cells that support your conclusion, and the third paragraph to indicate which differences also support your conclusion. End the essay with a concluding paragraph restating your topic sentence and adding closing comments about your main idea.

HIGH SCHOOL VERSION

This lesson can also be taught in high-school biology courses. Material from a high-school textbook or other source material used in high school can be used instead of the passages on plant cells and animal cells included in this lesson. A compare-and-contrast graphic organizer filled in by high school students is included at the end of this lesson as an example of what high-school students can do in comparing and contrasting these two types of cells.

ASSESSING STUDENTS' THINKING
WHEN THEY COMPARE AND CONTRAST

To assess this skill, ask students to write an essay to answer any of the application questions or others you select. Ask students to describe how they compared and contrasted the two subjects. Determine whether they are attending to each of the steps of comparing and contrasting, and whether they have determined significant similarities and differences between the two figures.

ANIMAL CELLS AND PLANT CELLS
MIDDLE SCHOOL SOURCE MATERIAL ON ANIMAL AND PLANT CELLS

ANIMAL CELLS

Cells are the basic units of life. Each cell can carry on the basic activities of living tissues. Animal cells are the building blocks of animal tissue. They are usually very small and require a microscope to be seen. They appear colorless and nearly transparent. Animal cells do many different jobs. For example, they can work as blood cells carrying oxygen or nerve cells conducting electric signals.

Animal cells are made up of many parts. They are surrounded by a cell membrane, which allows only needed substances like water and nutrients to pass through while it keeps important substances like genetic material inside. The cell membrane is flexible; therefore, animal cells can change shape.

Animal cells also have a nucleus, which controls the activities that take place in the cell. Inside the nucleus are chromosomes. These are rope-like structures made of DNA, a chemical that acts like a blueprint and carries instructions for making more cells.

The inside of the cell is filled with a jellylike fluid called cytoplasm. The cytoplasm is like a thick soup filled with small structures that have specific jobs to do in the cell. These structures are called organelles, and they work together to keep alive molecules used by the cell. Animal cells also have centrioles, a structure needed for cell reproduction.

Animal cells come from other animal cells by the process of cell division. During cell division, one cell makes a second copy of its genetic material, its nucleus and its organelles. The original cell divides in half, makes more cell membranes and becomes two smaller cells. Each cell grows until it reaches its original size.

Animal cells need energy to do all this work. They get the energy from food molecules. After food molecules are digested, they enter the cell and are used as fuel to produce chemical energy in a special organelle called a mitochondria. The process of converting food into chemical energy is called cellular respiration. This process requires oxygen and food and produces heat, carbon dioxide and water. The energy is then used to carry on activities like reproduction, growth, and movement.

PLANT CELLS

Plant cells are the building blocks of plant tissue. They are usually very small and require a microscope to be seen. They often appear green because many plant cells contain the green pigment chlorophyll. Plant cells can do many different jobs, including working as root cells absorbing water or as leaf cells collecting sunlight.

Plant cells are made up of many parts. They are enclosed in a two-layer covering made up of a cell membrane and a cell wall. The cell membrane allows only needed substances like water and nutrients to pass through and serves the dual purpose of keeping important substances inside. The cell wall is a rigid shell that surrounds the cell membrane. It is not very flexible. Therefore, most plant cells do not change shape. The cell wall gives plant cells a solid structure so that they can be built into structures like tree trunks.

Plant cells also have a nucleus, which controls cellular activities. Inside the nucleus are chromosomes. These are rope-like structures made of DNA, a chemical that acts like a blueprint carrying instructions for making more cells.

Cells are filled with a jellylike fluid called cytoplasm. The cytoplasm is like a thick soup filled with small structures that have specific jobs to do in the cell. These structures are called organelles, and they work together to keep alive molecules used by the cell. Plant cells also have a large central vacuole, which stores water.

Plant cells come from other plant cells by the process of cell division. During cell division, one cell makes a second copy of its genetic material, its nucleus and of its organelles. The original cell divides in half, makes more cell membranes, and becomes two smaller cells. Each cell then grows until it reaches it original size.

Plant cells need chemical energy to do all this. They get the energy in a process called photosynthesis. During photosynthesis the energy of sunlight is used to assemble food molecules in chloroplasts. The process requires carbon dioxide and water and produces oxygen and food molecules. Then the food molecules are used as fuel to produce chemical energy in mitochondria. The chemical energy is then used to carry on activities like reproduction, growth, and movement.

Sample Student Responses • Plant and Animal Cells • Middle School

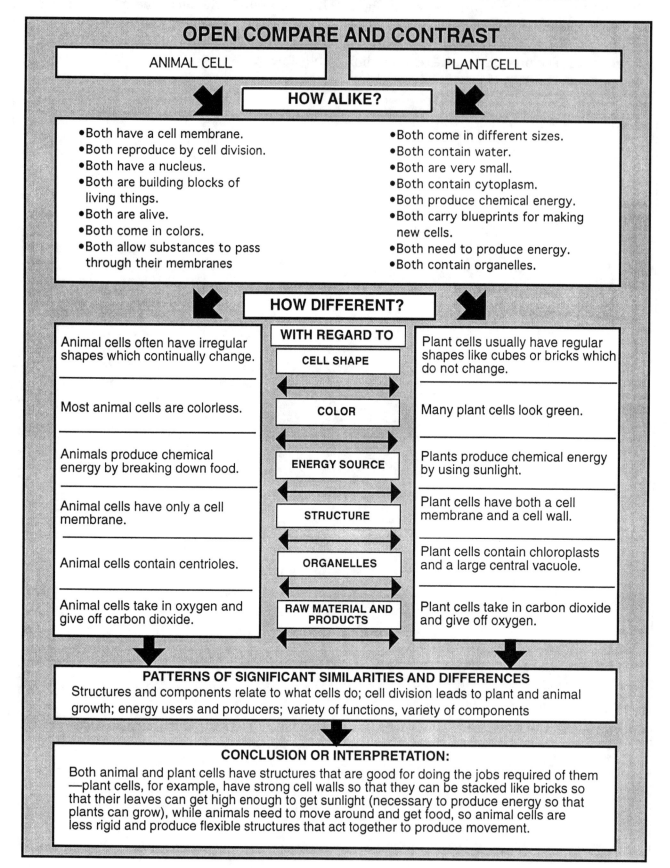

OPEN COMPARE AND CONTRAST

ANIMAL CELL	PLANT CELL

HOW ALIKE?

- Both have a cell membrane.
- Both reproduce by cell division.
- Both have a nucleus.
- Both are building blocks of living things.
- Both are alive.
- Both come in colors.
- Both allow substances to pass through their membranes

- Both come in different sizes.
- Both contain water.
- Both are very small.
- Both contain cytoplasm.
- Both produce chemical energy.
- Both carry blueprints for making new cells.
- Both need to produce energy.
- Both contain organelles.

HOW DIFFERENT?

	WITH REGARD TO	
Animal cells often have irregular shapes which continually change.	**CELL SHAPE**	Plant cells usually have regular shapes like cubes or bricks which do not change.
Most animal cells are colorless.	**COLOR**	Many plant cells look green.
Animals produce chemical energy by breaking down food.	**ENERGY SOURCE**	Plants produce chemical energy by using sunlight.
Animal cells have only a cell membrane.	**STRUCTURE**	Plant cells have both a cell membrane and a cell wall.
Animal cells contain centrioles.	**ORGANELLES**	Plant cells contain chloroplasts and a large central vacuole.
Animal cells take in oxygen and give off carbon dioxide.	**RAW MATERIAL AND PRODUCTS**	Plant cells take in carbon dioxide and give off oxygen.

PATTERNS OF SIGNIFICANT SIMILARITIES AND DIFFERENCES

Structures and components relate to what cells do; cell division leads to plant and animal growth; energy users and producers; variety of functions, variety of components

CONCLUSION OR INTERPRETATION:

Both animal and plant cells have structures that are good for doing the jobs required of them —plant cells, for example, have strong cell walls so that they can be stacked like bricks so that their leaves can get high enough to get sunlight (necessary to produce energy so that plants can grow), while animals need to move around and get food, so animal cells are less rigid and produce flexible structures that act together to produce movement.

ANIMAL CELLS AND PLANT CELLS
HIGH SCHOOL SOURCE MATERIAL
ON THE STRUCTURE OF ANIMAL AND PLANT CELLS

ANIMAL CELLS

The animal cell is the basic building block of animal tissue. It is usually microscopic. Animal cells have many specialized functions including those that provide structure, shape, nervous pathways, immunity, secretions, and reproductive machinery for the organism. Animal cells come in a variety of shapes and sizes ranging from the extremely small and round human egg to the extremely long motor neurons that connect our brains to our muscles. Animal cells can grow, respond to their environment, reproduce, and repair themselves.

A typical animal cell is made up of four major components. It has a cell membrane, a nucleus, a jellylike material called cytoplasm, and a set of structures suspended in the cytoplasm known as organelles. It also has an internal skeleton made up of tubes and fibers.

Many animal cell organelles are contained in their own cell membranes and include mitochondria, lysosomes, Golgi bodies, smooth and rough endoplasmic reticulum, and microbodies. Other cell structures, such as ribosomes and centrioles, do not have a cell membrane. Some animal cells, like the single cell of the protist *paramecium*, also contain small compartments called vacuoles, which collect excess water.

The animal cell nucleus is surrounded by a double cell membrane called a nuclear envelope, which is perforated by small passageways known as nuclear pores. The nucleus contains DNA, the cell's genetic material; DNA is found in long strands called chromatin that, during cell division, condenses into thicker coiled structures called chromosomes, which can be seen with the light microscope. The nucleus also contains a nucleolus, the site where ribosomes are manufactured, and a jellylike granular fluid called nucleoplasm.

The internal skeleton of the animal cell is a protein scaffold made up of three types of fibers: microfilaments, which can contract and cause cells to move or change shape; intermediate filaments, which are very strong and anchor organelles within the cytoplasm; and microtubules, which are hollow tubes used to maintain cellular shape. Animal cells have a pair of microtubular structures called centrioles, which are used during cell division.

Animal cells can also have several long, whiplike projections called flagella, which are used for locomotion by sperm and other kinds of cells. Cilia also extend from the surface of many kinds of animal cells. Single-celled organisms use these cilia for movement, and complex animals use them to move material over the surface of cells. For example, cilia move mucous out of lungs.

The cytoplasm of the animal cell is a soup of water, dissolved nutrients, minerals, electrolytes, small molecular building blocks like amino acids, macromolecules such as hormones and enzymes, and metabolic waste.

The animal cell membrane separates the cell from its environment. Like other cellular membranes, it is made up of a double layer of lipid molecules called a phospholipid bilayer. It functions to control the movement of substances into and out of the cell. It is considered a selectively permeable membrane because it allows only some substances to pass through.

Energy in the form of adenosine triphosphate (ATP) is produced in mitochondria by breaking down the food molecule glucose. This energy is used to power all the activities of the cell. The process is called cellular respiration. It requires oxygen and food, and it produces heat, carbon dioxide, and water.

Animal cells reproduce by cell division. The process by which one cell produces two identical offspring cells is called mitosis. When the organism reproduces, specialized cells called sex cells produce gametes (sperm, eggs). The process by which gametes are produced from sex cells is called meiosis.

Although animal cells are flexible and at times mobile, they maintain a very constant internal environment. Each of the many different kinds of animal cells contributes to the overall stable internal environment of the organism. Keeping the environment stable is a process called homeostasis.

PLANT CELLS

The plant cell is the basic building block of plant tissue. It is usually microscopic. Plant cells have many specialized functions including those that provide structure, shape, protection, secretions and reproductive machinery for the organism. Plant cells come in a variety of shapes and sizes. They can grow, respond to their environment, reproduce, and repair themselves.

A typical plant cell is made up of four major components. It has a cell covering made up of a cell wall and cell membrane, a nucleus, a jellylike material called cytoplasm and a set of structures suspended in the cytoplasm known as organelles. It also has an internal skeleton made up of tubes and fibers.

Many plant cell organelles are contained in their own cell membranes and include mitochondria, chloroplasts, a central vacuole, lysosomes, Golgi bodies, smooth and rough endoplasmic reticulum, and microbodies. Other cell structures do not have a cell membrane—for example, ribosomes. Nearly all plant cells contain one or several large membrane–enclosed compartments called central vacuoles that contain water and enzymes, can enlarge and shrink, and function in cellular digestion in changing the cell's shape.

The plant cell nucleus is surrounded by a double cell membrane called a nuclear envelope which is perforated by small passageways known as nuclear pores. The nucleus contains DNA, the cells' genetic material, and is found in long strands called chromatin. During cell division, the chromatin condense into thicker coiled structures called chromosomes which can be seen with the light microscope. The nucleus also contains a nucleolus, the site where ribosomes are manufactured, and a jellylike granular fluid called nucleoplasm.

The internal skeleton of the plant cell is a protein scaffold made up of three types of fibers: microfilaments which can contract and cause cells to move or change shape; intermediate filaments which are very strong and anchor organelles within the cytoplasm; and microtubules, which are hollow tubes used to maintain cellular shape.

The cytoplasm of the plant cell is a soup of water, dissolved nutrients, minerals, electrolytes, small molecular building blocks like amino acids, macromolecules such as hormones and enzymes, and metabolic waste.

In plant cells a cell wall made of cellulose surrounds the cell membrane. The cell wall functions to give plant cells rigidity for support. Plant cells communicate through adjacent cell walls via small channels called plasmodesmata. The plant cell membrane, like other cellular membranes, is made up of a double layer of lipid molecules called a phospholipid bilayer. It functions to control the movement of substances into and out of the cell. It is considered a selectively permeable membrane because it allows only some substances to pass through.

Energy in the form of ATP is produced in chloroplasts by converting the energy of sunlight into chemical energy. This energy is then used to produce the food molecule glucose. The process is called photosynthesis, requires carbon dioxide, water, and light energy, and produces glucose and oxygen. Plant cells then undergo cellular respiration in mitochondria where chemical energy stored in glucose made during photosynthesis is converted into ATP which is used to power all plant cell activities.

Plant cells reproduce by cell division. The process by which one cell produces two identical offspring cells is called mitosis. When a plant reproduces, specialized cells called sex cells produce gametes (sperm, eggs). The process by which gametes are produced from sex cells is called meiosis.

CROSS-SECTION OF AN ANIMAL CELL

A   Cell Membrane   E   Chromatin

B   Nuclear Envelope   F   Nuclear Sap

C   Nuclear Pore   G   Mitochondrion

D   Nucleolus   H   Golgi Complex

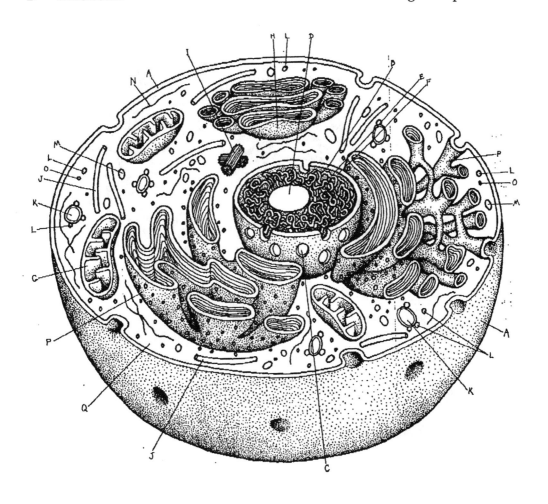

I   Centriole   N   Microfilament

J   Microtubule   O   Ribosome

K   Vacuole   P   Endoplasmic Reticulum

L   Lysosome   Q   Hyaloplasm

M   Microbody

CROSS-SECTION OF A PLANT CELL

A   Cell Membrane

B   Cell Wall

C   Plasmodesma

D   Vacuole

E   Crystal

F   Chloroplast

G   Leucoplast

H   Chromoplast

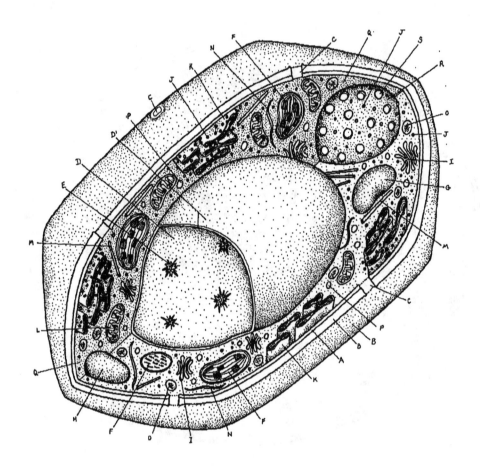

I   Golgi Complex

J   Ribosome

K   Endoplasmic Reticulum

L   Mitochondrion

M   Microtubule

O   Lysosome

P   Microbody

Q   Hyaloplasm

R   Nuclear Envelope

S   Nuclear Pore

Sample Student Responses • Plant and Animal Cells • High School

OPEN COMPARE AND CONTRAST

Animal Cell	Plant Cell

HOW ALIKE?

- Both are basic units of living material.
- Both carry on all processes necessary for life.
- Both are eukaryotes.
- Both are microscopic.
- Both contain a nucleus and nuclear envelope.
- Both contain chromosomes made up of DNA, which carries genetic information.
- Both can undergo cell division by mitosis & meiosis.
- Both can grow.
- Both are contained in a selectively permeable cell membrane.
- Both contain cytoplasm.

- Both produce chemical energy.
- Both can repair themselves.
- Both contain many identical organelles: mitochondria, ribosomes, Golgi complex, ER, microtubules, etc.
- They both form tissues.
- They share many common survival needs: water, minerals, electrolytes, etc.
- They both manufacture hormones and enzymes, proteins necessary for survival.
- They both can survive for long periods of time.
- Both are made of proteins, lipids, carbs & nucleic acids.
- Both require an energy input to carry on life processes.

HOW DIFFERENT?

Animal Cell	WITH REGARD TO	Plant Cell
Animal cells require nutrition as a source of energy, which is converted to chemical energy in the process of cellular respiration in the mitochondria.	**ENERGY**	Plant cells require sunlight as a source of energy which is converted to chemical energy in chloroplasts in the process of photosynthesis.
Animal cells are contained in a flexible selectively permeable cell membrane which maintains its shape using a network of microtubules & microfilaments.	**STRUCTURE**	Plant cells are enclosed in both a cell membrane and a cell wall. The cell wall, made up of cellulose, provides a rigid structure for the plant cell.
Animal cells contain centrioles. Single cellular animals are the only animal cells which contain plastids. Animal cells contain few, if any, small vacuoles.	**ORGANELLES**	Many plant cells contain chloroplasts & many other plastids in large numbers. Plant cells contain one or two large central vacuoles which store water.
Single cellular animals can move as a result of the whipping movement of flagella and the beating movement of cilia.	**MOBILITY**	Plant cells are immobile and lack flagella. They move in response to stimulation as a result of growth hormones acting on selected cells.
Animal cells function to maintain homeostasis as they generate enough energy to provide for the acquistion of food via movement.	**FUNCTION**	Plant cells function to maintain homeostasis by providing support and protection to provide for the acquisiton of light via position.

PATTERNS OF SIGNIFICANT SIMILARITIES AND DIFFERENCES:

Neither has organelles nor structures which are not used. Each has unique structures and organelles necessary for the acquistion and processing of energy. Both use the same equipment to maintain homeostasis, synthesize cellular products, and reproduce.

CONCLUSION OR INTERPRETATION:

A cell's structure is determined by how it functions to maintain homeostasis and how it acquires energy. While both kinds of cells maintain an internal environment the same way, at some point in the history of life on Earth cells evolved two processes for obtaining energy, thereby accounting for the few significant differences between plant and animal cells.

DNA AND RNA

Biology, Anatomy, Physiology **Grades 9–12**

OBJECTIVES

CONTENT

Students will learn the central dogma of molecular biology, the role of deoxyribonucleic acid (DNA) in the replication of living cells. They will differentiate between DNA and RNA (ribonucleic acid) and will discover the connection between the structure and function of DNA and RNA.

THINKING SKILL/PROCESS

Students will learn to compare and contrast skillfully by finding significant similarities and differences, by determining patterns in the similarities and differences, and by reaching a conclusion based on the comparison and contrast.

METHODS AND MATERIALS

CONTENT

Students use relevant textbook material to review the structure and functions of DNA and RNA. Selections from an article, "The Best Deals on Street Wheels," *Consumer Reports*, are used to introduce skillful comparing and contrasting.

THINKING SKILL/PROCESS

This lesson features structured questioning, a compare and contrast graphic organizer, a think-pair-share activity, and metacognitive reflection to develop a plan for comparing and contrasting skillfully.

LESSON

INTRODUCTION TO CONTENT AND THINKING SKILL/PROCESS

- I had to buy something I was unfamiliar with a short time ago. My sister asked me to get my nephew a pair of in-line roller skates for his 14th birthday. I knew absolutely nothing about in-line skates, except maybe for having noticed their popularity and the growing number of skaters zipping by me during my morning walks. If you had been in my shoes, what might you have done in order to learn more about in-line skates so that you could make an intelligent decision on which brand to buy and how much to spend? Have students brainstorm a list of recommendations. STUDENT RESPONSES: *Check with people you know who have in-line skates and ask them what to look for in a good skate and what to avoid in a bad skate; visit a shopping mall and shop around; get a roller skating magazine and read product advertisements; watch home shopping channels on television or check the newspapers to see who's got the best deal.*

- **What do all these recommendations have in common?** Students usually realize that getting a good deal requires some type of comparison shopping, whether it be at the mall, between television shopping channels, competitive newspaper advertisements, etc.

- Actually, I considered doing most of these things, but I really didn't have much time to invest before making the purchase; therefore, I took a shortcut, one I often take whenever I need to make an informed decision about an important purchase. I got a copy of *Consumer Reports*. The company that publishes this magazine, Consumers Union, assesses the quality and value of products so consumers can make informed decisions about what they choose to purchase and use. Their ratings and recommendations are valued highly by the public. Let's take a look at overall

quality scores given to twelve men's and seven women's models of in-line skates. Give students a copy of the rating chart at right. Ask them to examine the ratings table for in-line skates and make a list of the kind of information the table presents. Then, ask the students to report on one of their items. STUDENT RESPONSES: *Different manufacturers; how much the different makes and models of in-line skates cost; ratings of important characteristics such as wheels, brakes, and comfort; an overall score for each skate.* Be sure they know how to interpret the ratings table. Check for understanding. Ask: "Which men's skate is rated highest?" "Which lowest?"

- **Let's take a closer look at the way** *Consumer Reports* **determines which products are best to buy. Read the source material from the article "The best deals on street wheels." Then, in your groups, discuss the probable steps by which the professionals at Consumers Union determine their overall ratings of in-line skates.** Most students are veteran shoppers and fairly accurately lay out the process for product testing and comparison. STUDENT RESPONSES: *First, they determine what product they are going to evaluate; next, they go out and purchase all the different makes and similar models available; then, they determine all the characteristics of the product that they can compare; then, they select those characteristics they determine are important to users of the product; then, they test each product for each characteristic; then, they describe the similarities and differences between the products; next, they see if any patterns emerge in the products they are testing; then, they compare test results and determine an overall score; last, they make a recommendation based on the test performance of each make.*

Ratings & Recommendations — In-line skates

Overall Ratings *Within types, listed in order of performance*

Key no.	Brand and model	Price	Brake	Sizes	Overall score (P F G VG E, 0–100)	Rolling	Turning	Stability	Ride
	MEN'S								
1	Rollerblade Macroblade Maxxum	$279	heel	5½–12½ [1][2]	▬▬▬▬▬	●	◕	○	◕
2	Roces LAX ALF	285	heel	4–15 [1]	▬▬▬▬▬	●	◕	○	◕
3	Roces LAX	299	rear-wheel	4–15 [1]	▬▬▬▬▬	●	◕	○	◕
4	K2 Exotech Extreme Flight	240	heel	5–14 [1][2]	▬▬▬▬	●	○	◕	○
5	Ultra-Wheels Ultra-Extreme, **A BEST BUY**	139	heel	4–13	▬▬▬▬	◕	◕	◕	◕
6	Bauer F/4 FM	220	cuff	6–13	▬▬▬▬	◕	◕	◕	◕
7	Rollerblade Spiritblade ABT	159	cuff	5½–12½ [1][2]	▬▬▬	◕	○	◕	◕
8	Oxygen XE 01	200	heel/rear-wheel	3–14 [1][2]	▬▬▬	◕	○	◕	○
9	CCM Falcon	125	heel	6–12	▬▬▬	○	○	◕	◕
10	California Pro Gemini 1000	100	cuff	4–12	▬▬	●	◖	◕	◕
11	Roller Derby Phantom GT	50	heel	5–12	▬▬	◖	◖	◕	◕
12	Variflex Excell 6000	50	heel	4–12	▬	●	●	◕	◕
	WOMEN'S								
13	Roces LAX ALF	285	heel	6–10 [1]	▬▬▬▬▬	●	◕	◕	○
14	Rollerblade Bravoblade GLX	239	cuff	5¼–11½ [1]	▬▬▬▬	●	◕	◕	○
15	Roces LAX	299	rear-wheel	6–10 [1]	▬▬▬▬	●	◕	◕	◕
16	Rollerblade Lightning TRS	199	heel	4,5,6,6½–10,11 [1][2]	▬▬▬▬	●	◕	◕	○
17	K2 Exotech Extreme Flight	240	heel	4–11 [1][2]	▬▬▬	◕	○	◕	◕
18	Ultra-Wheels Ultra-Extreme LS, **A BEST BUY**	139	heel	5–10	▬▬▬	◕	○	◕	◕
19	Oxygen XE 01L	200	heel/rear-wheel	4½–11 [1][2]	▬▬▬	●	○	○	○
	CHILDREN'S								
20	Ultra-Wheels Impulse	109	heel	1–13	▬▬▬▬	◕	○	◕	○
21	Blade Runner (by Rollerblade) Pro 2500	70	heel	1–12	▬▬▬	●	○	◕	○
22	Roces STL Jr.	135	heel	13–7	▬▬▬	◖	◖	○	○
23	Rollerblade Microblade	79	heel	10–6	▬▬	◖	○	○	○
24	Bauer F/2	120	heel	10–12	▬▬	◖	○	◕	◖
25	Variflex Excell 4000	40	heel	13–6	▬	●	●	◖	●
	The following skate is Not Acceptable because its toe brake is very hard to use on hills.								
26	Seneca Kristi Street Style	50	toe	12–10	▬	◖	◖	●	○

[1] Available in half sizes. [2] Shell size changes with each full or half size, making skate respond better to your movements.

The tests behind the Ratings

The Ratings are based on lab tests and on the judgment of panelists, who skated the hills and flats around our Yonkers, N.Y., headquarters. **Price** is suggested retail. **Brake** is the system used for slowing or stopping. Brakes are described in detail on pages 22 and 23. Skates come only in full **sizes**, except where noted. **Overall score** is based mainly on the following elements: **rolling**, how much effort it takes to skate; **turning**, how easy it is to skate through a slalom course; **stability**, how well skates that are directed straight ahead keep from wandering side to side or tipping backward or forward; and **ride**, how smooth the ride is over rough pavement. **Warranty** information is in "Details on the models."

Legend:
● Excellent
◕ Very good
○ Good
◖ Fair
● Poor

Adapted with permission from CONSUMER REPORTS, July 1996. Although this material originally appeared in CONSUMER REPORTS, the selective adaptation and resulting conclusions presented are those of the authors and are not sanctioned or endorsed in any way by Consumers Union, the publisher of CONSUMER REPORTS.

- **When you note the similarities and differences between things, decide which are important, look for patterns, and then formulate an insight about them that the similarities and differences reveal, you are "comparing and contrasting." Here is a thinking map of the important questions to ask when you compare and contrast in this way.** Post the thinking map on the wall. (Thinking map is on the next page.)

OPEN COMPARE AND CONTRAST

1. How are they similar?

2. How are they different?

3. What similarities and differences seem significant?

4. What categories or patterns do you see in the significant similarities and differences?

5. What interpretation or conclusion is suggested by the significant similarities and differences?

THINKING ACTIVELY

• One of the first things human beings realized about life was just how many characteristics offspring have in common with their parents. Later, animal breeders discovered that they could mix and match adults to produce offspring with certain desired characteristics. By 1860, Gregor Mendel had discovered the fundamental principles of heredity using pea plants. He inferred that units of heredity—later named genes—were located in cells and transferred information about traits such as pea color from one generation to the next. What do we call this science that maps out how genes are passed on from parent to offspring? Students readily identify this as the science of genetics. For nearly a hundred years after Mendel's work, biologists understood that genes were responsible for heredity but were not able to describe their chemical structure in sufficient detail to explain how they worked. In fact, the race to be the first to explain the chemical nature of genetic material was prolonged and heated and took the first half of the 20th century to be run. Just how is it that the chemicals that make up genes can produce exact copies of themselves so perfectly that none of the critical information necessary to manufacture an organism is lost when it is transferred from parent to offspring?

• Today, we're going to study how this question was answered by the breakthrough that biologists made while investigating the molecular makeup of living cells in the 1950s. They discovered that DNA and RNA molecules were the key ingredients in every living cell enabling the cell to reproduce other cells of the same sort. Explain that DNA is a nucleic acid called deoxyribonucleic acid and that RNA is ribonucleic acid. **We are going to look closely at these nucleic acids. They are often confused with each other. We will use skillful comparing and contrasting to uncover similarities and reveal differences between DNA and RNA so that we can gain some insight about how this fundamental mechanism of life works. The way we will do comparing and contrasting will be similar to the way the skates in** *Consumer Reports* **were compared. We will look for similarities and differences, think about which are important, see if we find any patterns of similarity or difference, and then draw a conclusion about these two nucleic acids. Comparing and contrasting them well will help us to avoid confusing them. To help us organize our thoughts, we are going to use a compare and contrast graphic organizer.**

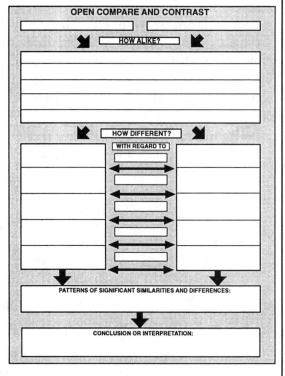

Give students copies of the open compare and contrast graphic organizer. Explain how it represents the process of questioning expressed on the thinking map of compare and contrast posted on the wall or chalkboard. If necessary, demonstrate its use with a familiar example like pickups and vans, trains, and planes, the army and the navy. This can be done by showing the students a transparency of a completed graphic organizer.

- Read the section in your textbook on DNA and RNA. As you read, be alert to ways that DNA and RNA are similar. Each time you discover a similarity, note it on a line in the "How Alike?" box on the compare and contrast graphic. Also, be on the lookout for ways that DNA and RNA are different. As you find these differences, note them to the left and right of the "with regard to" boxes in the "How Different?" section of the graphic. Think about in what way they are different and write it over the arrow. For example, the DNA in a cell lasts as long as that cell is alive, and RNA typically lasts less than an hour. What term describes that difference? Students usually answer "Lifespan." Allow students enough time to find at least four similarities and differences.

- Meet in your groups and together complete the "How Alike?" and "How Different?" sections of the compare and contrast graphic organizer. As each member contributes an idea, record that idea in the appropriate section of the graphic. Be prepared to report back to the class. Allow enough time for students to brainstorm in their groups and record at least five similarities and five differences.

- How are DNA and RNA alike? Have a spokesperson for each group report back one similarity to the class. As they do so, record their responses on a class transparency. As you record similarities on the graphic organizer, ask for clarification or extension of the answer by asking questions about the cause, effect, significance, or implications of the similarities that students cite. Discuss the response with the student in order to draw out his or her thought. RESPONSES OFTEN INCLUDE: *Both are found in cells; are polymers; have a helix structure; are made up of sequences of four different nucleotides, each of which is made up of a pentose sugar, a phosphate group and a nitrogen base; can form hydrogen bonds between complimentary base pairs; can act as substrates for enzymes; carry genetic information in the unique sequences of their nucleotides; take part in protein synthesis; have evolved from less complex molecules; and are made up of many carbon atoms.*

- How are DNA and RNA different? Ask for and record differences in the same way you did similarities, following them up with extending questions. Then use guided questioning to emphasize preciseness in identifying the kind of difference that has been identified and writing it in the "With Regard To" section of the graphic. Do not hesitate to use the chalkboard or transparency to clarify student responses; you might invite students to the board to illustrate their response. ANSWERS USUALLY INCLUDE BUT ARE NOT LIMITED TO THE FOLLOWING:

  ° *DNA has a double helix structure, RNA a single helix; DNA nucleotides contain the pentose sugar deoxyribose, RNA contains the pentose sugar ribose; DNA is made up of four nitrogen bases (guanine, adenine, cytosine, and thymine), RNA is made up of four different bases (guanine, adenine, cytosine and uracil); DNA is made from DNA templates using the enzyme DNA polymerase and is, therefore, self-replicating; RNA is made from DNA templates, is not self-replicating and requires the enzyme RNA polymerase to transcribe DNA into additional copies of RNA; DNA is always found in the nucleus of eukaryotes, RNA is found in the nucleus in nucleoli and as nuclear RNA but is also found in the cytoplasm as messenger RNA (mRNA), transfer RNA (tRNA) and ribosomal RNA (rRNA).*

  ° *During protein synthesis, DNA has two roles (acting as a library of genetic blueprints for building proteins and acting as a template for making (mRNA); RNA has three roles (acting as a messenger molecule between the nucleus and the cytoplasm (mRNA), transferring amino acids from the cytoplasm to sites on ribosomes where polypeptide chains are growing (tRNA), and serving as structural material found in ribosomes (rRNA);*

  ° *DNA exists as long as the cell does; most RNA is degraded inside the cytoplasm in less than an hour.*

- **Now let's think about the similarities and differences on your diagram. There are many true things that we could say about DNA and RNA that may not be very important. The fact that they both are made up of many carbon atoms does not add to our understanding of these nucleic acids—after all,** *every* **organic molecule is made up of carbon. Would it have made sense or been worthwhile for the staff at** *Consumer Reports* **to have considered skate color or product packaging when they were evaluating the performance of in-line skates? We want to base our understanding of nucleic acids only on factors that are important. Let's draw a line through any similarities or differences that are not important.** Review the class graphic and, after discussion, draw a line through those items the class agrees are not important. Students may predictably ask what is meant by "important." Any information, the omission of which would limit our understanding of the two nucleic acids, their molecular structure, and their functions in the cell, would be considered important.

- **Now let's look closely at the remaining entries in the "How Alike?" and "How Different?" sections of the diagram. Does there appear to be any theme or pattern emerging? For example, we might have compared carbon-based molecules with silicone-based molecules. You might then notice that the carbon-based molecule is associated with many organic functions and that the silicone-based molecule is associated with many inorganic functions. This would be a significant pattern of difference between the two.** After the students work together in their groups, ask the group spokesperson to respond to the class about the pattern(s) they detected. Students usually respond that structural differences come out in the comparison and contrast that seem to be related to differences in the overall function of these molecules within a cell.

- **Now meet in your groups and write one sentence that expresses a conclusion about DNA and RNA related to the patterns you have discovered within the similarities and differences between DNA and RNA.** Allow ample time for group discussion. Have each group report back to the class. Discuss each response, then fashion a consensus statement and write it in the "Conclusion or Interpretation" box on the class transparency. STUDENT ANSWERS TYPICALLY INCLUDE: *Small differences in the molecular structure of nucleic acids can create substantial differences in their function. The structure of the nucleotides in DNA—the double helix—helps it produce RNA, and the structure of RNA—a single helix—allows it to slip into the cell cytoplasm and work to translate the genetic plan carried by the DNA into proteins by making changes in the cytoplasm. Their location within the cell appears to be related to their function.*

THINKING ABOUT THINKING

- **Let's stop thinking about DNA and RNA now, and let's think about our thinking. How was the way you compared and contrasted these two nucleic acids different from the way you have compared and contrasted things before?** STUDENT ANSWERS INCLUDE: *When I compared and contrasted things before, I just listed similarities and differences. This time, I spent time thinking about the similarities and differences, identified the kind of differences, looked for patterns in the similarities.*

- **Was it helpful to compare and contrast this way?** STUDENT RESPONSES INCLUDE: *I will never confuse DNA and RNA again. It wasn't until I looked at the similarities and differences carefully that I realized how DNA and RNA are related and how the differences between the ways they are constructed makes them work differently inside a cell. I could understand that because RNA was a single strand, it could do things that DNA couldn't and also probably wouldn't last as long.* Most students agree that taking the time to think about what they are comparing and contrasting like this helps them to better understand what they are comparing and contrasting.

• **List some questions that can guide you in doing this kind of compare and contrast next time so that you don't slip back into the old way of comparing and contrasting.** Students usually list these questions, which you can put on the wall of the classroom or on the chalkboard so that it will be there to remind students. Explain that this is a "thinking map."

APPLYING THINKING

Not too long after the lessons, ask students to engage in the following activities. Leave the thinking map of comparing and contrasting on the wall or chalkboard and have blank graphic organizers available in the classroom.

Immediate Transfer

• **Compare and contrast meiosis and mitosis. Use the same approach that we used when we compared and contrasted DNA and RNA.** Don't guide the students as much as you did when they worked on DNA and RNA; ask them to work in groups and question each other about their similarities and differences like you did in class. When they have completed the comparison and contrast have some of them share their conclusions with the whole class.

• **What other things that you are studying in biology do you think it would be helpful to compare and contrast in this way?** STUDENT RESPONSES VARY BUT SOMETIMES INCLUDE: *Plant and animal cells, enzymes and hormones, evolution and natural selection, osmosis and diffusion.* Post a chart with "Items to Compare and Contrast" at the top, and list these topics on it. Ask students to add others as they think of them.

• **What other things are you studying in other subjects for which you think it will be valuable to compare and contrast in this way?** STUDENT RESPONSES VARY. Add these to the chart.

Reinforcement Later

Later in the school year, ask students to engage in the following activities.

• **Pick one of the sets of biology items from the chart, and compare and contrast them.** This can be assigned as homework. When students complete it, have a few of them share their results by going over their graphic organizers in detail. They can use a transparency of their graphic organizer to show the rest of the class.

• **Pick one of the sets of non-biology items from the chart, and compare and contrast them.**

• **Compare and contrast colleges to which you are considering sending applications.** (For 12th-grade students early in the year.)

• **Compare and contrast two items you are considering purchasing.** Use the strategy for focused compare and contrast.

WRITING EXTENSION

Use your conclusion about DNA and RNA as a topic sentence and write a compare and contrast essay about the two as if you had been asked to contribute this to a biology textbook.

ALTERNATIVE THINKING SKILL OBJECTIVE

Instead of using open compare and contrast, used focused compare and contrast (below).

FOCUSED COMPARE AND CONTRAST

1. What is the purpose of the comparison and contrast?

2. What kinds of similarities and differences are significant to the purpose of the comparison and contrast?

3. What similarities fall into these categories?

4. What differences fall into these categories?

5. What patterns of similarities and differences are revealed?

6. What conclusion or interpretation is suggested by the comparison and contrast that is significant to its purpose?

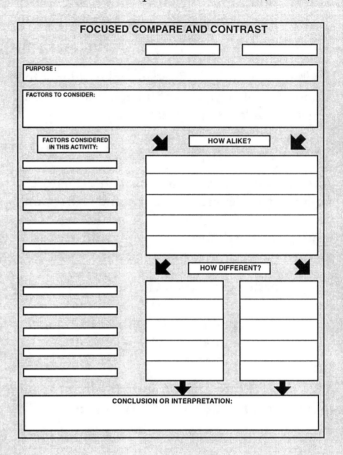

ASSESSING STUDENT THINKING WHEN THEY COMPARE AND CONTRAST

To assess this compare and contrast skill, ask students to write an essay to answer any of the application questions or others that you select. Ask students to describe how they compared and contrasted the two subjects. Determine whether they are attending to each of the steps of comparing and contrasting and whether they have determined significant similarities and differences between the two figures.

DNA AND RNA

SOURCE MATERIAL ON THE STRUCTURE AND FUNCTION OF NUCLEIC ACIDS

DNA

DNA, deoxyribonucleic acid, is the genetic material that organisms inherit from their parents. It is a long molecule, called a nucleic acid, that is a polymer made up of many smaller units. The monomers, the repeating units making up the nucleic acid, are called nucleotides. Each nucleotide is made up of three parts: a phosphate group, a nitrogen base, and a five-carbon sugar. The sugar is called deoxyribose.

Four different nucleotide monomers make up DNA: adenine, cytosine, guanine, and thymine. The nucleotides differ from each other only by small changes in the structure of their nitrogen base. Each base is made up of from 12 to 16 C, H, N, and O atoms.

DNA is made up of from one million to several billion nucleotides. The DNA strand is composed of alternating sugar and phosphate groups that form strong covalent bonds; the nitrogen bases project perpendicular to the DNA strand where they protrude and can readily form weak hydrogen bonds to other nitrogen bases.

However, the structure of each nucleotide allows it to hydrogen-bond to only one other nucleotide. For example, adenine bonds only to thymine; guanine bonds only to cytosine. These are called complementary base pairs.

It is this bonding of complementary bases that spontaneously produces the double helix structure of DNA.

DNA has two functions:

1) it reproduces itself (replication) and, as each cell divides, passes genetic information along to the next generation;

2) it acts as a template upon which RNA is assembled using suitable enzymes.

During the process of DNA replication, the surface architecture formed by DNA nucleotides acts to dock enzymes such as DNA helicase and DNA polymerase. These then proceed to build a second double helix on templates formed by the original DNA strands. In this way, each daughter cell resulting from a cell division ends up with a complete set of genes; in this manner, the genetic heritage of the new cell is transferred from its parent cell.

During the process of DNA transcription, a different set of enzymes, including RNA polymerase, assembles a different nucleic acid from the DNA template. RNA, which will be used to transfer information from where it is stored in DNA to where it is used in the cytoplasm of the cell, is also formed on DNA templates.

The set of unique surface features formed by the sequence and kind of nucleotides determines which enzymes will attach to the DNA.

Because DNA functions as a library of blueprints for making RNA, its lifespan is usually the cell's lifespan.

RNA

RNA, ribonucleic acid, is a single-stranded polymer made up of nucleotide monomers. RNA nucleotides are made up of a phosphate group, a five-carbon sugar (ribose), and a nitrogen base. There are four different nucleotides, each differing by small differences in the structure of the nitrogen base. RNA nucleotides include the bases adenine, guanine, cytosine, and uracil. Uracil and thymine both base-pair with adenine, but a small difference in their structure (uracil lacks a methyl group CH_3) defines their function in the cell.

RNA is made up of from several hundred to 20,000 nucleotides. The average eukaryotic RNA is about 6,000 nucleotides long. The RNA strand has alternating sugar and phosphate groups that form strong covalent bonds; the nitrogen bases project perpendicular to the RNA strand, where they can readily form weak hydrogen bonds with other nitrogen bases.

RNA can form base pairs with DNA: guanine bonds with cytosine and adenine bonds with uracil. RNA can also form base pairs with RNA.

RNA is made on DNA templates in a process called DNA transcription. The assembly of RNA requires the enzyme RNA polymerase.

RNA has three cellular functions:

1) acting as a messenger molecule (messenger RNA), carrying genetic information between DNA in the nucleus and ribosomes in the cyto-plasm. Here, it functions as a blueprint for the synthesis of proteins,

2) functioning to deliver amino acids to ribosomes (transfer RNA), where they are then assembled into polypeptides,

3) acting as the primary structural material in the formation of ribosomes (ribosomal RNA).

Messenger RNA is found in the nucleus and in the cytoplasm. Its lifespan, measured in minutes and hours, is relatively short.

Sample Student Responses • DNA and RNA • Open Compare and Contrast

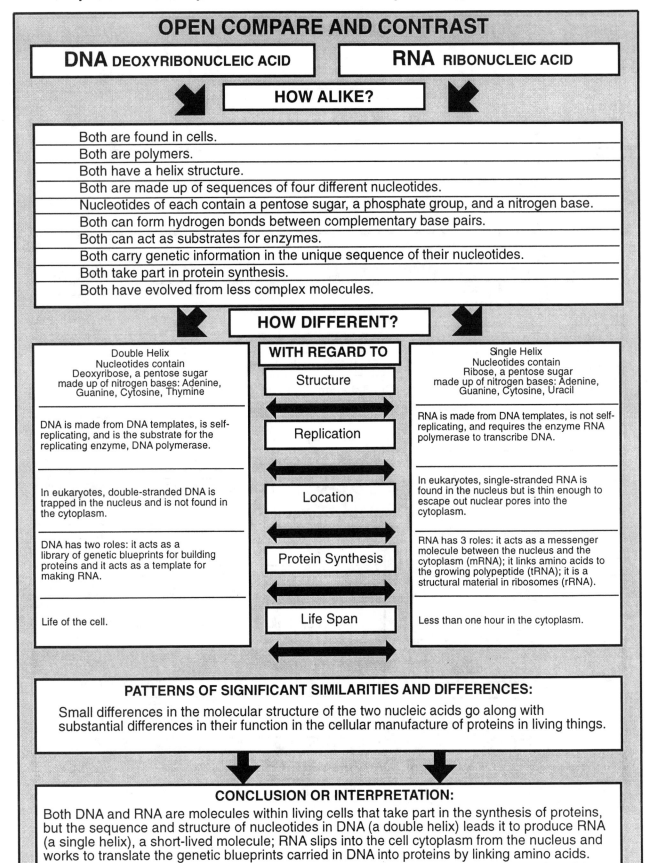

OPEN COMPARE AND CONTRAST

| DNA DEOXYRIBONUCLEIC ACID | RNA RIBONUCLEIC ACID |

HOW ALIKE?

Both are found in cells.
Both are polymers.
Both have a helix structure.
Both are made up of sequences of four different nucleotides.
Nucleotides of each contain a pentose sugar, a phosphate group, and a nitrogen base.
Both can form hydrogen bonds between complementary base pairs.
Both can act as substrates for enzymes.
Both carry genetic information in the unique sequence of their nucleotides.
Both take part in protein synthesis.
Both have evolved from less complex molecules.

HOW DIFFERENT?

WITH REGARD TO

DNA	WITH REGARD TO	RNA
Double Helix Nucleotides contain Deoxyribose, a pentose sugar made up of nitrogen bases: Adenine, Guanine, Cytosine, Thymine	Structure	Single Helix Nucleotides contain Ribose, a pentose sugar made up of nitrogen bases: Adenine, Guanine, Cytosine, Uracil
DNA is made from DNA templates, is self-replicating, and is the substrate for the replicating enzyme, DNA polymerase.	Replication	RNA is made from DNA templates, is not self-replicating, and requires the enzyme RNA polymerase to transcribe DNA.
In eukaryotes, double-stranded DNA is trapped in the nucleus and is not found in the cytoplasm.	Location	In eukaryotes, single-stranded RNA is found in the nucleus but is thin enough to escape out nuclear pores into the cytoplasm.
DNA has two roles: it acts as a library of genetic blueprints for building proteins and it acts as a template for making RNA.	Protein Synthesis	RNA has 3 roles: it acts as a messenger molecule between the nucleus and the cytoplasm (mRNA); it links amino acids to the growing polypeptide (tRNA); it is a structural material in ribosomes (rRNA).
Life of the cell.	Life Span	Less than one hour in the cytoplasm.

PATTERNS OF SIGNIFICANT SIMILARITIES AND DIFFERENCES:

Small differences in the molecular structure of the two nucleic acids go along with substantial differences in their function in the cellular manufacture of proteins in living things.

CONCLUSION OR INTERPRETATION:

Both DNA and RNA are molecules within living cells that take part in the synthesis of proteins, but the sequence and structure of nucleotides in DNA (a double helix) leads it to produce RNA (a single helix), a short-lived molecule; RNA slips into the cell cytoplasm from the nucleus and works to translate the genetic blueprints carried in DNA into proteins by linking amino acids.

Sample Student Responses • DNA and RNA • Focused Compare and Contrast

FOCUSED COMPARE AND CONTRAST

DNA	RNA

PURPOSE :
To determine the role of these two nucleic acids in the synthesis of proteins in living things.

FACTORS TO CONSIDER:
Their structure, internal composition, relation to proteins, life span, and location.

FACTORS CONSIDERED IN THIS ACTIVITY:

Location
Internal Composition
Structure
Relation to Proteins

HOW ALIKE?

Both are found in cells.

Both are made up of a sequence of four different nucleotides.

Both carry genetic information in the unique sequence of their nucleotides.

Both take part in protein synthesis.

HOW DIFFERENT?

Location
Internal Composition
Structure
Relation to Proteins
Life

Within the nucleus of the cell	Created in the nucleus, moves into the cytoplasm
Nucleotides contain deoxyribose, a pentose sugar	Nucleotides contain ribose, a pentose sugar
A double helix	A single helix
Contains genetic blueprint for building proteins and manufactures RNA	When in the cytoplasm it links amino acids into the growing protein
The life of the cell	Less than one hour in the cytoplasm

CONCLUSION OR INTERPRETATION:

DNA is the permanent genetic material in cells carrying the blueprints for synthesizing proteins; the "mastermind" behind the growth of specific types of proteins; RNA is a temporary agent, produced by DNA, that carries out the synthesis according to the DNA plan by linking amino acids in the cell's cytoplasm. RNA, the self-sacrificing "worker," gives up its life to carry out DNA's directions.

COMPARING AND CONTRASTING LESSON CONTEXTS

The following examples have been suggested by classroom teachers as contexts to develop infused lessons. If a skill or process has been introduced in a previous infused lesson, these contexts may be used to reinforce it.

GRADE	SUBJECT	TOPIC	THINKING ISSUE
6–8	Science	Human reproduction	Compare and contrast male and female reproductive systems to understand the physiology of sexual reproduction.
6–8	Science	Reproduction	Compare and contrast sexual and asexual reproduction to understand strategies for passing on traits.
6–8	Science	The solar system	Compare and contrast the Copernican and Ptolemaic solar systems in order to demonstrate the evolution of scientific thought over time.
6–8	Science	Eclipses	Compare and contrast solar and lunar eclipses in order to better understand the relative movements of the earth, moon, and sun.
6–8	Science	Cells	Compare and contrast animal and plant cells in order to demonstrate how structure determines function.
6–8	Science	States of matter	Compare and contrast liquids and gases in order to reveal seldom-thought-of similarities.
6–8	Science	Plants	Compare and contrast angiosperms and gymnosperms to clarify modes of plant reproduction.
9–12	Biology	Ecology	Compare and contrast ozone depletion and the greenhouse effect with regard to origin, effect, sollutions, and needed research.
9–12	Biology	Metabolism	Compare and contrast photosynthesis and cellular respiration to understand energy production in plants and animals.
9–12	Biology	Immune system	Compare and contrast the cellular and humoral immune systems to clarify the functions of each.
9–12	Biology	Cell division	Compare and contrast mitosis and meiosis to reveal distinctions between somatic and germ cells.
9–12	Biology	Plants	Compare and contrast spores and seeds to understand the differences between asexual and sexual reproduction.
9–12	Biology	Genetics	Compare and contrast genotype and phenotype to understand the relationship between genome and gene expression.
9–12	Biology	Biologic control systems	Compare and contrast negative and positive feedback mechanisms—for example, the control of body temperature and the control of contractions during childbirth.
9–12	Biology	Macro-molecules	Compare and contrast the monomers of carbohydrates, lipids, proteins, and nucleic acids to understand how the organic molecules of life are derived from simple building blocks.

COMPARING AND CONTRASTING LESSON CONTEXTS

GRADE	SUBJECT	TOPIC	THINKING ISSUE
9–12	Biology	Organisms	Compare and contrast prokaryotes and eukaryotes to understand the evolution of complex structure over time.
9–12	Chemistry	Acids and bases	Compare and contrast acids and bases with respect to pH, ions, reactivity with metals, indicators, taste, feel, nomenclature, and salts formed.
9–12	Chemistry	Chemical reactions	Compare and contrast oxidation and reduction reactions with respect to electron transfer, oxidation numbers, and processes leading to each, including: for oxidation, gain of oxygen, loss of hydrogen, loss of electrons, and movement of electrons away; for reduction, movement of electrons toward an atom, gain of hydrogen, gain of electrons, and loss of oxygen.
9–12	Chemistry	Heat	Compare and contrast exothermic and endothermic processes with respect to the system, the surroundings, the direction of heat flow, and the sign (+ or −) assigned.
9–12	Chemistry	Knowing the periodic table	Compare and contrast the noble gases (Group 0) with the alkali metals (Group 1A) with respect to reactivity, electron sublevels, ionization energy, and electronegativity to understand the organization of the periodic table.
9–12	Physics	Vectors	Compare and contrast vectors and scalars with respect to displacement (position), direction, and magnitude.
9–12	Physics	Machines and energy	Compare and contrast the conservation of energy in ideal and real machines to understand the energy needed to overcome friction.
9–12	Physics	Light	Compare and contrast the wave and particle theories of light to reveal the dual nature of its physical characteristics.

Chapter 6
Classification

Why is Skillful Classification Important?

Classifying things may seem like labeling them. Classification, however, carries with it much deeper significance than simply assigning a name. For example, when I identify a particular tree as an *oak tree*, I am noting that it has the specific characteristics of a tree, which distinguishes it from other plants, and of an oak, which distinguishes it from other trees. When I classify it as an oak tree, I attribute to it a cluster of important characteristics: It is a living thing; it is a plant; it has a root system, branches, and leaves. Moreover, it bears acorns that fall in autumn and can grow into offspring oaks. Through the seemingly simple act of classifying, a tree that I may never have seen before now becomes much more familiar to me.

Moreover, classifying this tree as an oak tree usually attaches to it an accumulation of much more information. For example, I recognize that it makes food by photosynthesis, that it can be a haven for birds, that its wood is hard and can be used for making furniture, and that it grows very tall. Classifications can be powerful shorthand devices for finding richer meanings. Our ability to classify individual things under general concepts provides us with an elegant way of organizing and expressing more complex forms of human knowledge.

Classification and definition. At its core, classification involves putting particular things in general categories because these things have certain characteristics that we use to define those categories. The terms that we use to label these categories usually mask the variety of characteristics that define them. When I call something an oak tree, I usually do not make explicit the wealth of characteristics that I think oak trees have.

Skillful definition of the term "oak tree" makes explicit the characteristics that we attribute to particular trees when we call them "oak trees." These characteristics may include some that, strictly speaking, do not define what an oak tree

is but that nonetheless are part of our conception of an oak tree — for example, that furniture can be made from its wood. The characteristics that we make explicit are all part of what we mean by the term "oak tree."

Hence, there is an intimate connection between the thinking involved in classification and the thinking involved in defining. Classification involves putting individual things into appropriate categories based on the defining characteristics of those categories. Definition involves analyzing the meanings of category names by making the defining characteristics of the categories explicit.

The purposes of classification. Classification serves many purposes according to our needs and interests. Here are a few of the diverse purposes that classification serves:

- *To help us select something we need.* For example, the specialty of a doctor is defined by characteristics of his or her practice (e.g., the type of treatment or the portion of the body about which the doctor has special knowledge). A urologist, for example, has expertise in ailments of the kidneys, bladder, etc. Hence, when we have such ailments, we can use our knowledge of how doctors are classified to select the appropriate doctor. In other examples, some of the ways that we classify tools or books help us select items that we may need. I may need something to shovel snow or a book to provide me with information about sights to see in Paris. Looking at the items classified "shovels" or "travel guides" can help me select what I need.

- *To protect things from harm or damage.* Classifying packages as fragile signals us to handle them carefully in order to protect them. Classifying foods as frozen foods helps us to determine where to store them so that they can be protected from spoiling.

- *To help us determine important properties and/ or relationships.* Classifying people by family relationships allows us to determine various forms of entitlements, rights, and duties important in our society. A father and mother, for example, have certain duties towards their children that a child who is not their son or daughter cannot claim. Classification systems used in special fields often call our attention to characteristics of things and important relationships between them. When an animal is correctly classified as a crustacean, we know that it is a shellfish that lives in water. Similarly, classifying elements in chemistry, weather patterns in meteorology, and minerals in earth science indicates certain features of these items that give us a deeper understanding of how these phenomena work.

Criteria for effective classification. It is often important to determine the best way to classify certain things. Suppose I need to classify all of my financial records for 1997. I do this effectively when the following occur:

1. My classification scheme indicates important features of what I am classifying so that it serves my purpose(s).

2. I understand what these features are.

3. I correctly classify things by noting that they have these features.

I may keep together all the pieces of paper that indicate how much I have paid for business expenses this year in an envelope marked "1997 Business Receipts." My purpose is to be able to locate them easily when I file my income tax return. Keeping them together in this envelope means that I won't have to sort them from non-business receipts when I need them. Of course, to sort them effectively, I have to understand what a 1997 business receipt is, and I have to identify correctly that the pieces of paper I put in the envelope are such receipts.

Problems that arise in the way we classify. We can, of course, group things in any number of different ways based on their common characteristics. For example, I could group my receipts by size. This might be a good way to classify if I were concerned about papers of different sizes, but it doesn't help me find my 1997 business receipts. This example illustrates one way that a classification scheme can fail; it may not serve the purpose we have for it.

A more frequent problem with the way we classify things is that the categories may be *too broad* or *too narrow* for our purposes. If I put all of my business expense receipts in a folder but did not discriminate among different years, I wouldn't be able to use this collection easily in doing my 1997 income tax. The classification would be too broad. Or I may just keep certain 1997 business receipts together — for example, my 1997 automobile expenses. Then I have to search for other receipts to be able to figure out my overall business expenses. In this case, the way I was classifying my expense statements would be too narrow to suit my purposes.

Another problem with classification occurs when we don't put things into the appropriate categories. Incorrectly classifying a poisonous snake as a harmless one can lead to a serious problem. We might make this kind of error for a number of reasons. The first is that we might be confused about what specific categories entail. For example, it may not be difficult to determine whether a document is for business and whether it is for 1997. However, if I am confused about what a business receipt is and put bills as well as receipts in the folder, that may cause problems. My classification scheme may be a good one; but, if I misunderstand the categories, I may misclassify the documents I put in those categories, negating the system. *Misunderstanding what categories signify* is a key source of misclassification.

There are other sources of misclassification. I may understand the defining characteristics of a class of things, like 1997 business receipts, and make a mistake about whether something I am classifying has these characteristics. I may think I see "1996" on a receipt and it may be "1997." This is a matter of *misperception*. I may also make the same error through *misjudgment*. If I classify a receipt for a meal as a business receipt but it fails to meet the requirements for a legitimate business expense, then I may have a problem with the Internal Revenue Service.

A summary of the problems with classification is contained in figure 6.1 (next page).

COMMON PROBLEMS WITH THE WAY WE CLASSIFY THINGS

1. The way things are classified may not fit the purposes of classifying them, e.g., the category may be too broad or too narrow.

2. We may have a superficial understanding of what makes something fall into a category. Hence, classifying it that way may lead to a superficial, rather than rich, understanding of what we classify.

3. We may not know what defines the category and may put the wrong things in it.

4. We may know what defines the category but misperceive or misjudge that an individual thing has those characteristics and, hence, misclassify it.

Figure 6.1

What Does Skillful Classification Involve?

When we classify something, we identify it as belonging to a certain category of things, and we understand the significance of its belonging to that category. We often recognize that something fits into a number of categories. Nothing falls into only one category. A tree is a living thing, composed of wood, cylindrical, leafy, a part of nature, and the subject of poetry. Sometimes identifying a range of different ways that a thing can be classified is called "classifying" it. However, this is not sufficient for most of the tasks that require skillful classification.

Bottom-up classification. Most natural thinking tasks involving classification are tasks in which we do more than just list ways that items can be classified. We often have to select, from among the variety of ways that we can classify something, a classification scheme that best serves our purposes. We then actively employ that classification scheme to sort things.

In the case of my business receipts, any one of the 1997 business receipts could be classified into a variety of categories; I might identify them as small pieces of paper, wood products, business receipts, 1997 business receipts, or simply receipts. There may, of course, be contexts in

which classifying them as wood products serves some purpose; but in this case, I want to use them to generate figures for my 1997 income tax return. So I can reject the categories "pieces of paper" and "wood products" because I am sorting for a business purpose. I can also reject the category "receipts" because I can't be sure that only *business* receipts are filed in this category, and I can reject "business receipts" because I can't be sure that the only business receipts filed here are from 1997. These categories are too broad. On the other hand, "1997 business receipts" seems to be a category that does serve my purpose. If I use it, I can total the receipts and arrive at a figure needed for my tax return.

This is the process of "bottom-up" classification. We initially determine many different ways that given objects can be classified by identifying a variety of their characteristics and noting the categories that these characteristics define. Then we select categories that serve our specific purposes. Figure 6.2 contains a thinking map of bottom-up classification.

BOTTOM-UP CLASSIFICATION

1. **What characteristics do the given items have?**

2. **What classifications do these characteristics define?**

   a. **What subclassifications fall under each main classification?**

3. **What purpose do we have for classifying the items?**

4. **What way of classifying the items best serves this purpose?**

5. **Which items fall into each category?**

Figure 6.2

Two graphic organizers can be used to guide us through this kind of thinking. The first (figure 6.3) is used after we have identified characteristics of the object(s) to be classified and we ask, "What categories do these characteristics define?" The categories defined by these characteristics are listed on the left; these categories are then grouped into broader types of categories. These are written on the lines in the upper right.

For example, suppose you are trying to organize items in your garage into useful categories.

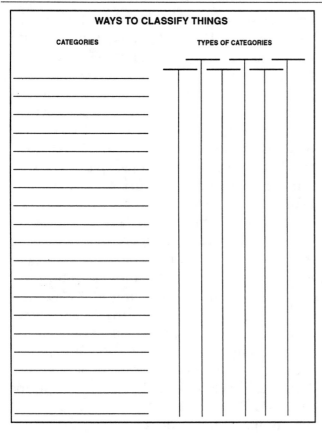

Figure 6.3

You find that some of these items have the following characteristics:

- They are made of metal
- They can be used to hammer nails
- They can be used to drive screws
- They are shiny
- They are heavy
- They are less than one foot long
- They are light in weight
- They are more than one foot long

They could be classified into categories like the following:

- Metal objects
- Tools that hammer nails
- Screwdrivers
- Shiny objects
- Heavy objects
- Objects less than one foot long
- Objects light in weight
- Objects more than a foot long

These categories should be written on the lines on the left of the graphic organizer. Then

we should determine if any of these categories are of the same type. Indeed, some are. Heavy objects and light objects are categorized by weight. "Weight" would then be written on one of the horizontal lines under "types of categories." "Heavy objects" and "light objects" should be connected to the vertical line leading to "weight." Similarly, "length," "use," and "material" can be written on the other lines for types of categories.

Determining types of categories in this way makes it easier to decide which way of classifying these objects best suits your needs or purposes. For example, you may choose "use" because you need tools to fix things. Or you may choose "material" because you are going to take these items to the city dump for recycling.

Next, determine which, if any, additional categories and subcategories should be added to your classification system. For example, the city dump might recycle metals and have bins for different types of metal. So you could further subdivide your categories under metal into steel, aluminum, copper, etc.

The second graphic organizer useful in classifying, a webbing diagram, shows categories and subcategories. It appears in figure 6.4.

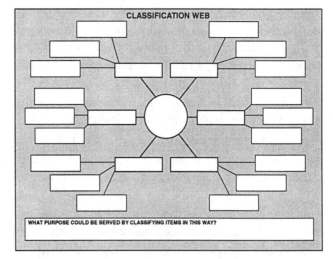

Figure 6.4

A variant on the webbing diagram appears in Figure 6.5. It is useful when the main category of items is divided into only two subcategories and then each subcategory into two others, etc.

To complete our classification project, we would then sort the remaining items in the

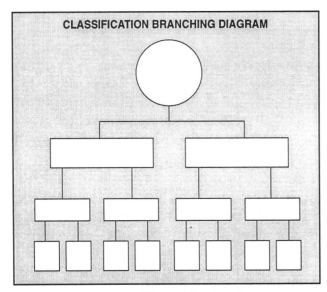

Figure 6.5

garage into the categories that we have determined. The strategy for top-down classification described in the next section provides us with a way of doing this skillfully.

We can also classify natural objects, as well as man-made products, using bottom-up classification. Animals can be classified by their body structure and functioning (e.g., quadrupeds) to help us understand life processes and evolutionary relationships. They can also be classified by habitat (e.g., animals that live in the oceans) to help us understand their environmental needs. They can be classified as predators or prey to understand the food chain and environmental balance. They can be classified by geographical area (the birds of North America) to enable us to predict what animals one might find where. How we choose to classify a group of animals is determined by our needs, interests, and purposes. Once we have chosen, the webbing diagram can help us to add useful categories.

Top-down classification. Sometimes, our classification task starts after the categories for classifying given objects have been determined. I may select "1997 Business Receipts" to classify certain receipts. I know what defines the category "1997 Business Receipts": An item that falls into this category must be a receipt for money spent, but it must also be from 1997, and it must be a business-related expense. Under that category, I may also want to sort my travel expense receipts, my home office expense receipts, my receipts for professional books, etc. I

also know what defines these subcategories. To put the right items into the right categories, I must sort through them and determine which have these characteristics. Then I can label them and sort them into the appropriate category.

This form of classification is "top-down." Top-down classification assumes a pre-established classification framework, one that we have developed bottom-up, or one that someone else has developed for us to use. For example, I may use categories already provided by the Internal Revenue Service for sorting tax records. In sorting other items for other purposes, I may use other established classification systems such as the periodic table, the Dewey Decimal System, or animal phyla.

Skill at this more restricted type of classification involves two things. We must know what defining characteristics are significant or commonly used for classifying things in the given categories. We must then detect the presence of those characteristics in individual things so that we can sort them correctly.

Figure 6.6 contains a thinking map for top-down classification.

TOP-DOWN CLASSIFICATION

1. What are the defining characteristics of the categories under which I want to classify things?

2. Which items have these characteristics?

3. How do I classify these items into the given categories?

Figure 6.6

The graphic organizer for top-down classification (figure 6.7, found on the following page) can guide the thinking.

How Can We Teach Students to Classify Skillfully?

We help students develop this skill by giving them many opportunities to classify, using both bottom-up and top-down classification. We should prompt students to differentiate between the two and guide them in using each strategy while performing classification tasks until they can guide themselves.

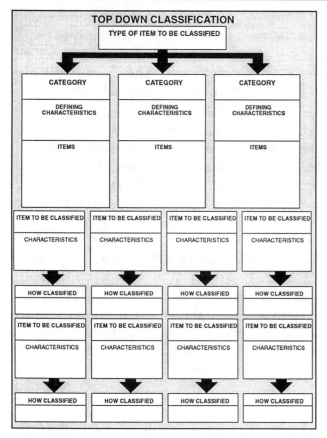

Figure 6.7

Classification and definition activities in standard curriculum materials. Classification is one of the traditional thinking skills that has been given attention in standard curriculum materials. Students are often asked to classify things, particularly in language arts and science. Be cautious about classification activities in common curriculum materials. Top-down classification is, by far, the most common form of classification students are asked to engage in. If this is the only form of classification that students are asked to do, supplement these activities with activities you design involving bottom-up classification.

Be cautious about activities in curriculum materials that only involve students in partial use of classification strategies. For example, bottom-up classification is often practiced by having students come up with a range of different ways to classify given objects. Sometimes that's *all* they are asked to do (or, if they are asked to do more, they are asked to arbitrarily select one way to classify the objects). You can supplement these activities by asking students the different purposes that classifying such ob-

jects can serve and then helping them make a selection based on specific purposes.

Similarly, students are often asked to engage in top-down classification by merely putting things into categories without explaining why. You can extend these activities so that students identify the defining characteristics of the given categories and look for these characteristics in the items they are asked to classify.

One of the most common tasks we ask students to do is to define terms. Often, however, students are simply asked to define words without any guidance about the connection between classifying and defining.

When students *classify skillfully*, they realize that things like ducks have characteristics which differentiate them from other things in the bird category and which delimit a subcategory into which these particular objects fall (ducks). Hence, to *define a word that can be used to classify something*, like "duck," the definition should include the category the thing falls into (bird), as well as the characteristics that make it different from other things in that category (other birds). Thus, instruction in skillful classification helps students understand the information they need to include in an adequate definition.

Contexts in the Curriculum for Designing Infusion Lessons on Classification

Classification is a type of thinking that permeates every field of study. Using the concepts of a field of study (e.g., force, energy, molecule, polygon) to identify specific things that fall under these concepts is a central ingredient in learning and understanding. Think of the way that works of literature are classified (e.g., tragedy, romantic, short story) and how important it is for students to understand the basis for these ways of classifying such works so that they can identify them in these ways.

Science offers a variety of conceptual frameworks that have been developed and are in constant use because of their utility in helping us identify things by various important characteristics. The concepts of different molecules help us to identify molecules by various characteristics related to their structure and properties of combining with other molecules. Insofar as the science curriculum helps students identify and understand the properties of individual things in terms of these concepts, contexts for classification lessons permeate these curricula.

Developing infusion lessons on classification in secondary school science can be particularly helpful in teaching lessons with these goals:

- *It is important for students to understand and differentiate various phenomena of the same or different types.* In science, scientific phenomena, organisms, geographic features, astronomical phenomena, etc. are differentiated.

- *Students are asked to organize various items that they are studying or using.* For example, in biology courses students may have to organize specimens, or in their laboratories they may have to store equipment.

- *Students are asked to define important concepts.* Students are asked to define operations of the human body like digestion, respiration, and reproduction; chemical processes like reduction, oxydation, or radioactivity; and physical processes like friction, magnetic attraction, and transverse waves.

A menu of suggested contexts for classification lessons is at the end of the chapter.

Tips for classification lessons. When you design lessons in one or the other of these contexts, make sure that the activities are authentic and not simply exercises in classification. Ask students to think through how to classify items in the natural contexts in which such classifications would be appropriate. One interesting type of activity involves students practicing bottom-up classification by creating files on a computer to store specific documents. You can, in fact, create a variety of analogies to classifying real objects in this way, thereby giving students valuable practice in using a computer and in learning important bottom-up strategies.

It is important to keep five points in mind when teaching classification:

- Always identify the task as classifying;

- Encourage students to recognize the purpose of classifying in each context;

- Make sure that your students express the

defining characteristic of the categories they are working with;

- Help them to relate the defining characteristic to the purpose of classifying; and

- Remind them to check the accuracy of their classifications of specific items.

Model Lessons on Classification

We include two model lessons in this chapter. The first is a middle school science/social studies lesson on animals. Bottom-up classification is taught in this lesson. This lesson illustrates how a multiplicity of different classification frameworks in science can be used to group the same set of individual things and how the choice of which way to classify things depends on the specific purposes and interests we have.

The second lesson is a challenging high-school chemistry lesson in which students engage in top-down classification of chemical reactions. This lesson serves as a model for a multitude of other lessons in which key concepts are used to classify chemicals or chemical operations. Such lessons can enhance students' understanding of these key relationships dramatically as they engage in classification activities.

As you read, ask the following questions:

- How would this compare to a lesson on the same topic in which students were engaged in top-down classification only?

- Which of the graphic organizers for classification is best suited for this lesson? Why?

- What other contexts can you think of for reinforcing this skill?

Tools for Designing Classification Lessons

Thinking maps and graphic organizers for both bottom-up and top-down classification are included on the following pages. The thinking maps and graphic organizers can serve as photocopy masters, transparency masters, or as models that can be enlarged and used as posters in the classroom. Reproduction rights are granted for use in single classrooms only.

BOTTOM-UP CLASSIFICATION

1. **What characteristics do the given items have?**

2. **What classifications do these characteristics define?**

   a. **What subclassifications fall under each main classification?**

3. **What purpose do we have for classifying the items?**

4. **What way of classifying the items best serves this purpose?**

5. **Which items fall into each category?**

WAYS TO CLASSIFY THINGS

CATEGORIES   TYPES OF CATEGORIES

_____ _____ _____

_____

_____

_____

_____

_____

_____

_____

_____

_____

_____

_____

_____

_____

_____

_____

_____

_____

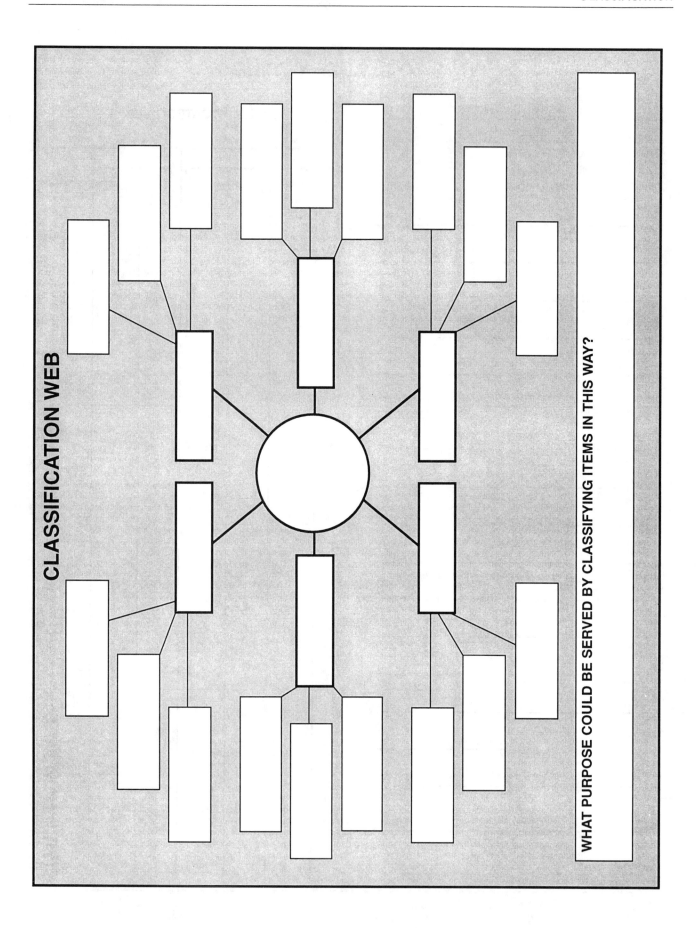

CLASSIFICATION WEB

WHAT PURPOSE COULD BE SERVED BY CLASSIFYING ITEMS IN THIS WAY?

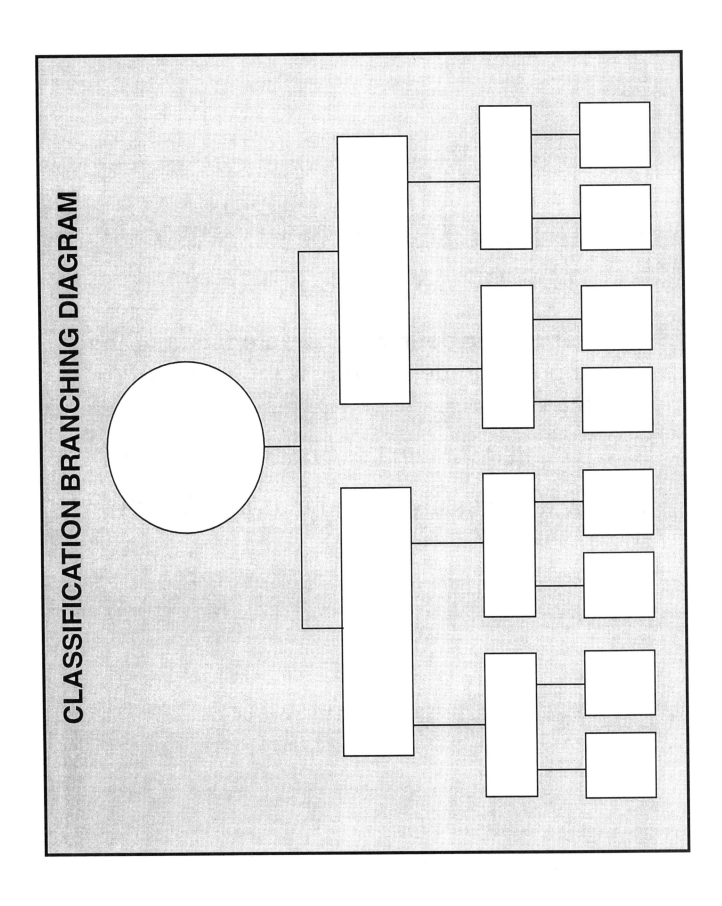

CLASSIFICATION BRANCHING DIAGRAM

TOP-DOWN CLASSIFICATION

1. **What are the defining characteristics of the categories under which I want to classify things?**

2. **Which items have these characteristics?**

3. **How do I classify these items into the given categories?**

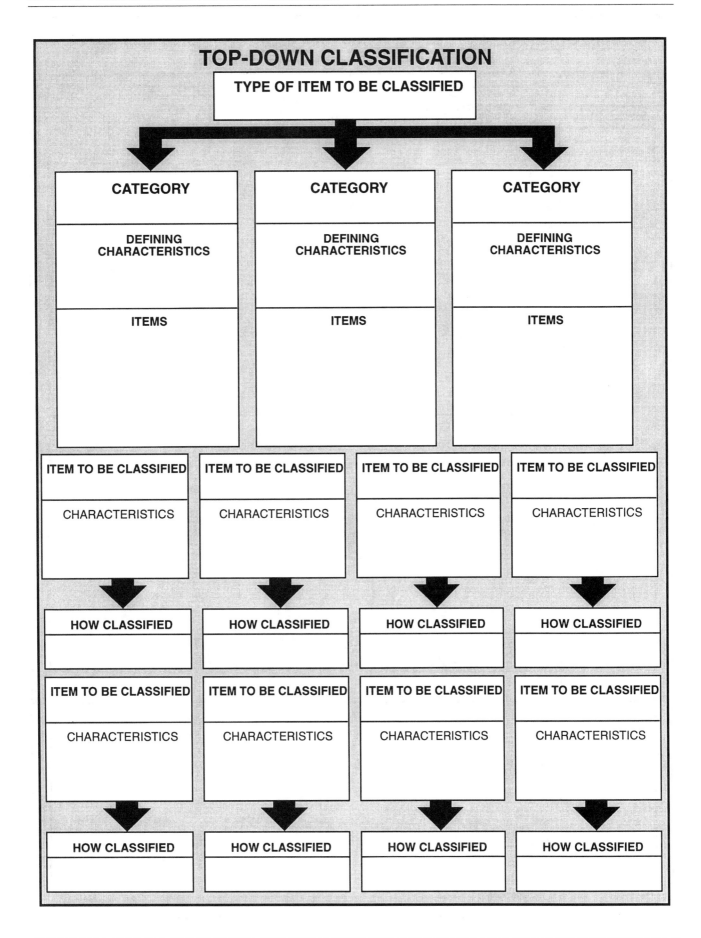

PURPOSES AND USES OF CLASSIFICATION SYSTEMS

WAYS TO CLASSIFY	PURPOSE OF THE CLASSIFICATION	WHO WOULD USE IT AND WHY

LIVING THINGS OF ALL SHAPES AND SIZES

General Science, Biology

Grades 6–9

OBJECTIVES

CONTENT	THINKING SKILL/PROCESS
Students will learn different ways of classifying organisms and recognize that different classifications yield different information about them (e.g., body structure, needs, habitat, etc.).	Students will identify defining characteristics of classes and recognize that the purpose for classification determines which characteristics are relevant.

METHODS AND MATERIALS

CONTENT	THINKING SKILL/PROCESS
Students will work together in collaborative learning groups of four to six. Each group will need index cards. Print resources such as an encyclopedia, biology texts, and other sources of information about living things should be available in class.	Students are guided by the teacher's questions and directions to classify a variety of living organisms, including plants, animals, protozoa, and bacteria. Students create a diagram to illustrate their classification scheme. A graphic organizer guides student thinking.

LESSON

INTRODUCTION TO CONTENT AND THINKING SKILL/PROCESS

- Think of a time when you classified or sorted various objects together. Work with two other students and describe this situation to them. **Explain the purpose of your classification. What characteristics of the things you classified were important to you for that purpose?** Have students work in groups of three and list their examples on the chart designed for this purpose. Then ask each group to report one example and write these on a copy of the chart on the chalkboard or on a transparency. Write the example, purpose, and characteristics in each case. The result might look like this:

EXAMPLE	PURPOSE	CHARACTERISTIC
Library Books	Find them easily Store properly	Alphabetical by author, Dewey decimal number, subject
Telephone Numbers	Connect two telephones	Area code Local exchange
Addresses	Deliver efficiently Find destination	Street number City, State Zip Code
Baseball Cards	Find easily Store properly	Team Player Position
Clothes	Find easily Store properly Keep track of	Type (socks, shirts, etc.)
Web sites	Find easily	Name Type (org, net, com)
Zoo Animals	Animal care Find easily	Habitat Country Biome

- **How did your purpose determine what characteristic you selected for your classification?** Students comment that classifying for different purposes requires paying attention to different characteristics.

- When we classify things, it is important to do it well. If we develop categories that don't serve our purposes or if we put things in our categories that don't belong there, we may not be able to do what we want to do with the things we are classifying. For example, if I organize my coats by the seasons they are made for and I need a winter coat, I know where to get it. On the other hand, if I just lump all of my coats together, then when I pick one from the group when I need a winter coat I may not end up with a coat that serves my needs. And if I am sloppy and then shirts and coats get put in the group that I label "Winter Coats," I may also end up with something that also does not serve my purpose. We're going to learn how to classify things carefully in this lesson. We will examine the things to be classified and then classify them in ways that will best serve specific purposes.

- Let's think in more detail about things that we might want to classify and categorize. Things that are collected, for example, usually have to be organized very carefully. They are grouped in categories to serve purposes like those we just considered. Basing our classification on characteristics that are important and that define these groups is what classification is all about. What kind of collections are you familiar with? Brainstorm a list of collectibles with the class and write them on the chalkboard as they are mentioned. STUDENTS USUALLY INCLUDE: *baseball and other sports cards, dolls, marbles, stamps, coins, paper currency, antiques, match book covers, college pennants, autographs, models, computer game software, music tapes, compact discs, and videos.*

- Let's consider one of these hobbies—stamp collecting. Here are some stamps that I copied for you to look at. Pass out copies of the diagram titled "Some U.S. and Foreign Stamps" found at the end of this lesson or alternately, distribute small numbers of a variety of real stamps to each group. If you were asked to make some sense of this batch of stamps—to organize them—how might you do this? Meet in your groups again and make a list of all the characteristics of stamps that might be useful in helping you to organize them into a meaningful collection. Students usually respond that they would have to know some basic information about the stamps. Using an overhead projector or the chalkboard, develop a list of characteristics by having each group contribute one characteristic at a time. TYPICAL STUDENT RESPONSES: *the country the stamp came from, the amount of postage, whether or not they were airmail or regular stamps, their size, shape, and color, when the stamp was printed, what was printed on it, how much the stamp was worth, its condition, whether it was new or used.* STUDENTS OFFER THE FOLLOWING RESPONSES LESS OFTEN: *whether or not the stamp was watermarked, how well the image was centered.*

- Some of these characteristics might be more important to you than others. Think about why you might be collecting stamps. Maybe you want to have all of the stamps of a particular country like the United States. Instead of keeping the ones you have all together in a box, what would be the *purpose* of organizing them into categories and classifying them? Students typically respond that the whole idea of organizing a stamp collection is so that you can quickly tell the ones you have and easily identify the ones you don't have.

- By which characteristics could you identify stamps to best serve this purpose? Most often, students respond that the stamps might be put into groups by country and year of issue. Some, however, identify the type of stamp—air mail, regular postage, etc.—as a basis for grouping. When asked why not by color or shape, they readily realize that having too few categories would make it difficult to find a particular stamp or locate the ones that are missing. Stamps are printed in only a few colors. Imagine trying to find a particular stamp in a catalogue that had categories like "rectangular" or "red." Ask students to design a classification system for grouping the stamps on the diagram containing U.S. and foreign stamps.

- Does this mean that grouping stamps into categories like "Red Stamps" and "Rectangular Stamps" or grouping them into categories like "Stamps with Pictures of National Leaders" would serve no purpose? Can you think of some purpose this type of categorization might serve? STUDENT RESPONSES VARY BUT SOMETIMES INCLUDE: *What if you weren't a stamp collector, but instead were an artist who used stamps as elements of montages; or what if you were a history teacher who liked to use stamps to show students portraits of famous people, events and places.* What determines which characteristics of the things to be classified are the most important? Students typically respond that the purpose or reason for organizing things determines which characteristics are important and that people might have many purposes for classifying the same objects. Additionally, these characteristics usually lead to different ways of classifying them.

- Let's put these ideas together and develop a thinking map for this type of classification. Here is a series of questions that grow out of this discussion: Notice that I've called this "Bottom-Up Classification." That is because in this type of classification, we start with specific items and develop a way to group them, and hence a system for classifying them, based on their characteristics and our purposes. This contrasts with "Top-Down Classification," in which we start with a given classification system that has already been developed and have to put individual things into the categories determined by that system (like when I ask you whether something is "Plant" or "Animal").

> **BOTTOM-UP CLASSIFICATION**
> 1. What characteristics do the given items have?
> 2. What classifications do these characteristics define?
>   a. What subclassifications fall under each main classification?
> 3. What purpose do we have for classifying the items?
> 4. What way of classifying the items best serves this purpose?
> 5. Which items fall into each category?

THINKING ACTIVELY

- We have been studying the origin of life on Earth. We already know that more than 30 million different species of living organisms populate this planet. Today, we are going to consider this diversity of life—the enormous variety of plants and animals—and think about ways of classifying them into intelligent and useful categories using **Bottom-Up Classification.** Post the thinking map for Bottom-Up Classification on the wall of the classroom or draw it on the chalkboard.

- In order to prepare for this lesson, I want each of you to take 3 index cards, and using your rulers, make each one have the same form as this. Draw a facsimile of the following graphic on the chalkboard or print the cards ahead of time using the form of the graphic as a master. **On each of these cards, place the name of an organism, like "Lion" in the example. Try to select three living things that have little in common with each other. That way, your group is likely to have some interest-**

ORGANISM:	**LION**		
CHARACTERISTICS			
1. lives in Africa	√	7. has sharp teeth	
2. has fur	√	8. runs fast	√
3. lives in a pride	√	9. has a mate	√
4. Is gold colored		10. is a carnivore	√
5. lives on land	√	11. has long tail	
6. sleeps daytime		12. hunts at night	√

ing selections. Then use what you know about these organisms and what you learn from your research to make a list of characteristics that describe each organism. Be sure to include at least eight characteristics. Remember, an organism is anything that is alive. Be sure students understand that organisms include animals, plants, bacteria and other single-celled organisms. Use your texts and the resources available in the classroom to get this information (for example, the animal cards shown at the end of this lesson, or one of the various computer programs available on living organisms). Alternately, set aside a period to visit the library, where students can benefit from more extensive resources. Resources typically found in high school libraries that have surveys of living organisms are listed at the end of this lesson. (A variation is to assign the completion of the cards as homework the night before.)

- Let's use these cards to generate a list of characteristics that we can use when we design a way of classifying these organisms. Meet in your groups and review each of the team's index cards. Put a check mark next to only those characteristics that appear at least twice on the index cards

of the group members. After students complete this task, develop a list of "Important Charac-teristics of Living Things" on the chalkboard or on a transparency. Do this by having each group report one characteristic. Go around the room, but be sure not to repeat similar characteristics. Then, repeat the process until you have a substantial list of characteristics (at least 30, if possible). Typically, the list includes information about physiognomy, habitat, niche, usefulness or harmfulness for humans, location, and range. Very often, students refer to the scientific class of an organism—for example, *it's a reptile.* Tell students to avoid such scientific terms at this point. Suggest that in their place, students should describe what they do that makes them fall into the scientific category—in this case, reptilian characteristics include: cold bloodedness, egg laying on land, and well developed lungs. POSSIBLE STUDENT ANSWERS: *organisms that live or grow in Africa, organisms inhabiting the sea, hostile and aggressive organisms, herbivores, organisms poisonous to humans, organisms that run.*

- **This list is long. How might we shorten it, make it easier to handle?** Students usually say many of the characteristics are similar and can be combined. Ask students to review the "Important Characteristics of Living Things" list and discuss those characteristics that can be combined.

- **We will now create categories that animals having these characteristics fall into. Write the category name for the organisms that have that characteristic on the left side of the "Ways to Classify" graphic organizer on the board, on a poster, or on a transparency by using the stem "Organisms that..." or the common name for the category, if one already exists (e.g., herbivore). Delete the original characteristic. Continue until all characteristics have been accounted for.** When students have completed this task, have four or five of them report one or two categories each.

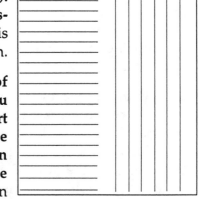

- **On the left side of the graphic organizer, we now have a list of categories of organisms representing the characteristics that you have identified. Some of these categories describe the same sort of feature. For example, organisms that are found all over the world and organisms found only in Africa describe the location or "range" of the organism. A term like "range" identifies the type of categories that have been identified. Write "range" on** one of the horizontal lines at the top of the diagram. Connect the appropriate subcategories (e.g., organisms found in Africa and organisms found all over the world) from the list on the left to the vertical line under "Range." **What other types of categories are identified in the column on the left?** As students note different types of categories, write these on the other horizontal lines and connect the categories they identify of that type with the vertical line that runs to the bottom of the page under that category type. Use different-colored chalk or pens to show the connections. POSSIBLE ANSWERS: *location (range), habitat, outer covering, what they eat, what eats them, benefit or harm to man, how they produce energy, niche, structure and functioning, how they bear their young, color, size, life span, population size, species stability.*

- **A common method for designing a classification system is to start with one group, one that includes everything to be classified, then to divide this group into several subgroups, and then to divide each subgroup into further subgroups and so forth until all categories of characteristics are accounted for. For example, one classification scheme for the stamps we discussed earlier began with an initial group of *Stamps of the World* and then this group was subdivided into two subgroups, *United States Stamps* and *Foreign Stamps*. The foreign stamps were then subdivided into stamps from each continent and then subdivided again**

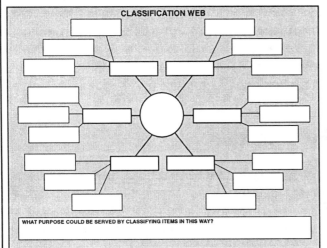

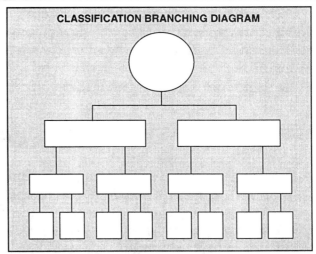

into stamps from countries within the continents. Each of these subgroups were further subdivided into groups for *Airmail* and *Regular Postage* and then *Date of Issue* and lastly *Postage*. Use the Classification Web or Branching Diagram graphic organizer to develop a classification system that falls under one of the types of categories identified in the chart, or develop your own freehand mind map of such a classification system. ANSWERS VARY.

- **We now have many different and interesting ways to classify living organisms. How do we decide which one is the best one?** Students typically respond that the way we choose the best way to classify organisms depends on our purpose. Ask students to meet in their groups and discuss the kinds of people who might be interested in classifying organisms by each of the types of categories they listed on the "Ways to Classify Things" graphic. As each group reports, add their response to a chart like the one included at the right. POSSIBLE ANSWERS: *Travelers and people who hunt animals for zoos would want to be able to locate particular animals; zoo and aquarium specialists creating artificial habitats would want to know the natural conditions in which specific animals survive; doctors of sick or injured animals would want to understand the various body parts and how they function.*

PURPOSES AND USES OF CLASSIFICATION SYSTEMS

WAYS TO CLASSIFY	PURPOSE OF THE CLASSIFICATION	WHO WOULD USE IT AND WHY

- **Because this is a science course, let's conclude this activity by focusing on a way of classifying living things relevant to our studies. Let's consider organizing organisms around their structure and function. Meet in your groups and using colored pencils, markers, graph or construction paper, develop a scheme for classifying organisms in this way. Use the webbing diagram or create a diagram or chart clearly showing the relationship between the different groups you have already named.** ANSWERS VARY.

- **Now that you have finished your classification systems, how do you think we might test them to show how well they work?** Students usually realize that if organisms they haven't thought about yet have a place in their system, then their scheme is useful. On the other hand, if they have to make a new category for their classification system each time a new organism is considered, then the system needs to be redesigned. Pass out copies of the illustration titled "Some Unusual Organisms" found at the end of this lesson and challenge students to find a category for each organism in their system. Alternatively, you could have each group send one student to the board to reproduce their chart. Then you can challenge each student to find a place for each of the unusual animals. The class could vote on which was most effective.

THINKING ABOUT THINKING

• **How did you go about classifying the organisms? What did you think about first, second, and third?** ANSWERS: *determined important information about the things to be classified; grouped, classified, and described these things according to common characteristics; chose a way to classify the objects that suits a specific purpose; classified the objects by developing a system of categories and subcategories.* The thinking map should include the steps in the chart for bottom-up classification.

• **How is the way you classified things different from using a classification system that is provided for you? Which way do you prefer? Why?** POSSIBLE ANSWERS: *You don't just sort them into categories you are given. When you do it this way, you have to observe the things, decide what is important, and decide what categories you are going to group them in based on a purpose. I prefer this type of classification because I can decide on the best way to classify for my purposes.*

• **How did classifying organisms this way help you to define the words for each category?** POSSIBLE ANSWERS: *I decided to group the animals together because of important characteristics, like eating only meat, that they had in common. These common characteristics define the category and can be used to define the category words (like "carnivorous animals").*

APPLYING THINKING

Immediate Transfer

• Suppose you had to organize the storeroom in the chemistry department at a local high school. What information might you need to know before you could get started? Consider the large variety of chemicals, glassware and apparatus you would be likely to find. Think through different ways you might organize the storeroom by using the thinking map for classification and recommend one, explaining why you think your classification system is a good one.

• Think about all the different courses and activities that are available in your school. It is your task to design an informative brochure which will be given to parents on "Back to School Night." It is important that the brochure organize the courses and activities of the school in an intelligent and useful manner.

Reinforcement Later

• Suppose you were the head of marketing for a large chain of department stores. You suspect that business is down because the layout of the store's departments and the organization of merchandise within each department has made it difficult for customers to shop. Consider the needs of your customers and then propose an overall design for this store describing how you would name and organize each department.

CONTENT EXTENSION

After students have finished their classification charts and diagrams, have them visit the library and locate a graphic or table showing the five kingdom classification schemes developed by biologists and traditionally found in biology textbooks and encyclopedias. Then have them compare and contrast this formal taxonomy with their class work. Have them report back on how their efforts measured up.

CONTENT EXTENSION

Ask students to consider the category of domestic organisms. Have them research a list of plants and animals that have been important to the emergence and maintenance of civilization. Then have them develop a Ways to Classify Things graphic organizer which listing those defining characteristics of each organism that have made it so useful to people. Students should then develop a chart or diagram showing the relationship between geography, culture, and domestic organisms.

ASSESSING STUDENT THINKING ABOUT CLASSIFICATION

To assess this skill, ask students questions like the immediate transfer questions above. The situation in which the students have to arrange things (e.g., clothing, books, or food in a store) is ideal for showing whether the students are thinking skillfully about how they classify things. Determine whether they are attending to each of the steps in the thinking map for classification. To determine their ability to monitor their own thinking about classification, encourage students to use terms appropriate to the skill of classifying (e.g., characteristics, groups, classification.)

SOME U.S. AND FOREIGN STAMPS

Sample Student Responses • Classifying Stamps

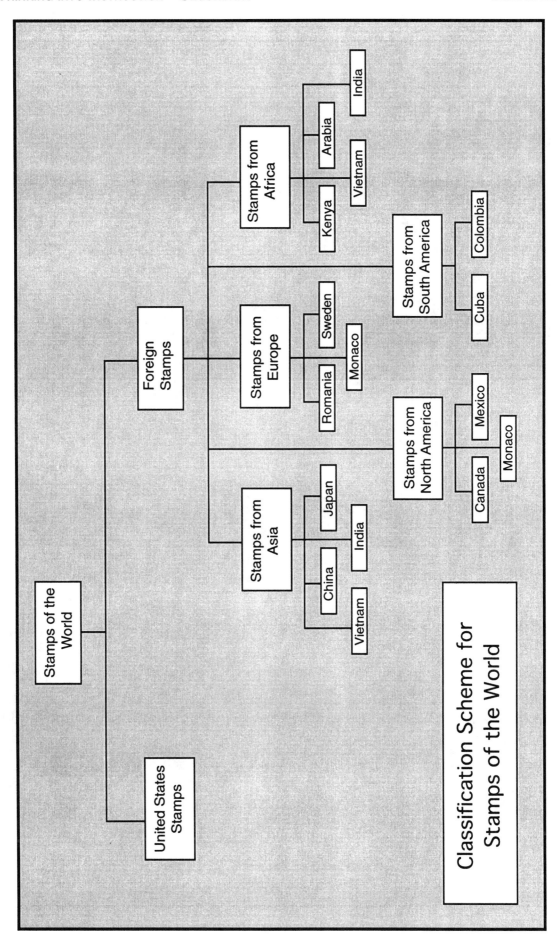

Classification Scheme for Stamps of the World

Sampling of Different Organisms That Students Identified in This Lesson

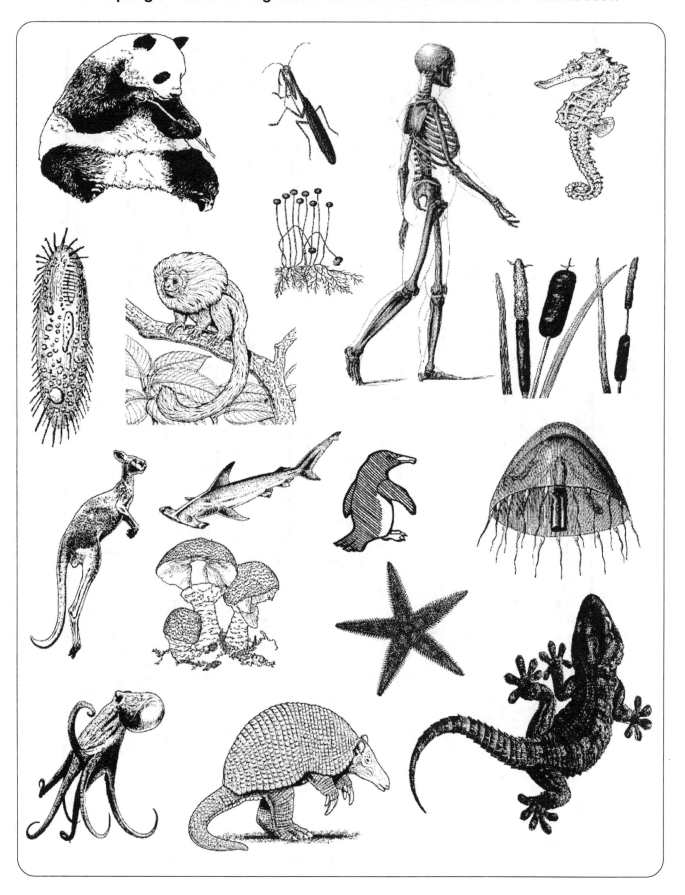

Sample Student Responses • Living Things of All Shapes and Sizes

WAYS TO CLASSIFY LIVING THINGS

CATEGORIES

	Lifestyle	Habitat	Niche	Reproduction	Structure	Life span
	Ecosystem component	Diet / Range	Source of energy	Appearance	Benefit/harm to man	

Organisms which live in Africa

Organisms that have fur

Organisms that live a long time

Organisms that live on land

Consumers

Carnivores

Organisms with four legs

Omnivores

Organisms that are eaten by people

Organisms that can eat people

Organisms that give live birth

Organisms with 4-chambered heart

Organisms that hunt

Organisms in temperate climates

Herbivores

Organisms that walk on two legs

Organisms that resemble humans

Organisms that live in small groups

Organisms that are intelligent

Organisms that live in salt water

Organisms that have lungs

Organisms that live all over the world

Organisms that live in fresh water

Organisms that lay eggs in water

Organisms that have smooth skin

Organisms that have wings

One-celled organisms

Organisms that cause disease

Organisms that reproduce sexually

Organisms that use sun for energy

Organisms that have leaves

Organisms that eat food for energy

Organisms that use carbon dioxide

Producers

Organisms whose cells have walls

Organisms that have roots

Organisms that produce seeds

Organisms that reproduce via pollen

Organisms that can be poisonous

Organisms used for medicines

Decomposers

Organisms that are parasites

Sample Student Responses • Living Things of All Shapes and Sizes

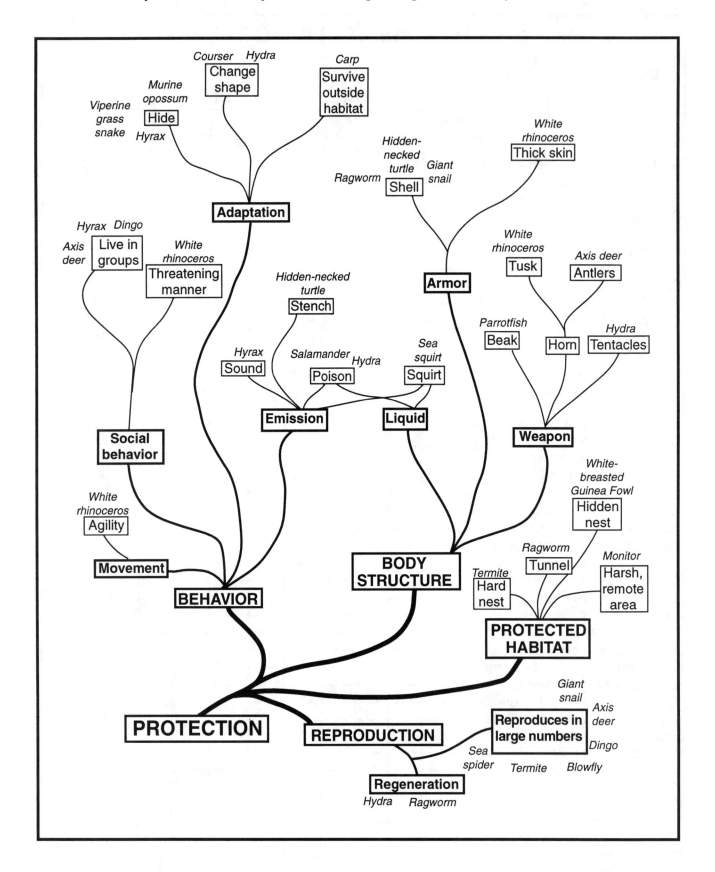

Sample Student Responses • Living Things of All Shapes and Sizes

CLASSIFICATION WEB

Animals that benefit and/or harm human beings

Animals that provide material for clothing

- Fur-bearing animals
- Animals that have durable skin
- Animals that produce fiber that can be spun and woven

Animals that can kill or harm humans

- Poisonous animals
- Animals that cause diseases
- Animals that treat humans as food

Animals that can be domesticated

- Animals that can be kept at home and that work for humans
- Animals that can be kept for protection
- Pets

Animals that serve as food

- Animals that provide edible protein
- Animals that contain food byproducts harmful to humans
- Animals that contain material desired for its taste

Animals that can carry and transport things

- Animals that can carry humans
- Animals that can carry goods for humans
- Animals that can pull vehicles

Animals that provide luxury items for humans

- Animals that provide materials for luxury items of clothing
- Animals that provide materials for luxury foods
- Animals that provide materials for jewelery

PURPOSES AND USES OF ANIMAL CLASSIFICATION SYSTEMS

WAYS TO CLASSIFY	PURPOSE OF THE CLASSIFICATION	WHO WOULD USE IT AND WHY
Location	To indicate which animals inhabit a particular region. To indicate where in the world a particular animal can be found.	Traveler who wants to know what animals can be seen in an area. People who find animals for zoos.
Ecosystem	To indicate what kind of environment the animal needs to survive. To indicate what kinds of animals are likely to be found in a type of environment.	Environmentalists and government officials who try to preserve animals. Naturalists who want to find and observe animals in their natural environment. Zoo and aquarium workers who keep animals healthy.
Habitation (nest, den, hive, shell, etc.)	To indicate what kind of environment the animal needs to survive. To indicate in what type of home the animal can be found.	Naturalists who want to find and observe animals in their natural environment. Zoo and aquarium workers who keep animals healthy. Architects who create homes and buildings based on natural principles of design. People who cultivate animals.
Outer covering	To indicate what kind of protection the animal needs to survive.	Environmentalists who try to preserve animals. Designers who create clothing based on natural design principles.
Body structure and functioning	To provide information about the bodies of animals. To indicate how animals function in their environment.	Biologists who explain diversity and evolution of animals. Doctors who treat ill animals. Zoo and aquarium personnel who keep animals healthy.
Benefit or harm to man	To indicate which animals can be used to benefit man. To indicate how various animals can benefit man. To indicate which animals we need to protect ourselves from.	Ranchers who cultivate animals for food. People who hunt animals for food (e.g., fishermen). Travelers who are going into the wilds. People who train animals.
Population/ species stability	To indicate size of population. To indicate which animals are endangered.	Environmentalists who try to preserve species of animals. People who hunt animals for food.

Sample Animal Cards

Cheetah
The fastest mammal on earth

The cheetah is the strangest of all the felines. Standing high off the ground, it has a narrow body and long tail. Its claws do not retract as do those of its cousins. It does not roar but emits an odd cry like a mewing hiss. Like a cat, it purrs to show pleasure. The cheetah is one of the rare diurnal felines. In fact, it hunts by day, pursuing the antelope, which it locates by means of its keen eyesight.

The cheetah lives in the open savanna and is never found in forests. It spends a large part of its time crouched on hillocks or large anthills, visually searching its territory for some kind of prey. Unlike the other big cats, which rely on surprise to take their victims, the cheetah overcomes gazelle and other antelope by running them down. Faster even than the greyhound, the cheetah can reach speeds of 90-98 km/hr (56-63 mph), which allow it to outstrip any other mammals on earth. Against this extraordinary performance must be set the fact that the cheetah has no endurance; it is exhausted after a run of 300-400 meters and has to rest for some time to get its breath back. It is a sprinter, not a distance runner.

The habits of the cheetah have remained mysterious until recent years. It steadfastly refused to breed in captivity until 1967, since zoologists did not know that the presence of several males is necessary before the female will mate. It is the female who selects her partner, always the strongest one of the group.

Long ago, the cheetah was used as a hunting animal in Asia and Arabia.

Gestation: 90 to 95 days Number of young: 2 to 5 Weight at birth: about 275 g (9 oz)	Adult weight: 30 to 65 kg (66 to 143 lbs) Sexual maturity: from 17 to 24 months	Longevity: 16 years Running speed: can cover 300 m (300 yards) at 70 to 90 km/hr (43 to 56 mph)		
Phylum: **Vertebrata**	Class: **Mammalia**	Order: **Carnivora**	Family: **Felidae**	Genus and species: **Acinonyx jubatus**

Tilapia
Of 'miraculous draught of fishes' fame ...

The tilapia is a large fish of African origin that has gradually spread throughout the tropical and subtropical regions of the world, for man has taken a hand and introduced it in his diet in areas where previously it did not exist. In the beginning, this fish was resident in tropical Africa, in the Nile basin and in Israel, Jordan, and Syria. It lives in slow-flowing lakes and rivers, as well as in estuaries and saltwater lagoons. It acclimatizes itself well to new habitats and its resistance is absolutely incredible. For example, Graham's tilapia does well in the exceedingly alkaline waters of Lake Magadi in Kenya, where the temperature reaches 28° to 45° C (80° to 112° F). The tilapia has a hundred different species.

Because of its ability to become used to different habitats, and above all for its food value to populations lacking protein, the tilapia has been introduced into many areas, both voluntarily and involuntarily. Thus, it appeared unexpectedly in Java in 1969, without anyone knowing how it traveled from East Africa to the East Indies. It then spread spontaneously throughout the Indonesian islands.

Recently, tilapias have colonized the waters of Texas and Florida and are becoming a serious menace to native species of fish. In 1951, the breeding of tilapia was begun in Madagascar. In several areas, tilapias are kept to clean the lakes and marshland from the dangerous mosquito larvae that infest them. Even by the time of the Ancient Egyptians, this fish was appreciated and it is certainly tilapias that are responsible for the Biblical 'miraculous draught of fishes.'

Oviparous Incubation: 8 to 20 days	Length: up to 45 cm (18 in) Sexual maturity: at 3 months	The tilapia can have 6 to 11 egg laying cycles a year.		
Phylum: **Vertebrata**	Class: **Osteichthyes**	Order: **Perciformes**	Family: **Cichlidae**	Genus and species: **Tilapia natalensis**

Safari Cards, Atlas Editions, Inc. Reprinted by permission of Atlas Editions, Inc., 33 Houston Dr., Durham, CT 06422. Cards shown are 67% of actual size. Only the outline of the animal has been reproduced. The actual animal cards show a full-color photo of the animal in its habitat.

SOURCE MATERIALS FOR LIVING THINGS LESSON

Cavendish, Marshall, *The Marshall Cavendish Illustrated Encyclopedia of Plants and Earth Science*, Bellmore, N.Y.: Marshall Cavendish Corp.

Classifying Plants and Animals, Deerfield, Ill.: Coronet Film & VideoGrizimek, Dr. H.C., *Grizimek's Animal Life Encyclopedia*, Van Nostrand Reinhold Co.

Herberman, Ethan, *The City Kids's Filed Guide*, New York: Simon & Schuster

Johnson, Hugh, *Encyclopedia of Trees*, New York: Portland House

Pearl, Mary Corless, *The Illustrated Encyclopedia of Wildlife*, Grey Castle Press

CHEMICAL REACTIONS

Chemistry   **Grades 10–12**

OBJECTIVES

CONTENT

Students will learn to recognize different types of chemical reactions and will identify these as either reactions involving combination, decomposition, single replacement, double replacement, or combustion based on the reactants and products in these reactions.

THINKING SKILL/PROCESS

Students will learn to classify by identifying defining characteristics of classes and will recognize that the purpose for classification is to determine those defining characteristics.

METHODS AND MATERIALS

CONTENT

Students will use their textbooks to identify the standard types of chemical reactions. Students will then use inspection and their background knowledge of chemical elements and compounds to identify reactants and products in sample reactions.

THINKING SKILL/PROCESS

Students are guided by the teacher's questions to classify various chemical reactions according to the defining characteristics of standard types of chemical reactions. Students use a graphic organizer for the top-down classification of single items, and the teacher uses the multiple-item graphic to record the results.

LESSON

INTRODUCTION TO CONTENT AND THINKING SKILL/PROCESS

- **Think about the different ways that products are grouped in the supermarket. When we group things together because they all have something in common, we are classifying them. With your partner, list as many ways as you can think of that food and other products in the supermarket are grouped. Give some examples of each way of grouping them.** List students' responses on a chalkboard or newsprint under the heading "How things are classified in the supermarket." POSSIBLE ANSWERS: *Type of food (fruit, vegetables, beverages, mixes, pasta), utensils for the home (food preparation utensils, storage containers, cleaning tools, waste containers), beauty or health products (products to clean or protect teeth, or skin, hair care products, first aid supplies, medications for minor ailments), foods that are used together (cake preparation [mixes, spices, flour, sugar], ethnic foods, foods for special diets), size and type of container (bulk or giant size packages may be grouped together on racks that are large enough to hold them), storage (food that doesn't spoil quickly is stored on shelves, perishable foods in a refrigerator, or frozen foods in a freezer), whether or not a salesperson has to get you a particular amount (bakery or deli products are bought by the amount the customer wants), whether or not a salesperson has to take special care of it (fruits and vegetables in the produce department must be watered and rotated and spoiled items removed each day).*

- **We always have a purpose for classifying things the way we do. Select one way of classifying things in the supermarket and describe to your partner why you think things are grouped that way.** List students' responses under the heading "Why things are classified this way." POSSIBLE ANSWERS: *Type of food (to find them easily when needed), utensils for the home* *(to find them easily and to display the type or size of utensil that is needed for a task), beauty or health products (to find them easily), foods that are used together (to remind us of products that are used in the same types of recipes, to show substitutes of some products, to display them the way people store them at home to find them easily), storage (to prevent spoilage), whether or not a salesperson has to get you a particular amount (to get only the amount of food that the customer wants), whether or not a salesperson has to take special care of it (to keep it fresh).* When the list is complete, add a summary statement at the bottom — for example: In the supermarket, most things are classified to help us find what we want to buy and to keep the food fresh.

- **Once things are classified a certain way, we have to understand what defines things in these categories in order to add other things to the same group. If you worked at the supermarket and stocked shelves every day, you'd have to know where to put things. Sometimes products come labeled and it's easy. For example, when you unpack a box that is labeled Macintosh Apples, it's easy to put them in the correct bin: the one marked Macintosh Apples. However, when things don't come labeled, you have to know what characteristics products should have in order to fall into the different categories, and you have to be able to identify items that have these characteristics so that you can put them in the right bin. To define a category, we identify its** *defining characteristics*: **the set of characteristics that all and only that kind of thing has. For example, when we think of fruit, knowing its defining characteristics helps us identify many kinds of foods that fall into that category. With your partner, list all the defining characteristics of fruit that you can think of.** After partners have had time to list characteristics of fruit, record their responses on the board. POSSIBLE ANSWERS: *Fruit grows on plants (trees, bushes, vines). It is the fleshy part of the plant that surrounds the seeds of the plant where a blossom used to be. The fleshy part can be eaten without being cooked.*

- **Knowing the defining characteristics helps us identify things correctly. It may be easy to identify fruit, but where in the fruit section should you put a specific type of fruit? That depends on what defines the kind of fruit it is.** Hold up the kiwi fruit. If one is not available, select a fruit or vegetable that students do not commonly buy. **What is this?** *A kiwi.* **How would you describe it to someone else?** *It is a fruit that is smaller than a pear and larger than a plum. It has a thin brown skin, sometimes looks almost hairy, and has green pulp inside. It tastes mildly sweet and comes originally from the South Pacific.* **Why would it be important to know the defining characteristics of the kiwi?** POSSIBLE ANSWERS: *If you have to put new fruit out, you'd know where to put kiwis. If someone sends you to the supermarket for one, you need to know what to look for or ask for. You also need to know which fruit the sign and the price refers to. You may also have to know what you are getting, if you are trying one for the first time, so that you can make sure that it is the right fruit.* **To describe a kiwi accurately or to be sure that we are selecting the right fruit when we look for a kiwi, we must be very clear of the defining characteristics of a kiwi.**

- **The thinking map to the right summarizes the questions to ask when you try to determine whether a specific item is classified correctly:**

> **TOP–DOWN CLASSIFICATION**
> 1. What are the defining characteristics of the categories under which I want to classify things?
> 2. Which items have these characteristics?
> 3. How do I classify these items into the given categories?

- **It isn't just in the supermarket that things are classified. We classify almost everything we know about in a variety of different ways, all for various** purposes. For example, when you studied mathematics in elementary school, you found that numbers are classified in various ways, such as into whole numbers or fractions. The shapes that you studied in geometry are also classified; in their case, they are classified in various ways for the purpose of understanding important facts about them. Like in the supermarket, once we understand what defines the way we classify different types of triangles, for example, we can identify individual ones that should be classified the same way. Then we know that what we can prove about the type of triangle applies to the individual triangles that are of that type as well. For example, when you classify triangles by the size of the angles, what are the different ways that they are classified in geometry? *Right triangles, obtuse triangles, and acute triangles.* **What are the defining characteristics of each of these types of triangles?** POSSIBLE ANSWERS: *Right triangles have one right angle; a right angle has 90 degrees. Obtuse triangles have at least one angle that is more than 90 degrees. Acute triangles have all three angles less than 90 degrees.* **When you sort them by sides, what small groups can you sort them into?** *Equilateral triangles, scalene triangles, and isosceles triangles.* **What are the defining characteristics of each of these types of triangles?** *In equilateral triangles, all three sides are the same length. In isosceles triangles two sides are equal. In scalene triangles none of the sides are equal length.*

• **Given these defining characteristics, how would you classify the following triangle?** Ask for responses from the class. *This triangle is a right triangle that is isosceles.* **What are the defining characteristics of the triangle that you noted on the basis of which you classified it this way?** *It has a right angle and two sides are equal in length.* **What are some theorems that you have learned about right triangles?** *The Pythagorean theorem: the square of the hypotenuse is equal to the sum of the squares of the other two sides in a right triangle.* **This is what I meant when I said that once you have classified a specific triangle a certain way—for example, as a right triangle—you know that what you've proved about the type of triangle it is (in this case the Pythagorean theorem) applies to this particular triangle. Can you think of anything else you know about this triangle because it is an isosceles triangle?**

THINKING ACTIVELY

• **In chemistry, you've been exposed to a number of classification schemes that are very useful in this field. Can you identify some of these?** STUDENT ANSWERS VARY BUT OFTEN INCLUDE: *Matter into elements and compounds; chemical elements in the periodic table; states of matter into solids, liquids, and gases; types of molecules into inorganic and organic; kinds of chemical reactions.* As students respond, write these classification systems on the chalkboard or on a transparency. **Work with a partner and select one of these classification systems in chemistry to review. Discuss the details of the classification system, including the major categories and subcategories, and discuss what purpose the classification system serves in chemistry.** After three or four minutes, ask for student reports about two or three of the classification systems they worked on, writing their views about the purpose for classifying things in these ways on the chalkboard or transparency next to the classification systems. STUDENT ANSWERS VARY.

• **We are now going to use skillful top-down classification to identify items that fall into the classification schemes of chemical reactions.** If students haven't identified this type of classification, introduce it and explain its details referring only to five of the basic categories that reactions fall into: combination, decomposition, single replacement, double replacement, and combustion.

• **Chemical reactions of these sorts are defined by the type of reactants and products in these reactions. In the box at the top of the graphic organizer, write "Kinds of Chemical**

Reactions." Use your text books and this graphic organizer to write statements of the defining characteristics of each of these types of reactions. After a few minutes, ask students to report on each of these types of reactions. Ask if anyone wants to add to the statement of defining characteristics in the report, then formulate the statement in a way that is acceptable to the student responding and write it in the appropriate box on a diagram of the graphic organizer on the chalkboard or on a transparency. STUDENT ANSWERS INCLUDE THE FOLLOWING: *Combination: The reactants are either two elements or two compounds, and the product is a single compound. Decomposition: The reactants are a binary compound or a ternary compound and the products are either two elements or two or more elements and/or compounds. Single Replacement: The reactants are a single element and a single compound, and the products are a different element and a new compound. Double Replacement: The reactants are two ionic compounds and the products are two new compounds. Combustion: The reactants are oxygen and a compound of carbon, hydrogen, and sometimes oxygen; the products include carbon dioxide and water.* If you have not already discussed with students what purpose is served by classifying reactions into these categories, do so now.

- Now, let's write a formula for a chemical reaction that you are all familiar with, the reaction representing the formation of rust. Ask for student responses. **The reaction is typically written like this:**
$3Fe + 2O_2 \longrightarrow Fe_3O_4$.

- Write this reaction formula in the first box labeled "Item to be Classified." What are the characteristics of this reaction in terms of the reactants and products? Write your description in the same box under "Characteristics." After a minute or two, ask for student responses. TYPICAL STUDENT RESPONSES INCLUDE: *Reactants: two elements (Fe and O). Products: a single compound (Fe_3O_4).* **Based on these characteristics, how would you classify this reaction?** ANSWER: *A combination reaction.* Write that in the box under "How Classified" and enter it in the "item" list under the category "Combination."

- Here is a list of other chemical reactions:

$$Mg + H_2SO_4 \longrightarrow MgSO_4 + H_2$$
$$2C_2H_6 + 7O_2 \longrightarrow 4CO_2 + 6H_2O$$
$$F_2 + 2NaBr \longrightarrow 2NaF + Br_2$$
$$2H_2O_2 \longrightarrow 2H_2O + O_2$$
$$CdBr_2 + Na_2S \longrightarrow CdS + 2NaBr$$
$$Sr + I_2 \longrightarrow SrI_2$$
$$Mg(ClO_3)_2 \longrightarrow MgCl_2 + 3O_2$$

• **Work with your partner again and write your assessment of each section's characteristics in the appropriate space on your graphic organizer; then add your classification of the reaction.** After four or five minutes, ask students to report. Accept a report on only one reaction from each team that reports. After each report, ask if students have any comments; then, formulate the response to accommodate appropriate comments from other students. Write the responses on the diagram on the chalkboard or transparency. As you field responses, follow the questioning strategy on the thinking map, asking first, "What is the item?" and then "What are its characteristics?" and then "Based on these characteristics, how would you classify this item and why?" STUDENT ANSWERS INCLUDE, FOR EXAMPLE: $Mg + H_2SO_4 \rightarrow MgSO_4 + H_2$: *Reactants: an element (Mg) and a single compound (H_2SO_4); Products: a different element (H_2) and a new compound ($MgSO_4$).* CLASSIFICATION: Single Replacement. $2C_2H_6 + 7O_2 \rightarrow 4CO_2 + 6H_2O$: *Reactants: Oxygen and a compound of carbon (C_2H_6); Products: CO_2 and H_2O.* CLASSIFICATION: Combustion

• **What is gained when you classify chemical reactions in this way?** Discuss with students the power of this classification scheme: once you identify the kind of reaction that will take place, you can predict the properties of the products and prepare correctly for the reaction.

THINKING ABOUT THINKING

• **How did you go about classifying the reactions? What did you think about first, second, and third?** POSSIBLE ANSWERS: *I examined them for important characteristics, defined groups by reference to some of these characteristics, identified each item according to whether it had these characteristics, and put the items into the defined groups. I described what I learned about chemical reactions by group-ing them in these ways.* If they are having trouble remembering, prompt their answers by referring to their graphic organizers. Their descriptions of classifying should include the steps on the thinking map for top-down classification (right).

TOP–DOWN CLASSIFICATION

1. What are the defining characteristics of the categories under which I want to classify things?

2. Which items have these characteristics?

3. How do I classify these items into the given categories?

• **We have classified specific chemical reactions by finding characteristics that made it possible to write them in forms we can identify. How did doing this help you understand the meanings of the words that describe the five chemical reactions we were studying? Why?** POSSIBLE ANSWERS: *It helps me define these words because the different reaction forms defined the type of reaction. It helps me understand how chemicals can interact with each other to produce predictable types of results.*

• **Is top-down classification a helpful way to learn more about the meaning of classification words and what they stand for?** POSSIBLE STUDENT RESPONSES: *I like top-down classifica-tion because it keeps me from guessing what to expect when putting together specific types of reactants. I have to think about what defines a type of thing and then see if the item I am trying to classify has a characteristic that matches those defining characteristics. Then, I'm certain what kind of thing the item is. Top-down classification, especially the graphic organizer, makes me go slowly in a step-by-step way to try to find out what sort of thing something is. It's much better than just defining words by other words because it gives me practice in finding out how to classify things by looking for characteristics they have.*

APPLYING YOUR THINKING

Immediate Transfer

Shortly after the main activity in this lesson and the students' metacognitive reflection, ask them to work on the following:

- **Another way to classify chemical reactions is as endothermic and exothermic. Use the same thinking strategy as you used in this lesson to define these categories and to identify at least three reactions that fall into each of these categories. Use your textbooks or other reference sources in chemistry as you engage in this activity. Finally, explain what purpose classifying chemical reactions in this way might serve.**

- **Classify algebraic symbols by their functions using top-down classification.**

Reinforcement Later

Sometime later in the school year, ask your students to work on the following:

- **How are books classified? Go to the school (or your town) library, select a dozen books at random, and determine how to classify each using the strategy for top-down classification. Does the way books are classified clarify the different purposes for reading them? Explain.**

- **In biology, living things are classified by their cellular complexity and evolutionary development. Physics, geology, and other special sciences also have their special ways of classifying things. Pick one of these systems of classification and do a top-down classification of a dozen or so items that fall under the classification system you have chosen.** ANSWERS VARY

- **Classify jobs by whether they provide goods or services to show how the community needs both.**

- **Classify land forms and bodies of water to clarify correct terms for each.**

ASSESSING STUDENT THINKING ABOUT CLASSIFICATION

To assess classification skills, ask students questions like the ones given in immediate transfer. You may want to ask students to classify given books into fiction or nonfiction to clarify the different purposes for reading them. Ask them to think out loud. Determine whether they are attending to each of the steps in the verbal map for classification. Encourage students to use the terms "classify" and "characteristics" and to state what the classification shows.

Sample Student Responses • Chemical Reactions

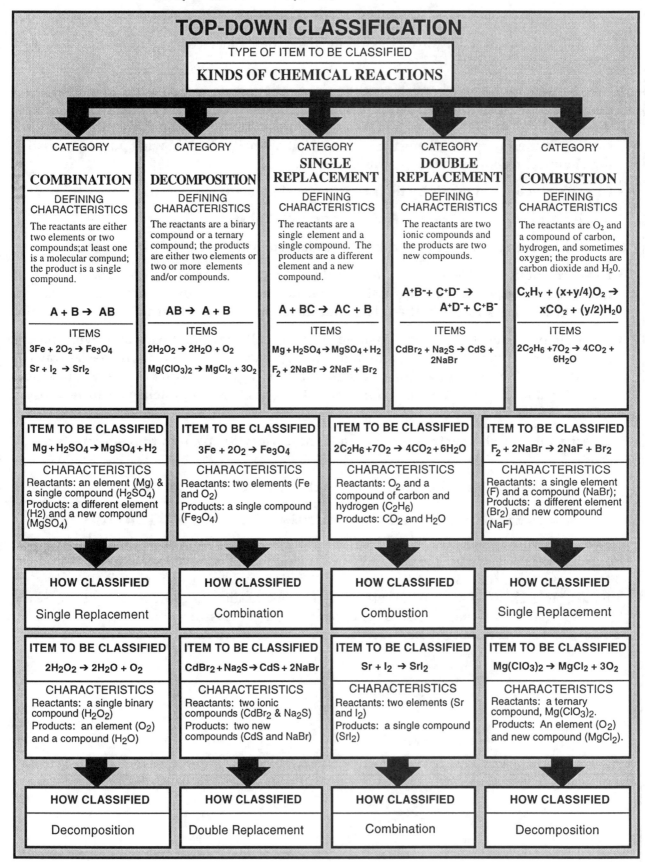

TOP-DOWN CLASSIFICATION

TYPE OF ITEM TO BE CLASSIFIED

KINDS OF CHEMICAL REACTIONS

CATEGORY	CATEGORY	CATEGORY	CATEGORY	CATEGORY
COMBINATION	**DECOMPOSITION**	**SINGLE REPLACEMENT**	**DOUBLE REPLACEMENT**	**COMBUSTION**
DEFINING CHARACTERISTICS	DEFINING CHARACTERISTICS	DEFINING CHARACTERISTICS	DEFINING CHARACTERISTICS	DEFINING CHARACTERISTICS
The reactants are either two elements or two compounds;at least one is a molecular compund; the product is a single compound.	The reactants are a binary compound or a ternary compound; the products are either two elements or two or more elements and/or compounds.	The reactants are a single element and a single compound. The products are a different element and a new compound.	The reactants are two ionic compounds and the products are two new compounds.	The reactants are O_2 and a compound of carbon, hydrogen, and sometimes oxygen; the products are carbon dioxide and H_2O.
$A + B \rightarrow AB$	$AB \rightarrow A + B$	$A + BC \rightarrow AC + B$	$A^+B^- + C^+D^- \rightarrow$ $A^+D^- + C^+B^-$	$C_xH_y + (x+y/4)O_2 \rightarrow$ $xCO_2 + (y/2)H_2O$
ITEMS	ITEMS	ITEMS	ITEMS	ITEMS
$3Fe + 2O_2 \rightarrow Fe_3O_4$ $Sr + I_2 \rightarrow SrI_2$	$2H_2O_2 \rightarrow 2H_2O + O_2$ $Mg(ClO_3)_2 \rightarrow MgCl_2 + 3O_2$	$Mg + H_2SO_4 \rightarrow MgSO_4 + H_2$ $F_2 + 2NaBr \rightarrow 2NaF + Br_2$	$CdBr_2 + Na_2S \rightarrow CdS +$ $2NaBr$	$2C_2H_6 + 7O_2 \rightarrow 4CO_2 +$ $6H_2O$

ITEM TO BE CLASSIFIED	ITEM TO BE CLASSIFIED	ITEM TO BE CLASSIFIED	ITEM TO BE CLASSIFIED
$Mg + H_2SO_4 \rightarrow MgSO_4 + H_2$	$3Fe + 2O_2 \rightarrow Fe_3O_4$	$2C_2H_6 + 7O_2 \rightarrow 4CO_2 + 6H_2O$	$F_2 + 2NaBr \rightarrow 2NaF + Br_2$
CHARACTERISTICS	CHARACTERISTICS	CHARACTERISTICS	CHARACTERISTICS
Reactants: an element (Mg) & a single compound (H_2SO_4) Products: a different element (H_2) and a new compound ($MgSO_4$)	Reactants: two elements (Fe and O_2) Products: a single compound (Fe_3O_4)	Reactants: O_2 and a compound of carbon and hydrogen (C_2H_6) Products: CO_2 and H_2O	Reactants: a single element (F) and a compound (NaBr); Products: a different element (Br_2) and new compound (NaF)

HOW CLASSIFIED	HOW CLASSIFIED	HOW CLASSIFIED	HOW CLASSIFIED
Single Replacement	Combination	Combustion	Single Replacement

ITEM TO BE CLASSIFIED	ITEM TO BE CLASSIFIED	ITEM TO BE CLASSIFIED	ITEM TO BE CLASSIFIED
$2H_2O_2 \rightarrow 2H_2O + O_2$	$CdBr_2 + Na_2S \rightarrow CdS + 2NaBr$	$Sr + I_2 \rightarrow SrI_2$	$Mg(ClO_3)_2 \rightarrow MgCl_2 + 3O_2$
CHARACTERISTICS	CHARACTERISTICS	CHARACTERISTICS	CHARACTERISTICS
Reactants: a single binary compound (H_2O_2) Products: an element (O_2) and a compound (H_2O)	Reactants: two ionic compounds ($CdBr_2$ & Na_2S) Products: two new compounds (CdS and NaBr)	Reactants: two elements (Sr and I_2) Products: a single compound (SrI_2)	Reactants: a ternary compound, $Mg(ClO_3)_2$. Products: An element (O_2) and new compound ($MgCl_2$).

HOW CLASSIFIED	HOW CLASSIFIED	HOW CLASSIFIED	HOW CLASSIFIED
Decomposition	Double Replacement	Combination	Decomposition

CLASSIFICATION LESSON CONTEXTS

The following examples have been suggested by classroom teachers as contexts to develop infused lessons. If a skill or process has been introduced in a previous infused lesson, these contexts may be used to reinforce it.

GRADE	SUBJECT	TOPIC	THINKING ISSUE
6–8	Science	Relationships among organisms	Classify these examples to clarify whether the relationship is parasitism, mutualism, or commensalism: fleas on a dog, ticks on a cat, bacteria in your intestine that help you digest your food, sheep farmer, clownfish and anemone, lichen, barnacles, hummingbirds and flowers.
6–8	Science	Simple machines	Classify common objects that change force by the type of simple machine to illustrate how each type of simple machine works.
6–8	Science	Ecosystem	Classify animals by the type of ecosystem they inhabit to clarify needs for survival.
6–8	Science	Nutrients in food	Classify foods by types of nutrients (proteins, carbohydrates, fats, minerals/vitamins, and water) to determine how much of each type of food is appropriate for good daily nutrition.
6–8	Science	Plants we eat	The American Cancer Society states that certain types of foods may help prevent cancer (foods containing vitamin A, vitamin C, or fiber and cruciferous vegetables). Classify the following fruits and vegetables to show which vitamins each offers in a cancer-prevention diet: apple, artichoke, onion, banana, strawberry, collard greens, papaya, lettuce, tomato, broccoli, cantalope, kale, kiwi, kohlrabi, grapefruit, Brussels sprouts. Which of these do you already eat regularly? Which might you add to your diet? Add additional examples.
6–8	Science	Electricity	Classify objects (wooden box, salt water, paper, copper wire, rubber band, styrofoam or cardboard take-out boxes, a wool sweater, air) according to whether they are conductors or insulators. Why? Add additional examples.
6–8	Science	Energy movement	Determine whether the way heat is transferred in each of the following situations fits into the categories of convection, conduction, or radiation: sand at the beach, hot air balloon, electric space heater, hot handle of a frying pan, melting snowball. Explain how the characteristics of each illustrates these types of heat transfer. Add other examples.
9–12	Biology	Linnaean classification of animals	Describe the sub-categories of the modern taxonomy for classifying animals developed by Linnaeus. How is it more useful than Aristotle's grouping of air, land, and water animals?
9–12	Biology	Fruit	Are the following fruit best classified as simple fruit, aggregate fruit, or multiple fruit: strawberries, walnuts, green peas, tomatoes, raspberries, pineapples, olives? Explain why. Add other items to these categories.

CLASSIFICATION LESSON CONTEXTS

GRADE	SUBJECT	TOPIC	THINKING ISSUE
9–12	Biology	Botany	Classify the following leaves by arrangement (simple, compound palmate, compound pinnate, compound doubly pinnate). The leaves are oak, maple, birch, clover, walnut, rosebush, elm, magnolia. Give additional examples of leaves and classify these as well.
9–12	Biology	Mendelian genetics	Determine, then classify the genotypes of F1 gametes resulting from a dihybrid cross of pure bread round yellow (RY) and wrinkled green (ry) pea plants.
9–12	Biology	Algae	Blue-green algae form chains of connected cells and carry on photosynthesis, but chorophyll is not contained in their choroplasts. To which kingdom of organisms do they belong? What other organisms belong to this kingdom? Explain these differences.
9–12	Chemistry	Periodic chart	Explain the basis by which elements are classified on a periodic table. How is the system of classification more descriptive than earlier classification systems?
9–12	Chemistry	Chemical bonds	Use a table of electronegativity difference to classify the following pairs of atoms by the type of bond that they are likely to form: H and O, Ca and O, Al and Cl, N and H, F and F, O and O. Why might it be important to determine the type of bond formed by a pair of atoms?
9–12	Chemistry	Organic molecules	Classify the following examples of organic molecules by type: sugars, starches, triglycerides, steroids, waxes, oils, enzymes, ribonucleic acids, and deoxyribonucleic acids. Draw the molecules with all covalent bonds indicated. What are the defining characteristics of each type of organic molecule? Why is it useful to classify such molecules?
9–12	Chemistry	Periodic chart	Explain the basis by which elements are classified on a periodic table. How is the system of classification more descriptive than earlier classification systems?
9–12	Physics	Electro-magnetic waves	Classify the following electromagnetic waves: gamma rays, X-rays, ultraviolet rays, sunlight, infrared rays, radio waves, electric waves. State the defining characteristics of each category.
9–12	Physics	Vector/scalar quantities	Classify the following quantities as either vector or scalar: velocity, speed, force, acceleration, pressure, weight, mass, volume, time, displacement. Give additional examples. Explain the relationship between motion and direction.

CHAPTER 7
DETERMINING PARTS-WHOLE RELATIONSHIPS

Why is it Important to Determine Parts-Whole Relationships Skillfully?

Everything around us is made up of parts. Man-made things, like automobiles and TV sets, depend for their functioning on the proper operation of their parts. Many natural objects, including the bodies of animals, the solar system, and great rivers, have parts that combine and operate together for the functioning of the whole.

Things that aren't physical objects also have parts. Stories, films, and human societies have component parts that give these items their distinctive character.

Whole objects or systems are not just collections of their parts. If the parts were combined together in different ways, something different would result. The special relationship between the parts and the whole that they comprise often makes the whole object or system what it is, allows it to function as it does, and permits it to retain its integrity.

The purposes of determining parts-whole relationships. Recognizing how parts contribute to the whole and how each part functions can help us better understand the world around us. In addition, analyzing parts-whole relationships can have some immediate practical applications. If we know what function each part serves, we are better able to sustain and maintain the whole. An automobile repair technician who knows the function of each part can usually fix a malfunctioning car because he knows what to repair or replace.

Indeed, our knowledge of how parts function can make us much more self-reliant. If the door sticks, we can sand or shave it. If the hinges are loose, we can tighten the screws. If we need a new lock, we know what to get. Being able to do these repairs efficiently ourselves depends on our knowing the parts of a door and how they function.

Knowing the function of parts can also contribute to our creativity. If we know what specific components can do, we may be able to combine them in new ways to serve certain purposes. New gadgets for use around the house, new organizational systems for managing work, original stories and works of art, and new economic systems connecting independent countries are all built on knowledge of how parts can contribute to the functioning of a whole.

Problems with the way we think about parts and wholes. People often have no difficulty identifying parts of things they see before them. Parts are usually smaller than the wholes they compose, and they usually look different from each other. The knob on a TV set, for example, looks quite different from the picture screen.

Often, when people think about specific parts of things that they see, they identify them only by their physical appearance. A person may see a series of controls on the front of a VCR, recognize them as parts of the VCR, but have no idea what each part does. The person may not even try to find out what the parts do as long as he or she can put a videotape in the VCR and play it. If we attend to appearance alone in characterizing a part, then our understanding of the part/whole relationship is *hasty and superficial*. Describing parts superficially blocks us from finding out how the part functions and recognizing the overall relationship between the parts and the whole they comprise.

A second problem about the way we think about parts is that we often do not consider subdividing parts into other parts in order to understand them better. This represents a *narrowness* in our conception of the parts of a whole that can limit our understanding of how the parts function. For example, I may identify the receiver on the phone as a part of the phone but need some prompting to separate it into a mouthpiece, an earpiece, and a connecting handle. Even if I did that, I would rarely go further and subdivide the mouthpiece and earpiece into their components.

A third problem is that, although we may identify many component parts of familiar things, we often think of them as *a scattered group of components* and do not consider how they may be connected to each other. I may, for example, notice that the vinyl padding on the back of a chair, its seat, the wooden chair back itself, and the arms on which my elbows rest are all parts of the chair, without thinking much about how they work together to make the chair useful and comfortable. I may not recognize, for example, that the padding, which supports my back, is held up at the right height by the chair back, which also supports the arms, which are positioned at the right height to rest my elbows on. Nor may I note that the arms are also supported by a strut that rests on the seat which, in turn, is supported by the legs. Although the separate parts of the chair are easily discernible, unless I think about how each contributes to the structure, function, and appearance of the chair, I may not understand their purpose or value.

Sometimes, we do think in richer ways about parts and the wholes that they comprise. I may identify the receiver on the telephone as a device that transmits my voice and receives the voice of others via the telephone lines and not just as a piece of plastic with a certain shape. I may identify the keyboard on my computer as something that transmits letters to the screen and contains numerous parts: keys, a space bar, a case, and the electronic circuitry that transforms pressure on a key into an electric signal, etc. I may also be aware that the keyboard, connecting wire, and inner circuitry in my computer transmit an image onto the screen. But many of us don't think enough about parts and wholes in this way. Awareness of the three problems with our thinking about parts-whole relationships can prompt us to try to think more skillfully.

Figure 7.1 contains a summary of these three common problems with the way we think about parts-whole relationships.

How Do We Think Skillfully about Parts-Whole Relationships?

Determining how parts function in relation to a whole. Determining the relationship between parts and a whole is a basic analytical thinking skill. In its most complete form, we strive to understand the basic parts of an object, an organism, a composition, or a system in

---

COMMON PROBLEMS WITH OUR THINKING ABOUT PARTS-WHOLE RELATIONSHIPS

1. We define parts based only on their appearances. (Our characterization of parts is **hasty**.)

2. We don't think of subdividing parts into other parts. (Our consideration of parts is **narrow**.)

3. We don't connect parts together in relation to the whole that they comprise. (Our thinking about parts is **scattered**.)

Figure 7.1

---

terms of how the parts function together in the structure or operation of the whole. The key to understanding parts-whole relationships is understanding what the parts *do* in relation to the whole, not just their immediate appearance.

A strategy for determining parts-whole relationships. A natural way to begin thinking about the parts of a whole is to identify as many parts as possible, describing them by their common names (e.g., computer keyboard) or, if these are not known, by using a description (e.g., the light on the front of the computer monitor). This may involve an investigation deeper than just looking at the object. We may have to take the cover off, for example, and look inside. We may also need a magnifying glass or even a microscope.

We then should consider these parts one-by-one to determine their function. Sometimes, this is easy. If a switch is marked "Off-On," then its function is to turn the computer on or off, presumably by opening or closing an electrical circuit. Sometimes, however, the function of a part is not clear. One important way to determine a part's function is to raise, and answer, the question, "What would happen if the object didn't have this part or if the part malfunctioned?" If there were no keyboard on my computer, I would not be able to type information into the computer memory. Therefore, I can conclude that the function of the keyboard is to transmit data to the computer. Finally, we should not think about just the function of each part. Rather, we should think about how the parts work together to make the whole.

The thinking map in figure 7.2 (next page) guides us in determining this relationship.

DETERMINING PARTS-WHOLE RELATIONSHIPS

1. **What smaller things make up the whole?**

2. **For each part, what would happen if it was missing?**

3. **What is the function of each part?**

4. **How do the parts work together to make the whole what it is or operate as it does?**

Figure 7.2

Sometimes we do not know what would happen if the object didn't have a specific part. For example, what would happen if the unmarked knob on the side of my portable radio were not there or did not function?

The two ways to answer this question are by seeking information from a reliable source or by investigating directly. In the first instance, I can determine the function of the radio knob by consulting the radio manual or asking a radio repair technician. Like other thinking skills, determining parts-whole relationships skillfully may require thinking about the reliability of sources of information.

We can also undertake some direct investigation. We could try the knob to see what happens and then remove it and see what happens. We might find that turning the knob changes the tone of the sound from the radio. When the knob is not there, it is more difficult to turn the metal rod that runs into the knob. Therefore, we realize that the function of the knob is to make it relatively easy to fine-tune the tone on the radio.

If we undertake this kind of direct investigation, we try to determine possible causal connections between the functioning of the part and the operation of the whole. Although determining the function of the knob on the radio is relatively straightforward, it might be necessary to engage in skillful causal explanation and/or prediction in order to be able to make a confident judgment about the function of a part. For example, does a specific bit of genetic material in the human body create the body's natural resistance to cancer? Although this is the same sort of question as finding the function of the

knob on the radio, in this case, a simple investigation would not suffice to answer this question. Research on the function of certain components of genes in the human body may involve collecting some very sophisticated evidence before hypotheses about gene functioning can be certified as well as supported.

Determining parts-whole relationships skillfully sometimes requires more sophisticated kinds of investigation, as in determining the function of specific genetic material. Fortunately, in everyday circumstances, like investigating the function of a radio knob, we can often engage in part/whole thinking skillfully without depending on such complex investigations.

Two graphic organizers to facilitate determining the relationship between parts and a whole. We may use graphic organizers to guide our thinking about parts and wholes and to record the results of our thinking so that we do not have to keep it in memory. The graphic organizer in figure 7.3 guides us to discriminate a number of parts and then to determine the function of *one* part to analyze further.

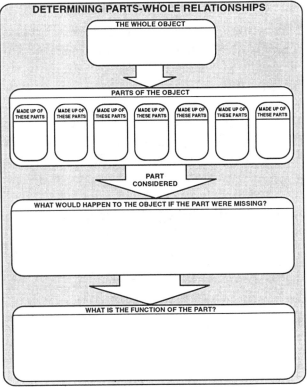

Figure 7.3

A similar graphic organizer in figure 7.4 can be used to record these results for all of the parts.

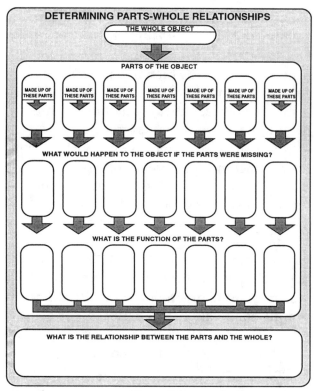

Figure 7.4

How Can We Teach Students to Determine Parts-Whole Relationships Skillfully?

Teaching skillful determination of parts-whole relationships explicitly. Teaching students the organized process of thinking depicted in figure 7.2 can help them improve the way they think about the parts of a whole. The approach we recommend is to teach skillful determination of parts-whole relationships *explicitly*. Explain the strategy directly and ask your students to practice it in a content lesson. As an alternative, you can guide your students to develop the strategy themselves as they examine the relationship between the parts and a whole that they are studying. In both cases, this should be followed by the use of a graphic organizer (figure 7.3) to reinforce the strategy.

Contexts in the Curriculum for Lessons on Determining Parts-Whole Relationships

The curriculum offers numerous opportunities to teach students to determine parts-whole relationships skillfully. Almost everything they study has component parts: plants and animals; systems like the solar system; social and political systems; poems, novels, and short stories; paintings and musical compositions; team sports; and machines.

Just as in other curriculum areas, parts-whole analysis is common and important in the sciences. In fact, the basic conceptual framework in biology and in chemistry quite explicitly treats much of what we see as compounds of smaller parts which have certain functions, each one of which is, itself, also a compound of smaller components, working down to the molecular and atomic levels. Hence, the understanding of the natural world typically presented in biology and chemistry curriculum materials is built on this idea. Contexts for parts-whole infusion lessons are replete throughout these curricula. Insofar as physics shares this basic conceptual framework, it too contains a multitude of similar contexts for such lessons. But physics also involves the study of complex physical systems of interacting forces (like the solar system). Systems, themselves, also are compounds, and their study in the physics curriculum serves as viable contexts for such infusion lessons. Finally, various machines, like a crane, can be subject to a parts-whole analysis that can enhance students' understanding of basic scientific principles (in this case, for example, having to do with pulleys, levers, weight, and force). In any parts-whole lesson of this sort, diagramming the way the parts connect together to yield the whole object often helps with this understanding.

When you choose instructional contexts in your curriculum for parts-whole lessons, make sure that there is a rich content objective to be achieved by the lesson. In general, there are two types of contexts in which determining the relationship between parts and a whole is important in the standard curriculum:

- *Understanding key concepts or processes depends on recognizing the function of component parts.* For example, understanding how certain machines operate by understanding the role of component parts like pulleys, gears, and levers, can provide a rich context for a parts-whole infusion lesson in physics, as can, of course, the study of the component systems that sustain human life, or the component chemical molecules that contribute to smog in the atmosphere.

- *Recognizing the function of component parts in a whole helps us understand changes in the whole.* A mass of cold air colliding with a mass of warm, moist air can bring about a severe storm like a tornado or a blizzard. Understanding how these two components of a weather system function to bring about such storms is a key to predicting weather.

A menu of suggested contexts for infusion lessons on parts-whole relationships can be found at the end of the chapter.

Reinforcing the process. Your goal in teaching these lessons is to help your students develop, remember, and internalize strategies for determining the relationship between parts and wholes so that they can guide their own thinking. This requires continued reinforcement. Make the graphic organizers for determining parts-whole relationships available in your classroom and encourage your students to use them on their own. Help them to practice this skill deliberately in curricular contexts other than the one in which you introduced it as well as in further applications within your curriculum.

You can also help students apply the parts/whole thinking strategy in non-curricular contexts in ways that have a science payoff. Simple items of more advanced technology found around the home, such as washing machines, microwave ovens, and computers, make excellent examples. The school building is a whole object whose smooth functioning depends on a variety of component parts: the electrical system, the plumbing, construction elements, etc.

Suggestions to students about how they might apply parts-whole analyses to everyday contexts that are non-scientific are not out of place in science and mathematics classrooms, or in any classroom. You might suggest that your students analyze how people in different roles contribute to the functioning of organizations, such as their supermarkets or schools. Families, communities, the scientific community, the community of mathematicians, nations, and even cooperative learning groups in the classroom are also systems made up of components that work together when these systems function well.

You will find that as the process of determining the relationship between parts and wholes becomes familiar, students will use the parts-whole thinking strategy without your guidance.

Model Lessons on Determining Parts-Whole Relationships

Two model lessons are included in this chapter. The first is a science lesson in which middle school students analyze the parts of a bird—the American kestrel—to determine their function with regard to the whole organism. In the Human Organism lesson, high-school biology students similarly analyze the systematic functioning of the human body, particularly the various systems (e.g., circulatory system) that sustain its integrity.

As you review these lessons, think about the following key questions:

- How does the thinking skill instruction interweave with the content in these lessons?
- Can you clearly distinguish the four components of infusion lessons in these examples?
- What differences are there in the way determining parts-whole relationships is treated?
- Can you identify additional transfer examples to add to these lessons?

Designing Parts-Whole Lessons

The thinking map on parts-whole relationships provides focus questions to guide students' finding of the relationship between parts and wholes in infusion lessons. It can be used as represented here or as modified by students.

The graphic organizers depicting parts-whole relationships supplement and reinforce the thinking map. The first two diagrams, one for five parts (designed especially for use in the primary grades) and the other for seven, are used to determine the function of a specific part. The third diagram is used to determine how multiple parts interconnect in the functioning of a whole.

The thinking maps and graphic organizers can guide you in designing the critical thinking activity in the lesson and can also serve as photocopy masters, transparency masters, or as models that can be enlarged and used as posters in the classroom. Reproduction rights are granted for use in single classrooms only.

DETERMINING PARTS-WHOLE RELATIONSHIPS

1. **What smaller things make up the whole?**

2. **For each part, what would happen if it was missing?**

3. **What is the function of each part?**

4. **How do the parts work together to make the whole what it is or operate as it does?**

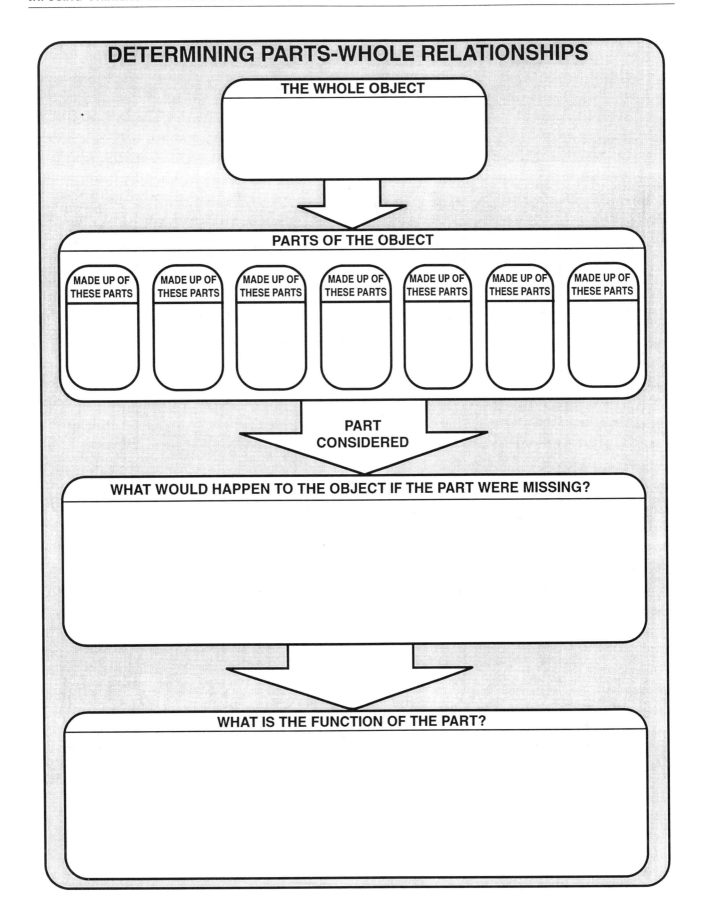

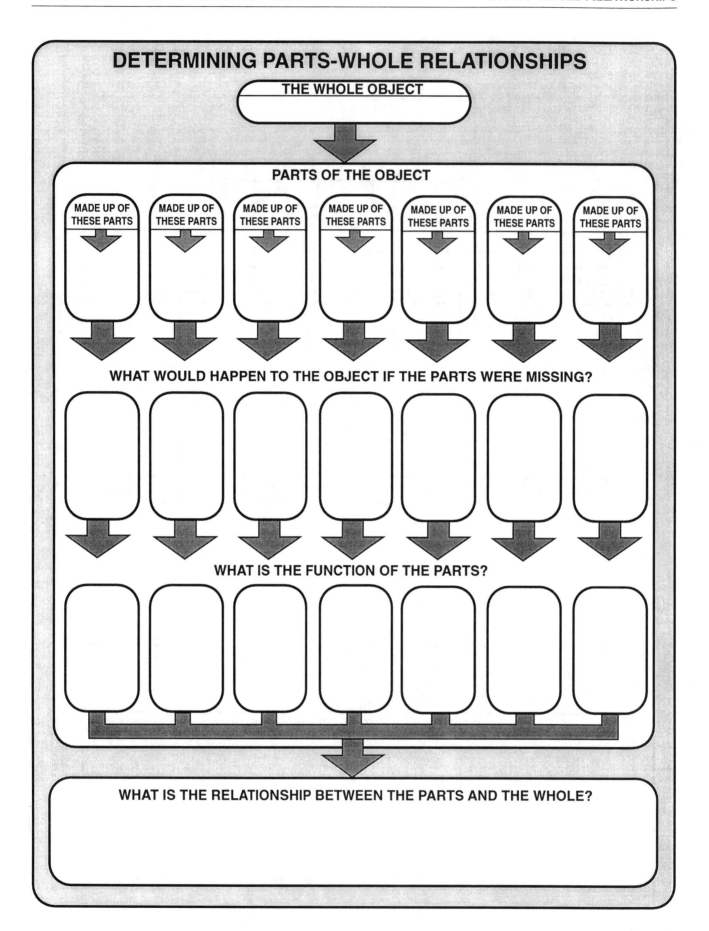

DETERMINING PARTS-WHOLE RELATIONSHIPS

THE WHOLE OBJECT

PARTS OF THE OBJECT

MADE UP OF THESE PARTS

MADE UP OF THESE PARTS

MADE UP OF THESE PARTS

MADE UP OF THESE PARTS

MADE UP OF THESE PARTS

MADE UP OF THESE PARTS

MADE UP OF THESE PARTS

WHAT WOULD HAPPEN TO THE OBJECT IF THE PARTS WERE MISSING?

WHAT IS THE FUNCTION OF THE PARTS?

WHAT IS THE RELATIONSHIP BETWEEN THE PARTS AND THE WHOLE?

THE KESTREL

General Science, Biology   **Grades 6–9**

OBJECTIVES

CONTENT

Students will learn the importance of the structure and use of different parts of a bird of prey, the kestrel, in sustaining its ability to survive.

THINKING SKILL/PROCESS

Students will identify the significant parts of a whole by identifying parts and determining their functions.

METHODS AND MATERIALS

CONTENT

Students will read a passage about the American kestrel and will examine pictures of it. If possible, a videotape of the kestrel and its behavior will be shown.

THINKING SKILL/PROCESS

A thinking map, a graphic organizer, and structured questioning prompt students to determine parts of a whole and how the parts function with regard to the whole. Collaborative learning enhances interchange of thinking.

LESSON

INTRODUCTION TO CONTENT AND THINKING SKILL/PROCESS

- We often notice some of the parts that make up a whole thing—for example, the engine and the body of a car. But sometimes, it is important to know more about the relationship between parts and the whole. For example, when something doesn't work properly, it is usually because one of its parts doesn't do what it was designed to do. If you know what the parts are supposed to do, it's easier to fix it. For example, if the door of the classroom doesn't close properly, what part might need fixing? POSSIBLE ANSWERS: *the door knob, the latch, the hinges, or the door frame.* The reason you can figure that out is because you know that these are parts of the door and that they make the door open, close, and stay shut.

- It is usually easy to pick out the major parts that make up familiar things. Determining the relationships between the parts and the whole may be more challenging. This often means figuring out how the component parts relate to one another to define a whole object or to enable it to do what it was designed to do. When we describe the relationship between a part and the whole, we usually do so in terms of the *function* of the part. For example, in addition to the engine and body, let's list some other parts of an automobile. POSSIBLE ANSWERS: *the transmission, the wheels, the frame, a radio, paint, a heater, a steering wheel, the instruments, the seats, the carpeting, and the windows.* **What are the functions of some of these parts?** POSSIBLE ANSWERS: *The engine functions to provide power for the car to move. The transmission delivers power to the wheels so that they can turn and propel the car. The body encloses the driver and passengers. The paint protects the metal and improves the appearance of the car. The radio and heater increase people's comfort.*

- The thinking map (at right) can guide you in identifying the parts of a whole and their relationships to the whole. Show a copy.

> ### DETERMINING PARTS-WHOLE RELATIONSHIPS
>
> 1. What smaller things make up the whole?
>
> 2. For each part, what would happen if it was missing?
>
> 3. What is the function of each part?
>
> 4. How do the parts work together to make the whole what it is or operate as it does?

THINKING ACTIVELY

- Let's try out these ideas in connection with something we've been studying in science: the predator-prey relationship. Predators depend on their prey for survival, but they also depend on their ability to capture their prey. Let's use our thinking strategy for determining parts-whole relationships to try to understand how one well-known predator, the American kestrel, has maintained a large population in both North and South America. Here is a passage about the kestrel from a guide to birds of America:

> *American Kestrel.* The American kestrel is one of the most common birds of prey in both North and South America. Its population has flourished. It is characterized by long, narrow, pointed wings and reddish brown and slate-gray markings. It is known for its habit of hovering in one place while hunting its prey, generally consisting of snakes, lizards, large insects, and small rodents. This evolved hunting skill involves the ability to detect any small movement on the ground, the ability to hover while scanning the ground for prey, and the speed and strength to plunge onto the prey from heights of 50 feet or more. All these skills depend on the kestrel's good sense of depth perception.

If possible, show a videotape of the kestrel in flight stalking and catching its prey.

- **Use the graphic organizer for determining parts-whole relationships to compile a list of parts of the American kestrel. Break some of these parts into their components and add these to your list. Work with a partner on this activity.** After a few minutes, solicit responses from the class. Ask each team to mention only one part so that as many teams as possible can report. POSSIBLE ANSWERS: *its beak, its head, the tip of its beak, its eye, a feather, a wing, its feet, a claw, its markings, its tail, its heart, its breast muscles, its neck, a pupil, the quill from a feather, a claw nail.*

- **Imagine what it would be like if the kestrel were injured. Pick a part and discuss with your partner how you think the bird would be affected if that part were injured. Write your ideas in the appropriate box on the**

Illustration by Kate Simon Huntley

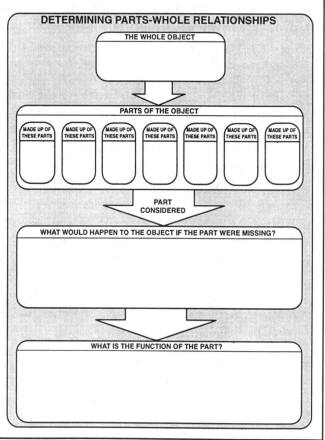

graphic organizer. After a few minutes, ask for responses from the teams. POSSIBLE ANSWERS: *If one of its eyes was injured, it would lose depth perception and would probably not be able to judge very well how far it must dive to reach its prey. This might affect its ability to get food, and it might starve. (Some students may know that, as a bird of prey, the kestrel has stereoscopic vision, which enables it to see objects as three-dimensional and to judge the distance of the prey.) If the kestrel was missing a talon, it might manage to catch its prey, since it has four talons on each foot, but it would be difficult. If the tip of the beak was injured, it would be difficult for the bird to kill and eat its prey. If the kestrel lost a feather, it could probably still fly unless the lost feather was a tail feather; in this case, the bird might not be able to dive accurately. A broken wing would be a disaster, since the kestrel depends on rapid wing flapping to hover over its prey.*

- **Now work with your partner to complete a statement about the function of the part in relationship to the whole bird.** After a few minutes, ask for a few responses. Write them on a large version of the graphic organizer "Determining Parts-Whole relationships." POSSIBLE ANSWERS: *The two eyes function together to give the kestrel a good sense of depth perception, allowing the bird to locate its prey accurately. The wings function to enable the kestrel to fly and to hover undetected above its prey. The claws function to enable the kestrel to grasp its prey.*

- **Using the information in the passage from the guide, describe how the different parts of the kestrel would function together in stalking and catching prey. As a pre-writing activity, fill in a new graphic organizer for determining parts-whole relationships with the responses of your fellow students in today's class work. Include it with your writing assignment.** POSSIBLE ANSWERS: *The eyes see the prey as the bird moves its head to scan the area and moves its wings rapidly (using its muscles) to hover. After the bird detects its prey and dives to capture it, changing the motion of its wings to do so, it uses its claws to grasp the prey and its talons to kill it. The bird moves its head so that it can use its beak to peck off morsels and its jaws and tongue to swallow them.*

THINKING ABOUT THINKING

- **How did you decide what parts to write in the "parts" boxes?** POSSIBLE ANSWERS: *I first looked at the large parts, like the head, and tried to find other things that were about the same size. I looked at the smallest parts I could see, like the nails on the bird's claws, and recorded them; then I moved to bigger parts, some of which were composed of these smaller parts (for example, the claws are partly made up of the nails). I started with a big part, like one of the bird's wings, then focused my attention on the parts of this part, like the feathers and markings; then I did the same for other large parts.* A variation on this activity can occur when students are working in pairs, listing the parts of the kestrel. Ask one student to think out loud about the parts when listing them; ask the other student to record how the first student is generating his or her list of parts. Then, the second student in each pair can report his or her observations.

- **Can you think of any other techniques for coming up with a list of the parts of the kestrel?** POSSIBLE ANSWERS: *I could find a real kestrel and look at it. I could dissect a kestrel. I could take a sample of its parts and further divide the sample; I could use a microscope.*

- **Work with your partner and describe how you thought about what would happen to the kestrel if one of its parts were injured.** POSSIBLE ANSWERS: *I imagined the kestrel trying, without the part, to do what the bird guide says it does. I then compared what I imagined to what I think happens when the kestrel has the part. That made me realize the function of the part.*

- **Sometimes, we can't tell what would happen without a specific part if the object is unfamiliar. If you don't know what would happen, how could you find out?** POSSIBLE ANSWERS: *I could* *observe the object operating without the part and compare it to the object operating with the part. I could observe the operation of a similar object that doesn't have the part. I could find someone who knows about kestrels and ask what would happen if the bird didn't have the part, or if the part were injured.*

- **What would you tell other people to think about to identify the part of a whole object and their relationships to the whole?** Students should suggest questions like the ones that appear on the thinking map for determining parts-whole relationships. For a variation on this activity, ask students to draw a flow chart that can be used by others to guide their thinking about parts and wholes.

APPLYING THINKING

Immediate Transfer

- **Stories are another example of wholes that have parts. For a story that you are reading, use your plan for determining parts-whole relationships to figure out how the different story parts function with regard to the whole.**

- **Select a machine that you are familiar with (like a washing machine) and analyze its parts and their functions.**

Reinforcement Later

Later on in the school year, introduce these additional transfer activities by saying the following:

- **In social studies you are studying complex societies. Pick a particular example of a society (for example, ancient Rome) to demonstrate your skill at determining parts-whole relationships. Explain the functions and roles of the different components that sustained the society over a period of time.** If students have a problem identifying components of a society such as Ancient Rome, tell them that the senate or the army are examples of such components.

- **Imagine that you are stranded on a desert island and want to make a device for signaling to passing airplanes. The only materials you have are a few long branches, a few pieces of clothing, some rocks, some matches, and a mirror. How could you put all or some of these materials together to construct a signaling device? Explain how each part would function.**

RESEARCH EXTENSION

Ask teams of students to use textbooks or school library resources to gather information from their textbooks or from resources in the school library about the parts of the kestrel (the feathers, the claws, etc.) and how they operate. Ask students to divide each of these components into parts and analyze how the subparts function in relation to the larger part being investigated.

ASSESSING STUDENT THINKING ABOUT PARTS-WHOLE RELATIONSHIPS

To assess student thinking about parts and wholes, give them a problem about an item that is familiar to them. Explain that it doesn't work because of a malfunctioning part and ask the students to figure out what part is causing the problem. As an alternative, you may have students construct a device from parts that you provide and ask them to explain how the parts function with regard to the purpose of the device. Ask the students to identify each step of the parts-whole strategy as they think about the problem. By attending to the questions they are asking, you may determine whether the students are attending to each of the steps in the thinking map for determining parts-whole relationships.

Sample Student Responses • The Kestrel • The Function of A Part

DETERMINING PARTS-WHOLE RELATIONSHIPS

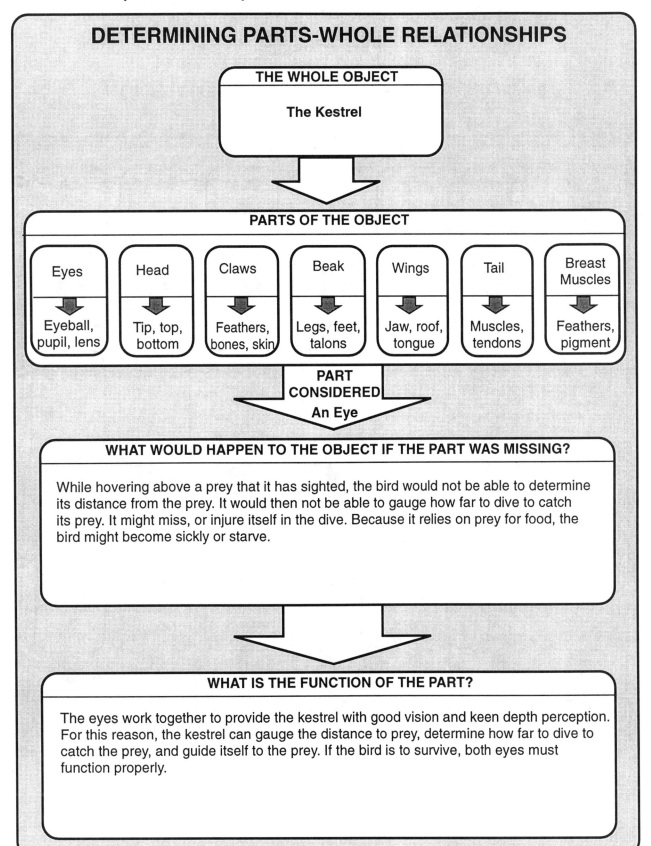

THE WHOLE OBJECT

The Kestrel

PARTS OF THE OBJECT

Eyes	Head	Claws	Beak	Wings	Tail	Breast Muscles
Eyeball, pupil, lens	Tip, top, bottom	Feathers, bones, skin	Legs, feet, talons	Jaw, roof, tongue	Muscles, tendons	Feathers, pigment

PART CONSIDERED

An Eye

WHAT WOULD HAPPEN TO THE OBJECT IF THE PART WAS MISSING?

While hovering above a prey that it has sighted, the bird would not be able to determine its distance from the prey. It would then not be able to gauge how far to dive to catch its prey. It might miss, or injure itself in the dive. Because it relies on prey for food, the bird might become sickly or starve.

WHAT IS THE FUNCTION OF THE PART?

The eyes work together to provide the kestrel with good vision and keen depth perception. For this reason, the kestrel can gauge the distance to prey, determine how far to dive to catch the prey, and guide itself to the prey. If the bird is to survive, both eyes must function properly.

Sample Student Responses • The Kestrel • Interrelationship Between Parts and Whole

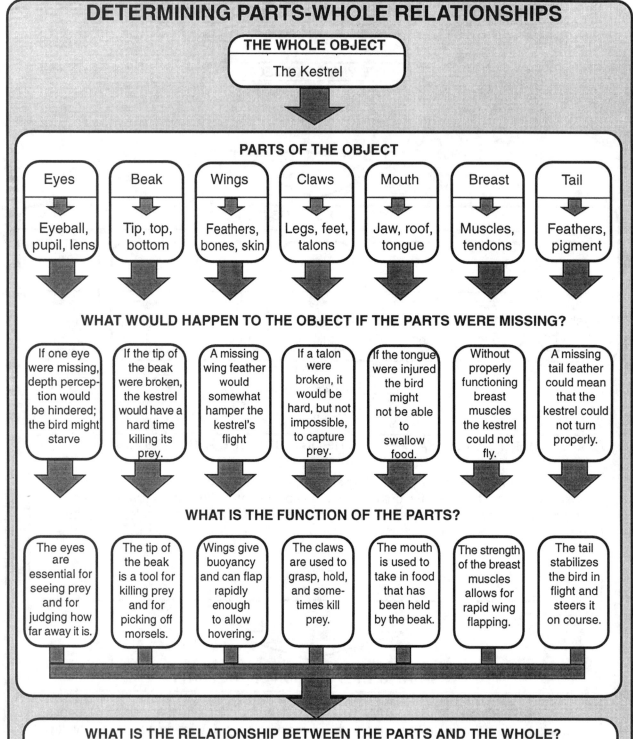

DETERMINING PARTS-WHOLE RELATIONSHIPS

THE WHOLE OBJECT

The Kestrel

PARTS OF THE OBJECT

Eyes	Beak	Wings	Claws	Mouth	Breast	Tail
Eyeball, pupil, lens	Tip, top, bottom	Feathers, bones, skin	Legs, feet, talons	Jaw, roof, tongue	Muscles, tendons	Feathers, pigment

WHAT WOULD HAPPEN TO THE OBJECT IF THE PARTS WERE MISSING?

| If one eye were missing, depth perception would be hindered; the bird might starve | If the tip of the beak were broken, the kestrel would have a hard time killing its prey. | A missing wing feather would somewhat hamper the kestrel's flight | If a talon were broken, it would be hard, but not impossible, to capture prey. | If the tongue were injured the bird might not be able to swallow food. | Without properly functioning breast muscles the kestrel could not fly. | A missing tail feather could mean that the kestrel could not turn properly. |

WHAT IS THE FUNCTION OF THE PARTS?

| The eyes are essential for seeing prey and for judging how far away it is. | The tip of the beak is a tool for killing prey and for picking off morsels. | Wings give buoyancy and can flap rapidly enough to allow hovering. | The claws are used to grasp, hold, and sometimes kill prey. | The mouth is used to take in food that has been held by the beak. | The strength of the breast muscles allows for rapid wing flapping. | The tail stabilizes the bird in flight and steers it on course. |

WHAT IS THE RELATIONSHIP BETWEEN THE PARTS AND THE WHOLE?

As the kestrel hovers, it turns its head from side to side and scans the ground with its eyes for the movement of recognizable prey. When it spots the prey, the kestrel uses its stereoscopic vision, which depends on the coordination of both eyes, to gauge the location and distance of the prey. The kestrel then changes the motion of its wings, dives, and uses its tail to guide it to the prey. When the kestrel reaches the prey, it grasps it in its claws, using its talons to immobilize and even kill the prey. The kestrel then uses its wings to fly to a spot where it puts the prey down and pecks off morsels, which it puts into its mouth with its beak and tongue. If the prey is still alive, the kestrel kills it with the sharp tip of its beak.

THE HUMAN ORGANISM

Biology, Anatomy, Physiology   **Grades 9–12**

OBJECTIVES

CONTENT	THINKING SKILL/PROCESS
Students will learn the structure and function of the twelve organ systems that make up human organisms.	Students will identify the significant parts of a whole by identifying specific parts and determining their functions. Students will then explore the relationship between all the parts and the operation of the whole.

METHODS AND MATERIALS

CONTENT	THINKING SKILL/PROCESS
Students will learn about the structure and function of organ systems by reading their biology text. Selected videotapes, for example relevant episodes of *Body Atlas*, will be made available for viewing.	A thinking map, graphic organizers, and structured questioning organize student thinking and prompt students to determine parts of a whole and how the parts function with regard to the whole.

LESSON

INTRODUCTION TO CONTENT AND THINKING SKILL/PROCESS

- We often recognize parts that make up a whole. For example, most of us can readily name the major parts of a bicycle. We also may know how these parts work. Some have obvious functions: the seat is for sitting, the handle bars are for steering. Other parts have a more complex operation. How do the caliper brakes on a bicycle work? POSSIBLE ANSWER: *They work by pinching both sides of the tire rim, producing friction, and slowing the wheel.*

- Just as a bicycle is a system of parts working together, each part is also a system of parts. To understand how the brakes work, we have to be able to identify and determine the function of the parts of the brake system. Bring a bicycle into the classroom, or draw a diagram of a bicycle on the chalkboard to illustrate these parts. Ask the students to identify the parts of the brake system. ANSWERS INCLUDE: *The brake pads, calipers, cable ties, tension nuts, springs, bolt and washers that hold the two sides of the calipers together.* Demonstrate how the parts of the brake system work together to stop the wheel or elaborate your diagram.

- Most of us already know enough about a bicycle and its parts to diagnose many of its problems and make simple repairs. For example, what is probably wrong with a bicycle that produces a rubbing sound every time it moves, and how would you correct the problem? POSSIBLE ANSWERS: *The brake calipers are not centered and the brake pad on one side is rubbing; fix by centering the brake calipers. The wheel axle is not centered within the fork and the off-center wheel is rubbing against one prong; fix by loosening the axle nuts and recentering the wheel between the fork prongs. The wheel rim is bent and rubs as it turns; fix by either replacing the wheel or adjusting the tension on its spokes.* When we figure out what part of a whole mechanism is malfunctioning, we are "troubleshooting." We can so readily troubleshoot the rubbing sound and identify the parts responsible because we understand their function. Knowing the function of the parts comes from knowing what would happen if the part was bad or missing.

- In order to *fully* understand the whole, *every* part must be named, its function defined, and its relationship to the whole expressed. Let's see how close you come to fully understanding how a bicycle works. Work in groups and list as many bicycle parts and their functions as you can. POSSIBLE ANSWERS INCLUDE: *brakes, to slow and stop the bicycle; handlebars, to turn the wheel in order to steer the bicycle; hand grips, provide nonslip surface to hold on to so that you can steer; gear shifters, pull the gear shift lines so the chain will move in and out of engagement with the gears; handbrakes, pull the brake lines in order to close the calipers; brake lines, close the calipers; gear lines, shift the chain guide; chain guide, changes gear ratios; axle, holds wheels; wheel rim, holds tires; tire, grips the ground so that when the wheel moves the bicycle moves along the ground; spokes, absorbs bumps, smooths ride, holds the wheel rim straight; frame, holds moving parts; seat, supports rider; reflectors, provide safety; calipers, force brake pads against wheel rim; brake pads, produce friction.* **Congratulations, it appears that you know a lot about a bicycle. You understand how its parts function. In order to have understood the relationship between the parts and the whole bicycle, you probably thought in a manner described by a thinking map similar to this one (at right). For example, after you identified the brakes as a major part of a bicycle, I asked you to think about their function. If you thought about question (2) and asked "What would happen to the bike if it had no brakes?" in order to respond to question (3), what came to your mind?**

  > **DETERMINING PARTS-WHOLE RELATIONSHIPS**
  >
  > 1. What smaller things make up the whole?
  >
  > 2. For each part, what would happen if it was missing?
  >
  > 3. What is the function of each part?
  >
  > 4. How do the parts work together to make the whole what it is or operate as it does?

  Solicit responses from the whole class. ANSWERS INCLUDE: *I wouldn't be able to slow or stop the bike when I wanted to. I might crash and hurt myself. The function of the brakes is to slow or stop the bike when the rider wants to in order to avoid crashing and hurting himself or herself.* **Thinking in this manner can also guide you when you are investigating a whole object whose parts you are less familiar with than a bicycle.**

- Organisms are also mechanisms that can be well understood by identifying their parts and understanding what would happen if the part malfunctioned or was missing. In biology, we have already explored the detailed composition of many plants and animals by naming the parts that make up the whole organism, then learning the function of the parts and the relationship between these parts, much like we did in the discussion of the bicycle. You were practicing skillful thinking. What are some other things you have studied that you now recognize you have studied in this way? ANSWERS VARY BUT MAY INCLUDE: *the components of the atom; the interactions of various types of atoms in forming the macromolecules of life; the cooperation between organelles necessary for cell life; the relationships between different tissues making up organs.*

- In order to cope with this increasing complexity, a somewhat more formal organization of our thinking guided by the thinking map of parts-whole relationships will be helpful. Seeing the "big picture" and appropriately determining parts-whole relationships of complex material is an ideal opportunity to explicitly practice skillful thinking in biology.

THINKING ACTIVELY

- Today, we begin a unit on the most complex of all living organisms, a human being. In fact, there are so many parts making up the human organism that one entire division of biology is devoted to naming them (human anatomy), another to describing how they work (human physiology), another to describing how they malfunction (pathophysiology), and another to troubleshooting and repairing the problems (medicine). The most direct way to study so complex a whole is to break it down into manageable collections of parts. In human biology, the major collections of parts are called organ systems. How many organ systems can you name? In your groups, make a list. ANSWERS INCLUDE: *the digestive system, the respiratory system, the nervous system, the reproductive system, the circulatory system, the skeletal system, the muscular system.* LESS OFTEN MENTIONED: *the endocrine system, the integumentary system, the lymphatic system, and the urinary system.*

- Let's first apply the steps in the parts-whole thinking map to the organ systems in learning about the human organism. Work in your groups and use this graphic organizer for determining parts-whole relationships to enter the list of organ systems you just developed in the boxes for parts of the object. Write "The Human Organism" in the box marked "The Whole Object." Use your text as a reference to add any others that may have been omitted so that you have a complete list of human organ systems. After a few minutes, solicit responses about the organ systems from the class and demonstrate on a transparency how to fill in the *parts of the object* section of the graphic with organ systems. Have each group name one organ system and return to that group after others have had a chance to contribute. Produce a class response composite on a transparency, including all twelve organ systems.

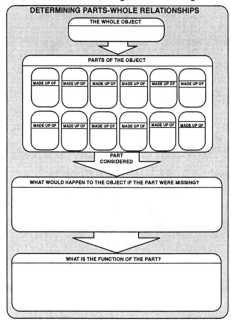

- Just as we did with the brake system in the bicycle, break each of these systems into their major component parts. For example, the digestive system is made up of which components? Ask for responses from the whole class but accept only one per student responding. ANSWERS INCLUDE: *the mouth and teeth, the esophagus, the stomach, the intestines, the rectum, the anus.* Work together in your groups again and list the major components of each of the systems of the body you have identified under "made up of" in the parts boxes in your diagram. Use your textbooks if you need them. After a few minutes, ask the groups to report by mentioning a set of parts of one of the organ systems. Write these on the graphic organizer. ANSWERS INCLUDE: *skeletal system: bones, joints, cartilage; muscular system: muscles, ligaments, tendons; nervous system: brain, receptors, spinal cord, nerves; cardiovascular system: heart, blood vessels, arteries, blood; urinary system: kidneys, bladder, urethra, ureter; reproductive system: male and female reproductive organs; respiratory system: pharynx, larynx, trachea, bronchial tubes, lungs; lymphatic system: thoracic duct, lymph nodes, lymph vessels; endocrine system: various glands, pancreas, testes, ovaries, hormones; integumentary system: skin, hair, nails.*

- Now that we have assembled a list of human organ systems and their parts, let's determine the function of each system. By learning what each system contributes to the biology of the entire organism, we should be able to uncover relationships between the systems. In other words, we should reach an intelligent understanding of how the whole human organism works by understanding the relationships between its parts. In order to determine the function of an organ system, let's ask the question that appears on the thinking map: What would happen to the whole organism if the particular organ system did not work? Remember how we did this when we asked what would happen if the brakes failed while you were riding a bike? Let's apply this tactic to the study of the human organism. In your groups, consider one organ system. Assign each of the organ systems to a different group. Ask them to write the name of their organ systems on the "Part Considered" arrow in the diagram. Discuss what might happen to the human organism if this organ system were to malfunction because one or more of its parts did not function, or if it didn't work at all. Use your text as a source of

information if you need to. For your assigned organ system, fill in the "What would happen to the object if the part were missing" box. If possible, have students view appropriate episodes of the video series *Body Atlas,* or other relevant source material. After a suitable inquiry period, have each group report back by identifying their organ system and then describing what would happen if it malfunctioned. POSSIBLE ANSWERS: *If the skin is missing, the body will become dehydrated. If the joints get arthritic, movement will be difficult. If muscles atrophy, strength will disappear. If the spinal cord is severed, paralysis will result. If the heart stops, it will no longer be able to pump blood to the brain and the body will die. If the lymph system fails, edema will result. If the lungs get damaged or fail to operate because of disease, suffocation might result. If the stomach isn't there, digestion of certain foods will not be possible and malnutrition or death might result. If the intestines stop working, food will not be digested and starvation will result. If the kidneys malfunction, blood-borne poisons will reach toxic levels. If reproductive organs fail, then no reproduction will occur. If the immune system fails, then our body will be defenseless against pathogens and we will get serious infections.*

- **Based on these ideas, take a few minutes and reflect on the function of the organ system in relationship to the organism and then write a statement summarizing your thoughts in the bottom box of the graphic organizer.** Have two or three groups report back, and ask each group to turn in copies of their graphic organizer. POSSIBLE ANSWERS: *The function of the skin is to protect other organs. The heart pumps and delivers oxygen and nutrients to tissues. The urinary system stores and then removes nitrogenous waste and excess metabolites. The lymphatic system removes excess liquids from tissues, and lymph nodes act to strain foreign materials from lymph and blood. The immune system provides defense against foreign organisms that might damage parts of the body, such as bacteria.*

- **Now that each group has determined the function of a major part of the human organism, let's put this information together in a way that will reveal the relationships between the parts and the whole, between the organ systems and the functioning of the whole human organism.** Record the results from each group on the summary chart of all the parts considered and give a copy to each group. **We will now put all this information about the function of organ systems together so that we can understand the relationship between organ systems and the human organism. Each group should compose a statement indicating the interrelationship between these twelve organ systems as they contribute to the functioning of the human organism.** Remind students to be sure to consider not only the relationship between organ systems and the organism, but also between various organ systems themselves. As examples, describe some of the interrelationships between the cardiovascular and respi-

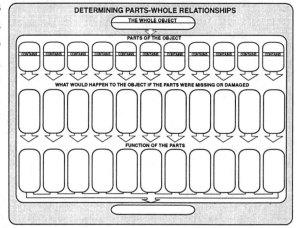

ratory systems; the immune and lymphatic systems; the nervous and muscular systems. Create a large version of this new graphic organizer with the parts, what would happen if the part were missing, and the function boxes filled in as they are on the copies given to the students. Post this on the wall or on the chalkboard. When the groups have composed their relationship statements, ask a few of the groups to report. Blend statements from all groups into one integrated statement from the whole class. This may take a day or two. POSSIBLE ANSWERS: *The human machine is built to procure and consume food and then process it into an absorbable source of energy that is used to power self-defense, propagation of the species, and metabolic housekeeping. Higher thinking centers of the brain work with the sensory apparatus for the nervous system to locate food and then coordinate muscle contractions, which produce movement of bones; as a result, food is acquired and ingested. Meanwhile, steady state body temperature, blood pH, hormone and enzyme concentrations, heart and respiratory rates, are adjusted in response to eating by the endocrine system. Nutrients absorbed are delivered to cells by vessels containing blood pumped by the heart. Oxygen, required for the metabolism of nutrients into life-sustaining energy, is acquired by the respiratory system and delivered to each cell by the circulatory system. Lymph collects leaked fluids and strains out foreign particles like bacteria while the kidneys remove excess electrolytes and nitrogenous wastes, which are then collected in the bladder as urine. Undigested food is removed during excretion. The human organism is protected from the external environment by skin and hair and defends against infection with the immune system. The organ systems of the human organism cooperate in order to maintain continuity of the species.*

- **Just as the organ systems are systems of parts working together, the whole human organism is a complex system as well. Based on your parts-whole analysis, work in your groups and come up with a definition of what makes a system like these** *systems.* **What is it for something to be a system?** ANSWERS VARY, but usually include the idea of parts working together in regular ways to bring about an overall result. Students sometimes say that there must be a purpose that they all have. This is an opportunity to discuss with students the difference between results, functions, and purposes, and to raise the question of purposes in the natural world.

THINKING ABOUT THINKING

- **What kind of thinking did we do in this lesson?** *Determining parts-whole relationships.*

- **How was the way you thought about the parts of a whole human being different from other ways that you've thought about parts of wholes before?** ANSWERS VARY BUT TYPICALLY INCLUDE: *In other school activities, like when we studied the atom, I was just asked to list the parts, like the nucleus, electrons, etc. This time, I had to think about what the parts I identified did—how they helped the whole organism of which they were a part.*

- **Is this a better way to think about parts and wholes?** Students typically say that it is because they understand what they are studying better. They say that they especially like figuring out how all the parts work together.

- **How did you decide what parts to write in the "parts" boxes?** ANSWERS VARY: In some cases, students identified these systems by asking themselves what the human organism does and then how it does it. Having a diagram of the human body helped, some say.

- **How did you figure out what would happen if the part you were considering was missing or defective?** Students often mention that they tried to remember cases in which someone lacked one of his or her body parts and what happened to the person. Or they say that they imagined they didn't have one of their body parts and then imagined what it would be like.

- **Is thinking about what would happen if the part were missing a good way to figure out what the part does?** Many students say that once they've said what will happen if the part were missing, it's easy to determine the function of the part. Students often say, however, that we should make sure we get reliable information because we could make a mistake if the information wasn't correct. Students will also say that when they have to look up what will happen if the part were missing, damaged, or diseased, it would be just as *easy* to look up the function of the part in a reference book. However, they say that by finding out what will happen if the part were missing, they understand the function of the part much better than if they just read about how it functioned.

APPLYING THINKING

Immediate Transfer

- Perhaps the most important parts of a living thing are its cells. Living cells are the basic building blocks of all living things. As you know from your studies, living cells have parts that function together to make a cell be able to do what it does in a living organism. Use the strategy for parts-whole analysis to determine the parts of a cell and how they function.

- Determine the parts of the whole that make up our solar system. What are each of the parts, and how do they function with regard to the whole?

- The school system in which you are learning is also a whole with parts that have specific functions. Use your skill at parts-whole relationships to determine the relationship between the parts of the school system and the whole system. Include specific consideration of individual people in the roles they have in the school system (e.g., your school principal).

Reinforcement Later

- Just as many natural things make up complex systems in which parts interact (as in the human body), so do many human-made objects. Identify some complex human-made objects or systems. ANSWERS VARY BUT SOMETIMES INCLUDE: *a jet airliner, an automobile, a computer, a telephone, the telephone system, a factory, the government.* Pick one of these and do a parts-whole analysis of it. When you're done, make a diagram to show parts interaction.

- Apply what you've learned about parts-whole analysis to something you've been studying in one of your other subjects in school—for example the parts of a story, of a piece of music, of a musical instrument, of a family, of a complex society, or of a language.

ART EXTENSION

Draw a systems diagram of one of the bodily systems showing how the parts interact to make up the whole system. Do the same for the human organism itself, showing how the different major systems in the body interact. Use arrows to indicate interaction, and label both the parts you show and the results of the interaction.

CONTENT EXTENSION

Consider one of the major systems of the human organism that we have studied in this lesson and identify the conditions under which it operates to its maximum efficiency. Then identify the ways it might break down. What are some of the common inhibitors of the working of the system?

INVENTIVE THINKING EXTENSION

How might you change the system you studied in the previous lesson extension so that it operated more efficiently than it does and so it didn't break down as frequently or easily? You may build into your redesign any resources available to you. Try to be as inventive as possible. Explain why your design would be better than the one we live with now.

ASSESSING STUDENT THINKING IN IDENTIFYING PARTS-WHOLE RELATIONSHIPS

To assess student skills at identifying parts-whole relationships, use any of the transfer examples and ask students to write in detail about their analyses. You can also ask students to select their own examples. Determine whether they are asking the questions identified in the thinking map of part-whole analysis. Also encourage them to use the language of the thinking skill.

DETERMINING PARTS-WHOLE RELATIONSHIPS

THE WHOLE OBJECT
The Human Organism

PARTS OF THE OBJECT

DIGESTIVE SYSTEM	SKELETAL SYSTEM	MUSCULAR SYSTEM	NERVOUS SYSTEM	ENDOCRINE SYSTEM	CARDIO-VASCULAR SYSTEM
CONTAINS	CONTAINS	CONTAINS	CONTAINS	CONTAINS	CONTAINS
Mouth Esophagus Stomach Intestines	Bones Cartilage Joints	Muscles Tendons Ligaments	Brain Spinal Cord Nerves Receptors	Glands Hormones Pancreas Thyroid	Heart Arteries Veins Blood

LYMPHATIC SYSTEM	RESPIRATORY SYSTEM	INTEGUMENTARY SYSTEM	URINARY SYSTEM	REPRODUCTIVE SYSTEM	IMMUNE SYSTEM
CONTAINS	CONTAINS	CONTAINS	CONTAINS	CONTAINS	CONTAINS
Thoracic Duct Lymph Nodes Vessels	Nose Pharynx Bronchial Tubes Lungs	Skin Hair Nails	Kidney Ureter Bladder Urethra	Male and Female Reproductive Organs	Glands White Blood Cells

PART CONSIDERED
The Digestive System

WHAT WOULD HAPPEN IF THE PART WERE MISSING?

If any of the organs that make up the digestive system—the esophagus, stomach, intestines, rectum and anus — were to become impaired or fail, then the absorption of nutrients and reabsorption of water into the blood would be either restricted or prevented. If the supply of energy-containing nutrients such as glucose are not absorbed into the blood, then they cannot be transported to tissues where they are absolutely essential for cellular metabolism. Without fuel and other essential molecules, cells die, tissues die, organs die, and ultimately the organism dies. Therefore, a malfunctioning digestive system can lead to malnutrition and death. Dehydration and death can result from a failure to reabsorb water.

WHAT IS THE FUNCTION OF THE PART?

The digestive system functions to provide essential nutrients and water to the cellular factory by processing food. It is composed of coordinated set of organs that process food by breaking it down mechanically and chemically into smaller and smaller particles; ultimately, usable nutrients are reduced to molecule-sized fragments, which are then absorbed into the blood system for delivery to every cell in the body. Undigested food is collected and evacuated from the anus.

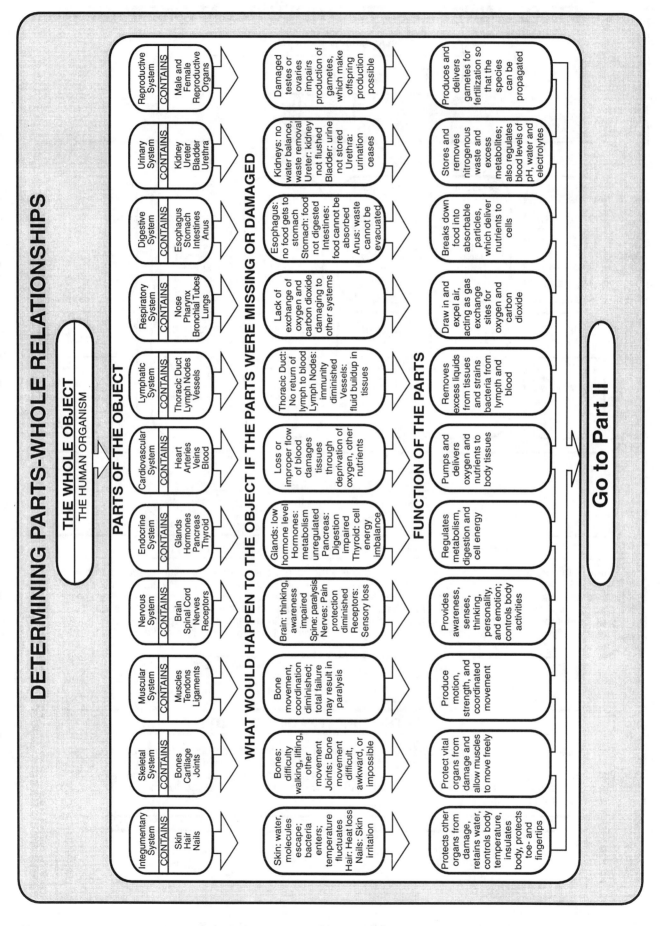

Sample Student Responses • The Human Organism • All Systems • Part I

DETERMINING PARTS-WHOLE RELATIONSHIPS

THE WHOLE OBJECT
THE HUMAN ORGANISM

PARTS OF THE OBJECT

System	CONTAINS
Integumentary System	Skin, Hair, Nails
Skeletal System	Bones, Cartilage, Joints
Muscular System	Muscles, Tendons, Ligaments
Nervous System	Brain, Spinal Cord, Nerves, Receptors
Endocrine System	Glands, Hormones, Pancreas, Thyroid
Cardiovascular System	Heart, Arteries, Veins, Blood
Lymphatic System	Thoracic Duct, Lymph Nodes, Vessels
Respiratory System	Nose, Pharynx, Bronchial Tubes, Lungs
Digestive System	Esophagus, Stomach, Intestines, Anus
Urinary System	Kidney, Ureter, Bladder, Urethra
Reproductive System	Male and Female Reproductive Organs

WHAT WOULD HAPPEN TO THE OBJECT IF THE PARTS WERE MISSING OR DAMAGED

- Skin: water, molecules escape; bacteria enters; temperature fluctuates Hair: Heat loss Nails: Skin irritation
- Bones: difficulty walking, lifting, other movement Joints: Bone movement difficult, awkward, or impossible
- Bone movement, coordination diminished; total failure may result in paralysis
- Brain: thinking, awareness impaired Spine: paralysis Nerves: Pain protection diminished Receptors: Sensory loss
- Glands: low hormone level Hormones: metabolism unregulated Pancreas: Digestion impaired Thyroid: cell energy imbalance
- Loss or improper flow of blood damages tissues through deprivation of oxygen, other nutrients
- Thoracic Duct: No return of lymph to blood Lymph Nodes: immunity diminished Vessels: fluid buildup in tissues
- Lack of exchange of oxygen and carbon dioxide damaging to other systems
- Esophagus: no food gets to stomach Stomach: food not digested Intestines: food cannot be absorbed Anus: waste cannot be evacuated
- Kidneys: no water balance, waste removal Ureter: kidney not flushed Bladder: urine not stored Urethra: urination ceases
- Damaged testes or ovaries impairs production of gametes, which make offspring production possible

FUNCTION OF THE PARTS

- Protects other organs from damage, retains water, controls body temperature, insulates body, protects toe- and fingertips
- Protect vital organs from damage and allow muscles to move freely
- Produce motion, strength, and coordinated movement
- Provides awareness, senses, thinking, personality, and emotion; controls body activities
- Regulates metabolism, digestion and cell energy
- Pumps and delivers oxygen and nutrients to body tissues
- Removes excess liquids from tissues and strains bacteria from lymph and blood
- Draw in and expel air, acting as gas exchange sites for oxygen and carbon dioxide
- Breaks down food into absorbable particles, which deliver nutrients to cells
- Stores and removes nitrogenous waste and excess metabolites; also regulates blood levels of pH, water and electrolytes
- Produces and delivers gametes for fertilization so that the species can be propagated

Go to Part II

Sample Student Responses • The Human Organism • All Systems • Part II

From Part I

WHAT IS THE RELATIONSHIP BETWEEN THE PARTS AND THE WHOLE?

Humans need all of their organ systems working together in order to survive. If one does not function normally, then the human organism could die. The organ systems work together to get and consume food, which keeps the organism alive. The energy that people get from digesting food is used by the body for keeping a steady body temperature, pH, hormone concentration and heart and breathing rates. The endocrine system speeds up and slows down these rates during diingested. Meanwhile, steady state body temperature, blood pH, hormone, enzyme concentrations, and heart and

respiratory rates are adjusted by the endocrine system in gestion. Energy from food is also used for movement, thinking, reproduction, self defense, immunity, and getting more food.

The heart pumps blood through vessels which deliver nutrients that are absorbed during digestion to each cell. Along with nutrients, blood carries oxygen to each cell. Oxygen is required to metabolize food into energy needed for the organism to stay alive. Lymph nodes mop up fluids that leak from cells and strain out foreign material like bacteria. protected from the external environment

by skin and hair and defend against infection with the immune system. The organ systems of human organisms cooperate in order to maintain continuity of the Kidneys remove excess water, salts and nitrogen containing wastes and transport them to the bladder as a fluid called urine. Undigested food is removed during excretion. Humans are protected from the environment by skin and hair and defend against infection with the immune system. Because humans can reproduce successfully, they have not become extinct. By producing male and female gametes, people sexually reproduce.

PARTS OF A WHOLE LESSON CONTEXTS

The following examples have been suggested by classroom teachers as contexts to develop infused lessons. If a skill or process has been introduced in a previous infused lesson, these contexts may be used to reinforce it.

GRADE	SUBJECT	TOPIC	THINKING ISSUE
6–8	Science	River	Describe a local river system. How does each part of the river contribute to the natural flow of the river and the welfare of the cities around it?
6–8	Biology	Eye	Describe each part of the eye and explain how each contributes to maintaining its healthy functioning?
6–8	Science	Coastlines	Describe the land and water parts of a coastline. How does each part contribute to sea life and the needs of people living in the area?
6–8	Science	Weather	Examine a weather map for your area. How do the parts of the map give you information about local weather?
6–8	Biology	Plants and animal cells	Describe how the parts of a cell function to support the life of the organism and allow us to discriminate between plant and animal cells.
6–8	Biology	Types of animal cells	Describe how the parts of nerve, blood, muscle, bone, and skin cells explain their function in that part of the body.
6–8	Science	Pollination	Explain how each part of the flower of an apple tree functions in reproduction.
6–8	Biology	Digestive system	Describe how the parts of the digestive system are interrelated in helping the body process and use food.
9–12	Biology	Organic molecules: nucleic acids	Explain how each part of an amino acid contributes to the functioning of nucleic acids.
9–12	Biology	Animal and plant cells	Explain how each organelle of an animal cell contributes to the functioning of the cell. Extend this to the plant cell.
9–12	Biology	Organs: The brain	Describe each major part of the brain and explain how each contributes to the coordination of sensory and motor nervous pathways. Discuss what would happen if one part was not functioning properly in terms of familiar pathophysiologies.
9–12	Biology	Heart functioning	Describe how each part of the heart functions. What does each part contribute to the function of the whole heart?

PARTS OF A WHOLE LESSON CONTEXTS

GRADE	SUBJECT	TOPIC	THINKING ISSUE
9–12	Biology	Blood composition	Describe the components of blood. Choose any part and describe the function of that part. What relation does this function have to the function of the blood?
9–12	Biology	Parts of a leaf	Describe how each part of a leaf contributes to photosynthesis.
9–12	Chemistry	Chemical formulas	Identify the parts of a given chemical formula and explain their functions.
9–12	Chemistry	Equipment	Examine a retort. What are its parts? How do these parts contribute to the proper functioning of the retort?
9–12	Chemistry	The periodic table: groups	Describe the physical and chemical characteristics of elements in each group of the periodic table (1-8A, 1-8B). Explain how each contributes to the organization of the periodic table. (Consider what would happen to its usefulness in predicting chemical interactions if the groups were ordered differently.)
9–12	Physics	Electricity	Examine and identify the parts of a simple motor and explain how each contributes to its normal functioning. Do the same for generators.
9–12	Physics	Simple machines	Select a simple machine and identify its parts. How does each function, and what is the relationship between the way the parts function and the function of the whole machine?
9–12	Physics	The atom	What are the parts of an atom? What is the function of these parts? How does this function contribute to the integrity of the atom?

CHAPTER 8
SEQUENCING

Why Is It Important to Sequence Things Skillfully?

Putting things and events in order is one of our most frequently practiced thinking tasks. We arrange names alphabetically in an address book in order to find an individual's address quickly. We arrange tools by size to be able to find a wrench or screwdriver that is just the right size for a particular task, to make efficient use of storage space, and to recognize quickly when a tool is missing or needed. We reconstruct a chain of events leading to an accident in order to determine who is responsible for it. We select a pet or plants for our yard based on size, the amount of care required, and the benefit that each brings to our lives and surroundings. We vote for candidates based on how well each fits our beliefs about government. In planning a work schedule, we prioritize tasks by importance and duration and fit them into time slots.

All forms of sequencing involve the same basic process—putting one thing or idea after another according to certain criteria. Any particular thing may occupy different positions in different sequences, however. Hurricane Hugo may be first on a list of the most devastating hurricanes of the decade, but it is eighth in order of hurricane occurrence in 1988. The characteristics you select to sequence things depend on your purposes.

Types of Sequences

We commonly employ many types of sequences when we put things in order. One of the first sequences that we learn is *alphabetical order*. Putting items in alphabetical order allows us to locate them quickly (i.e., finding the telephone numbers of individuals in a telephone directory). Sequencing alphabetically simply requires accuracy in complying with the correct order of letters.

We also use *time order* frequently. Memories of past experiences, our current schedule, and our future plans are all based on the order in which things happen. Time sequences some-times involve matching data about events to a timeline of given intervals. I may recall important events by relating them to other milestones, like moving to a new house or the birth of a baby in the family. Correlating national events with the leadership of particular individuals may help me understand whether they have been effective public officials and should be re-elected.

We commonly use three specialized types of time order: *operation analysis, causal chains, and cycles. Operational analysis* involves correctly setting up steps in a procedure. If I do not carry out the correct sequence of steps in a recipe for rolls, the bread may not rise properly. If I do not follow procedures in a given order, my word processing software may not give my computer the correct commands to check the spelling in what I am writing.

We employ time order to determine *causal chains* of events or actions leading to others. I must sequence events in the correct order to understand why something happened, who is responsible, and what I should do when similar conditions arise in the future.

Cycles are repeating sequences of events that follow in the same time order. The cycle of seasons is an example of the repetition of the same sequence of natural conditions at regular intervals. Understanding cycles allows us to predict recurring events and to realize points in the cycle in which change would affect the whole system. For example, massive rain forest cutting may interfere with the oxygen/carbon dioxide cycle in ways that affect the whole planet.

Ranking actions or things is a specialized form of sequencing. Ranking involves sequencing by quantity and/or by quality. Anything expressed in degrees can be ranked. For example, I may rank things by the *degree* to which they have certain properties (e.g., by how much sugar is included as an ingredient), by their *usefulness* for a certain task (e.g., how well various tools serve a particular purpose), and by their *value* (e.g., how beautiful they are or how well-qualified specific candidates might be for a job). Ranking can also be relative to a specific property (e.g., ranking foods by calories per fixed quantity), or it can be based on combining multiple criteria (e.g., ranking foods by their overall nutritional value).

Prioritizing is a form of ranking in which the highest-ranked item is the most important. We usually prioritize things that must be done. For example, I may regard paying my bills as a higher priority than going to the movies.

Problems in Sequencing

Although sequencing is a frequently used and important thinking skill, we do not always do it skillfully. Sometimes, we prioritize poorly because the criterion for sequencing does not fit the purpose. To store books, I might organize them in order of color or size, which might be visually attractive. However, if my primary purpose for keeping my books is finding information, I should arrange them by type in order to quickly locate books containing certain kinds of information. If I enjoy reading books by certain authors, I may want to arrange my books alphabetically by the authors' last names so that I can easily find a number of books by the same author.

Sometimes, we place items incorrectly in a sequence, either because of inaccuracy in the information or by inaccurately matching the data to the sequencing schema. For example, because of a misprint in a book, someone may think that John F. Kennedy died in 1973. If that information is placed on a timeline, then subsequent events, such as Lyndon Johnson's assuming the presidency, do not make sense. On the other hand, one may have the 1963 date correct, but because of inattention, enter it incorrectly on the timeline at 1973. In either case, one may mistakenly believe that Kennedy lived a decade longer than he did.

We may prioritize things poorly because we are not really clear about the criteria for ranking them. If I read "20% fat" on a box of cereal, I may mistakenly believe that this means that 20% of the weight of the cereal is fat. However, the percentage figure is really the percentage of calories provided by fat. The percentage of fat by weight may be much less than the percentage

of the calories that fat provides. If I buy cereal based on the wrong criteria for percentages, I may mistakenly believe that my fat intake is lower than it actually is.

Figure 8.1 provides a summary of these problems in sequencing.

COMMON PROBLEMS IN SEQUENCING
1. Picking a sequence that poorly fits the purpose of ordering.
2. Inadequate understanding of the criteria for putting things in a specific sequence.
3. Inaccurately matching items to a given sequence by misremembering, haste, or failing to attend to details.

Figure 8.1

How Do We Sequence Things Skillfully?

Sequencing falls into two categories: simple ordering (fitting items into a given sequence or one that we must figure out) and ranking (sequencing by degree or value), which includes prioritizing and is actually a more complex form of ordering.

Putting things in order skillfully. When I order things skillfully, I determine the purpose of the ordering and select an ordering schema that serves that purpose. If I want to find books in my collection by authors, alphabetical order is obviously a better schema to use for organizing my books than ordering them by size. I then have to place each item accurately according to the ordering schema. To organize my books, I identify the first letter of the last name of the author and put each book in the proper location.

Sometimes, determining the basis for ordering items is not so obvious. If a book is co-authored, whose name do I use? I can decide that I will use the name of the author whose name comes earlier alphabetically. I could also decide to use the name of the primary author to alphabetize the book, even though the primary author's name starts with a letter that comes later in the alphabet.

Deciding on the ordering characteristic and examining items carefully to determine accurately how each item fits in that order will allow me to place them properly. Skill in sequencing items involves accuracy, attention to significant details, and in some cases, sound memory.

More challenging types of sequencing activities involve sequencing the steps in a procedure according to cause-and-effect relationships (e.g., sequencing steps in a chess move, in a recipe, or in repairing electronic equipment), and reconstructing temporal sequences of events (e.g., reconstructing a crime).

The thinking map in figure 8.2 guides us through the process of sequencing skillfully.

SEQUENCING

1. What is the purpose of the sequencing?

2. What type of sequence best serves this purpose?

3. What criteria should be used to fit items into this type of sequence?

4. How does each item fit into the sequence based on these criteria?

Figure 8.2

We provide two different diagrams that are useful for fitting items into a sequence: a timeline and a flow chart. These diagrams represent two common sequencing patterns. The first type guides us in determining and recording the criteria for fitting items into a sequence. The second type provides an organizing pattern into which items fit.

The graphic organizer for determining sequencing criteria appears in figure 8.3. This graphic organizer focuses attention on the purpose of sequencing, the type of sequence that best serves that purpose, and the criteria for fitting items into that sequence.

The timeline diagram in figure 8.4 contains parallel time lines so that it can either show chronological occurrence based on a given time interval or illustrate concurrent events.

The flow chart diagram in figure 8.5 uses arrows to mark temporal connections.

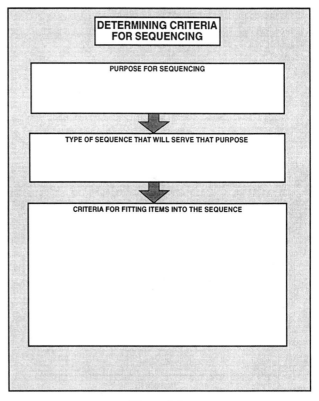

Figure 8.3

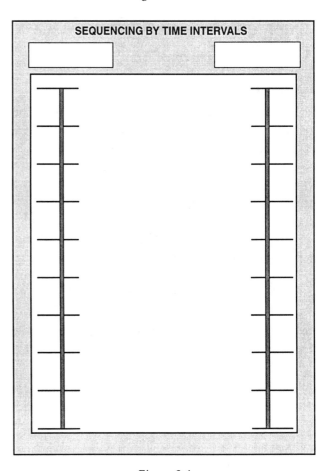

Figure 8.4

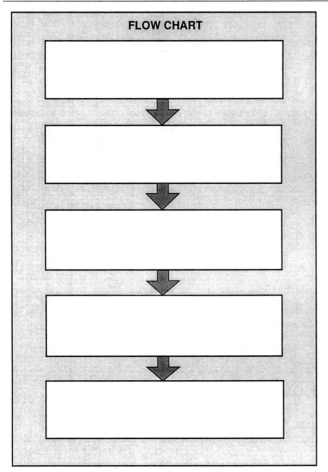

Figure 8.5

Ranking

When ranking things, we should clarify the purpose for ranking. If I am interviewing candidates for a new teaching position, I may want to rank the top five candidates. My purpose for ranking—selecting the best qualified candidate for the job—determines the criteria I will use. Experience in teaching, knowledge of the field, and personableness will be among my criteria. I may consider these qualities equally important, or I may consider one characteristic more significant than the others. If I then determine that one candidate has more experience, more expertise in her field, and is more personable than the others, she will be at the top of my list.

Ranking the job candidates may not be so easy, however. Perhaps one candidate has more experience and relates well with students but does not have as broad an understanding of his field as the other two, while another candidate has a broader base of knowledge in her field, is extremely well-liked by coworkers and yet has

had less experience teaching. How I rank these candidates may depend on subtleties in the relative importance of these criteria.

As demonstrated in this case, ranking is often based on our values and judgments about the relative importance of the criteria we use in determining how to rank specific items. Where these judgments play a role in our ranking, we should always justify them.

The thinking map for ranking (figure 8.6) is a variation of the map for sequencing.

SEQUENCING BY RANK

1. What is the purpose of the ranking?

2. What qualities must something have to serve this purpose?

3. Which of these qualities do each of the items being compared have, and to what degree do they have them?

4. Which has the largest degree of desirable qualities, the next largest, on to the least?

Figure 8.6

The graphic organizer for ranking (figure 8.7) can be varied for a different number of items.

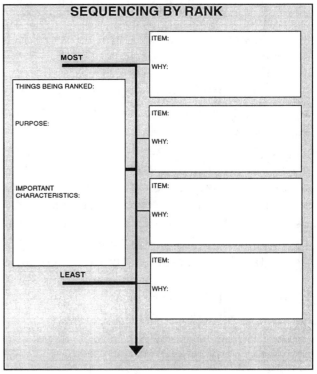

Figure 8.7

How to Teach Students to Sequence Skillfully

Students should learn both types of sequencing: ordering and ranking. They can develop family trees, reconstruct the major events in the plot of a story, and map out the water cycle to practice skillful ordering. They should also, however, practice ranking by engaging in activities such as ranking sources of air pollution by the amount and/or types of pollutants they put into the air or ranking United States presidents by their accomplishments. In teaching either kind of sequencing, it is important to introduce explicitly the specific form of sequencing being taught, to guide students through the sequencing process, and to involve them in reflecting about their thinking as they sequence things.

Often, students make impulsive decisions about the significance or usefulness of items without taking the time to compare the merits of each. For example, in selecting foods, many students base their preferences on taste, rather than ranking foods by caloric content or nutritional value. In teaching skillful ranking, it is important to include activities in which one issue is the relative usefulness of different items in achieving a certain goal. Students should start with a list of different items and think through the pros and cons of each item, ranking the items based on their analysis.

Because sequencing plays such an important role in our lives, it is also important to provide experiences for students in carefully setting up procedures, in recognizing the sequencing of actions, and in making decisions based on contingencies. Students can develop a plan of sequenced steps to obtain good reference material from the library; to figure out how much interest, overall, a loan will cost; to determine the trajectory of a space vehicle and when it will be the closest to the planet Mars; or to re-enact the investigative steps that led to the discovery of oxygen in the air in early science or the structure of DNA in more recent science. For simple sequences, students may use the flow chart diagram. When the plans involve responding to various contingencies, such as in scientific investigations, the basic flow chart diagram may be modified by adding branches.

Many sequencing exercises in textbooks involve matching items to a given order or continuing a given pattern. These exercises often provide students with information regarding the items that fall in the given sequence. For example, students are often given events in social studies texts, along with their dates, and asked to locate them on a timeline. Such exercises are valuable in teaching content. However, they do not help students learn to choose appropriate sequences for ordering items or to gain practice at figuring out where various items belong in a given sequence if this information is not evident.

Contexts in the Curriculum for Lessons on Sequencing

Sequencing plays a natural role in several types of contexts in curriculum materials:

- *Students learn about important sequences of events or conditions.* For example, students study the chain of causes and effects leading up to important natural events (e.g., the eruption of Mt. St. Helens). Clarifying the order of important changes in dinosaurs and in conditions on earth can help students understand what might have caused the extinction of the dinosaurs (e.g., "Did the diminution in plant life occur before, at the same time as, or after the extinction of the dinosaurs?"). Such causes and effects can be organized with a relatively simple flow chart. Sequences of certain other causes and effects studied in science can help students learn methods for controlling causal change (e.g., in nurturing the growth of plants or in understanding scientific procedures). Students also study sequences of organisms, chemical compounds, and forces. Sequencing events in a variety of ways (by their occurrence, importance, etc.) can be enlightening to students and serve as a vehicle for teaching skillful sequencing. Many of these sequences are quantitative (e.g., organisms by number of cells, organic compounds by number of carbon and hydrogen atoms). Although ranking may also be quantitative (e.g., ranking things by weight), some ranking is based on quantitative differences (e.g., how useful things are for certain purposes), and some on subjective properties (e.g., how sweet different liquids taste).

- *Students use sequencing to build important skills.* For example, students learn sequences of steps in solving physics problems, in testing certain chemicals in a laboratory, and they are taught to develop a sequenced plan in conducting their own investigative research.

Teaching students to sequence skillfully emphasizes order and the principles and purposes behind it. Incorporating such instruction into content lessons makes the sequences we teach —like timelines, cycles, and procedures—more relevant and interesting to students.

A menu of suggested contexts for infusion lessons on sequencing can be found at the end of the chapter.

Sample Lessons on Sequencing

We include two lessons on sequencing in this chapter. The first is a middle school/9th-10th grade lesson on ranking in which students rank the usefulness of various pieces of equipment according to specific needs.

The second sequencing lesson is one that has been used in the high school grades. It is on supermarket foods and involves students in working with the label of nutrition facts on packaged goods to determine a number of sequences, including a ranking of these foods for

dietary purposes. This is an important lesson because it not only shows how the science of nutrition studied in high school can be applied to everyday life situations, but also shows how the same items can be put into a number of different sequences depending on the purpose of the sequencing.

As you read these lessons, consider the following:

- How can we help students recognize what kind of sequencing serves a specific purpose?
- What variations of sequencing can be used to reinforce this skill?

Tools for Designing Sequencing Lessons

The tools on the following pages include the thinking maps and graphic organizers for ordering and ranking. The graphic organizers include one to guide students' thinking as they choose the best sequencing schemes to fit their purposes, two information-organizing graphics for the results of sequencing activities (the flow chart and timeline), and a thinking-activity graphic for ranking.

The thinking maps and graphic organizers can guide you in designing the critical thinking activity in the lesson and can also serve as photocopy masters, transparency masters, or as models that can be enlarged and used as posters in the classroom. Reproduction rights are granted for use in single classrooms only.

SEQUENCING

1. **What is the purpose of the sequencing?**

2. **What type of sequence best serves this purpose?**

3. **What criteria should be used to fit items into this type of sequence?**

4. **How does each item fit into the sequence based on these criteria?**

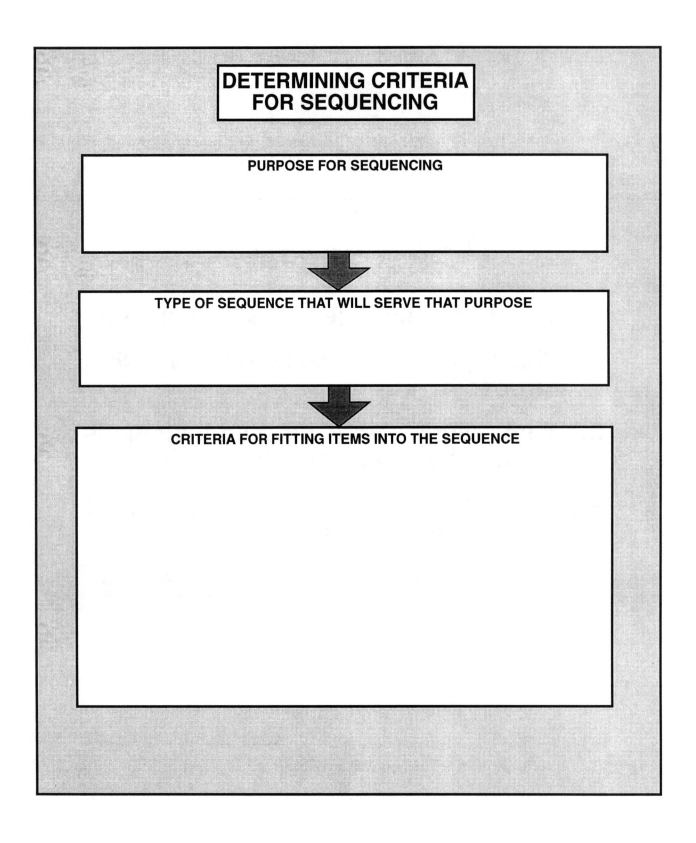

DETERMINING CRITERIA FOR SEQUENCING

PURPOSE FOR SEQUENCING

TYPE OF SEQUENCE THAT WILL SERVE THAT PURPOSE

CRITERIA FOR FITTING ITEMS INTO THE SEQUENCE

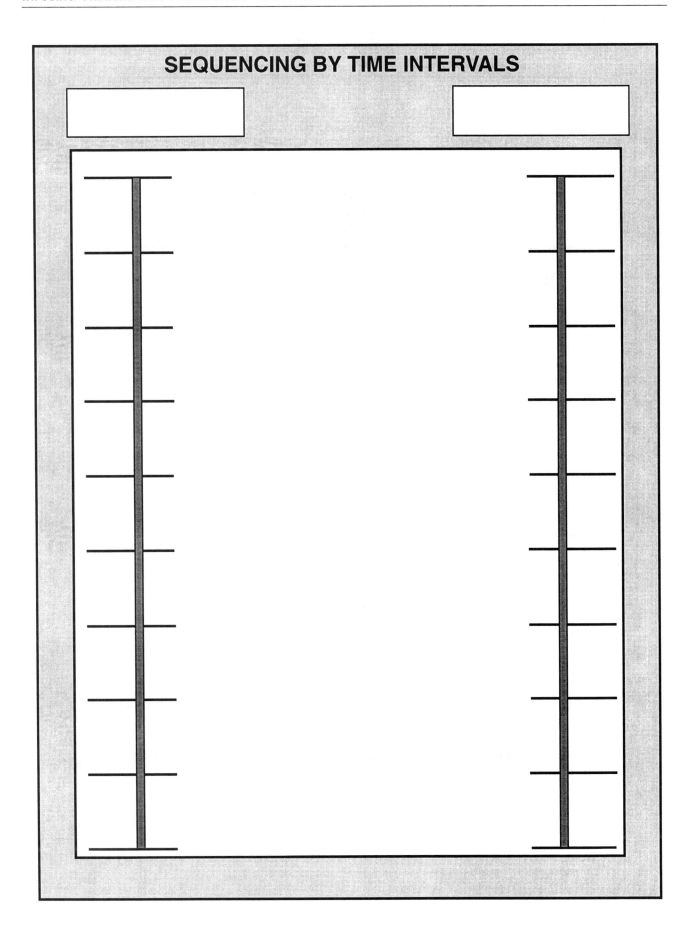

SEQUENCING BY TIME INTERVALS

FLOW CHART

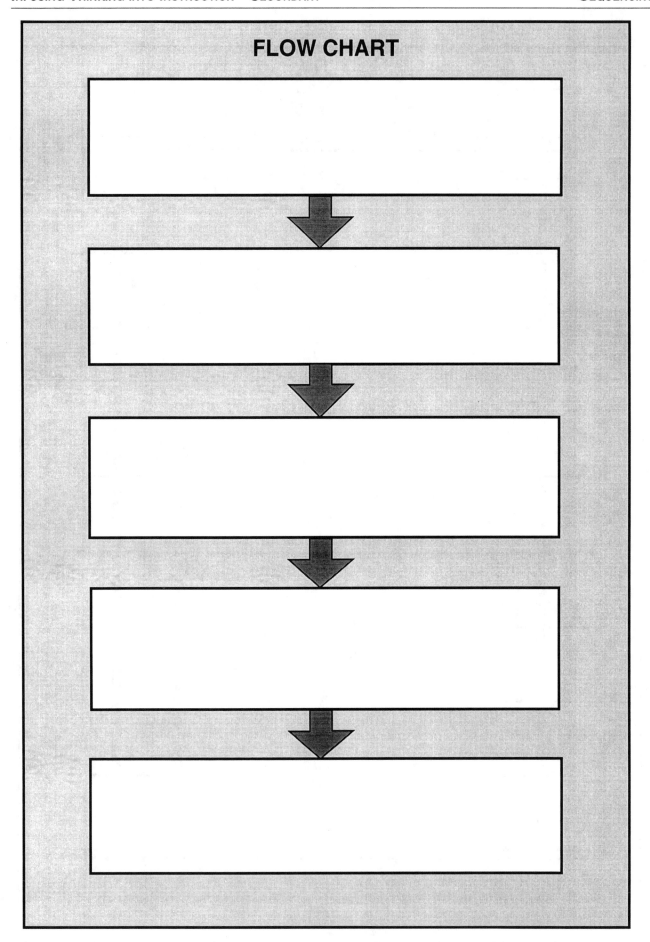

SEQUENCING BY RANK

1. What is the purpose of the ranking?

2. What qualities must something have to serve this purpose?

3. Which of these qualities do each of the items being compared have, and to what degree do they have them?

4. Which has the largest degree of the desirable qualities, the next largest, on to the least?

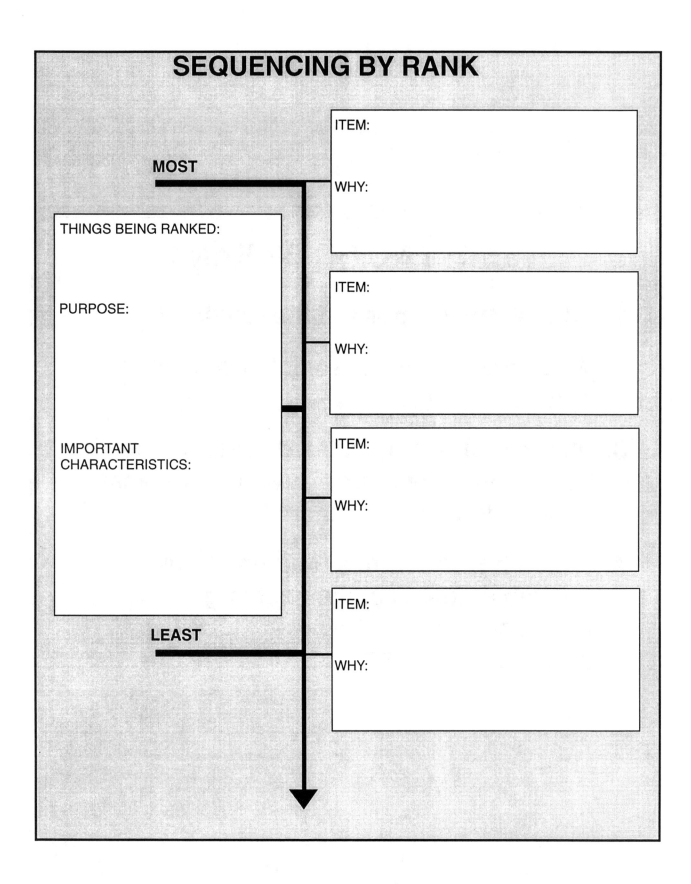

SEQUENCING BY RANK

MOST

THINGS BEING RANKED:

PURPOSE:

IMPORTANT CHARACTERISTICS:

LEAST

ITEM:

WHY:

ITEM:

WHY:

ITEM:

WHY:

ITEM:

WHY:

THE BEST WAY TO SEE IT

General Science, Biology

Grades 6–9

OBJECTIVES

CONTENT	THINKING SKILL/PROCESS
Students will clarify the characteristics of various types of observation instruments to decide which best fits different observation conditions or purposes. Students will differentiate between mosquitoes and flies.	Students will rank objects by how well they serve certain purposes. They will select the characteristic by which the items should be ranked and will accurately place items in order.

METHODS AND MATERIALS

CONTENT	THINKING SKILL/PROCESS
Examples of the observation equipment mentioned in the lesson or posters that show their features will remind students how each piece of equipment enhances observation. Use specimens of mosquitoes and flies, if they are available.	In cooperative work groups, students discuss the purposes of ranking the items and rank them correctly for a selected purpose. The thinking map guides the teacher's directions in describing the task and the process. Three posters, drawings, or transparencies of the graphic organizer for ranking are needed. Students also need three copies of the graphic organizer.

LESSON

INTRODUCTION TO CONTENT AND THINKING SKILL/PROCESS

- Think about things or events that we put in order. Our class schedule is the order in which we do things in class each day. A birthday chart is the order in which people's birthdays occur during the year. We think about coins in order of their value. We put things on shelves in order of their size. With your partner, list as many things as you can think of that you have put in an order during the last few days. Name the characteristic that you thought about to put them in order. On the board, list several examples and characteristics for ordering.

- When we put things in order, we sequence them in certain ways. One way that we sequence things is to rank them. Ranking means putting things in order by how much of a specific quality they have or by how well they serve a specific purpose. Ask four students to stand in front of the class.

- What are some ways that we might rank these four students? Make two columns on the chalkboard or on newsprint. The left column should be headed "Characteristic." Write each characteristic under the heading. POSSIBLE RESPONSES: *Although students usually include descriptions of physiognomy, like height, weight, and appearance, it is best not to focus on characteristics that, when attention is called to them, might embarrass students. More appropriate responses for further discussion often include: number of people in family, birthday, distance school is from home, distance from student desk to chalkboard.*

- Why would someone want to rank students by these characteristics? What would be the purpose of putting them in order this way? Label the right column "Why We Rank Them This Way." POSSIBLE RESPONSES: *distance from school (to arrange a carpool or plan rides home after after-*

school activities), birthdays (to make a birthday calendar so that the class can send birthday cards), family size (to discuss sharing or household chores), distance of student desk from chalkboard (to make sure everyone can see the board well, to reorganize seating so that everyone's sitting where they want to sit). Write students' responses in the right column. Ask the four volunteers to select one of the characteristics and to arrange themselves left-to-right according to that characteristic. After the group has reassembled themselves, ask them to clarify what they thought about to do it.

- **Describe what you thought about to place yourselves in this order.**
  Students' description of the thinking they used to rank themselves by birthday might include:

  (1) *We thought about who had the birthday closest to the beginning of the school year and we put this person on the far left.*

  (2) *We continued picking out the person with the next closest birthday to the beginning of the school year and placed that person next in order.*

  (3) *The person with the birthday furthest from the beginning of the school year was on the far right.*

  Now that the class sees the line of students in order of their birthdays, ask them for additional purposes for ranking people by their birth date. Then ask the students to arrange themselves by the size of their family. Note that the order has probably changed. Again, ask students when they think this ordering would be useful.

- **Sometimes we rank things according to their usefulness. With your partner, describe things that we rank in the order of how well they help us do certain tasks. What characteristics are important in deciding which items are the most or least useful?** Discuss several student examples. POSSIBLE ANSWERS: *Tools, drawing media (pencils, markers, pastels, tempera paint), means of transportation, outerwear (coats, sweaters, jackets, parkas).*

THINKING ACTIVELY

- **In this lesson, we are going to rank various tools for use in observing insects. When studying insects, we want to learn about where they live, what their bodies are like, and what they need to survive. To make observations of flies, we must choose the best tool to see different things about them.** Write "Observation tools" in the box under "Things Being Ranked." **Some tools for consideration are (1) a magnifying glass, often used to observe specimens in field where high magnification is not required, (2) the naked eye, which can provide more than 80 percent of the information received about the external environment, and (3) a compound microscope, which enables one to view an enlarged image of an organism by magnifying that organism up to 1000 times its original size.**

SEQUENCING BY RANK

MOST

THINGS BEING RANKED:

PURPOSE:

IMPORTANT CHARACTERISTICS:

LEAST

ITEM:
WHY:

ITEM:
WHY:

ITEM:
WHY:

ITEM:
WHY:

- **We will be ranking these three tools according to their usefulness in observing the patterns in a fly's wing.** Under the heading "Purpose," write "To see details on the wing of an insect." **Decide what characteristics of the observation tools are important in this task.** POSSIBLE ANSWERS: *Magnification (the wing does not need to be attached to the fly, so it can be mounted on a slide).*

- **Rank the three instruments from the most to the least useful. Write the order on the graphic organizer and explain why you ordered them this way.** After students have completed ordering the tools, ask students to explain why they placed the equipment in this order. List their responses on a transparency or a large drawing of the graphic organizer. POSSIBLE ANSWERS: *(1) The microscope has the greatest magnification, making it best for seeing the patterns. Although the wing is thin enough to be transparent, it has enough color for viewing it under the microscope without staining. The wing does have to be detached for viewing, however, destroying close examination of how it attaches to the body and how it works. The examination of the wing does not have to occur in the natural environment. (2) The magnifying glass offers enough magnification to see the wing but, perhaps, not in sufficient detail. It is portable and can show how the wing looks in relation to the fly's body without detaching it. The magnifying glass is easy to carry, allowing immediate examination of the specimen at the site. (3) The eye doesn't give enough magnification to see the details of patterns on the wing. The field of vision of the eye does allow the observer to see wing functioning.*

- **Suppose you wanted to understand how flies behave in their natural environment.** On a new copy of the graphic organizer, write "See how the fly behaves" under the heading "Purpose." **We will be ranking the three tools according to their usefulness in observing the activity of the flies in their habitat. Decide what characteristics of the observation tools are important for this activity.** POSSIBLE ANSWERS: *Wide field of vision, little interference with the flies.*

- **Rank the three instruments from the most to the least useful. Write the order on the graphic organizer.** POSSIBLE ANSWERS: *(1) Although the eye doesn't give enough magnification to see the details on the bodies of flies, its broad field of vision does allow the observer to see flies in relation to their environment. Using the eye means that the flies and their surroundings are undisturbed by handling or preparation. The eye is always ready for use, and its acuity is as good as the vision of the observer. (2) The magnifying glass allows close-up examination with little disturbance and no dissection of the fly. It is portable and easy to use, but its field of vision limits observing the surrounding environment. (3) Using the microscope requires preparation, which must be done in a laboratory environment, even if it is a field laboratory. Little examination of the habitat can be pursued other than examining samples of substances in the flies' habitat for microorganisms or chemicals. Some changes in the flies' bodies as the result of the environment might be detected, but the microscope's usefulness in understanding habitat is limited.*

- **Sometimes people mistake flies for mosquitoes. One way to differentiate them is to look carefully at their mouths. A female mosquito bites to get blood so that her eggs will develop. She pierces your skin with six little lances sheathed in a tube, interlocking to form a hollow needle. Before she sucks the blood, she injects an anticoagulant which causes the itchy, raised spot around the hole in a mosquito bite. Houseflies don't bite. Their mouthparts are spongy to absorb food. Biting flies have hard hooks above the mouth to shred its food before absorbing it. What characteristics of observation tools are helpful in examining the mouth?** POSSIBLE ANSWERS: *Some magnification; no damage to tissue in the fly's mouth.*

- **Rank the tools that would be most helpful in seeing the mouthparts of an insect in order to tell whether it is a mosquito or a fly.** POSSIBLE ANSWERS: *(1) The magnifying glass allows close-up examination with little disturbance and no dissection of the fly. Its magnification is adequate to see the details of the mouthparts of a fly or mosquito. It is portable and easy to use, and its field of vision is well-suited for seeing the head area. (2) The microscope requires preparation and must be used in a laboratory environment. Preparation destroys the connection between the mouthparts. Its magnification shows more details of the structure of the mouthparts than the identification task requires. (3) Unless one has very good near vision, it is very difficult to get a good look at the mouthparts. One can, however, observe the biting behavior and its effects and tell whether the insect is a fly or a mosquito.*

- **How did ranking these tools in different ways help you understand how to use them?** Students respond that they recognize that magnification is not the only factor in observing things. More magnification limits the field around the object. Too much magnification may obscure the whole structure that one is trying to see. This lesson can be extended by asking students to analyze the utility of other tools that might be helpful in observing flies, such as a video camera, movie camera, or photographic camera with a telephoto lens.

THINKING ABOUT THINKING

- **What did you think about to rank things accurately?** This thinking map contains the key questions in skillful ranking.

- **What advice would you give someone about the best procedure for ranking things and ideas?** Often, students make impulsive decisions about the relative usefulness of items without taking the time to compare the merits of each. In teaching the skill of ranking, it is important to have students start with a list of different items that may be useful for achieving a certain goal. In this way, students will gain practice in thinking through the pros and cons of each item and in ranking the items based on their analysis.

> **SEQUENCING BY RANK**
> 1. What is the purpose of the ranking?
> 2. What qualities must something have to serve this purpose?
> 3. Which of these qualities do each of the items being compared have, and to what degree do they have them?
> 4. Which has the largest degree of the desirable qualities, the next largest, on to the least?

APPLYING THINKING

Immediate Transfer

- List, in order of greater efficiency, which container would best fill a five-gallon aquarium that is five yards away from the water source: a one-gallon shallow pan, a 2 1/2-gallon plastic bucket with a handle, and a one-gallon glass jar. Explain why you ranked them this way.

- Rank the following items in terms of their usefulness in writing a term paper: a pencil, a typewriter, a word processor, a pen. Explain why you ranked them this way.

Reinforcement Later

- Find the calorie content of the snack foods available to you. Rank them by the range of their calories (0–100, 100–200, 200–300, etc.). Decide which ones fit your health goals and plan which ones you will select or reject or the frequency with which you will eat them.

- List all the tasks that you normally do between the time school is out and the time that you go to bed. Create a weekly schedule for yourself. Rank the five things that you do that are the least important and the five things that you do that are most important. Explain why you rank them this way. If you wanted to spend time on new things or to give more time to important things, how would you change your schedule?

ASSESSING STUDENT THINKING ABOUT SEQUENCING

Any of the transfer examples can be used for assessment tasks. Examples in which students prioritize are especially suited as assessment examples. Health textbooks are especially rich in contexts offered to students about ranking the relative importance of such things as foods and safety practices. The children's magazine *Pennywise* is produced by Consumers Union and features consumer information that often involves prioritizing. Ask the students to explain why they selected a given characteristic for determining rank and how they arrived at the order of items. Use the thinking map to help you determine whether your students are attending to relevant matters in ranking.

Sample Student Answers: The Best Way to See It!

HOW TO SEE THE DETAILS ON FLIES' WINGS!

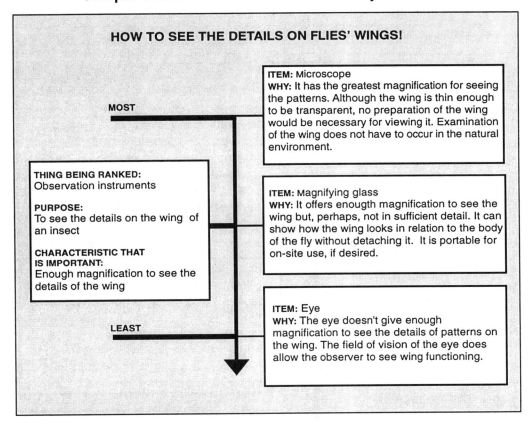

MOST

ITEM: Microscope
WHY: It has the greatest magnification for seeing the patterns. Although the wing is thin enough to be transparent, no preparation of the wing would be necessary for viewing it. Examination of the wing does not have to occur in the natural environment.

THING BEING RANKED:
Observation instruments

PURPOSE:
To see the details on the wing of an insect

CHARACTERISTIC THAT IS IMPORTANT:
Enough magnification to see the details of the wing

ITEM: Magnifying glass
WHY: It offers enough magnification to see the wing but, perhaps, not in sufficient detail. It can show how the wing looks in relation to the body of the fly without detaching it. It is portable for on-site use, if desired.

LEAST

ITEM: Eye
WHY: The eye doesn't give enough magnification to see the details of patterns on the wing. The field of vision of the eye does allow the observer to see wing functioning.

HOW TO SEE THE BEHAVIOR OF FLIES!

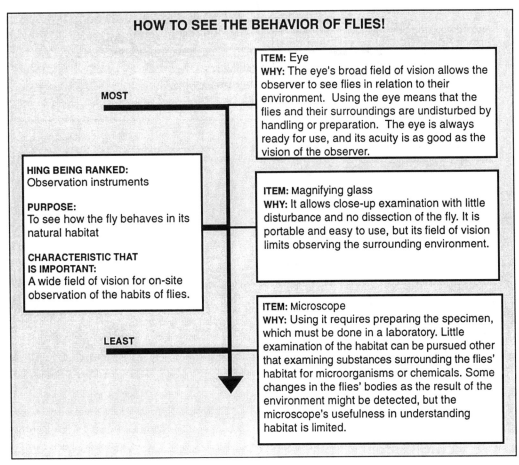

MOST

ITEM: Eye
WHY: The eye's broad field of vision allows the observer to see flies in relation to their environment. Using the eye means that the flies and their surroundings are undisturbed by handling or preparation. The eye is always ready for use, and its acuity is as good as the vision of the observer.

HING BEING RANKED:
Observation instruments

PURPOSE:
To see how the fly behaves in its natural habitat

CHARACTERISTIC THAT IS IMPORTANT:
A wide field of vision for on-site observation of the habits of flies.

ITEM: Magnifying glass
WHY: It allows close-up examination with little disturbance and no dissection of the fly. It is portable and easy to use, but its field of vision limits observing the surrounding environment.

LEAST

ITEM: Microscope
WHY: Using it requires preparing the specimen, which must be done in a laboratory. Little examination of the habitat can be pursued other that examining substances surrounding the flies' habitat for microorganisms or chemicals. Some changes in the flies' bodies as the result of the environment might be detected, but the microscope's usefulness in understanding habitat is limited.

Sample Student Answers: The Best Way to See It!

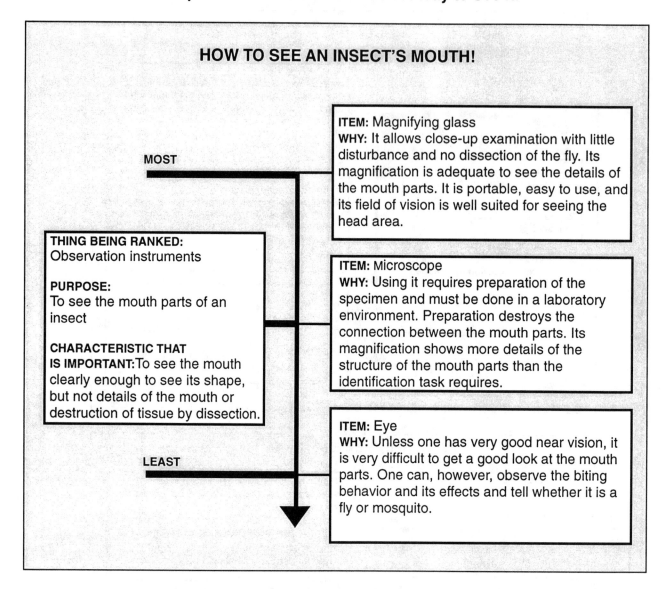

HOW TO SEE AN INSECT'S MOUTH!

MOST

ITEM: Magnifying glass
WHY: It allows close-up examination with little disturbance and no dissection of the fly. Its magnification is adequate to see the details of the mouth parts. It is portable, easy to use, and its field of vision is well suited for seeing the head area.

THING BEING RANKED:
Observation instruments

PURPOSE:
To see the mouth parts of an insect

CHARACTERISTIC THAT IS IMPORTANT: To see the mouth clearly enough to see its shape, but not details of the mouth or destruction of tissue by dissection.

ITEM: Microscope
WHY: Using it requires preparation of the specimen and must be done in a laboratory environment. Preparation destroys the connection between the mouth parts. Its magnification shows more details of the structure of the mouth parts than the identification task requires.

ITEM: Eye
WHY: Unless one has very good near vision, it is very difficult to get a good look at the mouth parts. One can, however, observe the biting behavior and its effects and tell whether it is a fly or mosquito.

LEAST

SUPERMARKET FOODS

General Science, Biology   **Grades 6–9**

OBJECTIVES

CONTENT

Students will learn nutritional information about various foods and relate this information to food production planning.

THINKING SKILL/PROCESS

Students will sequence items by how well they serve certain purposes. They will select the characteristic by which the items should be sequenced, determine the purpose for sequencing them that way, and accurately place items in order.

METHODS AND MATERIALS

CONTENT

Tables and graphs about important characteristics of foods are used. Some of these are obtained from labels used in the USA in accordance with the standards utilized by the federal government in revealing important nutritional characteristics of various foods.

THINKING SKILL/PROCESS

In cooperative work groups, students discuss the purposes of sequencing the items and sequence them correctly for a selected purpose. The thinking map guides the teacher's directions in describing the task. Three posters, drawings, or transparencies of the graphic organizer for sequencing are needed. Students also need three copies of the graphic organizer.

LESSON

INTRODUCTION TO CONTENT AND THINKING SKILL/PROCESS

- Let's think about things or events that we put in order. Our class schedule is the order that we do things in class each day. We think about coins in order of their value. We put books on library shelves by topic; and within each topic, they are ordered alphabetically by author. We list steps to go through in recipes in a specific order. With your partner, list as many things that you can think of that you have put in an order, or things that you've encountered that others have put in order. Name the characteristics that you thought about to put them in order. On the board, list the examples just mentioned and the characteristics used for ordering. Then add four or five examples from the class. **Discuss why we might want to put these things in that order.** Ask students to provide as many different purposes for these sequences as they can. Write them next to the appropriate item.

- When we put things in order, we *sequence* them in certain ways. How we sequence things depends on what purpose we have in sequencing them that way, which, in turn, determines the characteristics by which we determine the sequence that the things fit into. Here are some questions to guide skillful sequencing:

> ### SEQUENCING
>
> 1. What is the purpose of the sequencing?
>
> 2. What type of sequence best serves this purpose?
>
> 3. What criteria should be used to fit items into this type of sequence?
>
> 4. How does each item fit into the sequence based on these criteria?

THINKING ACTIVELY

- We've been studying food and nutrition in science. In this lesson, we are going to consider a variety of information about foods that can be purchased at any supermarket. This information is contained on the labels of these foods in accordance with rules established by the federal government about what information food manufacturers should provide to the general public about their products. Show students a typical food container with a label that contains a "Nutrition Facts" section. Explain what each percentage means (percentage of the normal allowable amounts for a standard diet).

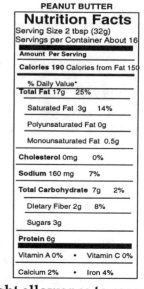

- The information contained on such labels can be used to put foods in various orders or sequences. For example, following the thinking map for sequencing, I may wish to sequence these foods according to the amount of sodium contained in them for the purpose of choosing foods for a low-salt diet. In this case, I might start with low-sodium foods and, consulting the "Nutrition Facts" label, sequence them by the amount of sodium in each serving. Or, because I have a limited amount of weight allowance to carry food with me on a trip, I might sequence foods by container weight. I'm going to give you a series of nutrition-fact labels from common foods. Work together with your partner and use this graphic organizer for sequencing to develop two sequences that you might put these foods into according to the information contained on the label. After three or four minutes, ask for some reports from the teams, taking no more than one from any given team. As the teams report, fill in a blank graphic organizer on a chart on the wall of the classroom, or on the chalkboard. (The chart should have room for at least five such sequences. In each, the items to be sequenced should be identified as "Supermarket Foods.") TYPICAL STUDENT ANSWERS IN-CLUDE: *Sequencing by fat content: cola, tuna fish, crackers, cream cheese, salad dressing, peanut butter. Sequencing by amount of sodium in a standard serving: cola, cream cheese, crackers, peanut butter, salad dressing/tuna fish. Sequencing by quantity in the container: tuna fish, crackers, cream cheese, salad dressing, peanut butter, cola. Sequencing by calories per serving: crackers, tuna fish, cream cheese/cola, salad dressing, peanut butter.*

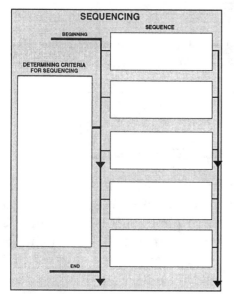

- Now suppose that you were a nutritionist and you wanted to produce a chart ranking these foods for their overall nutritional value. Ranking is a special kind of sequencing in which we sequence things by their value, usefulness, or importance. Instead of sequencing simply by quantity (from least to most), as we usually do, we sequence from worst to best. Work in groups of four now and rank these foods according to their overall nutritional value using the version of the sequencing graphic organizer called "Ranking." After four or five minutes, ask each team to report their results to the whole class. Give each team a blank transparency on which they write their results and use for their reports, or have a spokesperson for each team write their results on the chalkboard. If there is a difference of opinion expressed by the different groups, ask each group to explain why they ranked the foods as they did and give them a chance to revise their ranking based on this dialogue and discussion. If they do revise their ranking, ask them to change the graphic organizer on their transparency or on the chalkboard and explain why they changed their ranking. Show the class where they agree in their ranking and where they disagree and, if the class reaches a consensus, write the ranking that they accept on the board as the class's ranking. TYPICAL STUDENT ANSWERS INCLUDE: *Ranking by amount of protein for high-protein diets: cola/salad dressing, cream cheese/crackers, peanut butter, tuna fish. Ranking for low-fat diets: peanut butter, salad dressing, cream cheese, crackers, tuna fish, cola. Ranking for low-sugar diets: cola, salad dressing, peanut butter, cream cheese, crackers/tuna fish. Ranking for high-protein/low-fat diets: salad dressing, cream cheese, peanut butter, cola, crackers, tuna fish.*

THINKING ABOUT THINKING

- **What did you think about to sequence things accurately?** Students should identify the key questions from the thinking map for sequencing about the *purpose* of the sequencing and the *characteristics* that they based their sequence on as the basis for their sequenced lists.

- **Is this a good way to sequence things?** Students usually say that they didn't realize that the same things could be sequenced in a number of different ways and that if they are clear about the purpose of the sequencing and the characteristics according to which they are going to sequence the things they are working with, they will more likely put them in a useful and accurate sequence.

- **How would you describe the difference between ranking things and simply sequencing them?** STUDENT RESPONSES VARY, but usually include reference to the fact that ranking is a kind of sequencing in which things are sequenced according to their value, importance, or usefulness in performing a task. **Draw a thinking map for ranking that reflects these similarities and differences.** The thinking map should look like the following:

- **What advice would you give someone about the best procedure for ranking things and ideas?** Often, students make impulsive decisions about the relative value or usefulness of items without taking the time to compare the merits of each. In teaching the skill of ranking, it is important to have students start with a list of different items that may be useful for achieving a certain goal. In this way, students will gain practice in thinking through the pros and cons of each item and in ranking the items based on their analysis.

> **SEQUENCING BY RANK**
>
> 1. **What is the purpose of the ranking?**
>
> 2. **What qualities must something have to serve this purpose?**
>
> 3. **Which of these qualities do each of the items being compared have, and to what degree do they have them?**
>
> 4. **Which has the largest degree of the desirable qualities, the next largest, on to the least?**

APPLYING YOUR THINKING

Immediate Transfer

- Gather nutrition facts on 10 other common foods and add them to the sequences that you developed with your partner. Where would you place them in your ranking of foods by their nutritional value, and why?

- Make a list of 20 different types of animals. Sequence these in two different ways, according to two different purposes for putting these animals in a sequence. Now rank these according to their value or usefulness. You determine what kind of value or usefulness you will use as a basis for this ranking.

Reinforcement Later

• In the periodic table of elements that you have studied in science, chemical elements are sequenced. Analyze the way they are sequenced by determining the characteristic(s) by which they are sequenced. What purpose do you think scientists have in sequencing chemical elements in this way? Can you think of some other purpose someone might have for sequencing chemical elements? Explain what that might be and how it would change the way these elements are sequenced.

• List the kinds of objects that appear in our solar system. Sequence these according to some important purpose that a person might have for putting these in order.

• List all the tasks that you normally do between the time school is out on school days and the time that you go to bed. Create a weekly schedule for yourself by sequencing the tasks as you most commonly do them. Now rank the five things that you do that are the least important and the five things that you do that are the most important. Explain why you rank them this way. If you wanted to spend time on new things or to give more time to important things, how would you change your schedule?

ASSESSING STUDENT THINKING ABOUT SEQUENCING

Any of the transfer examples can be used for assessment tasks. Examples in which students sequence common items that they study in science, like types of cells, bodily organs, types of minerals, simple machines, pieces of communications technology, etc., are all well-suited for assessment purposes. Examples in which items are prioritized are especially suited as assessment examples for ranking. Health textbooks are usually rich in discussing items and practices that can be ranked in relative importance, such as safety practices, first-aid equipment, etc. The children's magazine *Pennywise* is produced by *Consumer Reports* and features consumer information that often involves prioritizing. Ask the students to explain why they selected a given characteristic for sequencing items, or for determining rank, and how they arrived at the order of items. Use the thinking maps to help you determine whether your students are attending to relevant matters in both general sequencing and ranking.

Nutrition Facts about Various Common Foods

CRACKERS

Nutrition Facts
Serving Size 2 crackers (14g)
Servings per Container About 12

Amount Per Serving

Calories 60 Calories from Fat 15

% Daily Value*

Total Fat 1.5g   2%

Saturated Fat 0g   0%

Polyunsaturated Fat 0g

Monounsaturated Fat 0.5g

Cholesterol 0mg   0%

Sodium 140 mg   6%

Total Carbohydrate 10g   3%

Dietary Fiber 1g   4%

Sugars 0g

Protein 2g

Vitamin A 0% • Vitamin C 0%

Calcium 0% • Iron 4%

SALAD DRESSING

Nutrition Facts
Serving Size 2 tbsp (30g)
Servings per Container About 12

Amount Per Serving

Calories 130 Calories from Fat 100

% Daily Value*

Total Fat 11g   17%

Saturated Fat 1.5g   8%

Polyunsaturated Fat 0g

Monounsaturated Fat 0g

Cholesterol 0mg   0%

Sodium 250 mg   11%

Total Carbohydrate 8g   3%

Dietary Fiber 0g   0%

Sugars 7g

Protein 0g

Vitamin A 0% • Vitamin C 0%

Calcium 0% • Iron 4%

PEANUT BUTTER

Nutrition Facts
Serving Size 2 tbsp (32g)
Servings per Container About 16

Amount Per Serving

Calories 190 Calories from Fat 150

% Daily Value*

Total Fat 17g   25%

Saturated Fat 3g   14%

Polyunsaturated Fat 0g

Monounsaturated Fat 0.5g

Cholesterol 0mg   0%

Sodium 160 mg   7%

Total Carbohydrate 7g   2%

Dietary Fiber 2g   8%

Sugars 3g

Protein 6g

Vitamin A 0% • Vitamin C 0%

Calcium 2% • Iron 4%

TUNA FISH

Nutrition Facts
Serving Size 2 ounces (56g)
Servings per Container 2.5

Amount Per Serving

Calories 70 Calories from Fat 10

% Daily Value*

Total Fat 1g   2%

Saturated Fat 0g   0%

Polyunsaturated Fat 0g

Monounsaturated Fat 0g

Cholesterol 25mg   10%

Sodium 250 mg   10%

Total Carbohydrate 0g   0%

Dietary Fiber 0g   0%

Sugars 0g

Protein 15g   27%

Vitamin A 0% • Vitamin C 0%

Calcium 0% • Iron 0%

CREAM CHEESE

Nutrition Facts
Serving Size 2 tbsp (29g)
Servings per Container About 8

Amount Per Serving

Calories 100 Calories from Fat 80

% Daily Value*

Total Fat 9g   14%

Saturated Fat 4g   20%

Polyunsaturated Fat 0g

Monounsaturated Fat 0g

Cholesterol 25mg   10%

Sodium 100 mg   4%

Total Carbohydrate 2g   1%

Dietary Fiber 0g   0%

Sugars 1g

Protein 2g

Vitamin A 8% • Vitamin C 0%

Calcium 4% • Iron 0%

COLA

Nutrition Facts
Serving Size 8 fluid oz (240g)
Servings per Container About 8

Amount Per Serving

Calories 100 Calories from Fat 0

% Daily Value*

Total Fat 0g   0%

Saturated Fat 0g   0%

Polyunsaturated Fat 0g

Monounsaturated Fat 0g

Cholesterol 0mg   0%

Sodium 35 mg   1%

Total Carbohydrate 27g   9%

Dietary Fiber 0g   0%

Sugars 27g

Protein 0g

Vitamin A 0% • Vitamin C 0%

Calcium 0% • Iron 0%

*BASED ON A DIET OF 2000 CALORIES PER DAY.

Sample Student Responses • Supermarket Foods

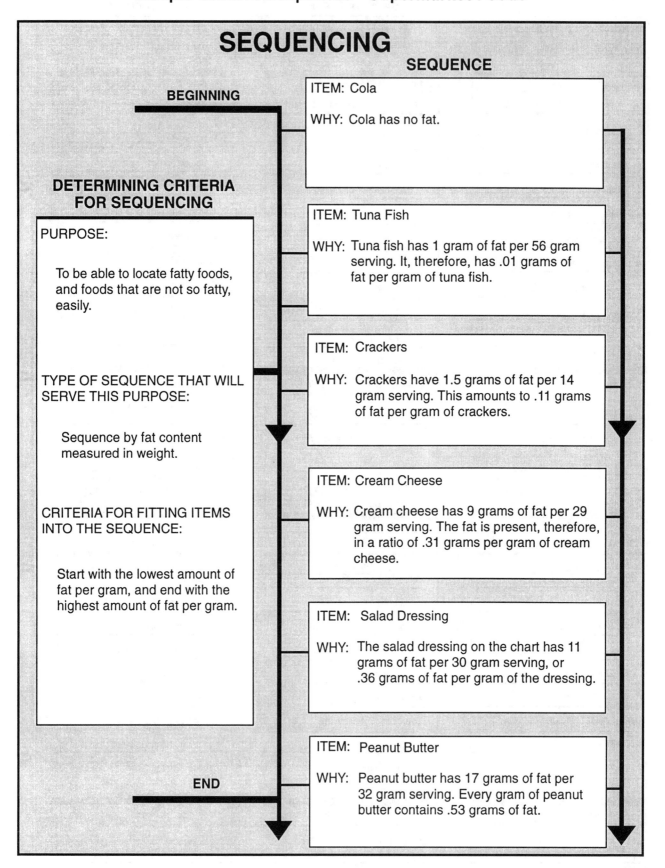

SEQUENCING

SEQUENCE

BEGINNING

ITEM: Cola

WHY: Cola has no fat.

**DETERMINING CRITERIA
FOR SEQUENCING**

PURPOSE:

To be able to locate fatty foods, and foods that are not so fatty, easily.

TYPE OF SEQUENCE THAT WILL SERVE THIS PURPOSE:

Sequence by fat content measured in weight.

CRITERIA FOR FITTING ITEMS INTO THE SEQUENCE:

Start with the lowest amount of fat per gram, and end with the highest amount of fat per gram.

ITEM: Tuna Fish

WHY: Tuna fish has 1 gram of fat per 56 gram serving. It, therefore, has .01 grams of fat per gram of tuna fish.

ITEM: Crackers

WHY: Crackers have 1.5 grams of fat per 14 gram serving. This amounts to .11 grams of fat per gram of crackers.

ITEM: Cream Cheese

WHY: Cream cheese has 9 grams of fat per 29 gram serving. The fat is present, therefore, in a ratio of .31 grams per gram of cream cheese.

ITEM: Salad Dressing

WHY: The salad dressing on the chart has 11 grams of fat per 30 gram serving, or .36 grams of fat per gram of the dressing.

END

ITEM: Peanut Butter

WHY: Peanut butter has 17 grams of fat per 32 gram serving. Every gram of peanut butter contains .53 grams of fat.

Sample Student Responses • Supermarket Foods

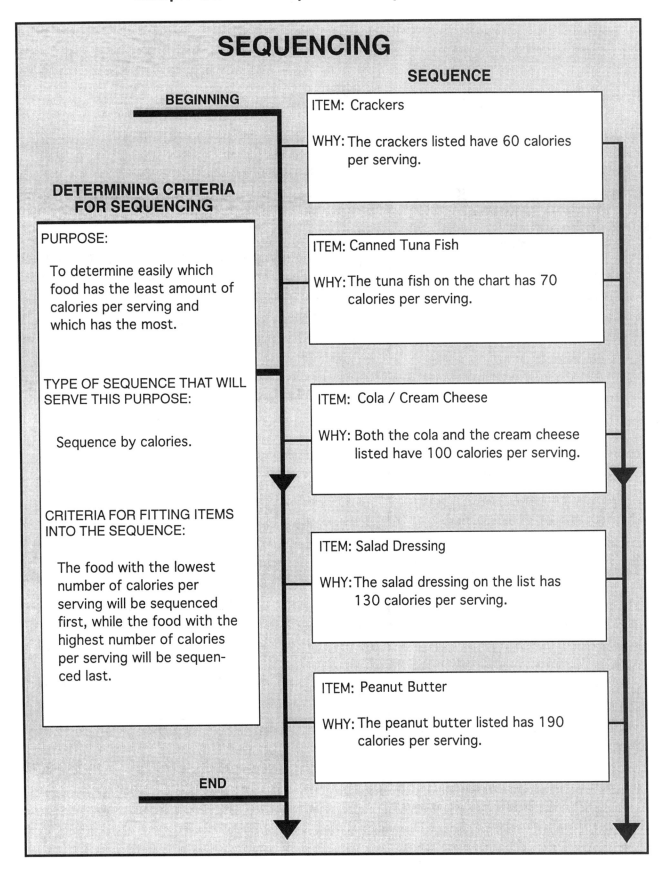

SEQUENCING

SEQUENCE

BEGINNING

ITEM: Crackers

WHY: The crackers listed have 60 calories per serving.

DETERMINING CRITERIA FOR SEQUENCING

PURPOSE:

To determine easily which food has the least amount of calories per serving and which has the most.

ITEM: Canned Tuna Fish

WHY: The tuna fish on the chart has 70 calories per serving.

TYPE OF SEQUENCE THAT WILL SERVE THIS PURPOSE:

Sequence by calories.

ITEM: Cola / Cream Cheese

WHY: Both the cola and the cream cheese listed have 100 calories per serving.

CRITERIA FOR FITTING ITEMS INTO THE SEQUENCE:

The food with the lowest number of calories per serving will be sequenced first, while the food with the highest number of calories per serving will be sequenced last.

ITEM: Salad Dressing

WHY: The salad dressing on the list has 130 calories per serving.

ITEM: Peanut Butter

WHY: The peanut butter listed has 190 calories per serving.

END

Sample Student Responses • Supermarket Foods

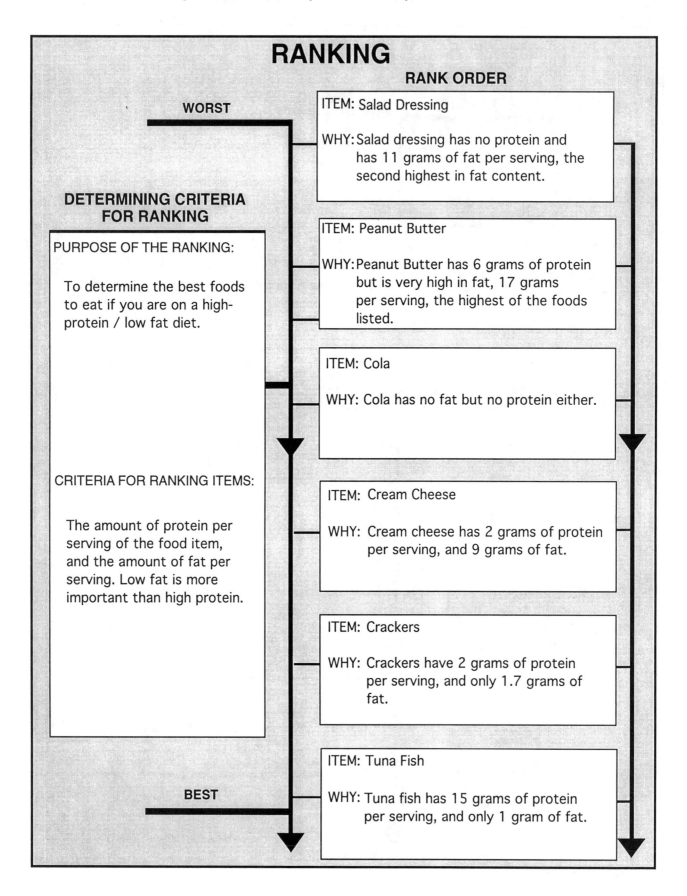

RANKING

RANK ORDER

WORST

DETERMINING CRITERIA FOR RANKING

PURPOSE OF THE RANKING:

To determine the best foods to eat if you are on a high-protein / low fat diet.

CRITERIA FOR RANKING ITEMS:

The amount of protein per serving of the food item, and the amount of fat per serving. Low fat is more important than high protein.

ITEM: Salad Dressing

WHY: Salad dressing has no protein and has 11 grams of fat per serving, the second highest in fat content.

ITEM: Peanut Butter

WHY: Peanut Butter has 6 grams of protein but is very high in fat, 17 grams per serving, the highest of the foods listed.

ITEM: Cola

WHY: Cola has no fat but no protein either.

ITEM: Cream Cheese

WHY: Cream cheese has 2 grams of protein per serving, and 9 grams of fat.

ITEM: Crackers

WHY: Crackers have 2 grams of protein per serving, and only 1.7 grams of fat.

ITEM: Tuna Fish

BEST

WHY: Tuna fish has 15 grams of protein per serving, and only 1 gram of fat.

SEQUENCING LESSON CONTEXTS

The following examples have been suggested by classroom teachers as contexts to develop infused lessons. If a skill or process has been introduced in a previous infused lesson, these contexts may be used to reinforce it.

GRADE	SUBJECT	TOPIC	THINKING ISSUE
6–8	Science	*Wonders of the Deep Sea*	Analyze the accounts of ocean exploration. Select one and describe how the research might have turned out differently if another sequence of events had occurred.
6–8	Science	*All About Famous Scientific Experiments*	Select one of the scientific experiments that had alternative sequences for carrying out the experiments. What difference in results would occur if a different order were followed?
6–8	Science	Acid rain	Describe the process of forming acid rain and explain where in the process measures can be taken to reduce it.
6–8	Science	Pollution	List four sources of air pollution in order regarding how much pollutant they put into the air.
6–8	Science	Plants	Prioritize five uses of plants from the most to the least important.
6–8	Science	Nuclear power	Trace the creation of nuclear energy from nuclear fission to an electrical generator to clarify nuclear safety concerns.
6–8	Science	Power	Rank, based on use, energy from different sources consumed in the United States.
6–8	Science	Air pollution	Select 10 major U.S. cities and rank them according to the risk of developing respiratory disease due to air pollution.
6–8	Science	Nutrition	Rank different foods according to the amount of cholesterol they contain and, then again, according to the amount of calories they produce.
6–8	Science	Buoyancy	Identify 20 bodies of water on Earth. Rank them according to their buoyancy. Explain why there is the difference that you indicate.
9–12	Biology	Protein synthesis and delivery	To clarify the order of events that occur during cellular protein syntheses, make a flow chart and sequence by time interval each step in the process (DNA transcription, messenger RNA nuclear processing, amino acid activation, ribosomal translation, polypepti folding, proofing, and vessicular export).
9–12	Biology	Evolution	Investigate, and then record the steps in the development of Earth's modern atmosphere. Consider what might have happened to the development of life on Earth if the sequence of steps were changed.
9–12	Biology	Diversity and evolution	Rank, by time interval, the origin of each of the body's organ systems. Consider what might have happened to the diversity of life on earth if they had evolved in a different sequence.

SEQUENCING LESSON CONTEXTS			
GRADE	**SUBJECT**	**TOPIC**	**THINKING ISSUE**
9–12	Biology	Taxonomy	Rank, by time interval, the origin of all major phyla on the arborum vitae. Consider the complexity of each taxon as it relates to the date of its emergence.
9–12	Biology	Developmental biology	Make a flow chart indicating the sequence of events leading to human birth and beginning with two gametes. Make a similar flow chart for a worm. Consider how gestation time and complexity might be related.
9–12	Biology	Cellular respiration	Make a flow chart showing the sequence of events that transpire during cellular respiration; include glycolysis, the citric acid cycle and the electron transport chain. Consider what might happen if the sequence were interrupted or an element were missing.
9–12	Biology	Photosynthesis	Make a flow chart showing the sequence of events during photosynthesis; include the light reaction and the electron transport chain, and the dark reaction and the Calvin cycle. Consider what would happen if any of the steps were missing or out of order.
9–12	Chemistry	Building atoms	Consider the steps in constructing diagrams of atoms (inventorying protons, neutrons, electrons, following octet rule in orbitals, placing electrons in pairs, etc.) and then make a flow diagram that would make the procedure clear.
9–12	Chemistry	Acids	Rank acids according to their corrosiveness.
9–12	Physics	Vector analysis	Given a large number of vectors, make a flow chart that would clearly express the sequence of steps in finding the resultant force. Consider how a change in the order of steps might affect your vector addition.

CHAPTER 9
FINDING REASONS AND CONCLUSIONS

Why is Accurately Finding a Person's Reasons and Conclusions Important?

Usually, we have many reasons for accepting the views that we hold or for doing the things that we do. I may conclude that I should buy a particular make of car because a friend has one, because I like a silver one that I saw, because an automobile magazine says that this make of car has a good repair record, or because an attractive person in a TV commercial told me to buy it.

A basic principle of critical thinking is that we should not accept a conclusion unless the reasons for the conclusion justify believing it. When I ask myself whether my reasons for buying the car are good ones, I should be prepared to withdraw my conviction that I should buy the car if I realize that my reasons are not good ones. Unless I can make explicit what my reasons are and state my conclusion clearly, I will be hampered in making these important judgments.

Analyzing and evaluating arguments. Other people sometimes try to convince us to act in certain ways or to adopt views that we may not currently hold. We sometimes try to convince others to adopt views that we hold. When people try to convince others by giving them *reasons* for accepting particular actions or views, they are offering an *argument*.

If we accept an idea or choice that someone else tries to persuade us to accept, we should likewise make sure that the reasons they offer justify the conclusion. When a salesperson tells me about the good features of a car to convince me to buy it, he or she is giving me an argument for buying. Before I become convinced that I should buy, I should make sure that the reasons justify the conclusion.

Arguments occur in a multitude of contexts, including political speeches, letters to the editor, and advertising. The same standards of critical thinking apply to all of these different forms of argument. The conclusion of an argument should not be accepted unless the reasons offered in support of it justify believing it.

Just as in our own case, to be able to decide whether the reasons others give us are good ones, we must first clarify what those reasons are. This fundamental *analytical* skill involves first noting the *conclusion* or *main idea* that is advanced and then looking for the *reasons that are offered to support the idea*.

When we *evaluate* the reasons, we use standards of critical thinking that enable us to distinguish good reasons from weak ones. If *there is some doubt about the accuracy of the reasons* or we're not sure whether *there is additional information available that counts against the conclusion*, then the reasons are weak. They are not, alone, sufficient for accepting the conclusion.

For example, an automobile salesman may try to convince me to buy an American-made car by pointing out that it costs less than any of the comparable foreign imports. Am I sure that this information is accurate? Are there any hidden costs? Even if the salesman is right about the cost, I have not yet investigated either the repair record or the cost of parts. I may still find that the car has a bad repair record and that spare parts are hard to get. This additional information may convince me that I should not buy the car, even though the initial information about its price is correct. In this case, the fact that the car costs less than any of the comparable foreign imports is not, in itself, a good enough reason for buying.

The type of analysis involved in *finding reasons and conclusions* is usually not difficult. We are told by reliable sources that eating lots of foods high in cholesterol can cause heart attacks. This fact is often used to try to convince us to change our dietary habits, specifically that we should not eat a lot of foods high in cholesterol. It is not difficult to recognize that "Eating foods high in cholesterol can cause heart attacks" is the reason being offered for not eating such foods. Not eating such foods is the conclusion proponents want us to accept.

There is a second, but unstated, reason in this argument: Heart attacks can kill and should be avoided. It is important to make unstated reasons such as this explicit when analyzing and evaluating arguments. When this unstated reason is revealed, it is easy to grasp why we find this argument compelling. The combined reasons clearly support the conclusion.

We can determine whether there is an unstated reason in this argument, and what it is, by asking whether there is an idea that the author believes we all accept. He accepts that heart attacks should be avoided, so he need not say it.

Difficulties in finding reasons and conclusions. It is not always easy to determine a person's reasons for his or her claims. In the cholesterol example, we are tipped off to the reason being offered by the word "because." Additionally, both the conclusion and his reason for it are stated clearly. Sometimes, neither the conclusion nor the reason are present.

Fiery speeches, dramatic rhetoric, the use of compelling visual images, and other appeals to our emotions often obscure the substantive reasons, if indeed there are any, that are offered to convince us of ideas or choices. Often, the devices used by advertisers distract us from finding good reasons to buy their products. Images and rhetoric, not reasons, may convince us.

Three common circumstances can lead to our accepting ideas without determining whether we can find reasons for doing so. The first is that many people do not think to ask for reasons. An idea may seem like a good one because of the way it is presented, but no reasons may actually be offered. If we do not seek reasons, we may base our acceptance on the idea's presentation, not its support.

The second circumstance is that we sometimes mistake appeal for reasons. The passion of a speech or the drama of a visual image in advertising may be enough to convince us to accept an idea. These devices are usually not offered as reasons but are, instead, used to make the idea seem appealing in and of itself.

Finally, we may take only stated reasons as the reasons for accepting an idea and not look for ones that are unstated. The unstated reasons may be ideas that are problematic. Unless we search for them, we can't gauge their credibility.

Difficulties in evaluating whether reasons are good reasons for a conclusion. Often, people are convinced by bad arguments because they hear some reasons supporting the conclusion and do not try to find out whether other unstated information counts against the conclusion. This happens frequently when people respond to advertising. You may buy a shampoo because you read in an ad that it will make your hair softer. Indeed, you find out that it does. But it also colors your hair, and you do not want that to happen. Unless you try to find out whether there are reasons against the conclusion and find that there are none, it is a mistake to take the reasons offered as conclusive.

There is a more subtle version of this problem in evaluating arguments. Sometimes, we find out whether there are reasons against a conclusion but do a hasty and superficial search. For example, I may ask the salesman for reasons to not buy the shampoo, and he may say that there aren't any. Thoroughness in a search for counter-reasons, however, is important. We should consult independent sources, such as *Consumer Reports* or someone who has used this type of shampoo, before being convinced of the absence of counter-reasons.

These circumstances illustrate the main problems in recognizing the conclusions that others try to convince us to accept and in identifying and evaluating the reasons that they give to support those conclusions. Figure 9.1 summarizes these problems.

**PROBLEMS IN FINDING
REASONS AND CONCLUSIONS**

1. We may not seek reasons for an idea someone is trying to convince us to accept.

2. We may mistake emotions or visual images for reasons.

3. We may not search for unstated reasons.

4. We may only consider the reasons in favor of a conclusion and not search for reasons against the conclusion.

5. We may not do a thorough search for reasons for and against a conclusion.

Figure 9.1

What Is Involved When We Skillfully Detect Reasons and Conclusions?

Breaking out arguments. How can we develop our analytical abilities so that we can cut through rhetoric, images, and imprecision to the heart of what others try to communicate? How can we determine clearly what ideas others are trying to convince us to accept and what reasons they give us for accepting these ideas?

We "break out an argument" when we determine a person's conclusions or main ideas and the reasons offered for these conclusions. It is always helpful to try to articulate both the conclusions and reasons as clearly as possible, detaching them from a particular context in which they appear. This is commonly done by stating the reason(s) first, then stating the conclusion using words or symbols that indicate this relationship. Figure 9.2 represents a possible argument about heart attacks and cholesterol.

ARGUMENT

(1) Eating a lot of foods high in cholesterol can cause heart attacks.

(2) Heart attacks should be avoided.

Therefore

(3) We should not eat a lot of foods high in cholesterol.

Figure 9.2

Sometimes "therefore" is replaced by three dots (∴). The reasons come before the "therefore"; the conclusion comes after it. While the reasons in an argument are often called its premises, we prefer to use the term "reasons."

Considering opposing viewpoints. One of the important dispositions of critical thinking is open-mindedness. This involves a willingness to consider all sides of an issue before making up one's mind. Often, people who are convinced about something are not willing to listen to other viewpoints or change their minds. We know, however, that sometimes new information or a fresh perspective can reveal that even firmly held views may need modification. Open-mindedness counters such narrow thinking.

Seeking and assessing arguments on both sides of an issue, whether or not we are disposed towards one viewpoint or another, fosters open-mindedness. Breaking out arguments and setting them side-by-side often helps to clarify an issue and paves the way for considering whether the argument on the other side has merit.

A graphic organizer and thinking map for clarifying reasons and conclusions. The graphic organizer in figure 9.3 depicts an argument. It illustrates the idea that conclusions should be supported by reasons.

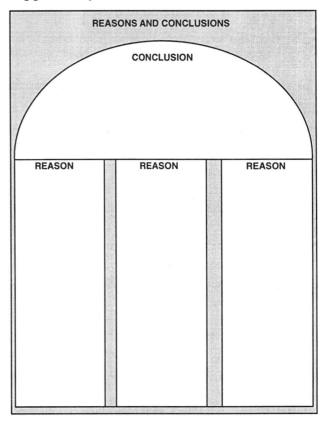

Figure 9.3

The number of reason "pillars" in the graphic organizer can vary, depending on the number of reasons offered in support of the conclusion. Both stated and unstated reasons can be written on the pillars.

Note that this graphic organizer is used for one argument only. If there are multiple arguments in a passage, a number of these diagrams should be used. To depict opposing viewpoints, two of these diagrams can be used, each with a contrasting conclusion.

The thinking map in figure 9.4 prompts the reasons and conclusions for the diagram.

FINDING REASONS AND CONCLUSIONS

1. What is the author trying to convince us to accept or do?

2. What reasons does the author provide to support accepting or doing that?

   a. Are there any words that indicate support (e.g., "therefore," "so," "because")?

   b. Does the author provide any other indication as to why he or she concludes what he or she does?

3. Is there anything that you think the author believes is common knowledge that he or she does not state but uses to support the conclusion?

Figure 9.4

This thinking map makes it clear that we should consider only the reasons that a specific person gives for a judgment or conclusion. We should not try to provide our own reasons. Our reasons may not be ones that the person accepts.

It is not difficult to identify reasons and conclusions skillfully. When someone offers reasons for an idea or action, that person often uses words like "therefore," "so," "because," etc. to indicate which statements are reasons and which are conclusions. When such signal words are not present, we can think about where we would put a "therefore" or a "because." Asking "Why does he think that?" can also lead us to the speaker's reasons.

Assessing whether reasons support accepting a conclusion. Finding a person's reasons for his or her conclusions is an important first step in thinking critically about whether or not a person has given good reasons to accept an idea or recommendation. To make this judgment skillfully, however, requires more thought than simply clarifying the reasons and conclusions. We must now use our critical thinking to assess whether the reasons are sufficient to support accepting the conclusion.

One way to make this assessment is to ask whether we need any other information beyond the reasons we've uncovered before we are willing to accept the conclusion. If we do, the given reasons may not be sufficient to justify the conclusion. If not, the reasons may be adequate.

In general, three basic standards should be applied to determine whether the conclusion is acceptable based on the reasons offered:

- The information stated as reasons is accurate.
- The reasons count strongly in favor of the conclusion.
- No other significant information counts against the particular conclusion.

For example, suppose a salesman tells me that I should purchase a car because it gets good gas mileage, does not cost a lot, and has the best repair record of all foreign cars. Because these are major concerns of mine, his reasons initially present a compelling case. I need much more information before I should be convinced to buy the car based on his reasons.

First, I should confirm that the information the salesman has provided is accurate. Sometimes, salespeople distort the truth to sell a product. Is that happening in this case? Maybe it is; maybe it isn't. Having the best repair record is a good reason for purchasing a car only if that claim is accurate. I may want to confirm what the salesman says by looking at another source like *Consumer Reports*.

Even if the information supporting an action or conclusion has proven to be accurate, the given reasons still may not be sufficient to support purchasing the car. Other factors may count against buying this particular car and override the benefits. I may find out that the car is not easily repaired and that spare parts are costly. I should not expect that a salesperson is going to tell me about these negative considerations. Until I have good reason for thinking that there is no overriding information, the salesperson's argument should be treated as inconclusive.

When a thorough and balanced argument lays out accurate reasons for and against a proposition *and* the reasons for accepting it are more compelling than the reasons against it, no additional information may be necessary. In this case, the conclusion to buy the car would be justified by the reasons.

In most cases, we more or less informally apply the standards just described for judging the viability of arguments. In some cases, however, our thinking should be more rigorous. For example, arguments about guilt or innocence in criminal matters often require more careful analysis and critical judgment. Such cases require careful evaluation involving critical thinking skills such as causal explanation or determining the reliability of sources of information.

The graphic organizer in figure 9.5 can be used to record the results of this inquiry into the soundness of an argument.

```
╭────────────────────────────────────────────────────────╮
│            ARGUMENT EVALUATION CHECKLIST                 │
│                                                          │
│  1. Is there anything you need to find out in order to   │
│     determine whether the reasons are accurate?          │
│                                          YES☐  NO☐       │
│                                                          │
│  2. If so, what information do you need to find out?      │
│     _____  │
│     _____  │
│     _____  │
│                                                          │
│  3. Given that the reasons are accurate, is additional   │
│     information needed before you can accept the         │
│     conclusion?                          YES☐  NO☐       │
│                                                          │
│  4. If so, what information do you need?                  │
│     _____  │
│     _____  │
│     _____  │
│                                                          │
│  An argument should be convincing only if you answer     │
│  "NO" to questions 1 and 3 above.                        │
╰────────────────────────────────────────────────────────╯
```

Figure 9.5

Using the argument evaluation checklist. After you have recorded reasons and conclusions in figure 9.3, you can analyze the argument using the checklist in figure 9.5. Record your response to whether you have enough information to determine that the reasons are accurate. If you do, check the "Yes" box; if not, check the "No" box and list the information you need to make this judgment. In cases in which more rigor is needed to determine the accuracy of the information, you may have to determine the reliability of the sources of this information more carefully. The strategies in Chapter 13: Determining the Reliability of Sources may be used to supplement this part of the argument evaluation checklist.

The second question on the checklist presupposes that the information given as reasons is accurate and asks you to consider whether you need any more information for the argument to be convincing. This question has to do with whether the reasons alone are sufficient to justify the conclusion. In the case of the auto purchase, for example, you would check the "Yes" box and write things like "Is this car easy to repair?" "What is its repair record?" "How do other owners rate this car?" etc.

If additional information is needed, either to show that the reasons are accurate or to supplement these reasons in order to justify the conclusion, then the given reason(s) *does not justify accepting the conclusion.* The reason(s) given in this case is not a good reason(s) for accepting the conclusion, and the argument is not viable.

There are limitations to using the evaluation checklist, however. Answering the questions on the checklist will enable us to make a judgment only about whether a specific argument provides good reasons to justify its conclusion.

If I determine, for example, that the argument offered by the car salesman is not good enough to support buying this particular car, this does *not* show that there are not other, better reasons for buying the car than those given by the salesman. There may well be. Indeed, using the checklist may help me find them. When I write the additional information that I need to be convinced of the conclusion, this can guide me in conducting further research about the car. When I do this research, I may find, for example, that people who own this type of car give it very high praise, that it has a good repair record, that it gets good mileage, etc. As I accumulate this additional information, I may be able to develop a much better argument for buying the car than the salesman gave me. If so, then when I use the argument evaluation checklist to assess this argument, it should show up as a good argument.

To summarize, using the argument evaluation checklist helps us clarify whether or not we should acquire additional information before we accept the conclusion. As we answer the questions raised in using the checklist, we may get information that can be used to construct another, more convincing argument.

Fallacies of reasoning. An alternative strategy for determining whether conclusions are well-supported by given reasons involves recognizing patterns of faulty ("fallacious") reasoning. There are specific types of fallacious arguments that occur with such frequency that they have been categorized for easy recognition. These patterns of faulty reasoning are usually called *fallacies*.

Once an argument is examined using the strategy for finding reasons and conclusions, it may be easy to identify it as an incorrect argument if it represents one of these patterns. For example, people often argue that something happens *because* of something else only on the basis of the fact that it occurred *after* that event. This pattern of argument is fallacious because it could be a coincidence that these two events occurred one after the other. The traditional name for this pattern of incorrect reasoning is the *post hoc, ergo propter hoc* ("after this, therefore because of this") fallacy. If someone offers an argument and you recognize it as an argument of this type, you can conclude that it is incorrect: the reasons are too weak to establish the conclusion.

Identifying informal fallacies is a technique used in many critical thinking textbooks, as well as in some language arts materials. Although learning to identify these patterns of faulty reasoning may be helpful in identifying some fallacious arguments, many other fallacious arguments do not fit these patterns.

In contrast to recognizing fallacies, using the argument evaluation checklist guides us in assessing whether *any* argument contains reasons supporting the conclusion. This strategy is employed in the lessons in this handbook because it is versatile, does not require technical language, and is easier for students to comprehend.

How Should We Teach Students to Detect Reasons and Conclusions Skillfully?

Giving students practice in analyzing arguments offered in the content they are studying and helping them to guide themselves in this process is a key to students' internalizing the strategy for finding reasons and conclusions. It is important that students extract arguments from natural contexts in which these arguments appear and restate them in a clearer form. Break-

ing out an argument, as represented in figure 9.2, makes an argument explicit, facilitating assessment of whether it is a good argument.

Contexts in the curriculum for finding reasons and conclusions. The curriculum offers many contexts for analyzing reasons and conclusions. Reasons for doing things are offered from character to character in fiction. Proponents of various points of view throughout history have offered reasons supporting their points of view, and reasons are frequently offered in subjects like writing for preferring one style of writing to another in certain contexts.

Offering reasons for conclusions is not just one of the things done occasionally in science and mathematics. Rather, there is a sense in which offering reasons in support of conclusions is what the activity of science is all about. Scientific principles, to be acceptable, must be supported by empirical evidence. Citing the evidence that supports a principle or judgment is what is usually offered as reasons for accepting that principle or judgment.

In general, infusion lessons on reasons and conclusions in science can be developed in three types of contexts:

- *Arguments are presented in textbooks for certain views or courses of action.* For example, in drug education programs (sometimes integrated into middle school science programs or ninth-grade biology courses), reasons are given to avoid substance abuse. It is important to help students to understand these reasons and assess their significance rather than just accept the conclusions. Similarly, when students study controversial topics dealing with the environment, such as genetic engineering, hazardous waste, nuclear power, or clear-cutting the rain forests, textbooks often include arguments on both sides of the issues. If not, you might find supplementary material in which such arguments are contained.

- *Arguments are offered by historical figures in support of certain points of view.* For example, Galileo initially offered a number of arguments for abandoning the geocentric view of the universe; Darwin, and other supporters of evolutionary theory like Thomas

Huxley, offered arguments in its favor, and Einstein published some books (that are quite accessible to high school students studying physics) in favor of the theory of relativity and in opposition to Newtonian mechanics. Such arguments are ripe for analysis, especially since opponents usually offer counter-arguments that can also be analyzed.

- *Students are asked to develop their own arguments.* For example, students are often asked in science classes to defend the conclusions they draw in their laboratory work or in their research. In these cases, the standards of good support are those inherent in scientific methodology: conclusions should be accepted only on the basis of observational data that is strong enough to support them. Asking students to write out their arguments provides raw material for a self-assessment lesson on reasons and conclusions (or other students can critique their arguments). This also helps students to learn to write persuasive prose, especially in science. They discover that being clear about their reasons for their conclusions can be a valuable asset. Similarly, debating skills can also be improved when other students analyze their arguments.

A menu of suggested contexts for infusion lessons on finding reasons and conclusions can be found at the end of the chapter.

Tips for Teaching Lessons on Reasons and Conclusions

Lessons that include arguments on both sides of an issue provide richer experiences for students than analyzing arguments on one side of the case. A variety of arguments should be considered on major issues so that students can develop a sense of the variety of reasons, some weak, some not, that people have offered to support their ideas.

It is especially important to help students apply this skill to school-related extracurricular activities and in other experiences outside of school. Helping them to analyze arguments provided in advertising, in political speeches, and on moral and social issues will put students in a much better position to assess these arguments.

Model Lessons on Finding Reasons and Conclusions

One lesson on reasons and conclusions is included in this handbook. It is a lesson in which students analyze and assess two opposing arguments about the dissection of animals in the science classroom. It is appropriate for middle and high school science students and provides a model on the basis of which similar lessons can be constructed in mathematics when supporting calculations are difficult or tricky.

As you read each lesson, consider the following questions:

- How does the graphic organizer in figure 9.3 help students become aware of the structure of an argument?
- How does using that diagram help students to clarify major and minor differences in two opposing viewpoints?
- How can the graphic organizer be used in the writing extension to help students outline arguments they will present?
- What are some other examples that can be used to transfer and reinforce this skill?

Tools for Designing Reasons and Conclusions Lessons

The thinking map for finding reasons and conclusions can be used to help students break out any argument they are considering, whether it occurs in written prose, orally, or it involves the use of visual images, like in advertising. The graphic organizer is similarly versatile.

The thinking map and graphic organizers can guide you in designing the critical thinking activity in the lesson and can also serve as photocopy masters, transparency masters, or as models that can be enlarged and used as posters in the classroom. Reproduction rights are granted for use in single classrooms only.

FINDING REASONS AND CONCLUSIONS

1. **What is the author trying to convince us to believe or do?**

2. **What reasons does the author provide to support accepting or doing that?**

   a. **Are there any words that indicate support (e.g., "therefore," "so," "because")?**

   b. **Does the author provide any other indication as to why he or she concludes what he or she does?**

3. **Is there anything that you think the author believes is common knowledge that he or she does not state but uses to support the conclusion?**

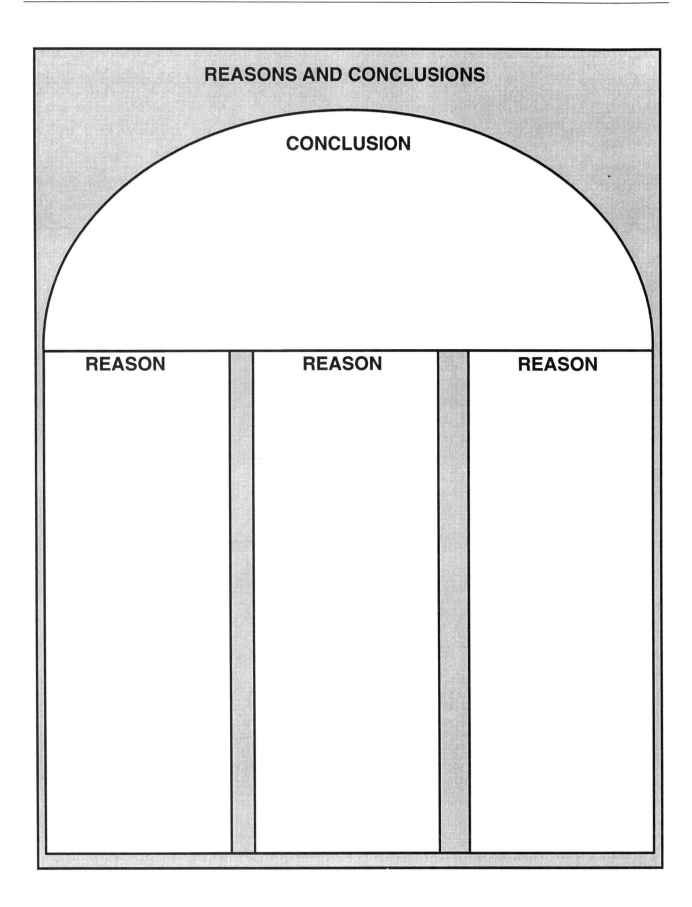

ARGUMENT EVALUATION CHECKLIST

1. Is there anything you need to find out in order to determine whether the reasons are accurate?

YES ☐ NO ☐

_____

2. If so, what information do you need to find out?

_____
_____
_____

3. Given that the reasons are accurate, is additional information needed before you can accept the conclusion?

YES ☐ NO ☐

_____

4. If so, what information do you need?

_____
_____
_____

An argument should be convincing only if you answer "NO" to questions 1 and 3 above.

TO DISSECT OR NOT TO DISSECT?

General Science, Biology, Anatomy, Physiology   **Grades 8–12**

OBJECTIVES

CONTENT

Students will learn reasons that opponents of dissecting animals in school classrooms offer for their position and reasons that proponents of animal dissection offer in support of their position.

THINKING SKILL/PROCESS

Students will develop skill at detecting conclusions and the supporting reasons offered for these conclusions.

METHODS AND MATERIALS

CONTENT

Students will read passages from *Objecting to Dissection*, by the National Anti-Vivisection Society, and from "What's Wrong With 'Animal Rights'" by Adrian Morrison. They will work together to reconstruct the arguments offered for and against dissection.

THINKING SKILL/PROCESS

Detecting reasons and conclusions is guided by structured questioning and by using a graphic organizer that highlights the conclusion and supporting reasons for an argument.

LESSON

INTRODUCTION TO CONTENT AND THINKING SKILL/PROCESS

- Has anyone ever tried to convince you of something important by giving you reasons why you should accept what he was saying? Write down two situations in which this has happened to you. Ask students for a few examples. Ask them to state exactly what the other person tried to convince them of and what reasons were offered.

- What the person was trying to convince you to believe or do was a *conclusion* he arrived at based on the *reasons* he gave you. The reasons together with the conclusion make up an *argument* he offered to convince you to accept his position. Use one of the situations you just wrote down; state the reasons and the conclusion, and write them on the reasons and conclusions diagram. See the chart at the right.

REASONS AND CONCLUSIONS

CONCLUSION

REASON   REASON   REASON

- Sometimes, people don't state some of their reasons because they think we already know them. For example, if someone says to you that you shouldn't eat a lot of sugar because it can make you gain weight, that person is presupposing that gaining weight isn't good for you. It is important to make these hidden reasons explicit and to write them on your diagram. Have you found hidden reasons in the situation you were working on? Discuss this with your partner. Add them to your diagram.

- Look at your diagram and think about the argument you see there. Discuss with your partner whether the reasons you have detected are good reasons for the conclusion. If so, why? If not, why not? Ask for two or three assessments of the reconstructed arguments.

- Biology is the study of life. Ever since this science first developed, biologists have learned a great deal from observing living things and their parts. To observe the parts of living things, the standard practice has been the dissection of such things after they have died. This practice has become a staple in most biology courses as well. Biology instructors have, for a long time, had good reasons for having students dissect animals—usually frogs, fetal pigs, and other small animals. Students have usually taken it for granted that their teachers were right. Did you know, however, that many people object to the practice of dissection and think *they* have plenty of good reasons for abandoning this practice? In fact, many people feel so strongly about this issue that they have engaged in highly emotional protests against dissection. To help us decide for ourselves which position to take, however, we should resist these emotional appeals and try to determine which side offers the best reasons in support of their position. To do this, we should determine what reasons are given on both sides of the issue first and then evaluate whether any of these reasons are good reasons for accepting their conclusions. We're going to look at two pieces of writing: one in support of dissection and the other against. We'll then extract the arguments from each by determining their conclusions and the reasons they offer in support of their conclusions. We will use the "Reasons and Conclusions" graphic organizer as well as this thinking map of important questions to guide us through this kind of thinking.

> **FINDING REASONS AND CONCLUSIONS**
>
> 1. What is the author trying to convince us to believe or do?
>
> 2. What reasons does the author provide to support accepting or doing that?
>
>   a. Are there any words that indicate support (e.g., "therefore," "so," "because")?
>
>   b. Does the author provide any other indication as to why he or she concludes what he or she does?
>
> 3. Is there anything that you think the author believes is common knowledge that he or she does not state but uses to support the conclusion?

THINKING ACTIVELY

- Imagine that you are a biology teacher. In the past, you've taught biology from a standard biology textbook supplemented by lab work, part of which involved dissecting a frog. However, a newly published pamphlet contains a discussion of whether animals should be dissected in biology classes. You haven't thought much about this question, but you, like many other people, are interested in what this pamphlet has to say because you are now not sure how you feel about dissection. Many students have been squeamish about this practice, and some have, in fact, refused to engage in it, though you have always felt that it was a good learning experience. You've just received a free copy of the pamphlet and are looking at the following passage. You want to see whether it provides good reasons for not dissecting animals in school classrooms. The following is from the National Antivivisection Society (53 West Jackson Blvd., Suite 1552, Chicago, IL 60604. Reprinted with Permission):

> "When dissection was introduced into educational curriculum in the 1920s, it was thought to be a good learning tool in the study of anatomy, physiology, and the theory of evolution. Today, more sophisticated teaching methods have been developed which can replace dissection and save animals.

> But dissection is big business. Millions of animals a year are killed to be dissected for educational experiments. Many of these animals, such as frogs, earthworms, crayfish, and perch are collected from their natural environments. In the process, their habitats are decimated and entire ecologies are threatened. A recent investigation of certain biological supply houses showed that cats obtained for dissection suffered very painful and cruel deaths.

> Even more destructive is the desensitizing effect of mutilating and dismembering animals in the name of education. Somehow the study of a "life science"—meant to instill

wonderment and respect for life—has become a science of death. Dissection teaches students that animal life is expendable and unimportant. As a result, some of the best potential scientists, who have a deep respect for animal life, may end up dropping out of a field they love because they refuse to take part in senseless killing."

- **What, overall, is the main idea of which the authors of this pamphlet are trying to convince you? Work with your partner, formulate the conclusion they are trying to convince you to accept, and write it in the "Conclusions" arch of a blank "Reasons and Conclusions" diagram.** After students have had a few minutes to read and reflect on this passage, ask for some statements of the conclusion. POSSIBLE ANSWERS: *The authors of the pamphlet are trying to convince readers that we shouldn't dissect animals. We shouldn't dissect animals in science classes. Dissecting animals in science classes is wrong.* Write students' responses on the chalkboard.

- **Work with your partner once again, asking yourselves why the authors of the pamphlet think their conclusion is a good one. Look for signal words like "therefore" that indicate reasons are being offered. Formulate the reasons they give in your own words. Using one pillar for each paragraph, complete the diagram with the reasons that are offered against dissecting animals in science classes.** Students offer three sets of reasons:

    Paragraph One: *There are better teaching methods available today that teach the same thing as dissection but don't involve killing animals.*

    Paragraph Two: *The habitats of animals used for dissection are destroyed when these animals are collected and killed. Some animals themselves—for example, cats—suffer pain and die in cruel ways when they are obtained for dissection.*

    Paragraph Three: *Dissecting animals in school classrooms makes students think that animal life is not important and can be dispensed with. Because of this, some students who could become good scientists don't continue in science.*

    Ask for reports from student groups by asking them to share one reason they have found. Go paragraph by paragraph. As students suggest reasons contained in each paragraph, have the class try to reach a consensus on how to state them. Write the reasons extracted by the students on the three pillars of the class diagram.

- **Before deciding that an argument is convincing, it is important to consider arguments for the opposite position. Here's a selection written by Adrian R. Morrison, a proponent of dissection. It first appeared in an article in the American School Board Journal, Jan. 3, 1992, and is reprinted here with permission. As you read it, see if you can answer the question, "What is this person trying to convince us to believe?"**

    "Some teachers argue that students can study animal parts just as easily using charts, models, or computer graphics. Were that the sole reason for dissection, I could accept that "Classics Comics" approach to biology instead of the use of actual specimens. But dissection is a valuable experience for more fundamental reasons.

    First, it's simple, direct science. The only pieces of equipment required are the student's own hands, eyes, and brain. Nothing else intervenes between the students and their observations, a rarity in modern sciences.

    In addition, from their observations, students learn that the statements in their textbooks—and the images on their computer screens—are not true or absolute but can only approximate reality because individual specimens differ.

    Finally, and most important, a student can better understand evolution—one of science's pivotal concepts—by comparing the anatomy of different species and observing how similar structures, such as arms and wings, are adapted to meet different environmental needs."

Ask students for their ideas about what the author is trying to convince us to do or to believe. Write each idea on the chalkboard. Clarify the author's conclusion the same way that you clarified the conclusion advanced by opponents of dissection. POSSIBLE ANSWERS: *Dissection is a valuable experience for students in biology classrooms. We should continue to dissect animals.*

• **Write this author's conclusion on a blank "Reasons and Conclusions" diagram. Think about the reasons he offers to convince us of his conclusion. Work with your partner and formulate the reasons in your own words. Ask what, in each paragraph, explains why he thinks we should continue to dissect animals in biology classrooms. Examine each paragraph.**

> Paragraph One: *Students can do science by direct observation when they dissect animals; they can't do direct science simply by using charts, models, or computer graphics of animals and their parts.*

> Paragraph Two: *Students can learn about differences between individual specimens of the same animal through dissection; they can't learn these things through their text books and computer graphics of animals and their parts.*

> Paragraph Three: *Students can better understand evolution through dissecting animals from different species and comparing how their parts are adapted to their environment.*

Discuss each paragraph with the class and try to reach consensus on how to state the author's reasons. Ask the students to write the reasons on their diagram.

• **What signal words did you find to indicate that a reason is being offered?** *"For more fundamental reasons" in paragraph one.* **Can you think of other signal words that might indicate that a reason is being offered for a conclusion?** *"Because," "so," and "therefore."*

• **You've now analyzed the passages and extracted the two arguments. They are each depicted on your two diagrams. Often, when people hear arguments on two sides of an issue like this, they both seem like good arguments. Then it is very easy for colorful rhetoric or the use of emotions to become the deciding factor. However, there are ways of thinking about arguments like these and evaluating whether they are good arguments. If an argument is a good one, independent of emotions and rhetoric, the information in the reasons has to be *accurate* and *adequate to support the conclusion*. If the information offered as reasons turns out to be false, we shouldn't accept the conclusion. If it turns out to be accurate and true, it still may not be strong enough to support the conclusion. Therefore, a person should determine whether it is necessary to find out any additional information in order to feel confident that the conclusion is acceptable. We're going to use an "Argument Evaluation Checklist" to write down our thoughts about whether each argument is a good argument based on this strategy. You are asked two types of questions on the checklist. First, do you have any questions you need to have answered before you believe *the reasons are accurate*? Second, even if the reasons are accurate, is there anything else you would need to know before you can *accept the conclusion based on the reasons*? Half of the class will work on the argument against dissection, and half of the class on the**

ARGUMENT EVALUATION CHECKLIST
1. Is there anything you need to find out in order to determine whether the reasons are accurate?   YES ☐ NO ☐
2. If so, what information do you need to find out? _____ _____ _____
3. Given that the reasons are accurate, is additional information needed before you can accept the conclusion?   YES ☐ NO ☐
4. If so, what information do you need? _____ _____ _____
An argument should be convincing only if you answer "NO" to questions 1 and 3 above.

argument in favor of dissection. Work in groups of four now and try to come up with as many questions as you can. What must be answered before you can make a decision about the argument?

POSSIBLE ANSWERS:

The argument against dissection: *Is it true that cats prepared for dissection felt a lot of pain and that those preparing them were cruel? Maybe the author is just saying this to make it seem like dissection causes pain and suffering for animals. And even if they do now, do they have to suffer? And is it true that students learn that life is expendable and unimportant? How did he find out? Did he ask students? Even if all these things are true, could there be other more important benefits of dissection that he doesn't mention?*

The argument in favor of dissection: *Is it true that we can learn what we are supposed to from dissection without instruments like microscopes? Can't individual differences between specimens be illustrated in textbooks and computer graphics? Can't evolutionary differences as well? Maybe he's exaggerating the difference to convince us. And even if these things are true, maybe there are other consequences of dissection that count against it that are more important than these. The author of this passage doesn't mention any negative effects of dissection. Are there any?*

Based on these argument assessments the students realize that neither of these arguments is a good one. Help them to understand that this does not mean that the conclusions of each are false. All it means is that these particular arguments are too weak to support the conclusions they purport to uphold. This still leaves open the question of whether a better argument could be produced that would adequately support either of these conclusions.

- **For further class work, try to answer these questions about the argument you worked on. Use the variety of source material about dissection in science that I have provided in the classroom. Then, for homework, develop an argument based on this research that you think is a better argument for one or the other of these positions. Use the "Argument Evaluation Checklist" to double check whether it is a good argument. Come in tomorrow with your arguments and be prepared to explain why you think your argument is a good one.** ANSWERS VARY. Ask four or five students who are in favor of dissection and four or five who oppose dissection to share their arguments with the rest of the class.

THINKING ABOUT THINKING

- **What were some questions you asked that helped you figure out what the conclusions were in these arguments?** POSSIBLE ANSWERS: *What did the author want to convince me of? What was the point of what he was saying? What was his purpose in telling us all of these things?*

- **To help you determine what conclusions and reasons people are offering, plan a strategy for thinking about things that people tell you.** Answers should resemble the steps on the thinking map for reasons and conclusions.

- **What do you call the result of this kind of thinking?** *An argument.*

- **Once you have extracted an author's argument from something he or she has said, what kinds of questions would you ask?** POSSIBLE ANSWERS: *Is this a good argument? Are the reasons correct? Is there anything else that I should know before I accept the conclusion?*

APPLYING THINKING

Immediate Transfer

- **Analyze and evaluate an argument that one of your classmates develops as homework for this lesson using the strategies you learned in the lesson and the accompanying graphic organizers.** To make this activity less personal, photocopy students' arguments, code them, and distribute them to other members of the class for their evaluation and analysis. Display the ones that class members evaluate as good arguments.

- **Locate an advertisement in a magazine. Determine what the authors of the advertisement are trying to convince us to do or to believe and what reasons they offer. Use the diagram for reasons and conclusions and state their argument. Think about whether it is a good argument.**

Reinforcement Later

- **Arguments about the environment—e.g., about whether or not we should restrict logging in the Pacific Northwest to save the Spotted Owl—appear frequently in the popular press and in magazines like** *Time*. **Select a passage that contains such an argument. Then analyze and evaluate it using the strategies you learned in this lesson. Finally, find an argument on the other side and do the same.**

- **In American History, you have studied how we elect presidents and members of Congress. Find a political speech made by someone running for political office and extract any arguments you can find in it. Then think about whether or not these are good arguments.**

WRITING EXTENSION

Write a letter to the head of the biology (or science) department in your school offering an argument you think is a good one in support of or against dissection in biology courses.

ASSESSING STUDENT THINKING ABOUT REASONS AND CONCLUSIONS

Any of the transfer examples can be used for assessment tasks. The examples in which students analyze political speeches and advertising are especially suited. However, you may wish to select other controversial arguments in science, for example, environmental issues, nuclear power, or genetic engineering. Ask students to extract the reasons and conclusions to reconstruct the author's arguments. Ask students to explain why they think that they have accurately stated the author's reasons and conclusions. Use the thinking map for finding reasons and conclusions to help you determine whether your students are attending to relevant matters in determining what these arguments are.

You may also want to ask your students to explore questions they would ask to decide whether or not this is a good argument. Use the lesson responses as examples by which to judge whether or not your students are asking relevant questions.

Sample Student Responses • To Dissect or Not To Dissect

REASONS AND CONCLUSIONS

CONCLUSION

We shouldn't dissect animals in science classes.

REASON

Other teaching methods teach the same thing as dissection but don't involve killing animals.

REASON

The habitats of animals used for dissection are destroyed when these animals are collected and killed. Some animals themselves -- for example, cats -- suffer pain and die in cruel ways when they are obtained for dissection.

REASON

Dissecting animals in school classrooms makes students think that animal life is not important and can be dispensed with. Because of this some students who could become good scientists don't continue in science.

Sample Student Responses • To Dissect or Not To Dissect

REASONS AND CONCLUSIONS

CONCLUSION

We should continue to dissect animals in science classes.

REASON

Students can do science by direct observation when they dissect animals; they cannot do direct science simply by using charts, models, or computer graphics of animals and their parts.

REASON

Students can learn about differences between individual specimens of the same animal through dissection, but they cannot learn through their textbooks and computer graphics of animals and their parts.

REASON

Students can better understand evolution through dissecting animals from different species and comparing how their parts are adapted to their environment.

Sample Student Responses • To Dissect or Not To Dissect? • Argument Evaluation

ARGUMENT EVALUATION CHECKLIST
THE ARGUMENT BEING CONSIDERED AGAINST DISSECTION

1. Is any additional information needed to determine whether the reasons are accurate? YES [X] NO [ ]

2. If so, what information is needed?

Is it true that cats prepared for dissection felt a lot of pain and that those preparing them were cruel? (Maybe the author is just saying this to make it seem like dissection causes pain and suffering for animals.) Even if they suffer now, do they have to? Also, is it true that students learn that animals are expendable and unimportant? How did he find out? Did he ask students?

3. Given that the reasons are accurate, is additional information needed to make the argument convincing? YES [X] NO [ ]

4. If so, what information is needed?

Even if all the information given is true, could there be other more important benefits of dissection that he doesn't mention? Are there things that students can learn from dissection that they could not learn in any other way and that are so important that it doesn't matter if some habitats are disturbed and some animals feel pain? Maybe there aren't, but I'd like to hear whether anyone thinks there are before this argument convinces me that we should abandon dissection.

An argument should be convincing only if no additional information is needed.

This argument is not convincing.

ARGUMENT EVALUATION CHECKLIST
THE ARGUMENT BEING CONSIDERED IN FAVOR OF DISSECTION

1. Is any additional information needed to determine whether the reasons are accurate? YES [X] NO [ ]

2. If so, what information is needed?

Is it true that we can learn what we are supposed to from dissection without instruments like microscopes? Can't individual differences between specimens be illustrated in textbooks and computer graphics? Can't evolutionary differences be as well? (Maybe he's exaggerating the difference between dissection and other ways of learning about the parts of animals in order to convince us.)

3. Given that the reasons are accurate, is additional information needed to make the argument convincing? YES [X] NO [ ]

4. If so, what information is needed?

Even if the information offered is true, maybe there are other consequences of dissection that count against it that are more important than these. The author of this passage does not menton any negative effects of dissection. Are there any? (For example, in the argument against dissection that I read the author said that animals collected for dissection suffered. Do they? What would he say about this? Why is this not more important than anything we could learn by dissection?

An argument should be convincing only if no additional information is needed.

This argument is not convincing.

FINDING REASONS AND CONCLUSIONS LESSON CONTEXTS

The following examples have been suggested by classroom teachers as contexts to develop infused lessons. If a skill or process has been introduced in a previous infused lesson, these contexts may be used to reinforce it.

GRADE	SUBJECT	TOPIC	THINKING ISSUE
6–8	Science	Energy	People in this country disagree about which energy source is the best for the country to rely upon. Analyze two of the arguments that have been presented on each side in this debate. What unanswered questions do you have before you accept or reject these arguments?
6–8	Science	Electricity	Ask students to suggest reasons why fuses and circuit breakers are connected to a circuit in series, whereas appliances are connected in parallel.
6–8	Science	Respiration	What reasons do people have for warming up before strenuous exercise? Are any of these good reasons? Why?
6–8	Science	Pollution	What are some of the reasons for and against having a landfill near homes? Are any of these good reasons? Why?
6–8	Science	Heredity	After Mendel had sprinkled pollen on the pistil of a pea plant, he covered it so that no other pollen could reach it by insects or wind. If you had asked him why he thought this was necessary, what reasons would he give? Was he right? Why?
6–8	Science	Heredity	Scientists studying genetics use fruit flies as subjects of their experiments. What reasons do they have for using flies instead of dogs, cats, or horses? Are these good reasons?
6–8	Science	Digestion	During digestion, starch is broken down into sugars. People who have diabetes suffer from too much sugar in their blood and must restrict how much sugar they consume. If a diabetic drew the conclusion that he or she must also restrict the intake of starches, what reason would people likely give? Is that a good reason? Why?
9–12	Biology	Diet	Consult an advertisment for dietary vitamin supplements. What are the reasons offered for taking these dietary supplements? Are these convincing reasons? Develop the most convincing argument you can for not taking these dietary supplements.
9–12	Biology	Genetic engineering	Many people have argued that genetic engineering is an extremely important practice that should be fully supported. Others have argued that genetic engineering should be banned. Research genetic engineering in the school library and develop the strongest argument you can for genetic engineering and the strongest argument you can against genetic engineering. Evaluate each argument.
9–12	Biology	AIDS	Suppose you were developing an AIDS education program. What reasons would you give to your students for becoming concerned about AIDS? What sorts of changes in behavior would you recommend?
9–12	Chemistry	Fluorides	Fluoride has been put into drinking water because it helps in preventing tooth decay by combining with the enamel of the teeth as they are being formed. Break out and evaluate this argument for fluoridation.
9–12	Chemistry	Sulfur	What arguments have been offered in favor of curbing sulfur emissions from industrial smokestacks? Do you have any questions about these arguments?

CHAPTER 10
UNCOVERING ASSUMPTIONS

Why is it Necessary to Attend to Assumptions Skillfully?

When we make choices, accept information from others, or draw conclusions, we often take certain ideas for granted. When I get in my car to drive to work, I assume that the road hasn't developed huge cracks that would make it impassable. We can't think of every possible contingency and check it out.

Fortunately, most of the time we don't have to check out our assumptions. Many of the assumptions we make are perfectly reasonable. Before taking a certain route to work, I have good reason for thinking that the road hasn't developed huge cracks. It was in good condition last night, and there have been no unusual occurrences since then, like an earthquake. Our ability to assume things and act on these assumptions without bringing them to mind can make our thinking and acting very efficient.

Problems in Not Attending to Or Questioning Our Assumptions

Relying on assumptions has advantages and disadvantages. We sometimes make assumptions that we either don't have reason to believe, we have reason to believe are false, or we can check out easily and discover to be false. Because we are often unaware of our assumptions, acting on them can sometimes lead to irreparable mistakes. If I decide to make a lengthy trip to a department store to buy a television set that they have advertised, I may be assuming that they haven't sold out of them. I may not be aware of what I am assuming until I get there and find, to my dismay, that they don't have any left. If I had been aware of this assumption beforehand, it may have prompted me to call the store to determine whether I could assume that the sets would be available when I got there.

We may be aware of what we are assuming but not consider whether our assumptions are justified. I may think that a friend whom I have lost touch with still lives in the same location. I

may not think to question this. When I go to visit him, however, I find that he has moved. Then, I realize that I had made an assumption. That he lived there five years ago was not a good reason for thinking that he would be there now.

The two main problems with assumptions that sometimes lead us to do things that are in error are recorded in figure 10.1. A skillful approach to examining assumptions involves asking and attending carefully to both.

COMMON PROBLEMS WITH ASSUMPTIONS
1. We often don't ask, "Am I taking anything for granted?"
2. Sometimes we are aware of assumptions that we make but don't try to find out whether they are reasonable.

Figure 10.1

Attending to Assumptions

Strategies for uncovering and testing assumptions. A simple way to uncover our assumptions is to ask, "What am I taking for granted in thinking this?" Sometimes, it is difficult to focus on specific assumptions by asking this direct question. Another way to clarify assumptions is to ask, "If what I think is correct, what must I also believe?" If I think that I can buy a television set on sale by going to the store, asking these questions should prompt me to list assumptions like "the sale is not yet over" and "there are still TVs like the kind in the advertisement available at this price." The skill of *uncovering assumptions* helps us to become aware of assumptions that we or others make and decide whether these assumptions are acceptable.

Once we uncover these assumptions, we can try to determine which ones are reasonable. To do this, we should first ask, "Do I have any reasons for thinking that this assumption is correct?" Additionally, we should ask whether there are any reasons for thinking the assumption is false. We may remember that this department store has a history of "bait and switch" advertising. There may be only one of the advertised television sets available, and it is likely to be sold quickly. When we find a questionable assumption, we should check it out.

When we have succeeded in checking out an assumption and find that there are reasons for or against it, it ceases to be an assumption. When I call the store to inquire whether there are plenty of TV sets still available at the sale price and am told by a clerk that there are, I no longer assume that the sets are available. I know it.

When we try to uncover the assumptions of others, it may be necessary to first list possible assumptions that lie behind the person's behavior and then determine which are best supported by the circumstances. Suppose you invited a friend for dinner on Saturday evening, and he arrived on Friday evening. He might have assumed that you had invited him for Friday, not Saturday, or he might have assumed that the day he arrived was Saturday, not Friday. Both would be mistaken assumptions that explained his actions. If you find that he wrote in his appointment book that the dinner date was Friday, you've got reason for thinking that he assumed that Friday was the date for dinner.

A Thinking Map and Graphic Organizer for Uncovering Assumptions

The thinking map in figure 10.2 can guide us in detecting assumptions and in determining whether or not those assumptions are justified.

UNCOVERING ASSUMPTIONS

1. What action, belief, or conclusion of a specific person might be based on assumptions?

2. What might that person be taking for granted in performing the action, accepting the belief, or drawing the conclusion?

3. Are there any reasons for thinking that any of these possible assumptions are being taken for granted? If so, what are these reasons?

4. Are these assumptions well-founded? Explain.

Figure 10.2

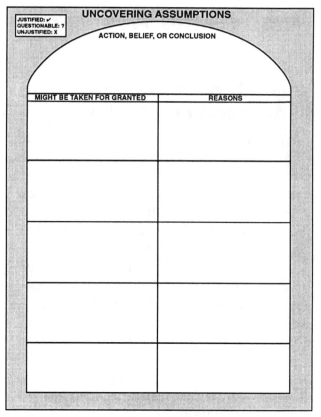

Figure 10.3

The graphic organizer in figure 10.3 can be used to record the answers to questions that help us understand whether it is reasonable to believe that assumptions are being made.

The graphic organizer in Figure 10.4 can be used to help people make explicit what they can do to check out their assumptions after they become aware of them:

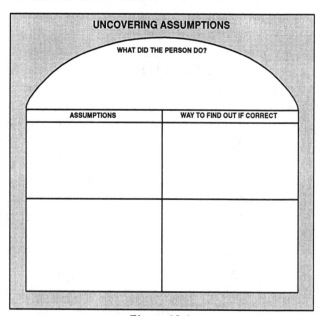

Figure 10.4

Assumptions in Arguments and Controversies

The second type of situation in which we must try to understand assumptions involves examining the reasons people give for a conclusion that they try to convince us to accept. Clarifying assumptions is especially important when people hold opposing viewpoints on an issue. Conflicting points of view often occur because each side's opinion is based on different assumptions. Examining the assumptions behind each side's arguments may help us to understand clearly why they are taking opposite sides. Uncovering assumptions can also help with conflict resolution. Once differing assumptions have been stated, the disputing parties can consider whether these assumptions are reasonable. This may not be so easy, especially when the assumptions are deeply held beliefs. Then the parties may at least "agree to disagree," respecting that they have differing assumptions.

Continued dialogue, even about the reasonableness of our most deep-seated beliefs, is a characteristic of good thinkers. Although two parties in a dispute may agree to disagree, this should not mean that they stop inquiring about how well-founded their own assumptions are. The ideal of true Socratic questioning is especially applicable in these situations. The Socratic questioner uses probing questions to prompt others to examine the bases of their beliefs and to accept only those that are well-founded. It may be that once the assumptions are revealed, one of the parties realizes that his or her assumptions are unwarranted, resolving the dispute.

UNCOVERING ASSUMPTIONS IN ARGUMENTS

1. **What action, belief, or conclusion is being recommended in this argument?**

2. **What reasons are being given for the action, belief, or conclusion?**

3. **Do the reasons suggest that the speaker has taken anything for granted? If so, what?**

4. **Are there good reasons for accepting what is taken for granted? Why?**

Figure 10.5

The thinking map in figure 10.5 guides uncovering assumptions in arguments.

The graphic organizer in figure 10.6 includes a "foundation" on which the speaker's reasons for the belief are based.

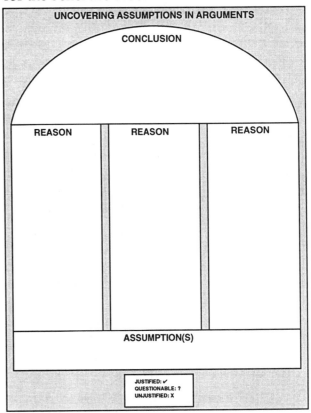

Figure 10.6

How Can We Teach Students to Attend to Assumptions Skillfully?

In teaching students to examine assumptions, it may be easier for them to focus on the assumptions that others make before thinking about their own. When you introduce the skill, for example, you can describe situations, like the one about the sale on television sets, and ask the questions on the thinking map (figure 10.2) to guide students in uncovering and reflecting on assumptions that are made. Then, you may wish to ask students to describe some assumptions they have made and ask them to assess them.

Contexts in the Curriculum for Infusion Lessons on Uncovering Assumptions

The curriculum is ripe with opportunities for infusion lessons on assumptions. People portrayed in stories often make assumptions that they later discover were incorrect. Sometimes these are costly mistakes. Similarly, historical figures have made assumptions that have been costly also (e.g., Moctezuma's assumptions about Cortés on the basis of which he did not try to prevent Cortés from entering his city). Any such contexts are opportunities for infusion lessons.

The need to uncover assumptions being made also plays a role in science. There are many examples of breakthrough research being done only after assumptions were challenged and the research took a new direction. For example, the theory of plate tectonics was revisited only after assumptions that the Earth was solid to its core made by geologists up to that time were made explicit and questioned. This theory, now viewed to be well-established, was not feasible given these assumptions. Assumptions about the way disease spreads (by inorganic agents in the air) blocked, for many years, the discovery of the microorganisms that cause many diseases. Only when these assumptions were examined and then rejected, did research on diseases progress.

The four contexts in the curriculum for infusion lessons on uncovering assumptions are

- *Historical scientists that students study make assumptions in what they say or do.* Scientists accepted the assumption that cells are spontaneously generated from matter for many years before that assumption was challenged by Virchow. ("All cells come from cells.") Similarly, Darwin's theory of evolution challenged assumptions that many made about the age of the Earth (e.g., it was about 6,000 years old). Such assumptions can be uncovered, and students can tell whether they were reasonable at the time.

- *Theories that students study rest on assumptions.* For example, the theory of spontaneous generation, long since discredited, rested on assumptions about the nature of organic matter. The Ptolemaic theory about the structure of the universe assumed the Earth as the center. Not all assumptions that lie behind theories are unreasonable. Some ancient thinkers (e.g., Democritus) advanced the theory that all matter was composed of small non-divisible particles,

an assumption that has been accepted as reasonable by contemporary atomic theory.

- *Principles or practices that students learn rest on assumptions.* Generalization rests on the assumption of the regularity of nature.

- *Students often base misconceptions of natural phenomena on assumptions.* Science instruction, in particular, has focused on misconceptions that students have about nature. For example, people often assume that weight determines speed, believing that heavier things fall faster than light things. Another common misconception is that when we exert pressure on an inanimate object (e.g., when you press on a table) the only force operating is the pressure of your hand. But Newton's first law says that for every action there is an equal and opposite reaction. The table reacts with equal force.

The thinking map for uncovering assumptions in figure 10.2 contains prompting questions adaptable for activities in each of these contexts.

When students have mastered this skill, they will ask the questions on the thinking map themselves in appropriate circumstances without your guidance. As in other infusion lessons, transfer is enhanced by helping students to think about and plan an effective strategy for attending to assumptions and then providing opportunities for them to carry out this thinking plan.

A menu of suggested contexts for infusion lessons on uncovering assumptions can be found at the end of the chapter.

Model Lessons on Uncovering Assumptions

One lesson is included in this chapter on uncovering assumptions. It is a chemistry lesson called "Dangerous Assumptions." It involves uncovering an assumption that led to an explosion at a chemical plant. This lesson focuses on a very common type of assumption made in science, one regarding background conditions, that people often do not think to examine.

As you read these lessons, consider the following questions:

- What prompting questions guide students to uncover assumptions in these lessons?

- How does uncovering assumptions enhance skillfully finding reasons and conclusions?
- What other contexts can be used to reinforce this skill during the rest of the year?

Tools for Designing Lessons on Uncovering Assumptions

The thinking maps contain guiding questions for thinking about assumptions that are made in a wide range of contexts and for assumptions upon which arguments rest. Please note that one set of thinking maps and graphic organizers is for uncovering assumptions in general and the other is used only when the assumptions behind an argument are considered.

The thinking maps and graphic organizers can guide you in designing the critical thinking activity in the lesson and can also serve as photocopy masters, transparency masters, or as models that can be enlarged and used as posters in the classroom. Reproduction rights are granted for use in single classrooms only.

UNCOVERING ASSUMPTIONS

1. What action, belief, or conclusion of a specific person might be based on assumptions?

2. What might that person be taking for granted in performing the action, accepting the belief, or drawing the conclusion?

3. Are there any reasons for thinking that any of these possible assumptions are being taken for granted? If so, what are these reasons?

4. Are these assumptions well-founded? Explain.

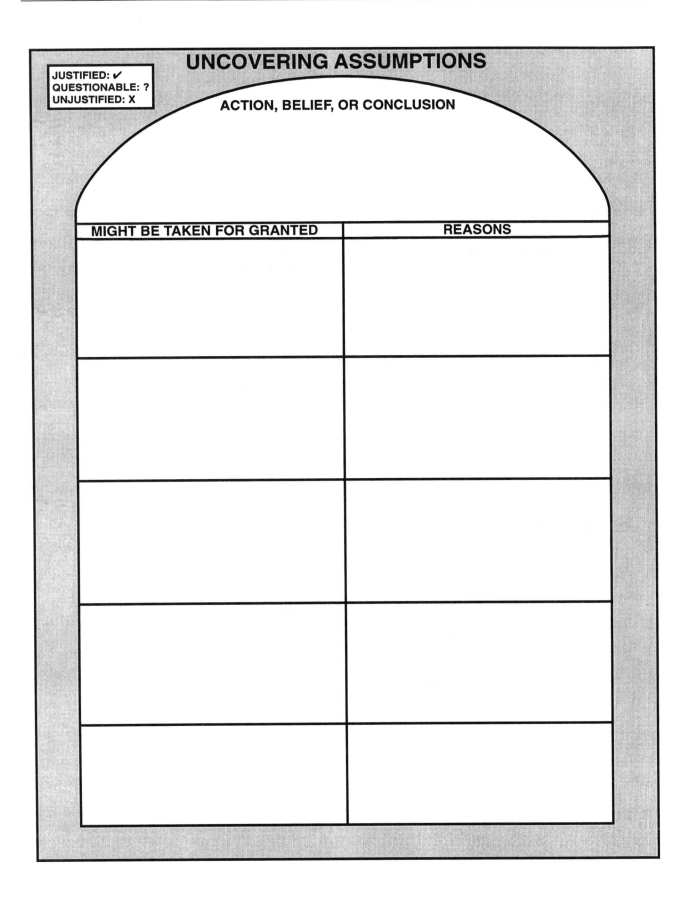

UNCOVERING ASSUMPTIONS IN ARGUMENTS

1. **What action, belief, or conclusion is being recommended in this argument?**

2. **What reasons are being given for the action, belief, or conclusion?**

3. **Do the reasons suggest that the speaker has taken anything for granted? If so, what?**

4. **Are there good reasons for accepting what is taken for granted? Why?**

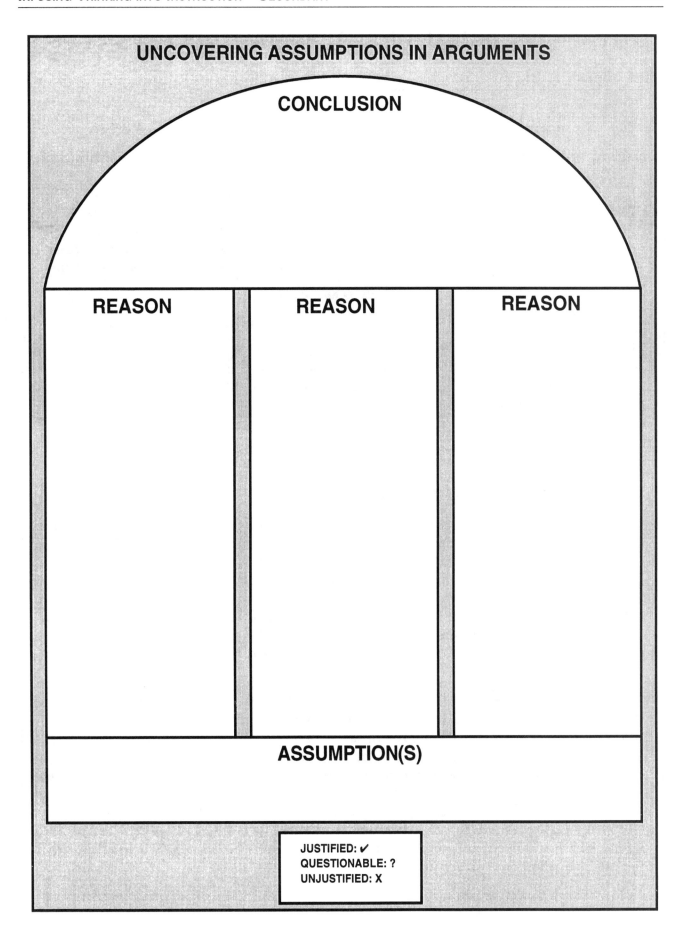

UNCOVERING ASSUMPTIONS IN ARGUMENTS

CONCLUSION

REASON   REASON   REASON

ASSUMPTION(S)

JUSTIFIED: ✔
QUESTIONABLE: ?
UNJUSTIFIED: X

DANGEROUS ASSUMPTIONS

Chemistry **Grades 9–12**

OBJECTIVES

CONTENT

Students will learn about the processes used to separate gases from a mixture by liquefaction and boiling off individual gases. Students will also learn the chemistry of ammonia.

THINKING SKILL/PROCESS

Students will learn to uncover assumptions by identifying what might be taken for granted by certain actions and determining whether there is evidence that the person actually made those assumptions. They will also decide whether the assumptions were supported by good reasons and devise tests for the untested assumptions.

METHODS AND MATERIALS

CONTENT

Students read the chapter in their chemistry textbooks about the liquefactions of gases and will separate gasses by liquefying them and then boiling off individual gases.

THINKING SKILL/PROCESS

An explicit thinking map, a graphic organizer, and structured questioning are used to uncover assumptions behind a person's actions. Collaborative learning enhances the thinking.

LESSON

INTRODUCTION TO CONTENT AND THINKING SKILL/PROCESS

- When we read about a sale at the mall, we often go to the store to buy things. But sometimes, that does not work out well. Once, I read in the newspaper that a store was having a sale on television sets. I needed one, so I went to the store to buy it. When I got there, however, I found that the store had no more of the television sets I wanted. It had sold out the day before and wasn't going to get new ones for another month. I was very disappointed. Then I realized that, if I had been thinking about what I was doing, I could have saved myself the trouble of traveling to the store. **What could I have thought about or done that would have saved me the trouble?** POSSIBLE ANSWER: *I could have called the store and asked if they still had the TV sets available.*

- I didn't think of calling the store because I was making a number of assumptions of which I wasn't aware. For example, I assumed that they still had television sets of the sort I wanted to buy. An assumption is something you believe without thinking about it or checking it. Assumptions sometimes lead you to do or say certain things. Finding out what a person was assuming helps you understand why he or she did a certain thing. Assumptions might be correct or mistaken. **When I set out to go to the store, what else do you think I was assuming?** POSSIBLE ANSWERS: *That the store would be open, that my car would work, that I had enough gas in my car to get to the store, that I had enough money to buy the television set, that I needed this particular television set.* **My assumption that they still had the television set I wanted to buy was mistaken. Were any of these other assumptions mistaken?** ANSWERS VARY. In this case, all of the other assumptions were correct.

- I didn't check out my mistaken assumption because I didn't think about it. I wasn't even aware that I was making that assumption. If I had asked myself, "Am I making any assumptions here?" I might have become aware that I was assuming all of these things. If I had questioned whether these were good assumptions, I might have checked them out and saved myself a trip to the store.

- Can you think of a time when you did something based on a mistaken assumption that you could have uncovered and checked out but didn't? Discuss it with your partner; help each other list all the assumptions that were made in this situation. Identify the ones that were mistaken and the ones that weren't. Write them down. Discuss how you could have checked out the mistaken assumptions if you had been aware of them. After a few minutes, ask for reports from the class. Ask the students reporting to identify their assumptions, those assumptions that were mistaken, and how they could have been checked. STUDENT ANSWERS VARY.

- Everyone makes assumptions. Not all assumptions are mistaken, however. For example, you are all assuming that the chairs you are sitting in are strong enough to continue to hold you. That's not mistaken, is it? And you all had been assuming it, hadn't you: No one thought to check out whether the chair was strong enough to hold them before they sat down, did they? Still, it's a good idea to become aware of your assumptions, or assumptions that other people make when they try to convince you of something, and to check them out before you act on them, especially when something important is at stake. That's because sometimes people do make mistaken assumptions, like you all discussed earlier. If you don't become aware of what you or someone discussing something with you are assuming, you'll never know whether anything mistaken is being assumed. Here is a thinking map for uncovering assumptions that can guide you in this process:

UNCOVERING ASSUMPTIONS

1. What action, belief, or conclusion of a specific person might be based on assumptions?
2. What might that person be taking for granted in performing the action, accepting the belief, or drawing the conclusion?
3. Are there any reasons for thinking that any of these possible assumptions are being taken for granted? If so, what are these reasons?
4. Are these assumptions well-founded? Explain.

- In the history of science, many assumptions have prevented real scientific advancement. For example, it was assumed for a very long time that because the earth didn't appear to move but the sun and stars did, that the earth was the center of the universe. Examining that assumption in an open way earlier than it was called to question in the history of science may have led to much earlier discoveries of the real nature of the solar system and the universe. Can you think of other assumptions that early scientists made that were abandoned once they were uncovered and tested? STUDENT ANSWERS VARY, BUT SOMETIMES INCLUDE: *Because of the appearance of maggots on decaying food, life could develop from nothing. Air does not enter into a combustion reaction, so there must be some substance responsible for burning (thereby leading to the positing of phlogiston as a special substance that was responsible).* If the people who made these assumptions had become aware earlier that they were making them and had then checked them out, the history of science may well have been quite different.

- We've been studying gases like oxygen, nitrogen, and hydrogen and have learned about where they occur in nature. Let's consider a real situation in which assumptions about such gases led to a near-tragic accident. Let's see if we can uncover what those assumptions were, how they might have been checked out, and which were mistaken, to determine whether, if the person making them had become aware of them earlier, the accident might have been prevented.

THINKING ACTIVELY

- A number of years ago, when the process of manufacturing ammonia by combining hydrogen and nitrogen was a new process, an engineer was given the task of designing a plant for making 300 tons of ammonia per day. What is the chemical formula for ammonia? *NH_3.* The standard form of hydrogen and nitrogen have which formulas? *H_2 and N_2.* Now let's return to our story. The engineer reasoned that the plant would have a large supply of H_2 because it was to be built next to a petroleum refinery. Where might the plant get the large quantities of pure N_2, the other essential raw material? *N_2 is present in air, so it is present all the time in the air around the plant.* Is it the only gas that is present in air? *Air also contains oxygen, carbon dioxide, argon, and other trace gases.* How could they get the N_2 from the air? *The air could be liquefied and the nitrogen extracted.* How could the air be liquefied? *The air could be compressed and chilled until it reaches the temperature at which it changes state and becomes a liquid.* How could the nitrogen be extracted from the liquefied air? *When liquefied, the N_2 could be distilled by boiling the liquid air.* How could you tell which of the gases that was distilled from the liquid air was nitrogen? *I could look up the boiling points of the different gases in air and determine the order in which they would boil off. Nitrogen would boil off first because it has a lower boiling point than the other ingredients in air.* After the extraction is complete, what by-product would remain? *Concentrated O_2 plus the other gases found in air—CO_2, argon, etc.* Next, ask the students to draw a chart or a graph indicating how much of the yield of nitrogen there would be in comparison to the quantity of oxygen, etc, in the by-product. Allow them to use their texts or other resources to get this information. STUDENT RESPONSES INCLUDE:

COMPONENT:	N_2	O_2	Ar	CO_2
VOLUME (%):	78.10	20.94	0.93	0.03

- The large quantities of nitrogen that could be produced by this process, together with the almost unlimited supply of hydrogen from the nearby petroleum refinery, made the prospects for a high-yield operation seem very good to the engineer. The process by which the ammonia was ultimately to be manufactured was based on the following reaction:

$$3H_2 + N_2 \longrightarrow 2NH_3$$

Based on all of this information, the engineer designed the plant to consist of an air separation unit, compressors, and a tower to contain the reaction that generated the ammonia. In the tower, a mixture of N_2 was to be passed over an iron catalyst at a temperature of 500°C and a pressure of 150 atmospheres, and then the ammonia produced cooled to standard temperature and pressure so that it would liquefy.

- What risks are involved in this process? Ask students to work in pairs and to make a list of any risks in this process. Students usually distinguish five phases in the process: (1) the liquefaction of the air, (2) the distillation and storage of the nitrogen, (3) the by-product storage and subsequent release into the atmosphere, (4) the chemical reaction whereby the ammonia is produced, and (5) the cooling and storage of the liquid ammonia. STUDENT ANSWERS INCLUDE: Phase (1): *the compression unit could malfunction and pressure buildup would cause the walls of the air separation unit to explode outward, liquid air could leak out and injure workers;* Phase (2): *the liquid nitrogen could become gaseous too rapidly and the buildup of pressure could cause the walls of the storage chamber to explode;* Phase (3): *the concentrated oxygen in the by-product might cause the combustion and explosion of flammable materials that might mix with the by-product by mistake;* Phase (4): *leaky valves in the reaction tower could lower the pressure and/or temperature, making the yield lower than expected and sticky valves in the reaction tower could cause a buildup of pressure and an increase in temperature that would cause an explosion;* Phase (5): *leakage of liquid ammonia could release noxious fumes into the surrounding atmosphere.*

- **None of these problems occurred in the first year of the plant's operation. In fact, the plant ran quite smoothly and produced a better yield of ammonia than expected because of greater efficiency in the separation process. But one day, for no apparent reason, there was a large explosion that created a huge fireball and destroyed the air separation unit. Luckily, no one was hurt. But large pieces of metal from the air separation unit were scattered as far as a quarter of a mile from the plant.**

- **Work in teams again and follow the thinking map for uncovering assumptions using this graphic organizer to help you record the assumptions you think the engineer made in designing this plant. What was the process engineer convinced of that was based on these assumptions?** *The ammonia plant, as he designed it, would operate efficiently and safely.* **Write this in the dome at the top of the graphic organizer. Write what you think he took for granted in accepting this belief in the appropriate boxes on the graphic organizer under the dome and explain why you think he made these assumptions in the boxes on the right.** STUDENT ANSWERS INCLUDE: *the data on the composition of air obtained in the engineering handbook is representative of the composition of the air near the oil refinery, no trace amounts of combustible materials (like hydrocarbons) would enter the air separation unit with the air and would accumulate sufficiently over time to become dangerous when mixed with the big concentrations of liquid oxygen that accumulates as a by-product in the unit; the oxygen is too cool for any ignition or detonation to take place;* hydrogen and nitrogen will react together to form ammonia; nitrogen can be separated from the oxygen in liquid air by boiling it; the compressor will operate properly; valves and other operating parts will be serviced sufficiently to keep them in good working order. After a few minutes, ask the groups to mention one assumption they identified and write it on a chart, diagram it on the chalkboard, or trace it on a transparency of the graphic organizer. Ask for additional responses until you have written all the assumptions the groups identify on the diagram you are using. As students report, ask them why they think he made this assumption.

UNCOVERING ASSUMPTIONS

JUSTIFIED: ✔
QUESTIONABLE: ?
UNJUSTIFIED: X

ACTION, BELIEF, OR CONCLUSION

MIGHT BE TAKEN FOR GRANTED	REASONS

- **Each group should select one or two assumptions from this list and use a modified version of the graphic organizer for uncovering assumptions. The group will write in the appropriate** box what the process engineer might have done to verify their validity if he had been aware that he was making those assumptions. Ask the groups to report on the assumptions they select and assign a different one to a group that selects an assumption already chosen by another group. After a few minutes, ask the groups to report and write these next to the assumptions on the chart, the diagram of the graphic organizer on the chalkboard, or on the transparency on which you had written the assumptions the groups identified. STUDENT ANSWERS VARY, but should include mention of laboratory tests that might have been conducted, where appropriate.

THINKING ABOUT THINKING

• You uncovered the process engineer's assumptions about the ammonia processing plant in this lesson. What questions guided you to determine what his assumptions were? POSSIBLE ANSWERS: *I first asked what the engineer believed or did that might be based on assumptions. I then thought about what the engineer might have taken for granted that would explain why he acted as he did. Then, I thought about what the engineer might have done to check to determine whether these assumptions were correct.*

• To guide you in uncovering assumptions in the future, construct a thinking map. The thinking map should contain the key questions for uncovering assumptions shown at the right.

• Is it helpful to be able to uncover assumptions? Why or why not? Many students respond that it is helpful to become aware of their assumptions and decide whether or not they are good ones. We should ask ourselves the questions for uncovering assumptions until we know that our assumptions are well-supported. This can often keep us from acting on mistaken assumptions that we are not fully aware of. Sometimes, acting on mistaken assumptions can lead to disasters, as in the case of the ammonia processing plant.

> ### UNCOVERING ASSUMPTIONS IN ARGUMENTS
>
> 1. What action, belief, or conclusion is being recommended in this argument?
> 2. What reasons are being given for the action, belief, or conclusion?
> 3. Do the reasons suggest that the speaker has taken anything for granted? If so, what?
> 4. Are there good reasons for accepting what is taken for granted? Why?

• How did working in a group help you reflect on what the engineer's assumptions may have been? Students comment that the group discussion helped them to come up with ideas by building on what others in the group said, much like the brainstorming process does.

APPLYING THINKING

Immediate Transfer

• Carbon-14, the radioactive isotope of carbon, is often used to determine the age of ancient objects. Carbon-14 decays into nitrogen-14 at a known, constant rate. The half-life of carbon-14 (the time it takes for one-half of a sample of the radioisotope to change to nitrogen-14) is 5,370 years. Living things take in both carbon -12 and carbon -14 when they are alive. Then, when an organism dies, the carbon-12 in its body remains constant, while the carbon-14 continues to decay. The ratio of carbon-14 to carbon-12 helps scientists estimate the age of an object that contains once-living material.

- Suppose a scientist finds that the ratio of carbon-14 to carbon-12 in the remains of a bird is one-fourth the ratio in the tissue of an organism living today. The scientist concludes that the remains of the bird are about 11,460 years old. What assumptions underlie the scientist's conclusion? Use the thinking strategy for uncovering assumptions to identify these and to sketch out how you would examine these assumptions. Then check them out and determine whether these assumptions are valid.

Reinforcement Later

- We've been studying energy sources. What, if any, assumptions are we making in our reliance on fossil fuels as the main source of energy in this country? Are these assumptions justified? How can we check them out? Based on what you have uncovered, what would you advise that we do about our sources of energy?

- What assumptions are being made by someone who says that by the year 2025, 100,000,000 people will be infected by the AIDS virus? How can these assumptions be verified? Based on what you conclude about these assumptions, what judgment would you make about this prediction about AIDS? Select a major decision that you have made recently. Use the strategy for uncovering assumptions to determine whether this decision is based on any untested assumptions. If so, how would you verify them? Do so and rethink your decision based on what you find.

CONTENT EXTENSION

Ask students to research the types of impurities that have been found in air. Have them design a technique for purifying air that contains these types of impurities.

THINKING SKILL EXTENSION

Ask students to use the strategy for causal explanation to make the most reasonable judgment about the cause of the explosion at the ammonia plant based on the information provided in this lesson.

Ask students to use the judgments they make about the most likely cause of the explosion to decide how to redesign the plant to avoid the same kind of problem again.

ASSESSING STUDENT THINKING IN UNCOVERING ASSUMPTIONS

To assess student thinking about assumptions, describe a significant action by a scientist or engineer. Ask students to think about what the people involved probably took for granted. Urge students to make their thinking explicit. Determine whether they attend to each of the steps in the thinking map for uncovering unstated assumptions and whether they are using the language of the skillful practice of uncovering assumptions.

Sample Student Responses • Dangerous Assumptions

UNCOVERING ASSUMPTIONS

WHAT DID THE PERSON DO?

The engineer believed that the ammonia plant would operate efficiently and safely.

ASSUMPTIONS	WAY TO FIND OUT IF CORRECT
No trace amounts of flammable material like hydrocarbons are in the air around the ammonia plant.	Air around the plant can be tested at different times to determine whether the air has anything other than the standard gases in it. Reliable tests for hydrocarbons, for example, can be researched and then used to make this determination.
Any traces of flammable material in the air around the ammonia plant will not accumulate in the air-separation unit enough to become dangerous in the presence of the oxygen in the by-product of the separation process.	Any traces of flammable material in the air can be mixed in various quantities with liquid oxygen in a laboratory to determine the amount that creates a volatile mixture, and projections can be made of the amount of time it would take for this amount of these materials to accumulate in the air-separation unit. It can also be determined whether steps will be taken to clean up accumulations of such materials before they reach critical amounts.

Sample Student Responses • Dangerous Assumptions

UNCOVERING ASSUMPTIONS

WHAT DID THE PERSON DO?

The engineer believed that the ammonia plant would operate efficiently and safely.

ASSUMPTIONS	WAY TO FIND OUT IF CORRECT
Nitrogen can be separated from oxygen in liquefied air because its boiling point is lower than that of oxygen.	Consult a table of boiling points of liquefied gases.
Hydrogen and nitrogen will react to form ammonia in sufficient quantities to make the plant viable economically.	Consult a chemistry textbook to determine how such a reaction works. Try the reaction in a laboratory, measure the yield, and note any special factors that must be taken into account in sustaining such a reaction. Do a cost analysis of the process and compare this to the projected gross income from the product.

UNCOVERING ASSUMPTIONS LESSON CONTEXTS

The following examples have been suggested by classroom teachers as contexts to develop infused lessons. If a skill or process has been introduced in a previous infused lesson, these contexts may be used to reinforce it.

GRADE	SUBJECT	TOPIC	THINKING ISSUE
6–12	Science	Scientific method	Describe the scientific method. On what assumptions is controlled experimentation based?
6–8	Science	Ptolemaic view of the cosmos	Why did it seem reasonable to assume that the earth was the center of the universe with the sun, planets, and stars rotating about it?
6–8	Science	Generating life	During the 1600s, scientists observed that whenever meat was left exposed to the air, maggots would soon appear on the meat. They also believed that fleas developed from sweat and that mice sprang fully formed from garbage. What were these scientists assuming about the origin of living things?
9–12	Biology	Specimen preparation	What is the assumption behind the fixation, embedding, sectioning, and staining of specimens?
9–12	Biology	Cell evolution	Currently, scientists hypothesize that the eucaryotic cells (the cells of higher organisms) originated as the result of one procaryotic cell (the cells of bacteria and blue-green algae) engulfing other procaryotic cells, eventually forming the nucleus and other structures of the eucaryotic cells. On what assumptions about the evolutionary origin of cells is this hypothesis based?
9–12	Biology	Mendelian genetics	A woman suffering from cystic fibrosis (CF), caused by a recessive trait and often fatal by age 30, gets married to a man who does not appear to have the symptoms of this genetic disorder. The woman decides to have a baby. What assumptions is the woman making?
9–12	Biology	The nature of living things	What assumptions are you making about the nature of living things when you determine that a virus is not alive?
9–12	Biology	Spontaneous generation	What mistaken assumptions did Jean van Helmont make three hundred years ago when he predicted that a dirty shirt placed in a box of wheat would produce adult rats in about three weeks?
9–12	Chemistry	Carbon-14 dating	What assumptions are made by scientists who rely on the use of the carbon-14 dating process to determine the age of ancient organic specimens? Are these reasonable assumptions? Explain.

UNCOVERING ASSUMPTIONS LESSON CONTEXTS			
GRADE	SUBJECT	TOPIC	THINKING ISSUE
9–12	Chemistry	Combustion	What did the early scientists who accepted the phlogiston theory assume about the role of air in combustion? How could they have checked out this assumption if they had become aware of it?
9–12	Chemistry	The extraordinary properties of water	Two open wood buckets of water are placed outside on a frigid winter day. One bucket is filled with cold tap water and the other is filled with boiling water. What mistaken assumptions might you have made about the behavior of water if you predicted that the cold water would freeze first?
9–12	Physics	The nature of light	Light reflects off surfaces, passes through glass, exhibits interference, can be polarized, and has frequency; these are all wave properties. One theory, therefore, suggests that light is a wave. What mistaken assumptions about the behavior of light does this theory make?
9–12	Physics	Astronomy: planetary motions	What mistaken assumptions did Ptolemy make about the motion of planets in order to develop the geocentric, or earth-centered, view of planetary motion?
9–12	Astronomy	Geocentrism	Early Greek astronomers observed that the stars all moved together in the sky, while the sun, moon, and the planets (which they called the "wanderers") each moved independently. They concluded that the stars must be contained on a sphere that moves around Earth, while the wanderers were not. What assumptions were these theories based on? What do you think this assumption was based on? Could they have checked it out? How?

PART 4

SKILLS AT GENERATING IDEAS: CREATIVE THINKING

Chapter 11: Generating Possibilities

Chapter 12: Creating Metaphors

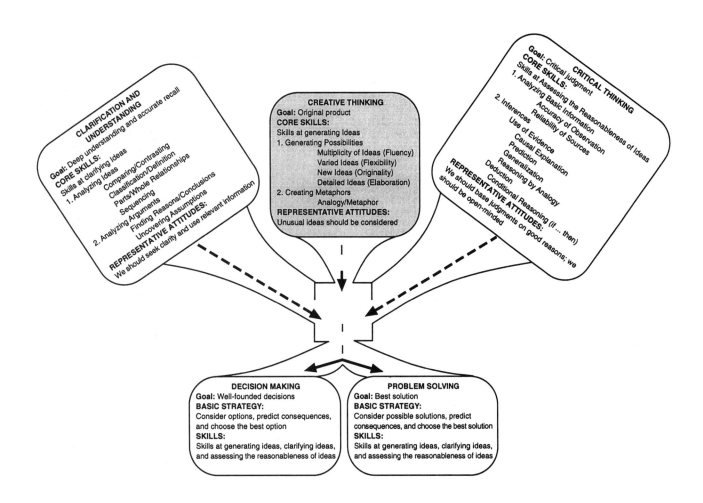

CLARIFICATION AND UNDERSTANDING
Goal: Deep understanding and accurate recall
CORE SKILLS:
Skills at clarifying ideas
1. Analyzing Ideas
  Comparing/Contrasting
  Classification/Definition
  Parts/Whole Relationships
  Sequencing
2. Analyzing Arguments
  Finding Reasons/Conclusions
  Uncovering Assumptions
REPRESENTATIVE ATTITUDES:
We should seek clarity and use relevant information

CREATIVE THINKING
Goal: Original product
CORE SKILLS:
Skills at generating ideas
1. Generating Possibilities
  Multiplicity of Ideas (Fluency)
  Varied Ideas (Flexibility)
  New Ideas (Originality)
  Detailed Ideas (Elaboration)
2. Creating Metaphors
  Analogy/Metaphor
REPRESENTATIVE ATTITUDES:
Unusual ideas should be considered

CRITICAL THINKING
Goal: Critical judgment
CORE SKILLS:
Skills at Assessing the Reasonableness of Ideas
1. Analyzing Basic Information
  Accuracy of Observation
  Reliability of Sources
2. Inferences
  Use of Evidence
  Causal Explanation
  Prediction
  Generalization
  Reasoning by Analogy
  Conditional Reasoning (if ... then)
  Deduction
REPRESENTATIVE ATTITUDES:
We should base judgments on good reasons; we should be open-minded

DECISION MAKING
Goal: Well-founded decisions
BASIC STRATEGY:
Consider options, predict consequences, and choose the best option
SKILLS:
Skills at generating ideas, clarifying ideas, and assessing the reasonableness of ideas

PROBLEM SOLVING
Goal: Best solution
BASIC STRATEGY:
Consider possible solutions, predict consequences, and choose the best solution
SKILLS:
Skills at generating ideas, clarifying ideas, and assessing the reasonableness of ideas

PART 4
SKILLS AT GENERATING IDEAS: CREATIVE THINKING

Imagine facing the following problem: Winters are cold where I live and my heating bills are very high. I usually just pay the bills and don't think much about them. But can I reduce them? Actually, I have no idea why they are so high. My neighbor pays only half as much as I do for heat. I've been told that the rates are going up. I can't tell what this increase will mean for my monthly budget. I don't know *what* to do.

To solve a problem ourselves without relying on others, we must come up with ideas that lead to workable solutions. Generating ideas will not, in itself, solve our problems. Unless we are able to generate a variety of possible solutions, however, we have little chance of finding a workable solution. In many ways, our ability to take charge of our lives depends on exercising this ability to generate ideas. When we can come up with no ideas about what to do, like the person described above, we are at the mercy of external circumstances.

Generating ideas is sometimes characterized as the active use of our creative imaginations. It is also described as an act of synthesis—putting thoughts together to get new ones. In either case, generating possibilities is recognized as the basic type of thinking that lies behind *creative thinking*, the generation of original ideas. Our ability to generate ideas is derived from two basic ingredients:

- our past experience, which furnishes the raw material of creative thought; and
- our ability to take apart and creatively combine ingredients from past experience.

Imagination plays a key role in this mode of thinking. We sometimes ask, "What would it be like if I tried to…?" For example, what would it be like if I tried to contain the heat in my house? Thinking about containing heat, I imagine a blanket. That idea may lead me to consider ways to "blanket" my house, as with weather stripping, storm windows and doors, and insulation.

Using active imagination to generate ideas relies on our ability to detect insightful analogies.

Two frequent manifestations of our ability to generate ideas deserve our attention:

- generating numerous, varied, and original ideas as possible answers to questions that arise in our personal and professional lives (generating alternative possibilities); and
- creating analogies that can help us gain and express insight (developing metaphors).

Thinking about possible solutions to problems, generating options to consider in decision making, developing alternative hypotheses to test when we try to find out why certain things are happening, and bringing to mind the possible consequences of our actions or of the events around us are just a few of the ways in which we consider alternative possibilities. Facility in generating possibilities can be developed by techniques such as brainstorming, used judiciously and enhanced by various other explicit strategies.

Poets are not the only ones who thrive on using analogies to develop metaphorical and figurative ways of thinking. Our everyday thinking is so permeated with metaphors that learning to generate them skillfully enhances the way we understand things, as well as our ability to solve problems. Think about how often we say things like "She's a dynamo," "It's a piece of cake," and "We're up the creek." Metaphors are powerful ways of emphasizing and communicating ideas that enhance both focus and deep insight.

Generating ideas helps us to gain insight and to take charge of situations. When we have active minds full of ideas, it becomes more likely that we will find options that work. When we do find such options and act on them, we enhance our ability to manage our own lives. The positive self-concept that many people have is a function of their ability to generate their own ideas. When you teach students skills at generating ideas, improved self-concept is one important benefit that you will bring them.

CHAPTER 11
GENERATING POSSIBILITIES

Why is it Important to Generate Possibilities Skillfully?

Considering possibilities is a major ingredient in much of our thinking. Could the high incidence of violence on TV be contributing to the increased incidence of violence in the streets? Could economic and social conditions be causing the violence? Finding out the causes of this national trend is essential if we have any hope of reversing it. Let's look for causes..

In general, when we are considering what caused some event, the hypotheses we generate for further testing are viewed initially only as possibilities—possible causes. Similarly, when we make decisions and try to solve problems, the options and solutions we develop are viewed as possibilities. Generating ideas and then deciding whether they work are crucial steps in engaging in more complex thinking processes such as decision making and problem solving. Unless we entertain a variety of possibilities as we wonder about causes, effects, options, and solutions, we will not be in a position to accept or reject them as good ideas.

Ideas that we consider can, of course, come to us from other people. Many times, the ideas of others prompt us to think of things that we would not have considered independently. Some people, however, depend on other people for their ideas. Nevertheless, they often face challenging situations that they have to deal with themselves. Learning to generate our own ideas frees us from dependence on others and is a major step in the direction of becoming autonomous, self-reliant thinkers. It broadens our thinking in ways that maximize our chances of finding the best solution to problems, making the best choices, determining the best explanation for why things happen, and even developing innovations that improve the quality of life.

Contexts in Which it is Useful to Think about Possibilities

In practical situations in which we are not sure about what to do, considering possibilities can be very helpful. Major decisions, like buying a house or changing jobs, are better made if we consider our options first. This is also true of more everyday decisions. In planning when to pay this month's bills, I may not be sure whether or not I should set aside time now to do this task. If I consider the possibility of delaying and paying them later, I may realize that such a delay would probably have adverse consequences. Since I have such a tight schedule in the next week, I probably *won't* have time to pay them after today. Because they are due in five days, it then becomes quite likely that I will have to pay additional finance charges if I delay. That's enough to make me decide not to delay any further. I would never have realized this if I had not first entertained the possibility of delaying and then asked what might result if I did delay. If we could only think about things as they are and were not able to bring to mind possibilities that may never happen, the kind of reasoning described would be impossible.

In addition, many of the technological advances that enhance the quality of our lives, like radio, TV, and the automobile, would simply not have been developed unless they had first been considered as possibilities. In one way or another, the great achievements people have made throughout history have had their origins in specific ideas. If others had not been innovators in developing these ideas, we would not enjoy their beneficial results.

Great works of creative imagination, like Melville's *Moby Dick*, also depend on this simple ability. These works spin out and elaborate ideas that come to their creators as possibilities. Nonetheless, the same processes of thinking are involved in generating these possibilities as are involved when we ask more practical "What-If" questions like "What might happen if I delay paying my bills?"

Problems with the Way We Generate Possibilities

When someone asks what he or she can do in a specific situation, what might have caused some disturbing occurrence, or what can result if he or she does something being considered, it is rare that no possibilities occur to that person. Often, though, the possibilities we think about are very limited. If I want to meet my friend for dinner at a restaurant across town, my first impulse may be to drive the same route I took the last time I went to the restaurant. Without much thinking, I may simply go that way. What is familiar and traditional often limits what comes to mind, so we often don't think further about it.

We may think in this routine way, but it also may not serve us very well. My route to the restaurant may not be the best one and may take me much longer than a more direct route. If I never think about another way to get there, I certainly won't choose it. The idea of improving things, whether it is how we get to a restaurant, how we prepare a meal, how we express ourselves in writing, or how we teach, will never occur to us unless we leave our ordinary ways of thinking and consider other possibilities.

Even when we consider other possibilities, they may also be limited. I realize that there are six different routes I could take to the restaurant. These alternatives allow me to choose the most desirable one, perhaps the one with the fewest traffic lights, less traffic, etc. However, when I get to the restaurant, I may find that the parking spaces are all filled and that the nearest parking garage is three blocks away. I realize that I could have taken a taxi or bus and avoided this problem. I didn't even think of that. As with other thinking processes, making decisions is often layered with assumptions. From the outset, when we spread a wide "net" for capturing ideas by considering a range of different possibilities, we often question many of our assumptions.

Narrowness in thinking suggests another shortcoming in how we generate and consider possibilities. I may, indeed, consider a number of possible ways of getting to the restaurant by alternate routes. I may think about other routine ways of getting there, such as taking a bus or taxi. The best way to get there on this occasion may be among a type of possibilities that I haven't considered at all—something entirely new for me at that time. Perhaps I can borrow my son's bicycle and get there quickly, less

expensively, and with easy access to the restaurant. Since I haven't ridden on a bicycle for many years, this means of transportation doesn't come to mind when I think about getting to the restaurant. Or perhaps I could call my friend and move our meeting place to another restaurant. She may not mind and may be relieved because she's having the same problem herself.

Moving from routine possibilities to novel ones is very important in considering ideas. Everyone has the ability to generate creative ideas, but few actually do. Breaking free of the constraints of routine or commonly accepted solutions to problems can lead to creative solutions. These alternatives challenge the assumptions we make in the way we typically solve problems, make decisions, and explain things that happen. Modifying aircraft hangars to make them expandable so that they can be moved around aircraft, rather than modifying an aircraft to make it fit the hangar, is a creative possibility for solving the problem of getting large jet planes into conventional hangars. This may not be the best solution to this problem, but we will never find out if we don't generate and consider this possibility.

The significance of considering unusual possibilities in such fields as medicine is also apparent. Researchers trying to isolate the cause of "Lassa fever" didn't consider the possibility that migrants from nearby urban areas were carriers transmitting the disease that was ravaging a rural African community. Believing that the source of the disease was local, they spent considerable time looking for its source in contaminated water, food, etc. within the community. While scientists were running into numerous dead ends in their investigation, many more people died of the disease, including some of the researchers themselves. Once the possibility that the disease was brought into the community by infected carriers occurred to them and was confirmed, they knew how to control the disease. If they hadn't thought of this possibility, many more lives would have been lost. However, if they had thought of it earlier, many more lives would have been saved.

Unusual possibilities that break away from routine ways of solving problems and explaining phenomena often provide breakthroughs that add considerably to human advancement and progress. Yet it is far too rare that people think of more than the ordinary ways of doing things when thinking about possibilities.

Finally, we often find fuzziness in people's thinking that blurs the distinction between the possible and the likely. Ideas that might work are just that: ideas that *might* work. Yet many people latch onto possibilities without realizing that they need to do more than just bring these ideas to mind in order to determine that they will work. In the restaurant example, because I think about getting there easily by driving on Elm Street, I may, without any further thought, get into my car and take the Elm Street route. This blurs the distinction between an option and a viable option. I run the risk that if I drive on Elm Street at a certain time of day, I could get caught in a rush-hour traffic jam, constantly stop and go because of the number of traffic lights, or have to take a detour because the road is blocked by construction. Taking another route may have been more satisfactory. Many people slip from "this may happen" to "this will happen" without much thought. When we fail to treat a possibility as only a possibility, we can easily make costly misjudgments.

Figure 11.1 summarizes the four major problems in the way we bring possibilities to mind for consideration.

COMMON PROBLEMS WITH THE WAY WE GENERATE POSSIBILITIES

1. We generate very few possibilities.

2. We generate the same type of possibilities only.

3. We generate only routine or ordinary possibilities.

4. We think that possibilities are desirable or likely without any reason.

Figure 11.1

How Do We Generate Possibilities Skillfully?

We can practice strategies for generating possibilities in order to develop the ability to gener-

ate a wide range of ideas as needed. The free association of ideas is sometimes mentioned as one such strategy. Although free association does generate many ideas, generating a range of possibilities for consideration is not the same thing as free association. Generating possibilities is always conducted in a context and with a purpose that constrains the possibilities that we consider by requiring that they must be possible ways of solving a problem, of making a decision, of explaining an event, etc. Since we think about possibilities in specific contexts, this suggests that a more organized strategy should be used to generate possibilities skillfully.

Attention Points in Generating Possibilities Skillfully

The first important thing to keep in mind when generating possibilities is *the purpose of these possibilities*. What is the issue that creates the need for a consideration of possibilities?

- I may face a problem: How can I finish my work before I leave on vacation?

- I may want to anticipate what will happen in a certain context so that I can plan well: What can I expect if I introduce this new sex education program into my school?

- I may want to find out what is causing something so that I can fix it: Why does the plaster keep flaking off my kitchen wall?

Identifying the context for considering possibilities is the first step in bringing these possibilities to mind.

To generate possibilities skillfully within a specific context, we should attend to four basic points: generating many ideas, generating different types of ideas, generating unusual ideas, and adding details to these ideas. We should take the time to develop a number of alternative possibilities that apply to this particular issue. But we should also make sure that we consider a range of different types of possibilities, that we strive for novel ones, and that we treat these ideas as mere possibilities not yet judged to be viable solutions.

Writers on creativity often distinguish these aspects of idea generation as involving *fluency* (generating a multiplicity of ideas), *flexibility* (generating ideas of different types), and *origi-*

nality (generating unusual ideas that have as yet not been tried in dealing with this particular issue). *Elaboration* (providing details of the ideas we are advancing) is another desirable feature of our ability to generate ideas. *Deferred judgment* (not yet taking a stand on which of these ideas is viable to meet the issue) is an attitude we should adopt about the ideas we generate in order to recognize them as possibilities and in order not to make premature judgments about which will work.

Brainstorming

Brainstorming is, of course, the technique commonly used to generate a rich set of possibilities. But research indicates that simple brainstorming is not enough. Just listing possibilities may not involve much variety or originality. Quantity of ideas does not always mean quality.

To enhance variety, we can deliberately make sure we have listed different kinds of possibilities. In the case of getting to the restaurant to meet my friend, I might deliberately ask myself, "Have I thought about other ways to get there?" A more strategic way to do this is to categorize the ideas I have come up with and ask two questions: "What other possibilities fall into these categories?" and "Are there any other categories that I might want to consider?" For example, I might classify the six routes for getting to the restaurant as ways of driving myself there. When I realize that I have not yet considered being transported by others, I can then easily come up with taking a taxi, a bus, or even a ride from a neighbor. Under the category of being self-propelled, I might then ask, "What are other ways that I can transport myself besides using a car?" A bicycle may come to mind.

Innovation and Invention: Strategies for Generating Original Ideas

Variety may not lead to originality. Have I thought of anything new and original? Simply asking myself this question is often enough to prompt new possibilities that are, indeed, quite original. Maybe my friend can drive to my location, and we can rent a horse-drawn carriage to take us to the restaurant. If I were thinking of a romantic evening, that would add some glamour to the way we get to the restaurant.

If original ideas don't come so easily, there are techniques that can be used to bring them out. For example, listing features of some of the possibilities already generated and then combining some of these features may lead to a new possibility. A bicycle has two wheels and is powered by the action of pedals. A carriage can transport me and my friend together. Perhaps we could take a two-wheeled vehicle powered by pedals that can transport two people: a bicycle built for two.

Some novel ideas might be wild and impractical; others might not be. We will never find out unless we think of them. We often have to dispense with original ideas because they are impractical, but it takes only one good idea to achieve our goal.

In this aspect of generating possibilities, we can use the full potential of our creative imaginations to develop new and original ways of doing things and solving problems. This is why generating possibilities, as a mode of thinking, falls within the domain of creative thinking. When we search for novel ways of solving problems, we draw on ingredients in our past experience and combine or connect them in new ways to develop new ideas. Novel solutions that work provide us with truly creative products that enrich and enhance the quality of our lives. Putting into practice original, viable ideas that serve our interests or needs is a mark of the truly creative thinker.

Emphasizing originality after working on quantity and variety in brainstorming yields the greatest potential for generating creative ideas. An effective strategy for doing this combines the ideas we have just discussed:

1. Categorize the possibilities derived from an initial brainstorming session;

2. Add new possibilities under these categories;

3. Suggest new categories; and

4. Combine categories to stimulate new possibilities that blend possibilities already on our list.

Suspending judgment during brainstorming makes brainstorming an excellent technique in carrying out this strategy.

This four-part strategy is especially useful if routine ways of doing things have their limitations. Improving products or practices that do not work well, composing and revising creative works of writing and art, as well as inventing new devices to meet needs that are not met by the present technology are common but significant contexts for generating new ideas. So, of course, are the more numerous opportunities for creative thinking in our own jobs and personal lives.

In generating possibilities through brainstorming, many people put a premium on original ideas. Indeed, we commonly stop short of the kind of thinking needed to create novel ideas when we are generating possibilities for consideration. However, when we decide which possibility to accept, it is inappropriate to *prefer* original ideas to routine ones just because they are original. Possibilities are only possibilities. If routine possibilities still give us the best options, these should be preferred to novel options that may not work as well. Renting a helicopter and lowering myself to the restaurant on a rope ladder may be a novel way to get a meal, but I shouldn't prefer it just because it is novel.

Assessing Possibilities

To assure that we are open to all possibilities until we have a reason for preferring one, we should think about what would make any of these possibilities preferable to others. What are the criteria for choosing one possibility among the ideas that we are considering? In the case of decision making, we should choose the option that leads to the best consequences. In the case of causal and predictive hypotheses, selection should be made on the basis of the evidence. Asking "What would make a possibility I am considering a viable one?" helps us to keep in mind that these possibilities require support to be accepted.

On the other hand, in contexts in which originality and creative elaboration are of utmost concern, creativity may be a determining criterion for choosing among the possibilities being considered. When our purpose is to develop creative products, such as stories or plays (especially fantasy) or visual works of art, seeking original ways to express ideas and insights is important. Then, all other things being equal, it is appropriate to put a premium on the more creative possibilities.

The strategy for brainstorming that we use in this chapter is effective in these contexts as well. It can help us to generate original ideas about how to express ourselves, just as it can help us to generate original ideas about our solutions to personal or professional problems. It prepares us to think through which modes of expression will be more effective.

Contexts for Brainstorming

Brainstorming can be done by individuals or in groups. Research on brainstorming indicates that ideas come faster, and with more variety, if the brainstorming is conducted within a group in an open exchange of ideas. Group members often get ideas by listening to the ideas of others in the group.

Thinking Maps and Graphic Organizers for the Skillful Generation of Possibilities

The thinking map in figure 11.2 shows a sequence of questions to prompt skillfully generating possibilities for consideration. Notice that the sequence of questions moves from a request for quantity to a request for variety and then to a request for originality. This sequence is extremely important in developing viable, original possibilities.

GENERATING POSSIBILITIES SKILLFULLY

1. **What is the task for which you are considering possibilities?**

2. **What possibilities can you think of?**

3. **What are some other types of possibilities?**

4. **What original or unusual possibilities can you generate by combining possibilities already listed?**

5. **What information would you need in order to decide which of these possibilities is best for the task?**

Figure 11.2

The graphic organizer in figure 11.3 supplements the thinking map.

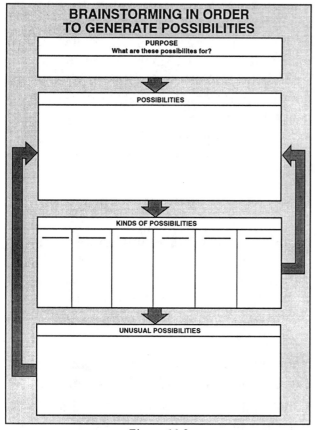

Figure 11.3

To use this graphic organizer, first state the purpose for the possibilities. For example, if they are for solving a specific problem, the problem should be mentioned. If they are determining the causes of something, that should be mentioned.

As free-wheeling brainstorming occurs, every possibility that is mentioned should be put in the next box. Suspending judgment about whether these are good ideas should be explicitly advocated.

The next box is used to organize the suggested possibilities into categories, to add other categories that are missing, and to add more possibilities under new and already established categories. Write the categories on the short horizontal lines, and list under each category the possibilities of that type. For example, if the possibilities are for getting to the restaurant, "In an internally propelled vehicle" might be one category. Under it I would list "by car," "by taxi," and "by bus." A new category that I might add is "In a

self-propelled vehicle." Under that I could add "by bicycle." Any additions should be circled and then added to the possibilities box.

A graphic organizer for the third step in this process is a webbing diagram, figure 11.4.

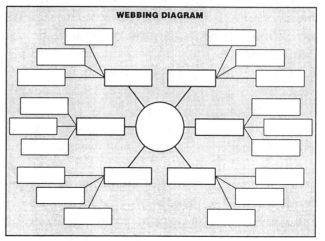

Figure 11.4

The purpose of the possibilities should be written in the circle. Types of possibilities already listed should be written in the boxes attached to the circle, leaving some of these free for new categories. Write possibilities that fall into these categories in the outer boxes. New possibilities should be added. All new possibilities should then be transferred to the main brainstorming diagram. The number of boxes on this diagram can be expanded as needed. A free-flowing cognitive map also serves this purpose.

The last step in using the brainstorming diagram (figure 11.3) involves generating original possibilities. We should transfer the original or unusual possibilities already listed in the top box to the lower one. Additional original ideas can be developed by combining features of other ideas from the original list and adding them to the lower box. The matrix in figure 11.5 (next page) can be used to combine ideas to develop more original and creative possibilities.

To use this matrix, list some of the categories that appear on the brainstorming diagram across the top and down the side. The ones listed down the side can be the same as or different from the ones across the top. Then, where feasible, combine them and develop a new possibility around the combination. For example, "Using a vehicle propelled by pedal action" might be one feature

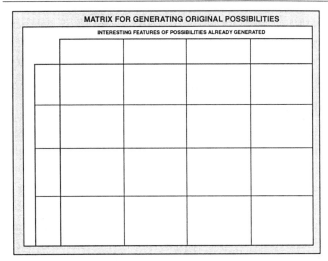

Figure 11.5

listed on the left and "Using a vehicle that can hold two people" on the top. Combining the two can yield "Using a bicycle built for two" in the intersecting cell.

Add the new possibilities generated in this way to the lower box of the brainstorming diagram (figure 11.3). They should be added to the possibilities box to create a large collection of varied possibilities, including a number of original ones, the goal of the first four steps.

Considering what criteria we should use to determine which possibilities are feasible, reasonable, or desirable rounds out this thinking skill. These criteria will, of course, be determined by the purpose of generating the possibilities, for example, whether these are options in decision making, possible causes, or possible uses of something.

How Can We Teach Students to Generate Possibilities Skillfully?

To teach students to generate possibilities skillfully, it is not enough to ask students simply to list possibilities for something. They should attend to variety in the types of possibilities that they come up with and should suggest original or unusual possibilities. Write down their suggestions as they think of them. With this guided practice, you should prompt your students to reflect on how they develop these ideas so that they can guide themselves in this strategy.

Encouraging original ideas through brainstorming is very important in the classroom. Even though some ideas may be very imagina-

tive, if not fantastic, and may be rejected later as not feasible, they should be both supported and encouraged. Permission to include even the most fantastic possibilities will allow students to start to develop broader, more creative habits of thought. Then, when highly imaginative ideas are needed (as in creative writing) these students will be adept at generating such ideas.

Contexts in the Content Areas for Generating Possibilities Lessons

Contexts in the curriculum in which students engage in decision making, problem solving, causal explanation, and prediction provide opportunities for teaching the strategy for generating ideas. Insofar as the science curriculum involves such contexts, it provides a rich array of contexts for generating possibilities to be considered. For example, you might design a lesson emphasizing generating options for a decision without guiding students in the whole process of decision making. What options might we have for cleaning up a hazardous chemical dump site? Or you may embed instruction in generating possibilities in the context of finding a cause for something or predicting the effects of a new process: What might be causing the upsurge in the gypsy moth infestations? What could have caused the asteroid belt? What are some possible results of the detonation of the first atomic bomb in the test in New Mexico?

Students particularly enjoy brainstorming the possible uses of something. Although this is commonly practiced in artificial, decontextualized activities (e.g., "What are some different uses of a brick?"), it is often an effective component in working through problems that call for new uses of a technology or creative solutions to hard problems. The various difficulties the Apollo 13 crew found itself in after part of their spacecraft was destroyed by an explosion provides a challenging context for students to generate possible ways of dealing with these difficulties. Finally, the uses of various pieces of apparatus in science classes also provides curriculum-robust contexts.

Although teaching students to generate ideas skillfully is important, it rarely is useful in isolation. Encourage students to reflect on occasions when it would be useful to engage in this type of thinking. This task is easy if generating possibilities is taught in the context of a decision problem, a problem about cause, etc. If not, it is important to ask students to suggest situations in which it would be useful to engage in this kind of thinking.

A menu of suggested contexts for infusion lessons on generating possibilities can be found at the end of the chapter.

Classroom Brainstorming Techniques

Brainstorming is a favorite activity in many classrooms today. Students quickly develop the risk-taking attitudes promoted by brainstorming ideas not judged to be right or wrong at the time. The enhanced version of brainstorming that we utilize in this chapter provides a generating-possibilities strategy in a simplified form easily mastered by students as they brainstorm.

Using a brainstorming strategy prompts students to draw on their knowledge and experience. Obviously, when students are asked to brainstorm in an area requiring background knowledge of a technical nature, their efforts will be hampered if they lack this background knowledge. Make sure that the topic that you ask students to brainstorm about is familiar enough to enable them to draw on an appropriate range of background knowledge. At the same time, you may be quite surprised at how much knowledge your students already have that can be brought out in brainstorming.

Brainstorming in a group is, therefore, especially important. It often leads to one student's ideas stimulating the thinking of other students—"piggybacking." In brainstorming possible uses for a graph, for example, one student may say "To record the growth of my family from my grandfather's day to the present." This may prompt another to say "To record world population growth from the beginning of this century to the present." The second student probably would not have thought of that possibility then if the first student had not mentioned the growth in size of her family. Encourage this type of cross-fertilization to enhance your stu-

dents' thinking. Perhaps the most direct way to do this is to explain the idea of piggybacking and then ask students to do it. They will easily.

Model Lessons on Generating Possibilities

We include two model lessons on the skillful generation of possibilities. The first is a middle school lesson in science/social studies on the use of natural resources and recycling what are considered waste products. It is very easy to extrapolate from this lesson to others in the science curriculum in which you want to help students develop more skill at brainstorming, especially when this involves developing creative and original ideas.

The second lesson included is a lesson in high school chemistry in which students tackle the question of the variety of ways of meeting the challenge of acidic pollution in the atmosphere. Here, the chemistry of acids is quite essential for students to master in developing possible ways of countering this pollution. Indeed, this lesson is so compelling that many students who would not otherwise have learned this material in chemistry become motivated to learn and use this information.

As you read these lessons, reflect on these questions:

- What specific strategies are included in this lesson to promote the generation of creative ideas by students?
- Are there other ways of helping students to recognize contexts for the use of this skill in their own lives?
- What additional examples can you find for reinforcing this skill after it is taught?

Tools for Designing Lessons on Generating Possibilities

The thinking maps and graphic organizers that follow can serve as photocopy masters, transparency masters, or as models that can be enlarged and used as posters in the classroom. Reproduction rights are granted for use in single classrooms only.

GENERATING POSSIBILITIES SKILLFULLY

1. What is the task for which you are considering possibilities?

2. What possibilities can you think of?

3. What are some other types of possibilities?

4. What original or unusual possibilities can you generate by combining possibilities already listed?

5. What information would you need in order to decide which of these possibilities is best for the task?

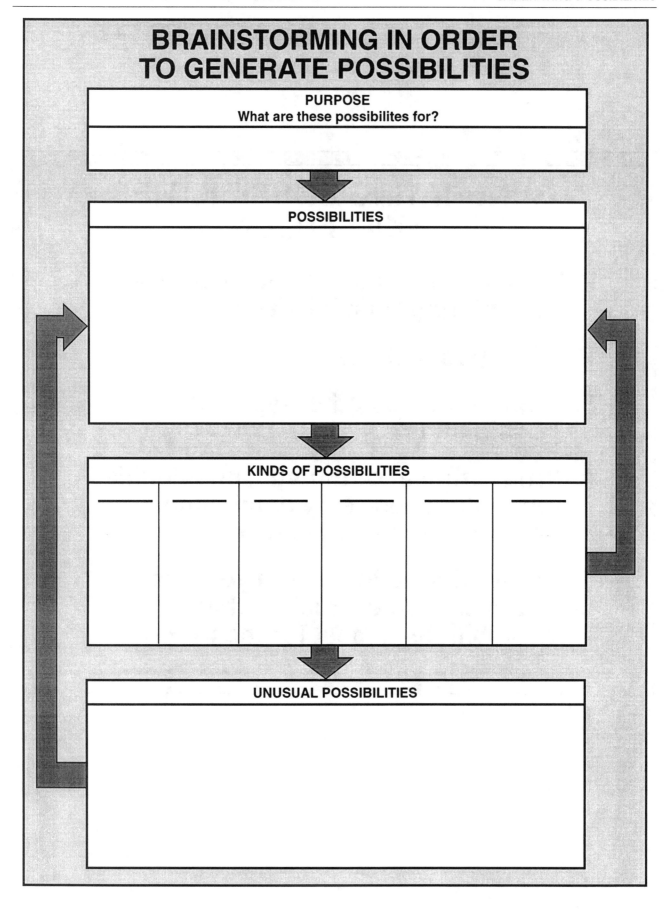

BRAINSTORMING IN ORDER TO GENERATE POSSIBILITIES

PURPOSE
What are these possibilites for?

POSSIBILITIES

KINDS OF POSSIBILITIES

UNUSUAL POSSIBILITIES

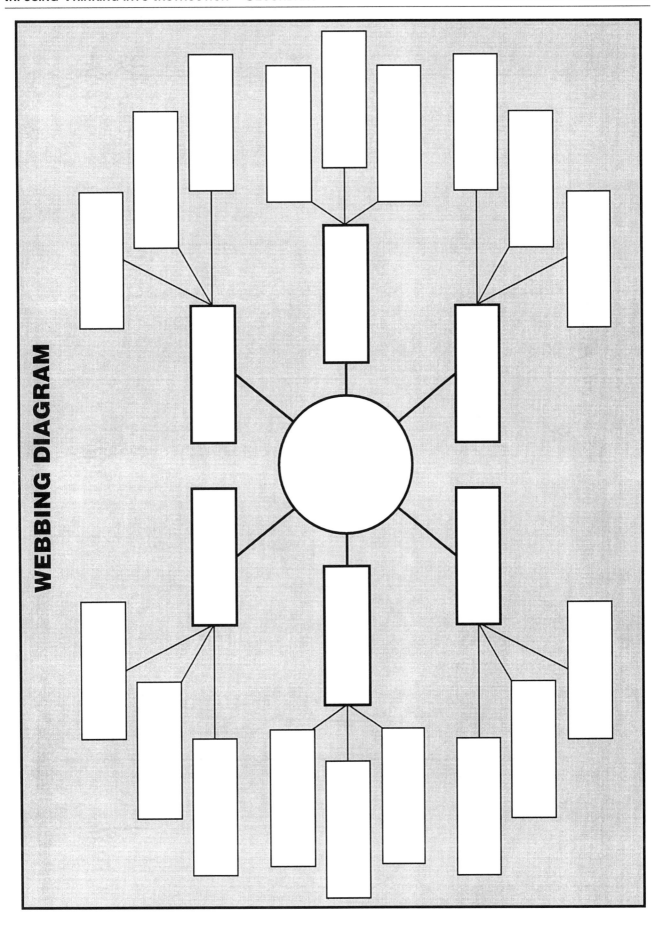

WEBBING DIAGRAM

MATRIX FOR GENERATING ORIGINAL POSSIBILITIES

INTERESTING FEATURES OF POSSIBILITIES ALREADY GENERATED

DIRT

General Science   Grades 6–9

OBJECTIVES

CONTENT

Students will learn that ordinary soil is a natural resource containing many components that have a variety of uses.

THINKING SKILL/PROCESS

Students will learn to generate possibilities skillfully in the context of solving a problem, by brainstorming a variety of ideas, and by combining ideas to generate new and original ones.

METHODS AND MATERIALS

CONTENT

Students will read in their textbooks about the composition of soil and will examine soil samples.

THINKING SKILL/PROCESS

An explicit thinking map, graphic organizers, and structured questioning facilitate a thinking strategy for generating uses for something. Collaborative learning enhances the thinking.

LESSON

INTRODUCTION TO CONTENT AND THINKING SKILL/PROCESS

- Have you ever solved a problem by coming up with an unusual way of doing something? For example, I dropped the key to my car into a storm drain on the street once, and luckily it landed on a ledge in the drain. However, when I put my fingers through the grate it was just out of my reach. And I realized that if I used a stick, that might knock it into the water. So I thought for a while. Then I realized how I could do it. I went into a drug store and bought some chewing gum. Then I chewed one piece. When it was soft, I stuck it to the end of a stick, pushed the stick through the grate, and got the key to stick to the gum. Then I had my key back!

- What I did was *problem solving*. I recognized that there was a problem, thought of a number of ways of solving it, thought about whether some of these solutions would work, and then figured out how to do it best. My solution was a creative solution because I used something that is not normally used for this purpose. Describe to your partner a time when you used something in a new way to solve a problem. After a few minutes, ask three or four students to describe what they had done. ANSWERS VARY.

- Whenever you come up with new ideas and put them into practice, you are doing *creative thinking*: You do something new that works that you haven't thought of before. Explain what you thought about that helped you discover a new way to use something to solve your problem. Students often say that they didn't have what they needed, so they started thinking about other things they could use or other ways to solve the problem. Some students say that they pictured using another object (like the coat hanger) altered in a way that would solve the problem. If using the object seemed to do the job, they tried it; if not, they considered another idea.

- We've discussed creative ways to solve a problem. You solved your problem because you thought of many ways to do it and picked a possible solution that worked. Whenever you come up with lots of new ideas, you are *generating possibilities*. This is a very important step in problem solving. You wouldn't solve your problem unless you thought of a good way to do it. We're going to try out a way to do this kind of thinking that will lead to many new and original ideas. These questions will guide us in this kind of thinking. Show the thinking map for generating possibilities skillfully.

- In this lesson, we're going to see if we can find new ways of using something that is very ordinary. We've been studying soil as an important natural resource, and how weathering erodes soil and washes it away. Soil is one of those natural resources that many people don't think is very valuable. But the discussion in our textbooks shows us how valuable it is for crops, as a bed for

> ## GENERATING POSSIBILITIES SKILLFULLY
>
> 1. **What is the task for which you are considering possibilities?**
> 2. **What possibilities can you think of?**
> 3. **What are some other types of possibilities?**
> 4. **What original or unusual possibilities can you generate by combining possibilities already listed?**
> 5. **What information would you need in order to decide which of these possibilities is best for the task?**

grasses and trees to grow in to keep it in place, and as a home for living things like earthworms that enrich it and, ultimately, contribute to its ability to support the growth of important crops. Let's review what we've learned about soil. Please break up into groups and each group work on one of these topics, writing down what you know about it:

> How does soil come into existence?
> What is soil composed of?
> How does soil support life?
> How is soil destroyed?

After a few minutes, ask each group to give a brief report. ANSWERS OFTEN INCLUDE: How does soil come into existence? *By mechanical and chemical weathering of rocks due to ice, frost, rain, etc. By chemical weathering due to the action of acids in rainwater, plant decay, etc. Oxidation of rocks and other materials. The deposit of dead organic matter like leaves, etc.* What is soil composed of? *Tiny mineral-laden rock particles, decayed organic matter, living microorganisms, and water.* How does soil support life? *Soil provides minerals, water, and anchoring for plants. Animals eat plants that grow in the soil.* How is soil destroyed? *Running water erodes it and washes it into rivers, wind blows it away when it is dry and has no trees or plants growing in it, excessive rain sometimes causes mud slides and landslides.*

- Sometimes soil, however, is thought of as dirt. Dirt is usually not considered to have any real value: We often get rid of it, clean it up, and throw it away. Let's consider a situation in which, typically, people have this attitude toward soil.

THINKING ACTIVELY

- Suppose someone you know is having a new house built and the workers have dug a large hole for the foundation. They've put the soil from the hole in a pile next to it. The pile of dirt is called a "by-product" of an industrial activity. Usually, a truck picks up the dirt and takes it somewhere to dump it, but these people don't like to throw away things that might be useful. They ask you to help them think of different uses for the pile of dirt. Then they'll try to figure out which is the best use for the dirt.

- One way to generate ideas is called "brainstorming." When you brainstorm, you try to think of as many ideas as you can about something. At this point, you don't try to decide whether these ideas will work. You can do that later. You just let your mind go and come up with as

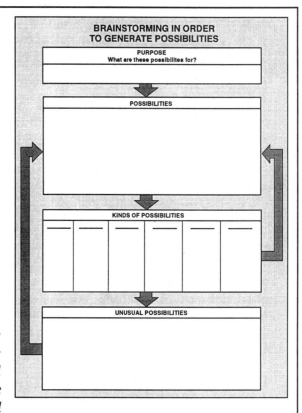

many ideas as you can. In this case, our purpose for brainstorming is to come up with a lot of ideas about different uses for the pile of dirt. **Write that purpose in the first box on the diagram for generating possibilities. Then brainstorm with a group of three other students. List the uses for dirt that you come up with in the possibilities box. Try to come up with five possibilities or more.** After about five minutes, ask the students whether they are finding it easy or difficult to come up with more ideas. Many say that they can't think of any more beyond the five or ten that their group has already listed on the diagram. Ask the groups to report their ideas. As they report, write their ideas on a large class diagram (on the board, on a transparency, or on a poster) for brainstorming to generate possibilities. POSSIBLE ANSWERS: *bring the topsoil to a farm and spread it over the farmland; spread the soil over the rest of the building site; advertise it and sell it to someone who needs it to fill in a hole; package it as potting soil for house plants; take the worms out and sell them for bait to fishermen; send it to schools to study life in the soil; leave it where it is for sledding in the winter; fill bags with it and give it to the city to use when the river is in danger of flooding.*

- **Many of you have said that it was difficult to come up with any more ideas. We're going to do something different now that will help you generate more possibilities. Work again in your groups and arrange the list of possibilities into categories. For example, you can put "Using it as a sled run" under the category "Used for Fun." Write these categories on the horizontal lines in the "Kinds of Possibilities" box. Under the horizontal lines, list the possibilities that fall into those categories. Try to think of new possibilities and add them under the appropriate categories. For example, under "Used for Fun," you might add "Bicycle run." Any new possibilities should be underlined and added to the possibilities box.** After a few minutes, ask for reports on some of the students' categories, as well as new possibilities that they've added. Write these on the class diagram, underlining the new ones. POSSIBLE CATEGORIES AND NEW POSSIBILITIES: *Moving the soil elsewhere (spread it over the building site, use it to fill a hole elsewhere, fill bags with it and give it to the city to use when the river is in danger of overflowing, bring it to the playground and make a hill with tunnels). Taking things out of the soil (extract the worms, have students in a science class extract the living things from the soil and study them, package the soil as potting soil for house plants, extract the rocks and use them in a rock garden). Used for fun (sled run, bicycle run, ski run, mud slide, mud castles, mud balls). Used for learning (students in a science class study the living things in the soil, test the impact of acid rain on the soil).*

- **Now work with your groups to add at least one new category. For example, a new category might be "Mixing it with other things." Next, list at least two possibilities under that category. For example, you could add "Mix it with water and make mud bricks" to this new category. What other things could you make by mixing the dirt with something else? Let's try this first in the whole class, then you add a new category to your diagram and list at least two new possibilities.** Ask students for new possibilities and add them to the diagram. Underline them.

SOME NEW POSSIBILITIES: *Mix it with water to make mud balls. Mix it with compost to make good top soil. Mix it with rocks to make dirt that can be used on a dirt road.* **I've underlined these new possibilities because they weren't on the original list; I will add them to the possibilities box. You do the same when you add possibilities under your new category.** Add the new possibilities to the possibilities box. After another few minutes, ask for some reports about other new categories and possibilities and add these to the class diagram. POSSIBLE NEW CATEGORIES AND NEW POSSIBILITIES: *Used as an environment for living things (grow grass on it and make it into a grassy hill; put a fence around it, put goats inside, and feed them regularly; make it into a huge ant hill; plant trees on it). Put things in it (bury trash in it, hollow it out and store apples there to keep them fresh, bury a treasure there for a treasure hunt). Use it for building (hollow it out and use it as a club house, put it on the roof of the house that is being built and make a sod roof). Use it for health and beauty aids (extract minerals from it, make a mud pack for a facial). Use it for decoration (make terraria with it, put it in flower pots and plant bulbs in the pots).* **Notice how many new possibilities are brought out by this way of organizing your thinking. Is this a good way to add to the variety of ideas you came up with in brainstorming? Do you think you would have thought of these possibilities if we hadn't categorized them first?** Many students say that they wouldn't have thought of these new ideas if we had only brainstormed. They reiterate that they were finding it difficult to add other ideas to their lists and say that categorizing helped them to come up with new ideas.

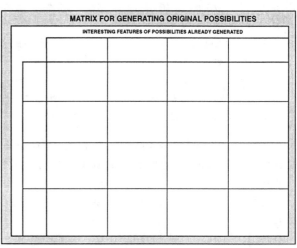

- **Now let's think about the ideas you've come up with that are unusual. These ideas involve your creative thinking. We're going to use a technique that will help you generate many more creative ideas. This time, work individually with your diagram for generating possibilities. From your possibilities list, choose all of the ideas that you think are unusual or original. Write these in the last box of the diagram. Now, let's generate some other unusual possibilities by using a different diagram. The matrix for generating original possibilities is used to combine ideas to generate new and original possibilities. Choose four of your categories and write them across the top and again down the side of the diagram. Then try to blend them together. Write in the boxes of the matrix any new possibilities that these combinations suggest.** After a few minutes, ask the students to select one of the newly developed possibilities that seems interesting or worth considering and share it with a partner. Then ask a few of the teams to report on one of their ideas. Ask them to explain why it seems worth considering. POSSIBLE ANSWERS: *Combining transporting the dirt with using it for something that is fun (take it up in a space shuttle and eject it into the upper atmosphere at night so that the particles will fall back to Earth and burn up to create a great meteoric display visible from the earth). Combining mixing the dirt with something else and using it in building (mix it with water and clay, make bricks, dry them, and use them to build a house). Combining moving it with using it for something that is fun (take it to the playground and make a hill with tunnels for children to play in).*

- **Now we've generated a list of many more interesting possibilities for using the pile of dirt. If you hadn't used this diagram for combining categories, would you have thought of these ideas?** Many students say they wouldn't have and that the diagram helped them to think of new possibilities.

- **When we began this activity, I said that people think that there is only one thing they can do with the pile of dirt left after digging a foundation: dump it. However, it took just a little** thinking to come up with a large number of ideas. We have to be careful, though. Sometimes, ideas that seem really great at the outset turn out to be not so good after we think about them a little. Discuss with your partner some circumstances in which it might make sense to use the soil in some of the ways now included in the possibilities box. For example, if the building site is damp and needs more fill, but fill is not easily obtainable in this region of the country, it might make sense to spread this dirt over the rest of the building site. With your partner, elaborate some circumstances for some of the other possibilities. Ask for some reports. ANSWERS VARY. For example, some students might say that taking the soil up in a space vehicle might make sense in the following circumstances: *the Fourth of July is coming soon and it is the 50th anniversary of the first space flight; some spectacular show is needed for the celebration and 50,000 people have contributed two dollars each to pay for the event.*

• **Work with your group. To help you decide which use of the soil is best, write the various things you would want to find out about each of them. Use the descriptions of the circumstances in which these uses make sense. Jot them down so that you can share them with the class.** POSSIBLE ANSWERS: *Whether there is a need to use the soil in this way, how much it would cost, whether it would please people if the soil were used in that way, whether anyone might get hurt if the soil were used in that way, whether there was a better way to do the thing without using the soil, whether it would take a long time to use the soil that way, whether the purpose for which we would be using the soil is an important one.*

THINKING ABOUT THINKING

• **What questions were important to answer as you did this kind of thinking?** Students should mention at least the questions that are on the thinking map for generating possibilities. If they are having trouble, point to some of these questions and ask if each was a question we asked. Students identify the following questions: *What are some possible ways to use the soil? What other kinds of uses are there besides the ones mentioned? Are there any unusual or original uses? How can you decide which use is best?*

• **What kind of thinking did these questions lead you to do?** POSSIBLE ANSWERS: *Generating possible uses of something. Generating possibilities. Brainstorming possibilities.*

• **How did you think of the ideas you first listed when you brainstormed?** POSSIBLE AN-SWERS: *I thought of one possible way of using the soil: by extracting the worms. Then I thought of other things that could be extracted, like seedlings for replanting. That led me to think about who does the replanting, usually a gardener. So I thought of how a gardener might use the soil, and I came up with planting a garden.*

• **Is this a good way to try to generate ideas? Why or why not?** POSSIBLE ANSWERS: *We can get a lot of ideas this way. We have a chance to develop some really original ideas that we can test out later.*

• **Consider the thinking map I showed you at the beginning of the lesson. Is this a good way to form questions to guide you for the next time you want to generate possibilities? If not, how would you change it?** Most students say that this is a good way to organize the questions.

• **In this activity, you worked in groups. Is this a good way to do brainstorming or would you rather work on your own? Why?** Most students say that they like to work in groups because they get ideas from other students. If it comes up, identify this as "piggy-backing." Some students, say, however, that they would rather work alone because other students don't let them talk or because what other students say makes them confused. If these problems come up, ask the class to treat it as a creative thinking/problem solving task: how can the group work be managed so that these problems do not arise?

- The next time you have to generate possibilities, how will you do it? Will the diagrams we used help you? What will you do with them? ANSWERS VARY.

APPLYING THINKING

Immediate Transfer

- When we studied ancient Egypt, we learned that the pyramids are a reminder of this great civilization. Imagine you are in charge of building the pyramids. The plan is to build them using massive stone blocks. These blocks must be quarried, transported to the site, and moved into place. Select one of these tasks, and do some creative thinking to generate possible ways of accomplishing it.

- Think about the long trip west to California that many people made in the 19th Century. Imagine that your class is trying to compose a musical that will represent the pioneering spirit as well as the dangers of the trip. You won't be composing any music yourselves, but you may select any popular or classical music that you think will convey the ideas and feelings of the California movement. Brainstorm different ways of starting the musical that combine music, stage setting, and a story.

Reinforcement Later

- Many wars have been fought because of disputes of different countries over territory, resources, etc. Pick a war that you have studied. Brainstorm alternatives to violence as a way of settling the dispute that caused the war. What might the countries have done to avoid fighting with each other?

- Suppose that a veterinarian has prescribed large capsules that your pet dog must swallow once a day. List as many different ways you can think of to get your dog to take the capsule. Include lots of creative ideas. What would you need to find out in order to choose the best way to get your dog to take the medication?

ASSESSING STUDENT THINKING ABOUT POSSIBILITIES

To assess the ability to generate possibilities, select examples that challenge students to generate a wide range of ideas. Any of the transfer examples can serve as assessment items for written or oral responses. Developing multiple uses for common objects is a task that is often used to demonstrate this kind of thinking. Challenging problem-solving tasks in which you emphasize generating alternative solutions are also excellent vehicles for this type of assessment. Use the thinking map for generating possibilities as a guide to check that students are focusing their attention on the three basic factors for generating good ideas: quantity, variety, and originality.

Sample Student Responses • The Pile of Dirt

BRAINSTORMING IN ORDER TO GENERATE POSSIBILITIES

PURPOSE
What are these possibilites for?

The use of a pile of dirt left after a foundation hole is dug at a building site.

POSSIBILITIES

Bring it to a farm and spread it over the farmland.	Use it as a hill for a bicycle run.
Spread it over the rest of the site.	Use it as a mud slide.
Put up a sign that says "Fill" and give it away.	Mix it with water to make mud balls.
Sell it to someone who needs fill.	Mix it with compost to make rich top soil.
Package it as potting soil for plants.	Mix it with rocks to put on a dirt road.
Fill the hole and build a house without a basement.	Grow grass on it.
Extract the worms and sell them.	Make it into a fenced goat hill.
Use it in school to study life in the soil.	Make it into a huge ant hill.
Use it as a winter sledding run.	Plant trees on it for a shady knoll.
Bag it and use it on river banks to prevent flooding.	Bury trash in it.
Extract the rocks for a rock garden.	Store apples in it.
Bury a treasure in it for a treasure hunt.	Use it for a sod roof on your new house.
Hollow it and use it as a clubhouse.	Start a "Soil Recycling" business.
Use it for landscaping.	Extract nutrients and sell it as "Natural Fertilizer."
Test the impact of acid rain.	Extract the living things and use the rest as sterile fill.

KINDS OF POSSIBILITIES

Moving it Elsewhere	Extracting Things	Conserving It	Using it for Learning	Sell It
Spread over site	Sell worms	Spread it on farmland	Study living things	Sell it for fill
Fill hole elsewhere	Study living things	Refill the hole	Test impact of acid rain	Put it in pots and plant bulbs
Bags for flooding	Potting soil	Spread it over site		
	Rocks for garden	Use if for landscaping		

Using it as a Habitat	Putting Things In It	Using it for Fun	Mixing something in	Building With It
Grow grass	Bury trash	Winter sled run	Water for mud balls	Hollow for club-house
Goat hill	Store apples	Bicycle run	Compost for top soil	Mix it with water and make mud bricks
Ant hill	Bury treasure	Winter ski run	Rocks for dirt for road	
Plant trees		Mud slide		

UNUSUAL POSSIBILITIES

Extract and sell worms.	Start a soil recycling business.
Hollow it for a clubhouse.	Extract and sell natural fertilizer from it.
Use it for mud packs for facials.	Refill the hole after extracting the richest topsoil and settle for a crawlspace.
Bury a treasure for a treasure hunt.	Package and sell it as a biology or earth science lab kit.
Extract rocks for a rock garden.	Extract rocks and pebbles and use these under drain pipes.
Make it into a fenced goat hill.	Remove the living things and sell it as sterile fill.
	Make a sod roof for your house with it.

Sample Student Responses • The Pile of Dirt

	Move It Elsewhere	Extract Things From it	Conserve It	Sell it for Profit
Sell It for Profit	Move it to a large space you have rented, advertise for other "left-over" soil, sort what you get into different categories representing quality, and set up a "Soil Recycling" business.	Package it in small containers and market it as a biology or earth science Lab Kit for the study of the mineral or organism content of soil.	Sell it to farmers to replace lost topsoil that was eroded away by the action of wind and rain.	X
Conserve It	Give it away to builders who need fill for their building sites.	Remove the richest layer of topsoil and save for landscaping; then, refill most of the basement hole and settle for a crawlspace.	X	Sell it to a nursery or a gardening shop to use as potting soil.
Extract Things From it	Extract rocks, pebbles and bolders and use as gravel for drainage basins under the drainpipes that draw off water from the roof gutters. Give the soil that is left away to neighbors as "free topsoil" to be used for filling in flower beds or patio planters.	X	Sift out rocks; then build an enclosure to store sifted topsoil and rocks for future gardening and landscaping around the house.	Extract phosphorus, nitrogen and organic material such as leaves and roots; then sell this material in bags as "natural" fertilizer.
Move It Elsewhere	X	Extract all the living things from it and bring them into science class for classification and study. Figure out how they all got there. Move the rest of it to a spot where sterile fill is needed.	Use it to make a sod roof for your new house. Move it to the roof and plant grass seed in it.	Barter it with the township for use in filling potholes in the neighborhood in exchange for the township paving your new driveway.

INTERESTING FEATURES OF POSSIBILITIES ALREADY GENERATED

ACIDS IN THE ATMOSPHERE

Chemistry/Environmental Sciences   **Grades 9–12**

OBJECTIVES

CONTENT

Students will learn about the chemical interactions between industrial discharges into the atmosphere, emissions from automobiles, and the environment.

THINKING SKILL/PROCESS

Students will learn to generate possibilities skillfully in the context of solving a problem by brainstorming a variety of ideas and by combining ideas to generate new and original ones.

METHODS AND MATERIALS

CONTENT

Students will review in their textbooks material about nitrogen and sulfur compounds and oxidation and reduction reactions. They will then examine source material on the chemistry and technology of acid rain management.

THINKING SKILL/PROCESS

An explicit thinking map, graphic organizers, and structured questioning facilitate a thinking strategy for generating possible uses for something. Collaborative learning enhances thinking.

LESSON

INTRODUCTION TO CONTENT AND THINKING SKILL/PROCESS

- Sometimes, the most difficult and challenging problem is solved only when it is approached in a novel manner. Finding unique and unusual solutions requires creative thinking. For example, during a period beginning in 1881 and lasting more than 15 years, the French Company that had successfully built the Suez canal labored unsuccessfully to construct a canal connecting the Atlantic and Pacific oceans. After having investigated possible sites, including Nicaragua, the company chose the Isthmus of Panama because it was the shortest route. The project, nonetheless, seemed hopeless when the company investigated possible routes in Panama. Each required removal of a gargantuan amount of dirt and rock, in some areas tantamount to moving mountains. As might have been expected, they failed miserably, hardly making an impression on the dense jungle and rough terrain blocking the 50-mile route. Ultimately, disease, construction problems, and inadequate financing ended the project.

- The United States took up the project—and the challenge—with the financial backing of the federal government in 1904. The jungle, disease, and construction challenges still remained. Those working on the project for the United States didn't want to make the same mistakes as the French, however. Rather than again attempting the conventional approach of trying to carve a canal through miles of sometimes mountainous jungle, the team decided to consider some unusual and unconventional possibilities. The stunning solution to this problem lay in viewing a canal not so much as a thin ribbon of water snaking its way between two oceans, but rather as *any* utilizable or buildable body of water along the route. Instead of just digging a trench for the canal, the Americans simply dammed up an existing river, and created an enormous lake. They then built short stretches of canal from the Atlantic Ocean to one end of the lake and from the opposite end of the lake to the Pacific Ocean. The Panama Canal, as we know it today, took 10 years to complete and most probably would have never been finished had it not been for the willingness of the team to break away from the norm and to consider different ways of creating a body of water that could support major oceangoing vessels traveling from one ocean to another. An option that can enhance the use of this example involves showing pictures and maps of the Suez Canal, basically a ditch in the desert. Then, show the jungle terrain typical of the Isthmus of Panama and the Panama Canal, including the open areas of lake that ships pass through.

- What led to the construction of the Panama Canal is a combination of the willingness to depart from the norm and try new things and the ability to generate ideas for solving problems that are original and different. This is the essence of *creative thinking.* **Can you think of any other ways of getting goods from the Atlantic Ocean to the Pacific without having to transport them around Cape Horn at the tip of South America?** Do a brief, open brainstorming session with them on this topic. POSSIBLE ANSWERS INCLUDE: *unloading ships on one shore, transporting the goods overland, and then loading up other ships on the other shore; lifting whole ships onto railroad cars and transporting them across; building a huge conveyor belt across Panama.* Students usually come up with a limited (often not more than three or four possibilities) but interesting list of possibilities, including some original ideas. **Of course, not all novel and creative ideas work, as you might suspect from your list. Good problem solving involves the ability to come up with new and original ideas, like you just did, but that is not enough. You must also have the ability to determine, through *critical thinking*, that these ideas will solve the problem well without serious unwanted side-effects.**

- **Can you think of an example from your own experience of doing something in a different way from the way it is usually done?** Ask students to jot down one or two examples on a piece of paper and then to discuss one of them with a partner. After three or four minutes of discussion, ask for volunteers to share an example with the rest of the class. STUDENT EXAMPLES VARY. **Now try to come up with some other ways you might have done what you did.** STUDENT ANSWERS VARY, but again are limited—though they may include some original ideas.

- **In this lesson we are going to learn, assess, and develop our own strategy for generating creative ideas. We will build on the technique that you just used to do this. Does anyone know what that technique is called?** Students readily recognize this as brainstorming. Write the word "Brainstorming" on the board. **How do we do brainstorming?** Once again, students usually can describe the techniques used in brainstorming: *We think rapidly of different ideas for something we are concerned with, sometimes suggested by the ideas of others, and we don't yet pass judgment on whether the idea will work.* **This lesson takes us beyond ordinary brainstorming. Often, in ordinary brainstorming, we run out of ideas pretty quickly. In this lesson, we will learn how to generate ideas more easily, in quantity, with great variation and, especially, we will make sure that some of the ideas we are generating are original, new, and creative ideas. Of course, once we have generated such ideas we will then be ready to critically scrutinize them. Here's a thinking map that will guide us to do this type of thinking skillfully:**

- **We will try out this new way to brainstorm ideas in this lesson as we think about a problem that mankind faces today that is a by-product of one of the basic ways that we have achieved progress**

GENERATING POSSIBILITIES SKILLFULLY

1. What is the task for which you are considering possibilities?

2. What possibilities can you think of?

3. What are some other types of possibilities?

4. What original or unusual possibilities can you generate by combining possibilities already listed?

5. What information would you need in order to decide which of these possibilities is best for the task?

in this century, through the invention, mastery, and use of new technologies which involve *chemical* processes. Some of these new technologies produce harmful by-products; and although people are protected from many of these harmful by-products by safety measures, some by-products do change our environment and can cause harm to it and to us.

• We will be thinking specifically in this lesson about *air pollution.* Although there are many different air pollutants, including arsenic (As), benzene (C_6H_6), cadmium (Cd), chlorine (Cl_2), fluoride (F^-), carbon monoxide (CO), formaldehyde (HCHO), mercury (Hg), ozone (O_3), lead (Pb), and silicon tetrafluoride (SiF_4), the oxides of nitrogen (NO, NO_2) and sulfur (SO_2, SO_3) are among the most environmentally dangerous. They react with the atmosphere to produce caustic acids that often fall to earth as acid rain. This causes damage to animal and plant tissues and to many man-made structures like buildings and monuments. The primary sources of these pollutants are (1) the burning of fossil fuels in power plants and heavy industry and (2) the exhaust gases of internal combustion engines. The benefits of the technologies that we use that produce these effects (for example the automobile) make us very reluctant to give up their use. So we face a problem: How might we reduce or eliminate the harmful effects of the acidic pollutants that enter the atmosphere from the use of technologies that rely on chemical processes that produce the ingredients for such acids as by-products? In this lesson, we will try to generate ideas about the problem of air pollution using brainstorming techniques.

THINKING ACTIVELY

• We have already studied nitrogen and sulfur and their common compounds. We have also investigated acids and bases and learned about oxidation and reduction reactions. The chemistry of acid rain involves reactants, chemical reactions, and products that should be familiar to you, all of which lead to the presence of harmful ions that cause the real damage to living and non-living things. Let's fill in this chart to map the chemical processes involved in the production of acid rain. Work in groups to provide information that we can use for the chart. Use your textbooks as resources. Break the students into groups of three or four and assign half the class sulfur-based reactions and the other half nitrogen-based reactions. After five minutes or so , ask for reports from the groups about the reactants, about the reactions, and about the products in each of these reactions. If the students have difficulty filling in this

CHEMICAL REACTIONS		
TOPIC: ACIDS IN THE ATMOSPHERE		
Reactants	Reactions	Products
Step I	**Summary of Chemical Reactions** Fossil fuels such as natural gas, oil, and coal (containing iron pyrite) are burned in industry and in automobiles to produce sulfur dioxides, nitric oxide, and other by-products.	
Step II	**Summary of the Chemical Reactions** Gaseous oxides combine with atmospheric water vapor and oxygen to form sulfuric and nitric acid.	
Step III	**Summary of the Chemical Reactions** Nitric and sulfuric acids dissociate into hydrogen, nitrate and sulfate ions.	
Step IV	**Summary of the Chemical Reactions** Nitric and sulfuric acids dissociate in atmospheric water vapor and droplets into hydrogen, nitrate and sulfate ions, which lower the pH in lakes and ponds, react with stone on buildings, etc.	

chart ,provide them with resource material on acid rain like that included at the end of this lesson. Enter the reports on a master-chart on the chalkboard or on a transparency.

The result should look like this:

CHEMICAL REACTIONS TOPIC: ACIDS IN THE ATMOSPHERE		
Reactants	**Reactions**	**Products**
Step I	**Summary of Chemical Reactions** Fossil fuels such as natural gas, oil, and coal (containing iron pyrite) are burned in industry and in automobiles to produce sulfur dioxides, nitric oxide, and other by-products.	
S, O_2	$S + O_2 + Heat \rightarrow SO_2$	SO_2
O_2, FeS_2 (iron pyrite)	$4FeS_2 + 11O_2 + Heat \rightarrow 2Fe_2 + 8SO_2$	SO_2, FeO_2
N_2, O_2	$N_2 + O_2 + Heat \rightarrow 2NO$	NO
Step II	**Summary of the Chemical Reactions** Gaseous oxides combine with atmospheric water vapor and oxygen to form sulfuric and nitric acid.	
SO_2, O_3 (ozone)	$SO_2 + O3 \rightarrow SO_3 + O_2$	SO_3, O_2
SO_2, O_2	$SO_2 + \frac{1}{2}O_2 \rightarrow SO_3$	SO_3 (slow reaction)
NO, O_3	$NO + O_3 \rightarrow NO_2 + O_2$	NO_2, O_2
Step III	**Summary of the Chemical Reactions** Nitric and sulfuric acids dissociate into hydrogen, nitrate and sulfate ions.	
SO_3, H_2O	$SO_3 + H_2O \rightarrow H_2SO_4$	H_2SO_4
NO_2, H_2O, O_2	$4NO_2 + 2H_2O + O_2 \rightarrow 4HNO_3$	HNO_3
Step IV	**Summary of the Chemical Reactions** Nitric and sulfuric acids dissociate in atmospheric water vapor and droplets into hydrogen, nitrate and sulfate ions, which lower the pH in lakes and ponds, react with stone on buildings, etc.	
H_2SO_4	$H_2SO_4 \rightarrow 2H^+ + SO_4^{2-}$	H^+, SO_4^{2-}
HNO_3	$HNO_3 \rightarrow H^+ + NO_3^-$	H^+, NO_3^-

- Now that we are confronted with this environmental problem, let's see if we can come up with ideas to reduce this kind of air pollution and damage caused by acid rain. Let's first do some simple brainstorming.

- Write the purpose of the brainstorming in the first box on this diagram for generating possibilities: Students write statements like: *"To find ways of reducing or eliminating acidic air pollution caused by the combustion of sulfur and nitrogen compounds, and/or the damage caused by the acid rain that results."* **Now brainstorm with a group of three other students. List the possibilities that you come up with in the possibilities box. Try to come up with at least five possibilities.** After about four or five minutes, ask the students whether they are finding it easy or difficult to come up with more ideas. Many say that they can't think of any more beyond the five or ten that their group has already listed on the diagram. Ask the groups to report their ideas. As they report, write their ideas on a large class diagram (on the board, on a transparency, or on a poster) for brainstorming to generate possibilities. POSSIBLE ANSWERS: *Make it illegal to burn fossil fuels. Use battery-powered vehicles. Make more buses available. Locate factories/power plants near dams. Use better gasoline (higher octane). Use better catalytic converters. Make it illegal to drive alone in a car. Bring back more passenger trains. Harness the energy of volcanos. Put all highways underground. Make cars that hold more passengers. Make better scrubbers on smokestacks. Make stiffer fines for polluting air. Educate consumers to stop wasting energy. Use geothermal energy. Use solar energy.*

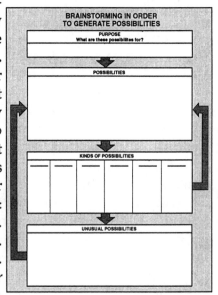

- Many of you have said that it was difficult to come up with any more ideas. We're going to do something different now that will help you generate more possibilities. Work again in your groups and arrange the list of possibilities into categories. For example, you can put "Use solar energy" under the category "Using an alternative source of energy." Write these categories on the horizontal lines in the "Kinds of Possibilities" box. Under the horizontal lines, list the possibilities that fall into those categories. Try to think of new possibilities and add them under the appropriate categories. For example, under "Using an alternative source of energy," you might add "Lasers" or "Nuclear reactors" Any new possibilities should be underlined and added to the possibilities box. After a few minutes, ask for reports on some of the students' categories, as well as new possibilities that they've added. Write these on the class diagram, underlining the new ones. POSSIBLE CATEGORIES AND NEW POSSIBILITIES: *Using a new alternative energy source (lasers, nuclear reactors, plasma fusion, flywheels). Improving existing technology and/or treatment of existing air pollution (more efficient engines in automobiles, seal houses, people should wear spacesuits, develop community-scale air scrubbers, stay indoors). Making new laws, levying taxes, providing subsidies (tax incentives for clean air, tax incentives for the use of solar power). Reducing demands for power (give out free bicycles).*

- Now, work with your groups to add at least one new category. For example, if one of these categories had been omitted, it could be added as a new category. Then you could list a whole set of new possibilities that fall into this new category. Underline any new possibilities that fall into this new category and move them up to be added to the possibilities box. After a few minutes, ask for some reports about other new categories and possibilities and add these to the class diagram. POSSIBLE NEW CATEGORIES AND NEW POSSIBILITIES: *Treat the acid rain that falls as a result of air pollution (add bases to lakes to neutralize the acids, use special acid-neutralizing fertilizer). Make things affected by acid rain acid-resistant (put an acid-resistant coating on buildings, put acid-resistant protective shields above crops, wear acid-resistant clothing).*

- Notice how many new possibilities are brought out by these two ways of organizing your thinking when you brainstorm. Is this a good way to add to the variety of ideas you came up with in brainstorming? Do you think you would have thought of these possibilities if we hadn't categorized them first or developed new categories? Many students say that they wouldn't have thought of these new ideas if we had only brainstormed. They reiterate that they were finding it difficult to add other ideas to their lists and say that categorizing helped them to come up with new ideas.

- Now let's think about the ideas you've come up with that are unusual. These ideas involve your creative thinking. We're going to use a technique that will help you generate many more creative ideas. This time, work individually with your diagram for generating possibilities. From your possibilities list, choose all of the ideas that you think are unusual or original. Write these in the last box of the diagram. Now, let's generate some other unusual possibilities by using a different diagram. The matrix for generating original possibilities is used to combine ideas to generate new and original possibilities.

MATRIX FOR GENERATING ORIGINAL POSSIBILITIES			
INTERESTING FEATURES OF POSSIBILITIES ALREADY GENERATED			

Choose four of your categories and write them across the top and again down the side of the diagram. Then try to blend them together. Write in the boxes of the matrix any new possibilities that these combinations suggest. After a few minutes, ask the students to select one of the newly developed possibilities that seems interesting or worth considering and share it with a partner. Then ask a few of the teams to report on one of their ideas. Ask them to explain why it seems worth considering. POSSIBLE ANSWERS: *Combine using an alternative energy source with treatment of existing air pollution or its effects. (Use solar collectors to cover the surface of smokestacks and supply power to run scrubbers to clean exhaust gas; use windmills on vehicles to produce power to operate an electrostatic exhaust gas scrubber). Combine reducing consumption on fossil fuels with treatment of existing air pollution and its effects. (Enclose cities in clear domes, purify the atmosphere under the dome, and produce heat by the greenhouse effect).*

• Now, we've generated a list of many more interesting possibilities for reducing or eliminating harmful acidic air pollution and its effects. If you hadn't used this diagram for combining categories, would you have thought of these ideas? Many students say they wouldn't have and that the diagram helped them to think of new possibilities.

• Even though it took just a little thinking to come up with a large number of ideas and some of them seem very interesting, we have to be careful. Sometimes, ideas that seem really great at the outset turn out to be not so good after we think about them a little, and ideas that seem doomed to failure turn out to be almost workable after some thought and perhaps some modifications. When we move beyond brainstorming possible solutions to a problem and start to sort out which ideas work best, we are combining important critical-thinking skills with the brainstorming to engage in skillful, creative problem solving. We will learn and practice this technique shortly. But for now, and to prepare for it, let's do one concluding activity about ways of dealing with acidic air pollution. Discuss with your partner some circumstances in which it might make sense to do some of the more unusual things now included in the possibilities box. For example, if there are powerful tidal currents in a coastal community that typically suffers from smog due to its heavy industries, it might make sense to undertake a project that generates energy from tidal sources. With your partner, elaborate some circumstances for some of the other possibilities. Ask for some reports. ANSWERS VARY. For example, some students might say that putting solar collectors on smokestacks might make sense in the following circumstances: *The climate is a sunny climate, solar collectors are relatively inexpensive, and solar energy technology has advanced to the extent that the energy demands of smokestack scrubbers can be handled by the energy produced by the solar panels alone.*

• Work with your group. To help you decide which ideas are the most promising, write the various things you would want to find out about each of them. Use the descriptions of the circumstances in which these uses make sense. Jot them down so that you can share them with the class. POSSIBLE ANSWERS: *Whether there is enough acidic air pollution in a region to be harmful, how much it would cost, whether it would please people if such a remedy were adopted, whether anyone might get hurt if the specific possibility was put into practice, whether it would take a long time to put the solution into practice.*

THINKING ABOUT THINKING

• What questions were important to answer as you did this kind of thinking? Students should mention at least the questions that are on the thinking map for generating possibilities. If they are having trouble, point to some of these questions and ask if each was a question we asked. Students identify the following questions: *What are some possible ways to reduce the harmful effects of acidic air pollution? What other kinds of ways to do this are there besides the ones mentioned? Are there any unusual or original uses? How can you decide which use is best?*

- **What kind of thinking did these questions lead you to do?** POSSIBLE ANSWERS: *generating possible solutions to a problem, generating possibilities, brainstorming possibilities.*

- **How did you think of the ideas you first listed when you brainstormed?** POSSIBLE ANSWERS: *I thought of one possible way of reducing the harm from air pollution—limiting the use of the fuels that pollute by making some illegal. Then I thought of other ways of limiting the use of pollutants—using more alternative sources of energy. That led me to think about who does the limiting, usually lawmakers. So I thought of how else a lawmaker might reduce the effects of acidic air pollution and came up with laws that require carpooling.*

- **Is this a good way to try to generate ideas? Why or why not?** POSSIBLE ANSWERS: *We can get a lot of ideas this way. We have a chance to develop some really original ideas that we can test later.*

- **Consider the thinking map I showed you at the beginning of the lesson. Is this a good way to form questions to guide you for the next time you want to generate possibilities? If not, how would you change it?** Most students say this is a good way to organize the questions.

- **In this activity, you worked in groups. Is this a good way to do brainstorming, or would you rather work on your own? Why?** Most students say that they like to work in groups because they get ideas from other students. If it comes up, identify this as "piggybacking." Some students say, however, that they would rather work alone because other students don't let them talk or because what other students say makes them confused. If these problems come up, ask the class to treat it as a creative-thinking/problem-solving task: How can the group work be managed to avoid these problems?

- **The next time you have to generate possibilities, how will you do it? Will the diagrams we used help you? What will you do with them?** ANSWERS VARY.

APPLYING THINKING

Immediate Transfer

- Study the history of the toxic waste problem at Love Canal in New York State. This toxic waste problem caused severe adverse health effects and ultimately drove people from their homes and from this community. Develop a chart about the specific chemicals involved in this case, where they came from (including the chemical reactions used to produce them), and their hazards. Use the enhanced brainstorming techniques we employed in the lesson on acids in the atmosphere to develop ideas about what might be done to "clean up" this site.

- Use the same techniques to brainstorm what might be done to prevent such calamities in the future while at the same time not compromising the need for the use of the types of chemical processes that ultimately caused this calamity.

- When we studied ancient Egypt, we learned that the pyramids are a reminder of this great civilization. Imagine you are in charge of building the pyramids. The plan is to build them using massive stone blocks. These blocks must be quarried, transported to the site, and moved into place. Select one of these tasks and generate possible ways of accomplishing it.

Reinforcement Later

- Brainstorm possible ways to keep pests from destroying crops without creating a human pesticide hazard.

• Many wars have been fought because of disputes over territory, resources, etc. Pick a war that you have studied. Brainstorm alternatives to violence as a way of settling the dispute that caused the war. What might the countries have done to avoid fighting?

• Select some dispute that you have been involved in. Brainstorm ways of settling this dispute using the techniques employed in the lesson on acids in the atmosphere.

CONTENT EXTENSION

Use the resource material included in this lesson to summarize the various methods that have been used to solve the problem of acid rain. When chemical reactions are involved, use the Chemical Reaction graphic to summarize and outline the chemistry of these solutions.

THINKING SKILL EXTENSION

Consider the various methods used to try to solve the acid rain problem and spell out their pros and cons. Of all the solutions presently available and in practice, decide which solution or combination of solutions to this problem is best. Write a recommendation to the Department of Energy in the federal government explaining your recommendation and why you think it is best.

ASSESSING STUDENT THINKING ABOUT POSSIBILITIES

To assess the ability to generate possibilities, select examples that challenge students to generate a wide range of ideas. Any of the transfer examples can serve as assessment items for written or oral responses. Developing multiple uses for common objects is a task that is often used to demonstrate this kind of thinking. Challenging problem-solving tasks in which you emphasize generating alternative solutions are also excellent vehicles for this type of assessment. Use the thinking map for generating possibilities as a guide to check that students are focusing their attention on the three basic factors for generating good ideas: quantity, variety, and originality.

ACIDS IN THE ATMOSPHERE
SOURCE MATERIAL ON THE IMPACT OF ACID RAIN

Acid rain is caused by the emission of exhaust gases into the atmosphere from the burning of fossil fuels. Combustion of coal and oil by the transportation, oil refining, power generating and heavy industries of the world all contribute to acid formation in the atmosphere. Gasoline-burning vehicles and coal-burning, electricity-generating power utilities are the main culprits. Electricity production accounts for about 30 percent of air pollution worldwide. Annually, automobiles and power plants release more than 125 million tons of acid-forming molecules into the atmosphere.

Water vapor in the unpolluted atmosphere is normally somewhat acidic (pH 5.6) because of oxidation of atmospheric carbon dioxide to form carbonic acid (H_2CO_3).

$$CO_2 + H_2O \rightarrow H_2CO_3 \rightarrow H^+ + HCO_3^-$$

However, heavily industrialized regions of the United States and Europe routinely produce a pH of precipitation of about 4. There are cases on record where rain has been as acidic as lemon juice (Norway, 1987) and in one incident at Kane, Pennsylvania, as acidic as vinegar.

Acid rain is made up of a mixture of nitric (HNO_3) and sulfuric (H_2SO_4) acids carried on airborne dust particles and in water vapor. These acids come from nitrogen and sulfur oxides produced as a by-product of the combustion of fossil fuels. For example, sulfur impurities commonly found in coal used to produce electricity in power plants are oxidized during combustion to form sulfur dioxides, which then combine with water vapor and air in the atmosphere to form sulfuric acid.

For example, iron pyrites (FeS_2) found in coal react with oxygen during combustion to produce sulfur dioxide:

$$4FeS_2 + 11O_2 \rightarrow 2Fe_2O_3 + 8SO_2$$

Sulfur dioxide also comes from the combustion of sulfur found in organic compounds:

$$S + O_2 + Heat \rightarrow SO_2$$

Sulfur dioxides are then further reacted several ways with atmospheric oxygen and ozone to produce sulfur trioxide:

$$SO_2 + O_3 \rightarrow SO_3 + O_2 \qquad Fast$$
$$SO_2 + \tfrac{1}{2}O_2 \rightarrow SO_3 \qquad Very\ slow$$

Sulfur trioxide then reacts with water to produce sulfuric acid:

$$SO_3 + H_2O \rightarrow H_2SO_4$$

Nitric oxide reacts in the atmosphere to produce nitrogen dioxide (NO_2). One mechanism for its formation involves ozone (O_3):

$$NO + O_3 \rightarrow NO_2 + O_2$$

Nitrogen dioxide is then converted in the atmosphere into nitric acid (HNO_3):

$$4NO_2 + 2H_2O + O_2 \rightarrow 4HNO_3$$

Precipitation of sulfate and nitrate ions can decrease the pH of a mountain lake to toxic levels. For example, half of the lakes in Sweden are acidic and half of the forests in West Germany have been damaged by acid rain.

When the pH of a mountain lake drops as a result of an increase in the acid content of the atmosphere, aquatic animals and plants sensitive to small changes in pH are adversely affected, often dying. It does not take a large change in pH to produce major effects. For example, at a pH of 5.4, the fish population drops off precipitously. At a pH of 5.2, snails disappear. Many plants also become threatened by the effects of acid rain on their leaves.

Acid rain has a very harmful effect on human artifacts, including buildings, and especially buildings and monuments made of marble. Sulfuric acid breaks down marble (crystallized calcium carbonate) into calcium sulfide, a much softer material, and releases carbon dioxide, according to the following reaction:

$$H_2SO_4 + CaCO_3 \rightarrow H_2O + CaSO_4 + CO_2$$

Pitting on many public buildings and monuments is often a sign of the effects of acid rain.

Reference: *Environmental Chemistry*, David E. Newton, J. Weston Walch, Publisher

ACIDS IN THE ATMOSPHERE

Source Material on Countermeasures to Acid Rain
for use in the lesson extension

There are a number of countermeasures now being used to fight acid rain. *Automotive catalytic converters* and smokestack *scrubbers* deal with preventing nitrogen and sulfur oxides from reaching the atmosphere. *Neutralization* deals with getting rid of the acid once it has been deposited on lakes, soil, and forests.

Neutralization One way to deal with the buildup of acid in lakes is to neutralize it by adding a base. Recall that a neutralization reaction is a double replacement reaction, one in which the mixing of an acid and a base produces water and a salt.

$$HCl + NaOH \rightarrow NaCl + H_2O$$
acid   base   salt   water

Acidified lake water can be neutralized by adding the base calcium carbonate ($CaCO_3$), also called limestone. Limestone is used because it is easily found in the large amounts typically used to treat a lake.

$$H_2SO_4 + CaCO_3 \rightarrow CaSO_4 + CO_2 + H_2O$$

Calcium oxide (CaO) is used to neutralize acidified soil. It is made from limestone at high temperature in a lime kiln.

$$CaCO_3 + heat \rightarrow CaO + CO_2$$

It is then mixed with water to form slaked lime ($Ca(OH)_2$).

$$CaO + H_2O \rightarrow Ca(OH)_2 + heat$$

Slaked lime is then applied to acidified soil.

$$Ca(OH)_2 + H_2SO_4 \rightarrow CaSO_4 + 2H_2O$$

Pretreatment and Scrubbing In response to regulations requiring cleaner air, industrial emissions from smoke stacks have been reduced as a result of treating fuels before they are burned, treating fuels as they are being burned, and treating flue gases after combustion. Sulfur dioxides, resulting from the combustion of sulfur impurities found in coal, is a chief contributor to acid rain.

Before combustion, coal can be cleaned of many impurities by a method called physical coal cleaning (PCC). By mixing crushed coal in water, the difference in their specific gravity allows coal to float and impurities such as sulfur sink. Sulfur removal approaches 35 percent by this method.

It is also possible to remove sulfur dioxide as it is being produced during the fuel burning process itself. In a process called FBC (Fluidized bed combustion) pulverized coal is mixed with powdered limestone and burned at high temperature (1000°C). Calcium sulfite is produced, and sulfur dioxide emissions are reduced.

$$SO_2 + CaCO_3 \rightarrow CaSO_3 + CO_2$$

However, to substantially reduce emissions, scrubbers are used. They reduce pollution after combustion by using centrifugal force, electricity, liquid sprays, chemicals, or other means to *scrub* toxic gases before they leave the stack. They can remove 95 percent or more of sulfur pollutants from smokestack flue gases and greatly reduce acid rain.

Look at Figure 1 below. It is a diagram that shows how three of these methods work.

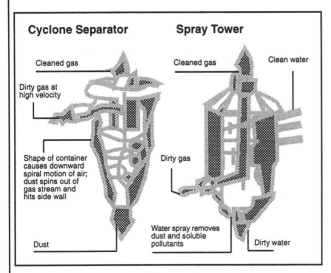

Figure 1

Cyclones Cyclones are a mechanical process used to remove solid and liquid particles from the waste gas stream before it enters the atmosphere. Dirty gas with suspended particles of waste enter the cyclone on top. As the stream is forced down, it spirals. Particles suspended in the gas stream are forced against the sides of the cyclone, where they fall by gravity to the bottom. The cleaner gas stream is then either fed into another cyclone or is sent to the smokestack. Multiple cyclone collectors can remove about 90 percent of solid and liquid particles from gas streams.

Spray Towers Spray towers are also known as wet scrubbers and work by bringing a liquid spray (usually water) into contact with a contaminated gas stream. When they mix, the water will dissolve soluble contaminants and the spray will capture stream-borne particles. Dirty water is collected at the bottom of the wet scrubber. Refer to Figure 1.

A specialized process of wet scrubbing can be applied to the desulfurization of flue gases. It combines the technology of the spray tower with neutralization chemistry to clear sulfur dioxide from smokestack emissions. The process is known as flue gas desulfurization (FGD). During FGD, the wet scrubber spray is made up of a slurry of water and either lime or limestone. Sulfur dioxide reacts chemically with the slurry in the presence of oxygen and is removed as calcium sulfate.

$$CaCO_3 + SO_2 \rightarrow CaSO_3 + CO_2$$
$$CaSO_3 + \tfrac{1}{2}O_2 \rightarrow CaSO_4$$

Electrostatic Precipitators Electrostatic precipitators are also used to remove small particles from the gas stream. They are used in nearly all power plants. Dirty gas enters in a horizontal direction and passes along high voltage electrodes that ionize small particles suspended in the gas stream. The gas stream then passes through a series of collecting plates that have the opposite charge. The plates collect the ionized pollutants, which can then be disposed of. The following figure diagrams this process.

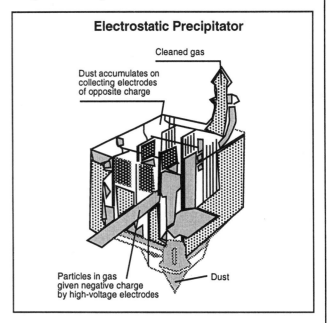

Figure 2

Catalytic Converters About half a billion automobiles are in the world, and they all contribute to air pollution. Auto emissions include pollutants such as hydrocarbons (HC) and carbon monoxide (CO). These are removed at one end of a typical automotive catalytic converter by combining with oxygen in the presence of heat and a catalyst (platinum or palladium) to produce carbon dioxide and water vapor in a catalyzed combustion reaction.

$$CO + HC + O_2 + Heat \xrightarrow{\text{catalyst}} CO_2 + H_2O$$

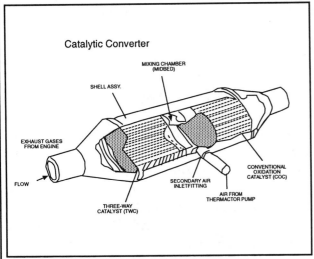

Figure 3

Refer to Figure 3 for a diagram of a typical automobile catalytic converter.

The automobile is also the main culprit in the production of airborne nitrous oxides, which end up reacting to form the nitric acid part of acid rain. Regulations now require the installation of catalytic converters to remove not only carbon monoxide and hydrocarbons, but also most of the nitrogen oxides present in automobile exhaust emissions. This poses a problem because the same process (combustion at high temperature) that gets rid of CO and HC increases the production of nitrous oxides. Therefore, a different catalyst, Rhodium, is used to remove nitrogen oxides, NO and NO_2.

A catalyst increases the rate of a reaction without being used up in the reaction. It does this by lowering the amount of energy necessary to make the reaction happen. In many catalyzed reactions, this is accomplished because the catalyst provides a surface on which the reaction can more easily occur. Rhodium works in this fashion. When exhaust emissions pass through the rhodium-covered honeycomb found at the other end of the catalytic converter, the reduction of nitrogen oxides to nitrogen gas and oxygen is catalyzed.

$$NO + NO_2 + Heat \xrightarrow{\text{Rhodium}} N_2 + O_2$$

Sample Student Responses • Acids in the Atmosphere

BRAINSTORMING IN ORDER TO GENERATE POSSIBILITIES

PURPOSE

Ways to reduce the harmful effects of atmospheric pollution by nitrogen and sulfur oxides resulting from the burning of fossil fuels in vehicles and industry.

POSSIBILITIES

Make it illegal to burn fossil fuels.	Give bicycles away free.
Use battery-powered vehicles.	Neutralize acids in lakes/lawns and gardens.
Make more buses available.	
Put factories/power plants near dams.	Acid-resistant coating on buildings/clothing.
Use better gasoline (higher octane).	
Use better catalytic converters.	Use dirigibles to transport cargo, people.
Bring back more passenger trains.	Design more efficient engines.
Harness the energy of volcanos.	Speed up development of plasma fusion.
Put all highways underground.	Use flywheel technology in vehicles.
Make cars that hold more passengers.	Give tax incentives for clean air, solar power.
Make stiffer fines for polluting air.	
Educate consumers to stop wasting.	Airlock houses and clean interior air.

KINDS OF POSSIBILITIES

Use alternate energy source	Better treatment of air pollution	Make new laws, levy taxes	Reduce demand for power	Treatment of acid rain
Batteries	Better gas	Make illegal to use fossil fuels	Put employees near work	Neutralize acids in lakes
Geothermal	Underground highways	Make illegal to drive alone	Enforce carpooling	Acid-neutralizing fertilizers
Solar	Engines	Levy stiffer fines for polluting	More buses	Buildings made acid-resistant
Lasers	Seal houses		More trains	Acid-resistant clothing
Nuclear Plasma fusion	Large air scrubbers	Tax incentives for clean air or solar power	Locate factories near dams	
Flywheels			Bigger cars	
			Free bicycles	

UNUSUAL POSSIBILITIES

Put highways underground.	Enclose cities in domes, purify air.
Give bicycles away free.	Give tax credits and no-interest loans for battery- or solar-powered cars.
Locate employees near work.	
Use airlocks on houses so that interior air is maintained clean.	Use windmills on vehicles to run exhaust gas scrubbers.
Use solar collectors on surface of smokestacks to power scrubbers.	Build factories and power plants on man made islands and use tidal currents.
Give utility rebates for using thermostat that cannot be set higher than 67.	Make it illegal to live more than bicycling distance from your job.

Sample Student Responses • Acids in the Atmosphere

MATRIX FOR GENERATING ORIGINAL POSSIBILITIES

INTERESTING FEATURES OF POSSIBILITIES ALREADY GENERATED

	Using an alternative source of energy	Reducing consumption of fossil fuels	Making new laws, levying taxes, or making subsidies	Treatment of existing air pollution or its effects
Treatment of existing air pollution or its effects	Use windmills on vehicles to produce power to operate an electrostatic exhaust gas scrubber	Enclose city in a clear dome, purify its atmosphere, and produce heat by the greenhouse effect	Make government subsidies available to start-up companies getting into the field of cleaning up the damage cuased by air pollution	X
Making new laws, levying taxes, or making subsidies	Give larger subsidies to individuals, universities, and companies for research and development of new energy sources	Give utility bill rebates for using a special thermostat that cannot be set higher than 67 degrees	X	Make stricter laws governing acceptable amounts of industrial air pollution, and make stiffer fines for noncompliance with the new laws
Reducing consumption of fossil fuels	Build factories and power plants on man-made islands and use tidal currents to generate power	X	Make it illegal to live more than bicycling distance from your place of employment	Reduce consumption by having everyone wear an "environment suit" that cleans the air you breathe and keeps you warm
Using an alternative source of energy	X	Replace fossil fuel-burning home heating units with a home or neighborhood-sized nuclear reactor	Give a tax credit and no-interest loans to car owners who switch to battery- or solar-powered electric vehicles	Use solar collectors to cover the surface of smokestacks to supply power to run scrubbers to clean exhaust gases

INTERESTING FEATURES OF POSSIBILITIES ALREADY GENERATED

GENERATING POSSIBILITIES LESSON CONTEXTS

The following examples have been suggested by classroom teachers as contexts to develop infused lessons. If a skill or process has been introduced in a previous infused lesson, these contexts may be used to reinforce it.

GRADE	SUBJECT	TOPIC	THINKING ISSUE
6–8	Science	Asteroids	How might we use asteroids?
6–8	Science	Space exploration	How might orbiting observatories be used in ways not yet undertaken?
6–8	Science	DNA	Create an amphibian/reptile that results from research on recombinant DNA. List the advantages and disadvantages of the new creature as compared to the two original species.
6–8	Science	Fumeroles	To what new uses could the energy from fumeroles be put?
6–8	Science	Plankton	Brainstorm new uses for plankton and/or seaweed.
6–8	Science	Weather	Brainstorm possible ways of making the weather in Antarctica less severe so that the continent could be habitable by humans.
6–8	Science	Dinosaurs	Imagine that some dinosaurs survived. Select a specific dinosaur. Brainstorm possible ways of keeping the dinosaur as a pet.
6–8	Science	Robotics	Brainstorm different ways that robots can be used in the following situations to save human beings time and/or diminish risk or danger to humans: food production, automobile manufacture, home maintenance, school use, space travel (pick one).
6–8	Science	Pollution	Brainstorm possible ways of making your community aware of pollution problems.
6–8	Science	Smoking	What are some possible ways to help people make good choices when it comes to smoking?
9–12	Biology	Conservation of natural resources	Make a list of possibilities for protecting remaining old-growth forests from destruction by humans; consider global threats from the logging industry, slash-and-burn farming, and urban sprawl.
9–12	Biology	Air pollution	Brainstorm ways of controlling air pollution in industrial areas. Consider sources of sulfur and nitrous oxides, carbon monoxide, ozone, soot, ash, and smog. Emphasize new and unusual ideas.
9–12	Biology	Properties of water	Think about water and living organisms and then brainstorm a list of all the ordinary and extraordinary properties of water that might be necessary for the evolution and maintenance of life.

GENERATING POSSIBILITIES LESSON CONTEXTS

GRADE	SUBJECT	TOPIC	THINKING ISSUE
9–12	Biology	Natural selection and adaptation	Design an extraterrestrial environment complete with weather, atmosphere, and surface features like oceans, volcanos, and mountains. Then, brainstorm the kinds of organisms that might have evolved within it. Draw pictures of each organism and indicate and explain its major adaptations to the alien environment.
9–12	Chemistry	Heterogeneous mixtures	Consider being asked to determine whether an unknown liquid is a solution, a colloid, or a suspension; brainstorm as many ways as you can to determine which heterogeneous mixture you have. Then, experiment to find out which idea works best in terms of time, materials, and expense.
9–12	Physics	Simple machines	Think about living in Egypt 2,500 years ago and the great challenges that were met in order to build the pyramids. Brainstorm different ways that simple machines were used or might have been used to accomplish these tasks. Draw diagrams showing how the machines might have been used.
9–12	Physics	Simple machines	Think about medieval Europe and the problems that armies encountered when they tried to storm a castle. Brainstorm different ways that simple machines were used or might have been used to lay siege to and defeat a castle's defenses (e.g., its moat, its walls, and its defenders). Extend by thinking about how simple machines might have been used to defend the castle from invasion.

CHAPTER 12
CREATING METAPHORS

Why Is It Important to Create Metaphors Skillfully?

Metaphors accentuate characteristics of objects, people, and events using an image of something analogous that highlights those characteristics clearly and dramatically. If a co-worker is described as a "loose cannon," that metaphor suggests that this individual is directionless, unguided by purpose or principle, unfocused, and capable of doing great harm with one impulsive, unexpected discharge. The image of an unsecured cannon on the rolling deck of a ship, primed and ready to go off, communicates these ideas very effectively.

The economy of language in using metaphors is striking. Two words, "loose cannon," call up an image that communicates several important perceptions about the co-worker. Comprehending and discussing those perceptions depends on the background of the person hearing the metaphor. If the listener doesn't know what a cannon is, the communication is not effective.

Developing effective metaphors is a form of creative thinking, connecting two things that are not ordinarily associated to give us a new insight. Using metaphors is common in creative expression and is especially important in creative writing and the visual media.

In problem solving, creating metaphors is also effective for suggesting new ideas for unusual problem solutions. By creating a metaphor for the problem, solutions that are not initially apparent may emerge. Creative problem solving based on metaphoric thinking can help people solve personal, professional, and technical problems.

Uses of Metaphors

Images in poetry often communicate ideas through various types of metaphor, such as figures of speech or allusions. In saying "All the world's a stage and all the men and women merely players," Shakespeare used the stage as a metaphor to give us insight into key relationships and actions in different periods of life. In "Dream Deferred," Langston Hughes used five metaphors in eleven short lines, one of which is "a raisin in the sun," to express the frustration of African-Americans in continually postponed justice and opportunity. In William Blake's "To See A World in a Grain of Sand," he used four metaphors in four lines to describe how understanding significant ideas is found in observing small things.

Metaphors communicate images and insight in other forms of literature as well. Plato's cave allegory, Aesop's fables, the parables of Jesus of Nazareth, the African trickster tales of Anansi the Spider, the humor of Abraham Lincoln and Will Rogers, and the homilies of Ben Franklin employ metaphors to express spiritual, moral, or political principles.

Metaphors are sometimes used to persuade. Franklin Roosevelt used the metaphor of an epidemic to support resisting the expansion of the Nazi movement and of lending a garden hose to a neighbor whose house was on fire to justify the lend-lease policy. Churchill's metaphor of an "Iron Curtain" became the common term to describe the division of Europe after World War II, an image that conveyed a "we-they" relationship of threat and impenetrability. "The rising tide of Communism" during the early 1950s suggested that political leaders should take precautions against a destructive and uncontrollable flood.

Metaphors also convey humor in jokes or puns. The Volkswagen ads of the '60s used humorous visual metaphors to draw attention to the special features of the "beetle." Soon after the 1969 moon landing, Volkswagen used a picture of the lunar landing module in its ad with the caption, "It's ugly, but it gets you there." Poking fun at itself and the charges that the "beetle" was ugly, Volkswagen also called upon American pride in the moon landing to sell its cars. Political cartoons similarly rely on metaphors for meaning and humor.

Metaphors may help people not familiar with certain objects, actions, and functions understand new information more easily. The computer industry uses metaphors like "virus," "mouse," "trash," "file," "byte," etc. to convey abstract, technical ideas or operating procedures in language that people not familiar with computer technology can understand.

Metaphors are also used in scientific investigation to develop and express sometimes complex and abstract scientific principles. Mendeleev's dream of a table on which all elements had a place or Kekule's image of the benzene ring as a snake with its tail in its mouth expressed ideas that these scientists had been incubating. Scientific models, like the double helix, serve as metaphors for abstract structures, properties, or complex causal relationships.

Problems in Creating Metaphors

Although metaphors can communicate ideas well, poorly selected metaphors can be misleading and/or confusing. One limitation in the way that people create metaphors involves using a metaphor that is too narrow to communicate what we are trying to convey, making the message superficial. To describe a starched tablecloth as "stiff as a board" is a fairly useful description. Both are materials that have a rigid appearance and do not bend easily. To describe someone as "stiff as a board" suggests that the person seems inflexible but does not convey the person's behavior or characteristics very meaningfully. Does the person have an overly polite or inexpressive manner, a rigid posture, dogmatic views, a stilted way of expressing ideas, or unimaginative ways of thinking? This metaphor is too thin to convey the complexity of human personality.

Metaphors may also convey incorrect ideas about something. We must be sure that the metaphor suggests what we want to describe and does not convey unintended meanings. If I notice how a person walks and comment that he is "squirrelly," I may be trying to convey that he makes scampering movements like a squirrel. I may be unintentionally communicating that he seems fearful, easily distracted, or that he hides things away. In this case, the metaphor is not effective because it carries ideas that suggest misimpressions, rather than insight. Most of us can think of a time that we used a metaphor jokingly and offended someone, either because it didn't fit the situation in a humorous way or because it carried some other message that was offensive.

Metaphors may also create confusion rather than clarification. Mixed metaphors, for example, often confuse us or seem humorous because they convey conflicting ideas. Consider the mixed metaphors in this advice, "To get ahead, keep your nose to the grindstone, your shoulder to the wheel, your ear to the ground, and your eye on the ball." The resulting image is one of motionless contortion, rather than progress.

Malapropism sometimes result from mixing metaphors, an error that the speaker often does not intend. If a pessimistic friend comments "Do not count your chickens before they come home to roost," the speaker is mixing a proverb about not expecting something good before you know it with a proverb about bad things always coming back to their origins. The listener does not know whether the speaker is confused, starting to say one thing and ended up in another proverb, is trying to make a joke, or is suggesting that the outcomes can be only bad ones. Malapropisms confuse the listener because the mixed metaphor blurs together several sets of connections, some of them conflicting.

Figure 12.1 summarizes these problems in creating metaphors.

PROBLEMS IN CREATING METAPHORS

1. We select metaphors with narrow meanings or that have only superficial connection with what they stand for.

2. We select metaphors that are inaccurate or misleading.

3. We mix metaphors in ways that convey conflicting or contradictory ideas.

Figure 12.1

Two additional problems arise in using metaphors: inappropriateness and overuse. A metaphor must fit the background of the listener for its communication to be complete. We are sometimes "put off" by "inside jokes" used in a group in which others share a common language and experience. We may feel slightly ignorant if we must look up allusions to understand a writer's metaphors or confess that we do not know what a speaker is referring to. We may be critical of a person using an obscure metaphor, believing that the speaker is being insensitive or "showing off" by using an analogy that excludes us from enjoying the humor or sharing the insight.

The esoteric metaphors of T. S. Eliot or Ezra Pound may not mean much to readers unfamiliar with the allusions. Since these metaphors are profound to those familiar with the background, the insights of these authors warrant the research that might be necessary to comprehend what the metaphors are supposed to convey.

Overuse also detracts from the effectiveness of metaphors. Skill in creating and using metaphor lies in connection with what we are expressing and in originality. Overuse of the same metaphor or using only hackneyed ones adds little interest or insight to our ideas. Poets and writers poke fun at themselves and others in overusing metaphors. Sylvia Plath's poem "Metaphors" and Ogden Nash's poem "Very Like a Whale" satirize overuse and misuse of metaphors. Shakespeare's sonnet "Shall I compare thee to a summer's day…" describes the limitations of using them.

How Do We Create Metaphors Skillfully?

When we draw an analogy between two things of different types, we recognize that they have important similarities. We use analogies for a number of purposes. When we reason by analogy, we extend something we know about one of the two things to another based on the analogy. In creating metaphors, we use analogies to convey something we already know about one thing by using an image of something else that is literally quite different from the original object. Creating metaphors is a creative thinking activity because metaphors clearly express ideas or insights by matching two things that are not ordinarily associated.

Creating metaphors is not just free association, although it almost seems to be that intuitive when we hear people who are good at metaphoric thinking spinning off metaphors. Some individuals hear and express metaphors spontaneously, but all of us can create metaphors skillfully with some careful, organized thought: determine what you want to say with a metaphor, identify the characteristics that suggest that meaning, and select a metaphor that is analogous enough to convey the same image or meaning effectively.

Creating metaphors can be facilitated by using a variety of creative-thinking techniques. Visualization or imagery prompts analogies that are used to create solutions to business or technical problems. Free association techniques that awaken stored memories or sharpen observation can generate metaphors for creative expression or practical problem solving.

In whatever way we get ideas for metaphors, it is important to think about whether these metaphors communicate what we intend, offer deep understanding or original expression, and do not carry misleading associations. The strategy described in this chapter is based on explicitly using analogies between things to give us ideas for metaphors and then analyzing whether or not the metaphors are a good fit. It is a strategy that we all can use, regardless of our background or experience with other creative thinking techniques.

Attention Points in Creating Metaphors Skillfully

We set the stage for creating new metaphors by asking, *"What is the purpose of this metaphor? What am I trying to communicate that a metaphor would express?"* Identifying the context for creating the metaphor is the first step in bringing possibilities to mind for consideration.

Next, we should list the key characteristics that explain what we want to communicate about the given object. These important characteristics are the "idea bridges" that allow us to search our memory to find something else that expresses in a new way the idea that we are working with. Clarifying these key characteristics allows us to put forward connections to a new object that is analogous. For example, I may want to communicate that a co-worker is dedicated to her job, very hard working, and quite productive. Key characteristics that express these ideas include that she is always active, puts out a lot of energy, and accomplishes many tasks.

A metaphor will work only if a number of characteristics connect it to the given object. We should consider a range of different types of characteristics related to the ones we start with so that the metaphor is a rich and useful one. Thinking about the details that elaborate each characteristic of the given object helps us to be very clear about what we are trying to convey.

With a variety of characteristics before us, we are ready to brainstorm a number of images, objects, or ideas that have these key characteristics. For example, when thinking about many types of characteristics of my co-worker, I think of a locomotive, an army, a dynamo, and a basketball player.

From this metaphor menu, we can select the ones that have the most significant characteristics in common with the given idea. When one object, person, or event seems most promising for our purpose, we can check how well the key characteristics of one fits the other, noting whether any significant differences might interfere with the effectiveness of the metaphor. A basketball player is similar, but there are too many breaks in the action in a basketball game. A locomotive is too overbearing. An army suggests a complex of individuals working together. A dynamo, on the other hand, does not have similar disadvantages.

The last step in creating metaphors is determining whether the characteristics and details between the two things are strong enough to make the metaphor a good one. "She's a dynamo" seems to convey what we wanted to describe about her very well.

The thinking map for creating metaphors (figure 12.2) summarizes these steps.

The graphic organizer for creating metaphors in figure 12.3 is an adaptation of the compare/contrast graphic, redesigned for the creative development of metaphors.

Add any additional characteristics boxes that may enhance the richness of your metaphor.

The graphic organizer follows the questions on the thinking map, moving vertically down the diagram. After writing in the upper left box

CREATING A METAPHOR

1. What do I want to describe about an object, person, or event that a metaphor would express?

2. What are the details of the key characteristics of what I am trying to describe?

3. What other things (objects, people, or events) have these key characteristics?

4. Which of these things might make a good metaphor?

5. What details of the metaphor fit the characteristics of what I am trying to describe?

6. Are there differences that make the metaphor misleading? If so, why?

Figure 12.2

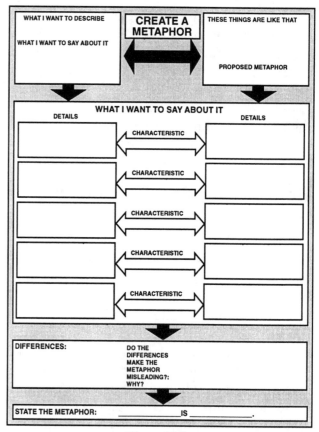

Figure 12.3

what you want to describe about the thing or idea, write these characteristics on the arrows in the middle of the box. Think about whether there are other related characteristics of the given object that occur to you as you consider the key characteristics. For example, I may want to describe my co-worker as active, energetic, and able to do many tasks. Then I realize that she is also forceful. I add this somewhat related characteristic to the list of things that I want to say about her and to the arrows in the middle.

To have a clear image of what you want to convey, brainstorm details about the given object and write them in the "Details" boxes. Figure 12.4 shows how the diagram may look at that point in the process.

After surveying the details and the characteristics of the thing that you wish to describe, suggest several things that have many of the same characteristics and write the list of possible metaphors in the top box on the right. From this list, select an object that seems like a good fit for many of the characteristics on the left, and write it in the "Proposed Metaphor" section of the same box.

Now brainstorm details of each of the characteristics as that characteristic applies to the "Proposed Metaphor." Add characteristics that are important in describing the proposed metaphor

to find out whether there are any similar features in the original object that we may have overlooked earlier. For example, a dynamo keeps on going for long periods of time without breaking down and so does my co-worker. I now add "keeps on going" to the characteristics and think of details about how both my co-worker and a dynamo keep on going. Figure 12.5 shows how the graphic organizer may look.

Some differences between the metaphor and what you are trying to describe give richness to the metaphor. The difference in the degree of energy, power, and productivity between a dynamo and my co-worker adds desirable meaning to the metaphor. Using the metaphor of a dynamo exaggerates those qualities in her in a complimentary way. However, I must check to see that the image of the dynamo does not carry with it some undesirable connotation that I do not want to ascribe to her.

Seeing the details and characteristics side-by-side, I can now decide whether the proposed metaphor has enough of the characteristics of the thing that I wish to describe to convey that

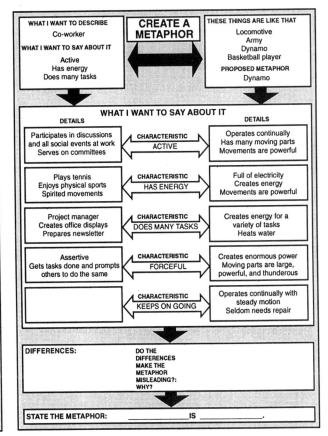

Figure 12.4   Figure 12.5

thing's important characteristics. Then, I write the metaphor: My <u>co-worker</u> is a <u>dynamo</u>. I may then use information in the "Characteristics" and "Details" boxes of the diagram to write about my co-worker as if she were a dynamo. Figure 12.6 shows a completed graphic.

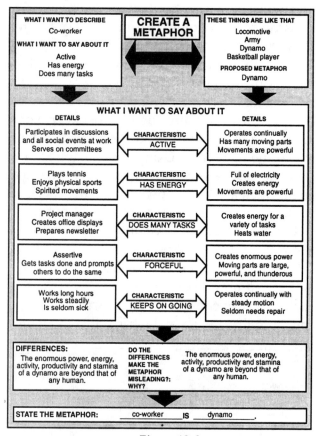

Figure 12.6

The metaphor web diagram in figure 12.7 is used to summarize the metaphors that have been selected to describe the same object or idea. This activity shows that many metaphors can be used to describe the same thing, each adding different understanding and emphasizing different characteristics. The web diagram can be used by an individual to compare metaphors that he or she has developed or in a group activity to summarize the metaphors developed by several group members.

Extended uses of metaphors in writing. To use a metaphor in a poem or persuasive prose, one should ask, *What details about one thing can I use to describe details about another?* In describing a co-worker as a "loose cannon" or as a "dynamo," the metaphor is a focused one and need not be elaborated. However, the force of the

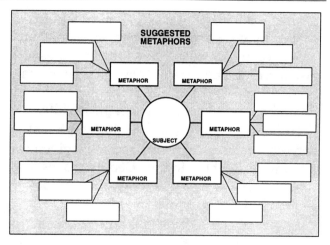

Figure 12.7

metaphor can be extended in continued dialogue by using details about a loose cannon or a dynamo to further describe the co-worker. I might say of her, "She never shuts down," or "She's full of electricity." This is an important device in poetry. In creating a metaphor like the cat in Carl Sandburg's poem "Fog," details are very important to express and extend the image of the cat to convey meaning about fog. The fog is described as creeping in "on little cat feet," for example. The details boxes in a completed graphic organizer like that in figure 12.6 contain the "raw material" from which a poem, some other literary work, or a pictorial representation can be created. The next step in the creative process is executing the images, poetry, or prose for which the metaphor is created.

Fluency and originality are important in any form of creative thinking. The three steps in this strategy that involve brainstorming (listing characteristics, creating a menu of possible metaphors, and generating many details for expressing the connection) should provide original material that can be used in the creative expression of the metaphors developed.

How Can We Teach Students to Generate Metaphors Skillfully?

It is not enough to ask students to "be creative" in developing and using metaphors. The strategy outlined in the thinking map helps students who may not view themselves as creative thinkers become skillful and confident about generating metaphoric images. Emphasize the importance of the purpose of the metaphor to give students a sense of why we develop and use metaphors and to establish a clear set of standards for selecting good metaphors from among the ways of expressing their ideas.

Creating metaphors reduces students' insecurity and confusion when asked to interpret symbols, figures of speech, or analogies in prose. In addition to its cognitive value, creating metaphors promotes self-esteem by confirming for students that they are creative thinkers who can suggest workable, original ideas.

As in the case of teaching other thinking skills and processes, you should guide students' practice and prompt them to reflect about how they developed these ideas so that they can guide themselves in using this strategy.

Contexts in the Curriculum for Lessons on Creating Metaphors

Metaphors that students are exposed to as well as contexts in which students can create their own metaphors appear throughout the curriculum. Metaphors have great value in science. The abstractness and complexity of many scientific concepts and principles make metaphors powerful learning tools in science that can lead to scientific understanding. In fact, some metaphorical language has worked itself into everyday scientific terminology already. We talk about the "flow" of electricity, using a water metaphor; electrons "orbiting" around a nucleus, using the solar system as a metaphor; chemical "bonds"; an electric "current." Models used in science are visual metaphors and can help not only in communicating more abstract concepts, but also in understanding how the phenomenon operates. For example, a model of a DNA molecule is a metaphorical representation that emphasizes key features of the molecule so that we can focus on them, like any good metaphor does. One should not expect that if you looked at a DNA molecule through an electronic microscope it would look like the models we make.

In general, there is one primary context in which infusion lessons on creating metaphors can be developed and taught.

- *Metaphors are useful in explaining complex or abstract phenomena.* Certain concepts in science can be understood by using visual

representations of them. For example, electromagnetism is sometimes compared to someone pulling a heavy object with a rope, so a positive and negative magnetic charge would be like a tug of war. Plate tectonics is often visualized as a flexible piece of plastic strung under another piece. Molecular structure is often exhibited by balls of different color being connected by rods to other balls of different color. These and other contexts in which students might create visual metaphors can provide contexts for infusion lessons on metaphor. Similarly, verbal metaphors can be useful in dramatizing certain abstract qualities of things like the metaphor of water flowing through a pipe does for an electric current.

A menu of suggested contexts for infusion lessons on creating metaphors can be found at the end of the chapter.

Model Lesson on Creating Metaphors

The model lesson on creating metaphors skillfully asks students to create visual metaphors or models for the atom bringing out crucial aspects of its structure. As you read the lesson, reflect on these questions:

- What specific strategies promote creating metaphors?
- How do the graphic organizers promote understanding and creating metaphors?
- What additional examples can you identify to reinforce this skill?

Tools for Designing Lessons on Creating Metaphors

The tools on the following pages include thinking maps and graphic organizers that can be used in designing infusion lessons on creating metaphors for grades 6–12.

The thinking maps and graphic organizers can guide you in designing the critical-thinking activity in the lesson and can also serve as photocopy or transparency masters or as posters in the classroom. Reproduction rights are granted for use in single classrooms only.

CREATING A METAPHOR

1. **What do I want to describe about an object, person, or event that a metaphor would express?**

2. **What are the details of the key characteristics of what I am trying to describe?**

3. **What other things (objects, people, or events) have these key characteristics?**

4. **Which of these things might make a good metaphor?**

5. **What details of the metaphor fit the characteristics of what I am trying to describe?**

6. **Are there differences that make the metaphor misleading? If so, why?**

7. **Is this metaphor a "good fit"? Why?**

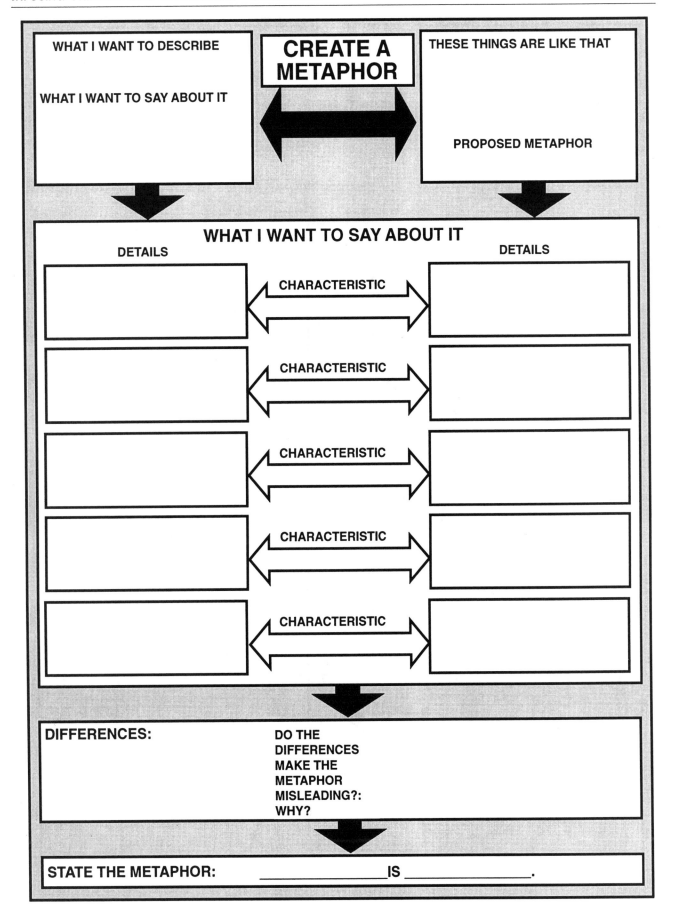

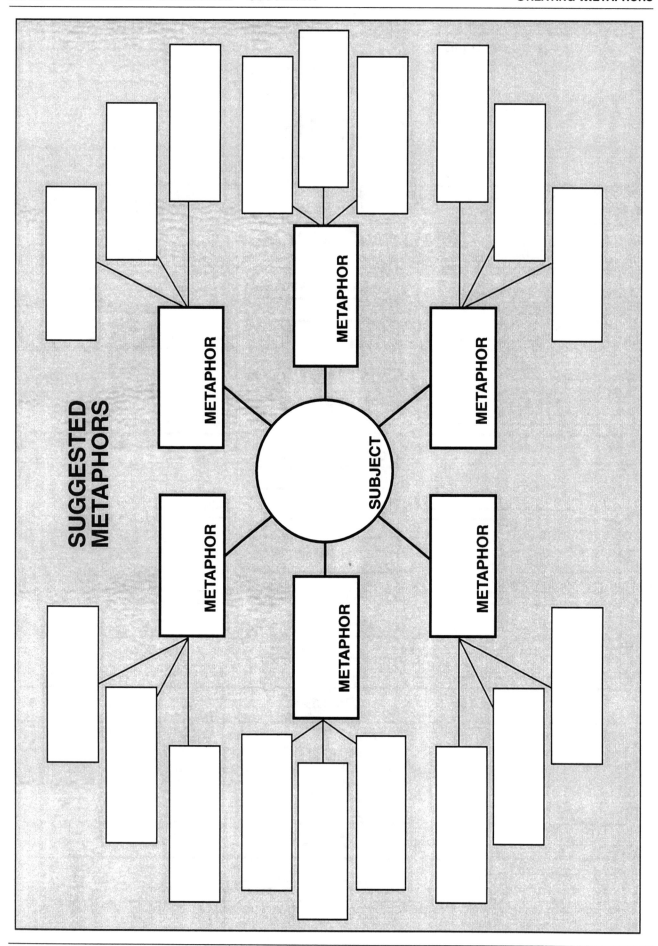

SUGGESTED METAPHORS

SUBJECT

METAPHOR
METAPHOR
METAPHOR
METAPHOR
METAPHOR
METAPHOR

A MODEL FOR A MOLECULE

Biology, Chemistry   **Grades 9–12**

OBJECTIVES

CONTENT

Students will learn the nature of chemical bonding and will determine which features of molecules are important to understand in order to understand their role in chemical reactions. They will also learn about the role of models in science.

THINKING SKILL/PROCESS

Students will create metaphors (models) effectively by doing the following: stating the idea that the metaphor will convey, listing the thing's characteristics, brainstorming other things with important similarities, and selecting an item whose details convey the idea well.

METHODS AND MATERIALS

CONTENT

Background reading will be done by the students on molecular structure in their chemistry texts.

THINKING SKILL/PROCESS

Creating metaphors is guided by structured questioning and the use of a graphic organizer. Brainstorming is featured twice in the lesson: first, to suggest possible metaphors, and, second, to generate details to express them.

LESSON

INTRODUCTION TO CONTENT AND THINKING SKILL/PROCESS

- Carl Sandburg's poem "Fog" begins:

  "The fog comes
  on little cat feet.
  It sits looking
  over harbour and city
  on silent haunches ..."

In this poem, what idea does Carl Sandburg use to tell us about fog? *The idea of a cat.* **When a writer or speaker uses one thing to tell us about another, he or she is using a metaphor. In the poem, what characteristics of a cat does Carl Sandburg use to tell us about fog?** *The cat's creeping quietly and the cat's being very still as it sits and looks around.* **What does that tell us about fog?** *The fog comes in quietly, and, as it hovers over a place, it does so silently and is very still (unlike wind).*

- The poet decides that a metaphor conveys what he wants to convey well by recognizing a clear and obvious analogy between the metaphor and the thing he describes. The metaphor makes those analogous features stand out in a striking way. **Can you think of other metaphors like that?** STUDENT ANSWERS VARY. TYPICAL ANSWERS INCLUDE: *calling a dull person a "wet blanket"; calling an exciting city with a lot to do a "hub"; calling a flashy car a "chariot."*

- In science, many metaphors are also used, especially because what we learn in science is abstract or technical. We talk about electricity "flowing" through a wire, about cells in the body as "building blocks," and about white blood cells as "soldiers." Science also uses two- and three-dimensional pictures and constructions—models—as metaphors. For example, an atom is sometimes pictured as an object with a ball at the center around which smaller balls, sometimes of different colors, rotate. A graph of the rise and fall of temperatures treats lines or bars as metaphors for degrees of temperatures. Each of these metaphors emphasizes certain features of what the metaphors stand for in striking ways—the strength and combination of cells to yield a whole living organism, the attack and destroy tactics of white cells when they detect a "foreign invader" in the blood, the structure of an atom, and the movement from "high" temperatures to "low" temperatures, and vice versa.

- It isn't too difficult to choose good metaphors to express an idea about something in this way. If we know what characteristics we want to stress through the use of a metaphor, we should first think about possible metaphors by coming up with things that have the same or similar characteristics. For example, if we want to stress that a person is nice, it isn't too hard to come up with possible metaphors like a philanthropist, a saint, etc. Then, we should pick the metaphor that works best—that conveys this characteristic and does not mislead us. Here is a thinking map:

CREATING A METAPHOR
1. What do I want to describe about an object, person, or event that a metaphor would express?
2. What are the details of the key characteristics of what I am trying to describe?
3. What other things (objects, people, or events) have these key characteristics?
4. Which of these things might make a good metaphor?
5. What details of the metaphor fit the characteristics of what I am trying to describe?
6. Are there differences that make the metaphor misleading? If so, why?
7. Is this metaphor a "good fit"? Why?

- In this lesson, we are going to create visual or three-dimensional models as metaphors for one of the important ingredients in the universe—the molecule.

THINKING ACTIVELY

- Suppose that you wanted to show someone what certain characteristics of molecules were in order to help them understand (1) how molecules can be built of atoms and yet combine with other molecules to make solids, gases, and liquids, and (2) how molecules react with other molecules to create molecules different from their source. Let's use this diagram to first take an inventory of what we know, what we think we know, and what we don't know but would like to know about chemical molecules. Give each student a copy of the graphic organizer and ask them to write appropriate information or questions in the columns. When students have had a few minutes to do this, ask them to divide into groups of four or five and share the items they put in each column. Then ask each group to list the five most important items on a group organizer. When they finish (take no more than 5 minutes), ask the groups to share their items orally. Write "Need to know" items on the board. ANSWERS VARY.

Molecules		
WHAT DO I KNOW?	WHAT DO I THINK I KNOW?	WHAT DO I NEED TO KNOW?

- In order to answer the need-to-know questions on the board, consult the chapter on chemical bonding in your chemistry textbook. Work in your group and write answers to your questions. After five minutes, ask students to report their findings. Ask for one per group until you have exhausted the questions. ANSWERS VARY.

- Let us now consider a simple molecule that we are all familiar with, the water molecule. Ordinarily, this is written H_2O. Actually, this is a model for the water molecule. Look through the chapter in your text on chemical bonding and identify any other models for molecules you find there. Add to these any others you think might be good models as well. Work in your groups and come up with a list of these. Students usually identify four types of models:

(1) The molecular formula, H_2O

(2) Various Lewis structural formulas:

(3) Two-dimensional diagrams showing orbitals as electron clouds:

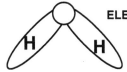 **ELECTRON ORBITAL MODEL**

(4) Three-dimensional models:

SPACE-FILLING MODEL   **BALL-AND-STICK MODEL**

- **Use the graphic organizer for creating metaphors now and choose an important piece of information about the water molecule that you want to express using a model. Write it in the appropriate box at the top left on the diagram (under "What You Want To Say About It") and "The Water Molecule," at the top of the box.** Students usually pick relatively obvious facts about the water molecule, such as that it is made up of two hydrogen atoms and one oxygen atom, or that two hydrogen atoms are bonded to one oxygen atom. Characteristics picked less frequently include the following: *The two hydrogen atoms are bonded to the oxygen atom by electrons. The electron bonding of the hydrogen and oxygen atoms in water creates a three-dimensional orbit of electrons from the oxygen to the hydrogen atoms. In water, hydrogen and oxygen molecules have covalent bonds.*

- **Now use your graphic organizer to elaborate the analogy between the model you think will work best and a water molecule. Do this by filling in one of the characteristics you want to elaborate on the arrow, for example "Bonding by Electrons." The arrows are "idea bridges" from the water molecule to your chosen model. On the left, include details describing how water displays this characteristic; on the right, include details describing how your model displays this characteristic. Do this with as many characteristics of water molecules as are expressed in what you want to say about it.** ANSWERS VARY.

- **Does your metaphor suggest characteristics of a water molecule that you didn't think of earlier? Add any additional characteristics and details to the arrows and boxes on your diagram.** POSSIBLE ANSWERS: *asymmetry, polarity.* As student groups report, write the names of their models on the webbing diagram for metaphors, indicating at least three important characteristics of a water molecule that are expressed by the metaphor.

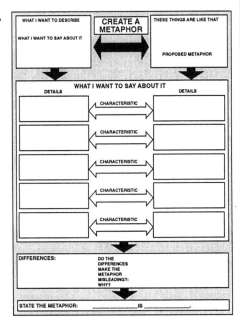

- **If there are any significant differences between a water molecule and your model, write them on the lower box.** POSSIBLE ANSWERS: *My model is two-dimensional (the covalent bond model), but the water molecule is three-dimensional. My model is visible, but the water molecule is invisible to the naked eye. My model (H_2O) suggests that the two hydrogen atoms are joined, but they are not in a water molecule. Only the valence electrons are shown in my model (the electron dot model), but oxygen has two electrons not shown. My model (a three-dimensional ball-and-stick model) is solid, but a water molecule is made up of much empty space.*

- **Are the differences between a water molecule and your suggested model so significant that the reader would be misled? Explain why or why not in the lower box.** Usually, students remark that the standard H_2O model, its chemical formula, is really misleading because of the suggestion that the two hydrogen atoms are combined. However, they often note that it can be used effectively in writing equations since these represent reactions and H_2O conveys the chemistry of the reaction. At the same time, it does not provide any information about chemical bonding, so one cannot use this model alone to predict a particular reaction. The differences between the other models and the water molecule, students say, are not significant enough to be misleading. For example, even though oxygen contains two additional electrons not shown, these electrons do not enter into chemical reactions so are not significant in models designed to emphasize how chemical reactions with water are possible. **If the differences are too misleading, pick another model and find out whether its characteristics and details are better suited to a water molecule.**

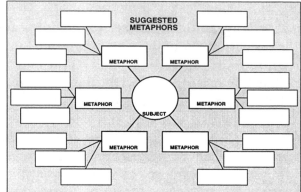

- **Write your model in the box at the bottom. Explain whether and how you think that your model is an effective one to describe a water molecule.**

- **Now, let's combine all of the different models for a water molecule in one graphic organizer to see what they show and what they don't show about this molecule.** Use the webbing diagram on a transparency or on the chalkboard to record student responses.

THINKING ABOUT THINKING

- **What did you think about to come up with your models?** Prompt students to recall the steps in the process. Record their strategies on the board or use a transparency of the thinking map "Creating a Metaphor," uncovering each step as students identify it. Discuss each step.

- **Do you think that this is a valuable way to think when you try to find one thing to describe another? Why or why not?** Most students comment that using this strategy is more organized than using the first idea that comes into your mind. It helps them find a good metaphor or model.

- **How does listing the characteristics of the water molecule on the graphic organizer help you "see" the traits it shares with the model?** POSSIBLE ANSWERS: *Listing the characteristics of a water molecule reminds me of details that I can use later to write or draw. Writing them down jogs my memory about other details or characteristics and makes selecting and using a model easier.*

APPLYING THINKING

Immediate Transfer

- Using the same strategy, create a model for the atom.

- **Suppose you were teaching students the following topics in science. Pick one and develop a verbal metaphor you can use to help your students understand key ideas under your topic: Plate tectonics, oxidation, nuclear fission or fusion, natural selection, the immune system.**

Reinforcement Later

- Develop a metaphor for one of the main periods in human history (e.g., the Enlightenment, the Industrial Revolution, the Space Age).

- Develop a metaphor for your science class.

ASSESSING STUDENT THINKING ABOUT CREATING METAPHORS/MODELS

To assess the skill of creating metaphors, ask students to answer any of the application questions or others that you select. Ask students to describe how they created the metaphor. If they write a poem using metaphors, ask them to show how they selected their metaphors. Determine whether they are attending to each of the steps on the thinking map.

Sample Student Responses • A Model for A Molecule

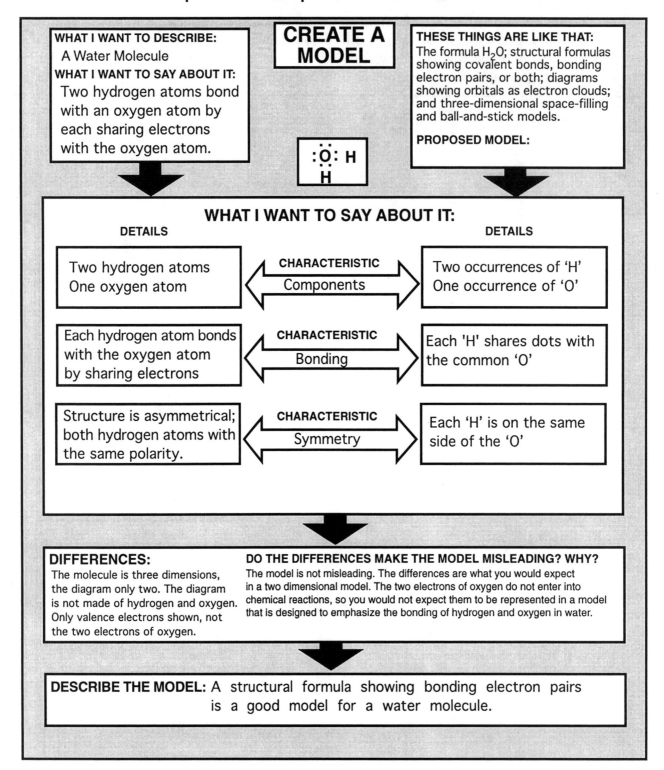

WHAT I WANT TO DESCRIBE:
A Water Molecule

WHAT I WANT TO SAY ABOUT IT:
Two hydrogen atoms bond with an oxygen atom by each sharing electrons with the oxygen atom.

CREATE A MODEL

:Ö: H
H

THESE THINGS ARE LIKE THAT:
The formula H_2O; structural formulas showing covalent bonds, bonding electron pairs, or both; diagrams showing orbitals as electron clouds; and three-dimensional space-filling and ball-and-stick models.

PROPOSED MODEL:

WHAT I WANT TO SAY ABOUT IT:

DETAILS   DETAILS

Two hydrogen atoms
One oxygen atom

CHARACTERISTIC
Components

Two occurrences of 'H'
One occurrence of 'O'

Each hydrogen atom bonds with the oxygen atom by sharing electrons

CHARACTERISTIC
Bonding

Each 'H' shares dots with the common 'O'

Structure is asymmetrical; both hydrogen atoms with the same polarity.

CHARACTERISTIC
Symmetry

Each 'H' is on the same side of the 'O'

DIFFERENCES:
The molecule is three dimensions, the diagram only two. The diagram is not made of hydrogen and oxygen. Only valence electrons shown, not the two electrons of oxygen.

DO THE DIFFERENCES MAKE THE MODEL MISLEADING? WHY?
The model is not misleading. The differences are what you would expect in a two dimensional model. The two electrons of oxygen do not enter into chemical reactions, so you would not expect them to be represented in a model that is designed to emphasize the bonding of hydrogen and oxygen in water.

DESCRIBE THE MODEL: A structural formula showing bonding electron pairs is a good model for a water molecule.

Sample Student Responses • A Model for A Molecule

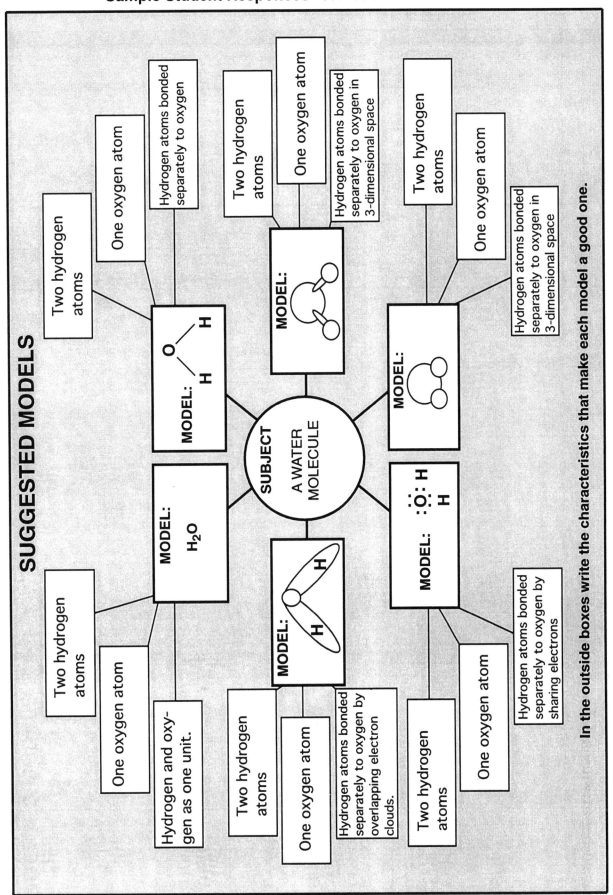

SUGGESTED MODELS

In the outside boxes write the characteristics that make each model a good one.

SUBJECT
A WATER MOLECULE

MODEL: H₂O

MODEL: O / H H

Two hydrogen atoms

One oxygen atom

Hydrogen atoms bonded separately to oxygen

Two hydrogen atoms

One oxygen atom

Hydrogen atoms bonded separately to oxygen in 3-dimensional space

Two hydrogen atoms

One oxygen atom

Hydrogen atoms bonded separately to oxygen in 3-dimensional space

Two hydrogen atoms

One oxygen atom

Hydrogen and oxygen as one unit.

Two hydrogen atoms

One oxygen atom

Hydrogen atoms bonded separately to oxygen by overlapping electron clouds.

Two hydrogen atoms

One oxygen atom

Hydrogen atoms bonded separately to oxygen by sharing electrons

CREATING METAPHORS LESSON CONTEXTS

The following examples have been suggested by classroom teachers as contexts to develop infused lessons. If a skill or process has been introduced in a previous infused lesson, these contexts may be used to reinforce it.

GRADE	SUBJECT	TOPIC	THINKING ISSUE
6–8	Science	White blood cells	Create a metaphor for the function of white blood cells in the human body.
6–8	Science	Circulatory system	Create a metaphor for the circulatory system that conveys important information about it.
6–8	Science	Volcanoes	Use metaphors to create a myth that would show how ancient people would explain volcanoes.
6–8	Science	Oceans	Use metaphors to create a myth or mural to show how the oceans were created.
6–8	Science	Crustaceans	The names given to crustaceans often fit their appearance, such as the horseshoe crab. Select three mollusks that you are just learning about and give them a name based on their appearance.
6–8	Science	Animal protection	Many animals have unusual means of disguising themselves. Create a metaphor based on a disguise so that you can remember how various animals disguise themselves.
6–8	Science	Earth science	Create a metaphor for how the layers of the earth were formed and write a story or draw a mural to show it.
9–12	Biology	Protein synthesis	Create a metaphor for the process of protein synthesis that describes its sequential nature. Then, draw a flow diagram of the process and its metaphor and evaluate the metaphor's effectiveness.
9–12	Biology	Cellular transport systems	Create a metaphor for the transport of raw material and cellular product in and out of a cell. Extend the metaphor to include shipment of raw material and delivery of product.
9–12	Biology	Cellular respiration	Create a metaphor for the metabolism of glucose into heat, energy, and water. Then, draw a diagram of a mitochondria and indicate on it the raw materials and products necessary to make ATP during cellular respiration; compare it with a similar diagram of the metaphor. Is the metaphor effective?
9–12	Biology	DNA	Create a metaphor for the role of DNA in a living cell. Explain the characteristic of DNA that the metaphor is designed to convey.
9–12	Biology	Botany	Create a metaphor for photosynthesis, the process of turning light energy into chemical energy. Compare diagrams of each and evaluate.

CREATING METAPHORS LESSON CONTEXTS

GRADE	SUBJECT	TOPIC	THINKING ISSUE
9–12	Biology	The digestive system	Create a metaphor for the breakdown of food into usable fragments.
9–12	Biology	Ocean life	Create a metaphor that expresses Sir Alister Hardy's theory of the vertical migration of plankton and their use of ocean currents to travel.
9–12	Biology	Symbionts	Create metaphors for the various symbionic relationships in nature: mutualism, commensalism, and parasitism.
9–12	Chemistry	The atom	Design a model for the atom that conveys important characteristics of the atom. Explain what those characteristics are.
9–12	Chemistry	The periodic table	Create a metaphor for the periodic table of elements.
9–12	Chemistry	Mixtures, etc.	Design models that convey the difference between a suspension, an emulsion, a solution, and a mixture.
9–12	Physics	Simple machines	Use simple machines like the lever and wedge as metaphors for human behaviors.
9–12	Physics	Nuclear fission	Nuclear fission creates an extremely rapid chain reaction. What are some possible metaphors for this dramatic event? Which would be good ones to convey the idea of a rapid release of tremendous energy?
9–12	Physics	Nuclear fusion	Create a metaphor for nuclear fusion that conveys the idea of energy generated by bringing nuclear particles together.
9–12	Physics	The solar system	Create a metaphor for the solar system. Extend by narrowing the characteristics of the metaphor to include motion as described by Kepler's laws.

PART 5

SKILLS AT ASSESSING THE REASONABLENESS OF IDEAS: CRITICAL THINKING

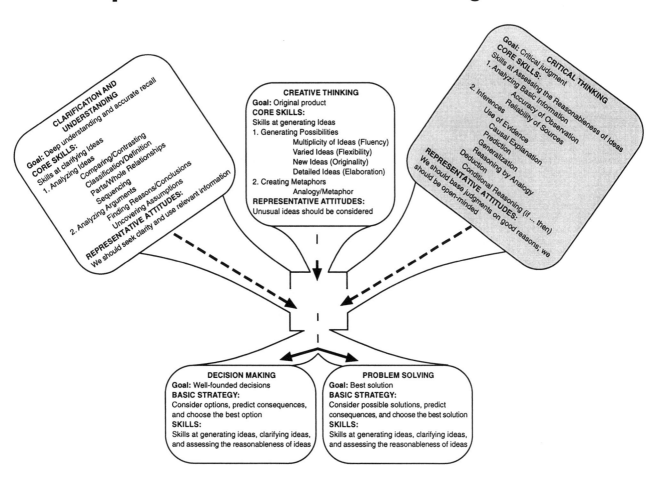

CLARIFICATION AND UNDERSTANDING
Goal: Deep understanding and accurate recall
CORE SKILLS:
Skills at clarifying Ideas
1. Analyzing Ideas
   Comparing/Contrasting
   Classification/Definition
   Parts/Whole Relationships
   Sequencing
2. Analyzing Arguments
   Finding Reasons/Conclusions
   Uncovering Assumptions
REPRESENTATIVE ATTITUDES:
We should seek clarity and use relevant information

CREATIVE THINKING
Goal: Original product
CORE SKILLS:
Skills at generating Ideas
1. Generating Possibilities
   Multiplicity of Ideas (Fluency)
   Varied Ideas (Flexibility)
   New Ideas (Originality)
   Detailed Ideas (Elaboration)
2. Creating Metaphors
   Analogy/Metaphor
REPRESENTATIVE ATTITUDES:
Unusual ideas should be considered

CRITICAL THINKING
Goal: Critical judgment
CORE SKILLS:
Skills at Assessing the Reasonableness of Ideas
1. Analyzing Basic Information
   Accuracy of Observation
   Reliability of Sources
2. Inferences
   Use of Evidence
   Causal Explanation
   Prediction
   Generalization
   Reasoning by Analogy
   Deduction
   Conditional Reasoning (If ... then)
REPRESENTATIVE ATTITUDES:
We should base judgments on good reasons; we should be open-minded

DECISION MAKING
Goal: Well-founded decisions
BASIC STRATEGY:
Consider options, predict consequences, and choose the best option
SKILLS:
Skills at generating ideas, clarifying ideas, and assessing the reasonableness of ideas

PROBLEM SOLVING
Goal: Best solution
BASIC STRATEGY:
Consider possible solutions, predict consequences, and choose the best solution
SKILLS:
Skills at generating ideas, clarifying ideas, and assessing the reasonableness of ideas

PART 5
SKILLS AT ASSESSING THE REASONABLENESS OF IDEAS: CRITICAL THINKING

Should I accept the idea that eating red meat regularly is a health hazard, or should I believe that it is risk-free? Should I endorse site-based management as a way of improving the quality of education, or should I reject it as ill conceived? Should I embrace the idea that teaching thinking is important, or should I concentrate on teaching students just the facts? Accepting ideas and acting on them is the primary way we progress through our lives and in our professional work. Unless we can assess which ideas are reasonable and which are not, we run the risk of acting on ideas that are incorrect, often leading to personal, professional, and even social harm.

When we engage in critical thinking, we assess the reasonableness of ideas. Critical thinking is crucially important to insure that the judgments we make are more likely to be correct than incorrect. We cannot expect to make sound judgments if we accept everything we hear and read or every idea that pops into our heads.

Critical thinking has been described in different ways: as the evaluation of reasoning and argument; as reasonable, reflective thinking directed at deciding what to believe or do; and as the application of standards to our judgments. What these conceptions of critical thinking have in common is that before we accept a judgment, we should be sure that it is supported by good reasons. If it is not, we should not accept it.

Skillfully assessing the reasonableness of ideas requires that we utilize acceptable standards. For example, when we appeal to the views of experts and well-respected authorities about matters we know little about, we use standards that are generally acceptable. However, a critical thinker does not simply accept standards that are in vogue but instead subjects the standards themselves to critical thinking. For example, when you reflect on appealing to an expert, you realize that even experts are some-

times wrong. We can think of many reasons to trust experts, but getting a second opinion may be more acceptable than trusting the word of one authority.

Critical thinking also involves important *attitudes and dispositions*. Critical thinkers search for reasons. They do so with open minds, looking for all of the reasons, both pro and con, before accepting an idea as feasible. They are willing to suspend judgment if they can find no reasons that support or counter an idea. They are willing to change their minds if they get new evidence. They are willing to submit even the most basic beliefs to critical scrutiny; for critical thinkers, everything they believe and do should be based on good reasons.

Critical thinking is as much a way of life as it is a set of abilities. Teaching for critical thinking must address the attitudinal and dispositional aspects of critical thinking, as well as the skills and abilities in good judgment that critical thinking involves.

The critical thinking skills in this handbook are the most frequently needed skills in our personal and professional lives. They fall into two categories:

- skills related to *basic information* that we get from a variety of sources, including media, textbooks, other people, and even our own observations (determining the accuracy of observation, determining the reliability of sources); and

- skills related to *inferences* in which we draw conclusions that we do not verify directly from information offered as *evidence to support* them (causal explanation, prediction, generalization, and reasoning by analogy), or inferences in which we *deduce conclusions* (conditional reasoning).

Each type of critical thinking skill involves the common theme of searching for reasons that support ideas and accepting those ideas only if they are based on good reasons. However, different types of supporting reasons are needed depending on the type of ideas being considered. For example, the support needed to determine whether a source is reliable (usual reliability, possible bias, expertise, etc.) is different from the type of evidence needed to support a prediction about tomorrow's weather, a generalization about voters' tendencies, an explanation of the causes of an accident, or a judgment about humans based on information about how mice behave in a laboratory. Critical thinkers can discriminate these different types of critical thinking and engage in each of them well.

Thinking skillfully about causal explanation, prediction, generalization, reasoning by analogy, conditional reasoning, and the reliability of sources of information is essential in our lives and professional work. Students will quickly develop the habits of mind necessary to become good critical thinkers when these thinking skills are taught across the curriculum and suitably reinforced.

CHAPTER 13
DETERMINING THE RELIABILITY OF SOURCES

Why is Evaluating Sources Skillfully Important?

We find out about many things without witnessing them ourselves, i.e., we rely on others as sources of information. We read about things in newspapers and magazines that we did not observe ourselves. We hear about events on television. Other people tell us about things that they've witnessed directly but we didn't.

We also rely on other sources for more specialized information. We sometimes consult sources for technical and/or general information that we may not be in a position to verify, like the usefulness of a medication in treating an illness or the correct translation of a foreign language passage.

The reliability of such information is extremely important. We often make key decisions based on it. We get certain medications for illnesses based on what a physician says. We make travel plans based on information we get from others. Misinformation in these contexts can be costly. People are often surprised to find out how much information we get from other sources and how much we depend on its being reliable.

Problems with Sources of Information

Unfortunately, we sometimes get misinformation from others. Deliberate deception or distortion is one source of misinformation. However, even well-meaning people who are unaware of the inaccuracy of their information can give us incorrect information. They may make hasty judgments, or their sources may not be reliable. Sometimes, a person's biases transform what they hear into subtle changes of meaning that are then unwittingly communicated from person to person. Rumors are extreme examples of misinformation transmitted by a number of people who are often influenced by such factors.

Problems with How We Think about Sources

One common problem in the way we handle information is that we often don't question our sources. Even if we do question a source, we may rely on just one or two factors to establish the source's credibility. In some instances, we accept a person as a reliable source because he or she seems honest. In other cases, if the source has expertise in a field, we assume that the information he or she provides must be accurate. However, we know that people who seem honest or who have expertise in a field may have biases or vested interests. They may be uncritical in accepting information they pass on to us. Often, if a witness describes an event at which he or she was present, people accept that witness as credible. An account published in a newspaper or book is considered reliable by some people just because it appears in print. These reasons for accepting reliability are risky oversimplifications.

These problems in our thinking about sources of information are summarized in figure 13.1.

COMMON PROBLEMS IN OUR THINKING ABOUT SOURCES OF INFORMATION

1. We accept information from sources without asking whether they are reliable.

2. We use information about only a small number of factors to make judgments about the reliability of sources.

Figure 13.1

To gather accurate information, we must exercise skillful discrimination about the information we are given. We can try to do this by confirming the information ourselves. However, when we have no direct access to the information, we must judge the credibility or reliability of the sources of the information in order to be able to judge its accuracy. Evaluating a source can, indeed, be done with care and skill.

How Do We Determine the Credibility of a Source Skillfully?

Types of Sources. Any person, publication, or other medium of communication can be a source of information. When the information we receive is important, we should take note of who or what the source is. The source could be a family member, a friend, a teacher, a salesperson, an event witness, a newspaper article, an encyclopedia, a new book, an advertisement, or a TV show.

We classify such sources in basic ways that have bearing on their credibility. The information may be secondhand (secondary), or it may be firsthand (primary). Generally, we tend to trust primary sources more than secondary sources. This, however, does not mean that secondary sources are untrustworthy. If the source is a secondary source, we should also find out where he or she got the information and try to trace it to its primary origins. Then, we should inquire about the primary source.

Determining Reliability

What else can we find out that would support or count against a source's credibility? Important factors include reputation, past history in providing reliable and accurate information, the person's expertise, the datedness of the information, the procedures used in gathering it, and the vested interests of those relating the information.

Eyewitness accounts are a special case of primary source information based on direct observation. In this instance, the conditions of observation, the use of observation-enhancing instruments, the person's expectations of what he or she is seeing, and when the report was recorded are all important.

We should apply the same principles to ourselves as sources of information. When we intend to communicate information to others, we can and should plan out how we acquire the information to be sure of its accuracy. When we plan and prepare observation reports, for example, we should be concerned with the same factors we take into account when we assess the reliability of others as sources of information. We may attend to some of these factors naturally, like putting our glasses on before we observe things, but we should also think about other factors that we often don't consider. For example, we should make sure that we know enough about what we are observing beforehand so that we don't misname or misdescribe it. We should also make sure that we write down

or otherwise record what we observe at the time, rather than waiting until later, when we might misremember.

Others develop confidence in us as reliable sources when we explain how we acquire our information. Describing how we make our observations or explaining any experimental procedures we use helps others to judge our reliability as observers. References to other reliable sources from whom we got our information also builds credibility. Indeed, these are standard procedures in the natural and social sciences for establishing a source's credibility.

Sometimes, we get information about one of the factors that influence credibility. We may find out, for example, that an individual who provides some important information on a current matter may not have given accurate information in the past. His past record casts some doubt on whether he is a reliable source for the information he is providing now. However, it would be a mistake to conclude that he is an unreliable source for the present information and that his report is inaccurate. An error in the past is not sufficient support for unreliability in the present.

Similarly, it is a mistake to think that because someone is an expert in a field, his or her judgment must be accurate. Experts often disagree and may have vested interests or biases (whether they acknowledge these biases) that can color and distort the information they provide. To have good reasons for thinking that a source is reliable or unreliable on a specific occasion, a good critical thinker must weigh all factors and establish a pattern of support for or against a source's reliability.

When all factors have been considered and the credibility of a source remains uncertain, one way to elevate the credibility of the source is to seek out other sources. If these additional sources are reliable and independently corroborate what the initial source tells us, that consensus can raise the credibility of the information.

Even without independent corroboration, we can still make judgments of relative reliability. If five peopled witnessed an automobile accident, determining which factors were present and which were not may suggest that one of these witnesses is more likely to give us accurate information than the others. One person may have been close to the accident and in full view of it; others may have had an obstructed view or have been attending to something else at the time. Although this may not guarantee that the first person is giving us an absolutely accurate description of what happened, there is a better chance that his or her account is more accurate than the others', all other things being equal. Of course, corroboration by additional witnesses can provide further support for the credibility of one or another of these accounts.

Tools for Determining the Reliability of Sources

A good way to make informed judgments about the reliability of sources is to develop a checklist of important factors to guide our search for information about the source. To exercise care in our judgments of reliability, we should think about all the factors on our checklist. Such a checklist can also be used to plan what we ourselves should do to become a reliable source of information.

The thinking map in figure 13.2 for evaluating the reliability of sources contains a checklist of this sort. It is useful when we don't know much about the source. If the information is an observation report, the more specialized map for determining the accuracy of an observation (depicted in figure 13.3) can be used.

These thinking maps can guide us in making judgments about sources in any field. This, however, does not imply that if you are a good source of information or a good observer in one field, you'll be a good source or observer in others. You wouldn't expect a person with expertise in medicine to be able to observe a malfunction in an automobile engine with the reliability of an auto mechanic, unless the medical expert is also knowledgeable about automobile repair. Expertise in a field is a criterion of accuracy, and reliability as a source and can vary from field to field. In using the checklists on these maps, interpret "Expertise" and "Background" to mean "in the field within which the information is contained."

The graphic organizer in figure 13.4 can be used to assess either secondary or primary sources of information. Write relevant questions from the thinking map in the boxes under

---

EVALUATING THE RELIABILITY OF SOURCES SKILLFULLY

1. What is considered the source of the information being considered?

2. List the factors present that are relevant to the reliability of the source in the following categories:

   Published?
   Date?
   Reputation of publication?
   Kind of publication (e.g., report, fiction)?
   Author?
   Expertise?
   Bias or distorting point of view?
   Special interest?
   Primary or secondary?
   If secondary, the reliability of any other sources the information is derived from?
   If primary, the other relevant factors, e.g., equipment used?
   Corroboration/confirmation?

3. Weigh the factors present and make a judgment of reliability based on them.

Figure 13.2

"Questions" to guide you in searching for information about the source. The items that you put in the "Questions" boxes can be derived from either the reliable sources or accuracy of observation checklists, depending on whether or not the source is an observer.

As you gather information about the source, write it in the "Information" box. Put a plus or minus next to it, depending on whether the information counts in favor of or against reliability. This process creates an informed profile of the source. Based on this profile, make your determination of the reliability of, unreliability of, or your uncertainty about the source. Usually, lack of information about corroboration and a mixture of other relevant information about the source, some of which counts for and some of which counts against reliability, should lead you to be uncertain about a source's credibility. Corroboration by an independent source of good repute can eliminate this uncertainty. On the other hand, in cases of such uncertainty,

DETERMINING THE ACCURACY OF AN OBSERVATION

1. Which of the following features of the observer, observation, and report are present in this case?

    Observer:
    Background?
    Qualifications?
    Usual reliability?
    Free of bias?
    State of mind?
    Physical ability to observe (eyesight, etc.)?
    Capacity to observe (proximity, direction, free of distraction)?
    Expectations/point of view?
    Vested interest in having audience believe the report?

    Conduct of the observation:
    Frequency?
    Equipment?
        Strength or accuracy?
        Condition?
        How operated?
    Date and location?
    Replicated?
    Observation conditions?

    Report:
    How soon after the observation?
    Details (drawings, photographs, graphs)?
    Language and findings expressed objectively?

    Corroboration:
    By others?
    By me?

2. When you weigh these factors, how reliable would you judge the observation to be?

Figure 13.3

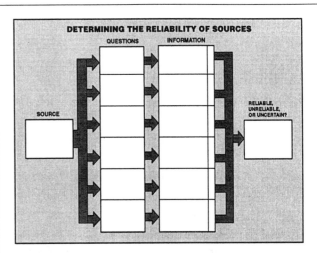

Figure 13.4

if other credible sources conflict with the one you are considering, this should lead you to judge that the source is not reliable.

How Can We Teach Students to Evaluate Sources Skillfully?

The goal of teaching students to evaluate sources skillfully is to encourage them to raise questions about reliability when appropriate and then to attend to the significant factors in judging reliability. It is not enough simply to ask students which sources among a given sample are most reliable. Challenge students to articulate their reasons for accepting or rejecting a source's reliability in the context of their understanding of the kind of information they need to make that judgment. Make the process of evaluating sources explicit through the frequent use of the language of the skill, the thinking map, and graphic organizers to guide your students as they respond to these challenges.

Contexts in the Curriculum for Lessons about the Reliability of Sources

Textbooks, teachers, and other resource people that your students use to learn about topics embedded in the secondary science and mathematics curriculum are sources of information that can serve as the focus of a reliable sources lessons. These are readily identifiable across the secondary school curriculum. In many ways, however, the best contexts for lessons on the reliability of sources are those in which an authentic issue about reliable information relates to curricular content. For example, when textbooks disagree about the facts or when there is a controversy about research data on a specific topic that students learn about in the curriculum, such contexts are particularly rich in both content-exploration potential and potential for developing skill at determining the reliability of secondary sources of information and of basic observations.

In the sciences, making sure that accumulated information is derived from reliable sources is of great importance. This is true of primary sources as well as secondary sources, and even to ourselves as observers.

In general, four sorts of instances provide rich contexts for reliable sources lessons in science:

- *Sources disagree about information students are learning.* For example, different sources may disagree about whether certain foods have nutritional or medicinal value or whether certain chemicals are harmful to the environment.

- *Observation reports and measurements sometimes conflict, reflect questionable procedures, or reveal something surprising.* For example, laboratory reports from students in science classes or measurements made in applying mathematics in science often differ widely. Lessons on the accuracy of observation reports are quite appropriate in such contexts.

- *A variety of sources of differing quality are available for student writing, reference, or research (e.g., in the school library).* For example, the library may contain popularized, outdated, and more recent scientific information on topics you ask your students to do library research on and to write about. Lessons built on these situations have great transfer potential across the whole secondary school curriculum.

- *People that students learn about have relied on other sources of information to advance their ideas about the natural world.* When students study the history of science, they learn about numerous figures who build on the work of others. Who are their sources, and was the person whose work is described justified in accepting those particular results as reliable and accurate?

Two Approaches in Teaching Students to Determine the Reliability of Sources

You may choose to use a direct approach in teaching this skill. For example, you can give students the thinking maps for reliable sources and accurate observation and ask them to use these as a guide, making a list of relevant factors they discover about a particular source. You can prompt students' consideration of these factors

by direct questions about the source, which you formulate yourself based on these maps.

Whenever possible, however, we recommend deriving maps for determining the reliability of sources from the students' own comments about a specific example. This inductive approach to learning the skill is featured in the sample lessons in this handbook. In conducting a lesson in this way, the teacher asks students to generate questions they would ask about sources and to categorize the questions into types of questions that one would ask about any source. This checklist can then be transformed into a thinking map for the skill. The graphic organizer in figure 13.5 is useful in conducting an inductive lesson of this sort.

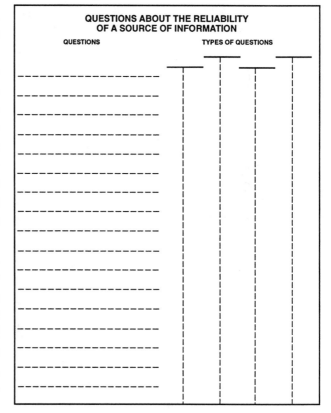

Figure 13.5

On the left side of this graphic organizer, students should list questions they would want to have answered in order to determine the reliability of the source. For example, to assess the accuracy of information in an article from a science magazine, students may list such questions as "What is the background and training of the author?" "Is the author usually reliable in giving information?" and "What is the reputation of the magazine in which the article appeared?"

Next, the students should group the questions into types, which they label on the short horizontal lines. They then connect the lines that the questions are on with the vertical line under the label. For example, "About the Author" might be listed on the first horizontal line under "Types of Questions." Then, all of the questions about the author, including the questions about the author's background and usual reliability, can be connected to the first vertical line leading to this heading. When this is completed, the types of questions can be reorganized into a checklist to develop a thinking map like the one we have included. If you use this technique for making the skill explicit, you should use the thinking map that the students generate, rather than the one included.

Reinforcing the Skill

After students have worked through a curricular example and have developed a checklist for the reliability of a source, you should give them more practice in using this checklist. When you ask them to apply the checklist to new examples, prompt students to use the terms from their checklist often and appropriately. Evaluating sources is particularly useful in resolving student conflicts. Your goal is to encourage students to use this skill deliberately whenever appropriate.

Model Lessons on Determining the Reliability of Sources

We include two model lessons on this skill: a middle school science lesson on astronomical observations and a lesson that has been used both in middle school life sciences and high school biology concerning information about life in the oceans. Each illustrates a different variation on teaching this skill mentioned in the commentary. The lesson on the oceans guides students in making judgments about the reliability of resources in science reference material. The other science lesson focuses on the accuracy and reliability of an observation report. As you review these lessons, try to answer the following questions:

- How is the importance of determining the reliability of sources of information introduced to students in these lessons?
- What techniques are used to prompt an inductive development of students' understanding of these skills?
- What are the near and the far transfer activities introduced in these lessons?
- Can you think of other ways of reinforcing students' skillful determination of the reliability of sources of information after these lessons have been completed?

Thinking Maps and Graphic Organizers for Determining the Reliability of Sources

The thinking maps provide questions to guide students' thinking in evaluating the reliability of sources of information skillfully. They can be used as written or modified by students as they reflect on how they engage in evaluating sources skillfully. One thinking map is for determining the reliability of any source, the second is for determining the reliability of an observer as a source of information.

Tools for Designing Lessons on the Reliability of Sources

Two graphic organizers are included. The first graphic organizer is for generating and categorizing questions to be asked in gathering information relevant to the reliability of a source or the accuracy of an observation. The second graphic organizer is for evaluating specific sources and reinforces the sequenced questions on the thinking maps.

The thinking maps and graphic organizers can guide you in designing the critical thinking activity in the lesson and can also serve as photocopy masters, transparency masters, or as models that can be enlarged and used as posters in the classroom. Reproduction rights are granted for use in single classrooms only.

EVALUATING THE RELIABILITY OF SOURCES SKILLFULLY

1. What is the source of the information being considered?

2. List the factors present that are relevant to the reliability of the source in the following categories:

   a. Published?
      Date?
      Reputation of publication?
      Kind of publication? (e.g., report, fiction)?

   b. Author?
      Expertise?
      Bias or distorting point of view?
      Special interest?
      Primary or secondary?
         If secondary, the reliability of any other sources the information is derived from?
         If primary, other relevant factors, e.g., equipment used?
      Corroboration/confirmation?

3. Weigh the factors present and make a judgment of reliability based on them.

DETERMINING THE ACCURACY OF AN OBSERVATION

1. Which of the following features of the observer, observation, and report are present in this case?

    Observer:

    Background?

    Qualifications?

    Usual reliability?

    Free of bias?

    State of mind?

    Physical ability to observe (eyesight, etc.)?

    Capacity to observe (proximity, direction, free of distraction)?

    Expectations/point of view?

    Vested interest in having audience believe the report?

    Conduct of the observation:

    Frequency?

    Equipment?

    Strength or accuracy?

    Condition?

    How operated?

    Date and location?

    Replicated?

    Observation conditions?

    Report:

    How soon after the observation?

    Details (drawings, photographs, graphs)?

    Language and findings expressed objectively?

    Corroboration:

    By others?

    By me?

2. When you weigh these factors, how reliable would you judge the observation to be?

QUESTIONS ABOUT THE RELIABILITY
OF A SOURCE OF INFORMATION

QUESTIONS

TYPES OF QUESTIONS

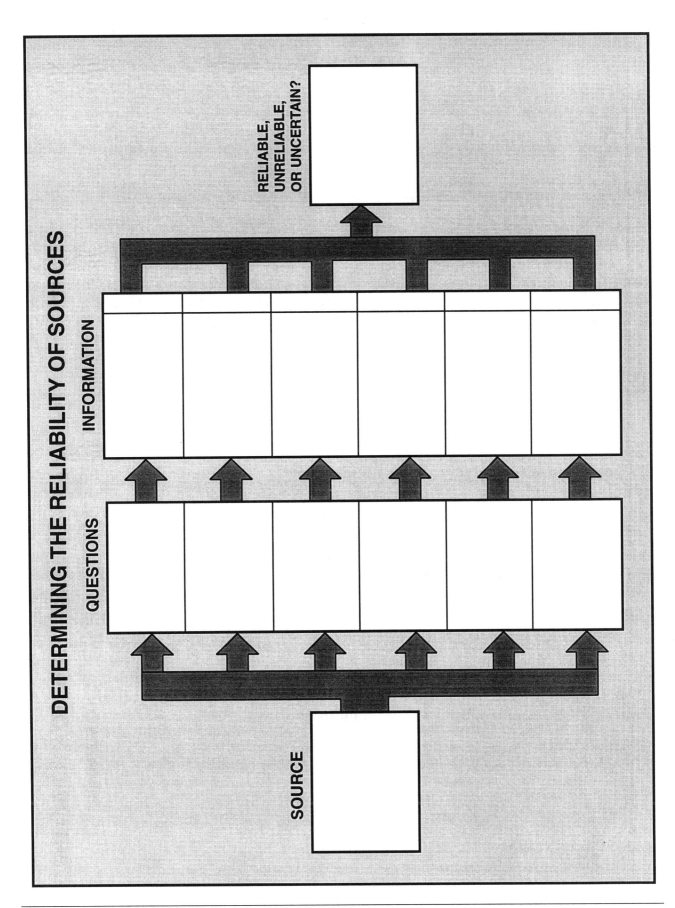

DETERMINING THE RELIABILITY OF SOURCES

RELIABLE, UNRELIABLE, OR UNCERTAIN?

INFORMATION

QUESTIONS

SOURCE

WHAT IS THE BOTTOM OF THE OCEAN LIKE?

General Science, Biology, Oceanography · **Grades 6–11**

OBJECTIVES

CONTENT

Students will learn how to select references for scientific topics. They will also learn about the ocean floor and about deep sea exploration.

THINKING SKILL/PROCESS

Students will learn how to make judgments about the accuracy and reliability of sources of information based on the presence or absence of relevant factors.

METHODS AND MATERIALS

CONTENT

Students read information about authors and publications that describe the bottom of the ocean. They interpret charts and read descriptions of the ocean floor. Library research is part of this lesson.

THINKING SKILL/PROCESS

This lesson makes use of structured questioning and a graphic organizer to aid students in identifying factors which influence the reliability and accuracy of sources of information. Collaborative learning enhances the thinking.

LESSON

INTRODUCTION TO CONTENT AND THINKING SKILL/PROCESS

- **We can't discover everything ourselves. To find out important information, we often have to rely on what other people tell us. Can you think of times when you have relied on other people for information?** Sources may include teachers, textbooks, radio talk shows, television documentaries, television commercials, parents, friends, salespeople, advertisements, and the movie page in the newspaper.

- **Sometimes, the information we get from others is accurate. Sometimes, it is not. It's important to be sure that we get good information when we have to rely on others. Can you think of a time when you got information from someone else and it turned out to be inaccurate? Discuss one with your partner.** Ask for a few examples from the class.

- **Finding information that we can rely on depends on how well we check out our sources of information. This means that we have to think about the many factors involved: who the sources are, how much they know, how they know about what they are telling you, what they believe about it, when they found it out, etc. Sometimes, we even have to evaluate whether we ourselves are good, reliable sources of information.**

- **Suppose a friend told you that the best bicycle to buy was a "Fantastic Flyer." What would you want to find out to decide whether this was reliable information?** POSSIBLE ANSWERS: *Did the friend learn it from a magazine or another person who owned a Fantastic Flyer? Was the information taken from an ad for the bicycle? Was it from an article about bicycles written by someone who didn't favor one particular brand? Does the friend have a good memory? Can you read the same article yourself?*

- **We've been studying the oceans in science. Much of the earth's surface is covered by the oceans. We get food from the oceans. That's why it is important to us to find out as much as we can about the oceans.**

The Bottom of the Ocean		
WHAT DO I KNOW?	**WHAT DO I THINK I KNOW?**	**WHAT DO I NEED TO KNOW?**

- Did you ever wonder what the *bottom* of the ocean was really like? You can't go there yourself, and an ocean map doesn't tell you much. Let's think about what you might find at the bottom of the ocean. In order to think about this in an organized way, I'm going to ask you to respond to three questions: What do you know about the bottom of the ocean, what do you think you know, and what do you need to know? As you respond, I will write what you say in three columns under each of these headings. On the chalkboard or a transparency, draw a diagram like the one shown. Ask for responses to the questions, one by one, from the class, and write student responses in the columns. Students often comment that they *know* things like the following: *It's very dark. It's very cold. There is a lot of pressure from the water. There are deep ravines.* Students say they <u>think they know</u> things like the following: *There are sunken ships. The animals will be very small because of the pressure. There will be debris thrown overboard from ships. It will be hard to see anything. The bottom of the ocean is miles from the surface.* Students often say that they <u>need to know</u> things like the following: *What kind of animals live there? How deep is it? Can humans survive there? Exactly how cold is it? Has anyone ever been there? Can we get food there?*

THINKING ACTIVELY

- Suppose you went to a library and found a number of books and articles about the oceans. You might want to use them to find out whether the class's ideas about the bottom of the ocean are accurate and to answer some of our questions. Here are some examples of books and articles:

  "Man's New Frontier," by Luis Marden, from *National Geographic*, April 1987.

  The Sea and Its Living Wonders, by Dr. G. Hartwig, 1860.

  Twenty Thousand Leagues Under the Sea, by Jules Verne, 1870.

  "Monsters Under the Sea," in *Great Science Fiction Stories*, 1955.

  "Incredible World of Deep Sea Rifts," by Robert Ballard and J. Frederick Grassle, from the *National Geographic Magazine*, November 1979.

- Is there anything about these works that might suggest that some of them are more reliable than others? What? POSSIBLE ANSWERS: *Since one of them is a science fiction story, perhaps its descriptions of what lives in the ocean are fiction, not fact. However, maybe the science fiction story is based on science facts.* Twenty Thousand Leagues Under the Sea *and* The Sea and Its Living Wonders *came out in the 1800s. People didn't have submarines then and may not have had any way to see what was at the bottom of the ocean. Also, wasn't* Twenty Thousand Leagues Under the Sea *a science fiction novel? The* National Geographic Magazine *has articles about real explorations. The articles written in the 1970s and 1980s were at a time when we could see what was there. Robert Ballard discovered the* Titanic *by going to the bottom of the ocean in a small submarine with lights and windows.*

- What other questions would you want to have answered to help you decide which sources are likely to give you the most accurate and reliable information about the bottom of the ocean? Work in your groups. Use the graphic organizer for questions about the reliability of a source. Write your questions on the dotted lines on the left. Compile a class list of questions on the chalkboard or on a transparency of the diagram by asking for one or two questions at a time from each group. POSSIBLE ANSWERS: *Was the article written by a person who actually explored the ocean? Does the magazine have a reputation for accurate information? Are the authors experts on the oceans? If the authors explored firsthand, what equipment did they use? Why did they write these articles? Does the article have photographs that show the bottom of the ocean?*

QUESTIONS ABOUT THE RELIABILITY OF A SOURCE OF INFORMATION

QUESTIONS   TYPES OF QUESTIONS

• We could try to answer these questions one by one, but consider a more systematic way. If we group them, we can make a checklist of types of things we want to find out to decide whether a source is reliable and accurate. Group these questions together into four or five basic categories, such as questions about the author. Write these categories on the short lines at the top of the diagram and connect the appropriate questions to the dotted line that leads to it. Call for responses and fill them in on the class diagram. POSSIBLE ANSWERS: *Questions about the author (whether or not he was there, his expertise about the oceans), about the publication (its reputation, special interests), about the article (when the article was published, corroboration by photographs).*

• In order to create a checklist for making these kinds of judgments, let's use these categories as part of a thinking map for determining the reliability of a source. The first question we should have on our thinking map is, "What is the source we are using?" The next question has to do with factors that count in favor of or against the source's reliability. Then, we should have a list of the factors we should attend to. You should fill in these factors yourselves. For example, one of your question categories concerned the author. Put "Author?" on your checklist of factors relevant to reliability. That will guide you to find out things about the author that are relevant to his or her reliability as a source of information. A subcategory you mentioned was the author's expertise about the oceans. That is something specific about the author that you'll want to find out. If he's an expert on the oceans, that will count in favor of his reliability; if not, that will count against it. What the students develop should resemble the thinking map shown.

EVALUATING THE RELIABILITY OF SOURCES SKILLFULLY

1. What is the source of the information being considered?

2. List the factors present that are relevant to the reliability of the source in the following categories:

   a. Published?
      Date?
      Reputation of publication?
      Kind of publication? (e.g., report, fiction)?

   b. Author?
      Expertise?
      Bias or distorting point of view?
      Special interest?
      Primary or secondary?
        If secondary, the reliability of any other sources the information is derived from?
        If primary, other relevant factors, e.g., equipment used?
      Corroboration/confirmation?

3. Weigh the factors present and make a judgment of reliability based on them.

• **Where can you get information to answer some of these questions?** You or the school librarian should show students where they can find the date on books and magazines and how they can find out about the author from the dust jacket of books, the title page, and biographical sketches in magazines, etc. Ask the students to make a list of these research strategies and add any others they find.

• Let's take a close look at one of the sources. Use your checklist to gather relevant information about whether it is a reliable source of information. Write your results on the graphic organizer for determining the reliability of sources. **Make a judgment about how reliable the source is for accurate information about the bottom of the ocean. Each group should explain its judgment to the class.** Use the article "The Incredible World of Deep Sea Rifts," find one or two others, or ask the students to go to the library and work there. The two *National Geographic* articles are usually identified as very reliable. When students explain why, their explanations should be thorough and should show an awareness of how relevant information indicates reliability. For example, the authors of "The Incredible World of Deep Sea Rifts" are

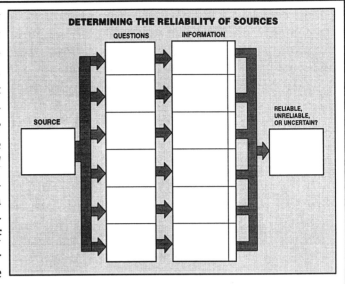

identified in the magazine as "a marine geologist and biologist" from the Woods Hole Oceanographic Institution. They are firsthand observers. Their submersible (the *Alvin*) is described in detail in the article as having observation ports, lights, mechanical arms, and the capability of traveling to the deepest parts of the ocean. Many colored photographs accompany the article. *National Geographic* has a reputation for publishing scientific works of discovery and exploration. The article does not indicate whether the authors have a reputation as good researchers or whether they have any vested interests or scientific biases. Endorsement of them as reliable sources is justified, but with the understanding that these questions remain unanswered. As each group reports, discuss with the class what the information on their graphic organizer reveals about the reliability of the source being considered.

- **Each group should also make a list of the interesting information it has gathered about the bottom of the ocean.** Ask each group to report one interesting piece of information it has found. As the groups report, circle information in the "I know" column of the original diagram if the information is confirmed. If it is not there, add it. Cross out any information it conflicts with, and if it answers any questions in the "I need to know" column, cross these out. POSSIBLE ANSWERS: *Worms six to eight feet long can be found in some of the deep-sea trenches at the bottom of the ocean. These trenches also have foot-long clams with red blood. Crabs and other sea life live near these trenches. The water is hot around these trenches because of lava, which is being ejected into the water. The water is rich in food for the plants and animals that live there. The pressure is very great. People can explore these deep-sea trenches in specially designed submersible craft like the Alvin.*

THINKING ABOUT THINKING

- **Let's put aside these questions about the bottom of the ocean and think about how you developed your checklist for determining the reliability and accuracy of a source of information. What did you think about first, and how did you proceed so that you had a good checklist?** Answers should refer to listing the questions, categorizing them, and then transforming the categories into a checklist. Some students may also comment on the collaborative nature of the activity.

- **Is that a good way to develop a checklist? Is the checklist you developed for judging the reliability and accuracy of a source of information a good and helpful one to use? Why?**

- Plan out how you will gather the most reliable information on the next topic you have to research. Describe what you will attend to, how you will get the information you need about the sources, etc. Students should indicate that they will use their checklist to guide their search for relevant information about reliability and accuracy. They should also describe where they will seek relevant information and mention the school librarian as a resource.

APPLYING THINKING

Immediate Transfer

- Think about something you might like to buy. Plan what sources you will use to get the most reliable information about it. Why will you use those sources? School libraries may feature *Penny Wise*, the *Consumer Reports* magazine for children, designed to provide consumer information. If they use this periodical, make sure they explain why they believe it to be reliable. Discuss other sources such as advertising, a salesperson, and someone who owns the item.

- We've been studying ocean pollution. Gather information about the kinds of pollution you find in the oceans, the problems the pollution causes, and what people have been trying to do about it. Explain where you got the information and why you think it is reliable.

Reinforcement Later

When the class studies the following topics later in the school year, use the given dialogue.

- We are now studying the effects of diet on health. We're going to look at various reports about diet and health in advertising, from health clinics, on milk cartons, etc. Apply your checklist to determine which of these is likely to be more reliable and accurate than others. What might you do to verify the information from these sources?

- We are studying the movement west in this country. One of the topics you study is the way of life of the Native Americans. Go to the school library and gather four sources of information about the Native Americans. Use your checklist to decide which of these sources is likely to provide the most accurate and reliable information. Explain why. Compare these descriptions to your textbook and to a movie you've seen about the Old West. Describe two important things you've found out about Native Americans in the Old West that you didn't know before and that you think are accurate. List some things that you think are not accurate. Explain.

ASSESSING STUDENT THINKING ABOUT RELIABLE SOURCES

Any of the transfer examples can serve as an assessment item. The examples about a new purchase, ocean pollution, diet, and the Native Americans are good ones for this purpose. The example about a new purchase is particularly suited to be used as pre- and post-tests on this skill. Decide beforehand which items from the checklist a student should mention in order to show improvement.

If your students are collecting information on a topic for a period of time, you can use portfolio assessment techniques to supplement other forms of assessment. You can ask students to develop a research plan and to keep a portfolio of references they research, each with an explanation of why the references are or are not reliable sources. You can also ask them to keep a journal in which they record information from the sources they think are reliable. This can be used in a future writing assignment.

IT IS EASY to become jaded these days, but there is a powerful antidote to that feeling. I experienced it last June in France as I witnessed the excitement and jubilation that followed the successful flight of *Gossamer Albatross*. That flimsy dragonfly made of graphite tubing, plastic, and tape, designed by aeronautical engineer Dr. Paul B. MacCready and driven by a wiry young biologist named Bryan Allen, had crossed the English Channel on Allen's leg power alone. The flight won not only £100,000 prize but also the plaudits of a world waiting to thrill to such exploits.

A few weeks later, I talked with a group of our editors who had been huddled around scientists at the Jet Propulsion Laboratory in California when the spacecraft Voyager 2 began sending back its breathtaking portraits of Jupiter's moons (pictures that will appear in our January 1980 issue). Again, high excitement was in the air, as a new frontier was unveiled by a superb team effort.

While Voyager 1 and 2 probed the outer world of our solar system, a manned deep-sea research vehicle named *Alvin*, using image-making equipment of high sophistication, brought back startling scenes from the inner world far below the surface of the Pacific Ocean. There, along rifts where earth's crustal plates are separating, were revealed life forms and geologic processes never seen before. As the scientists reported their findings to the National Geographic, which helped support the exploration, they were equally excited and jubilant.

Such moments of achievement and discovery have been occurring all during the 91 years we have been publishing, and they will continue as long as human beings find challenges to meet. Those on the vast scale of the planet have given rise to vast accomplishment. When we read Rick Gore's report on world desertification in this issue, we should remember that the earliest civilizations grew out of the deserts, that both advanced technology and societies were formed in response to their challenge.

It is not time to become jaded, when man had probed only the innermost fringe of the universe, and when he had seen only one mile in a thousand of the seafloor rift system. We expect to be publishing great adventures for a long time to come.

Gilbert M. Grosvenor

NATIONAL GEOGRAPHIC

THE NATIONAL GEOGRAPHIC MAGAZINE VOL. 156. NO. 5 COPYRIGHT© 1979 BY NATIONAL GEOGRAPHIC SOCIETY WASHINGTON, D.C. INTERNATIONAL COPYRIGHT SECURED

November 1979

The Desert: An Age-old Challenge Grows 586

It covers a third of earth's land, bringing hardship and suffering to a sixth of all people — and it is spreading. Rick Gore and photographer George Gerster report on a globe-circling survey.

Winged Victory of "Gossamer Albatross" 640

Its pilot, Bryan Allen, tells of his dramatic pedalpowered flight across the English Channel, a milestone in aviation history.

Which Way Oahu? 652

Buffeted by a tourist boom, inflation, land shortage, and urban sprawl, Honolulu's home island is a paradise in peril, Gordon Young and Robert W. Madden find.

Incredible World of Deep-sea Rifts 680

Marine geologist Robert D. Ballard and biologist J. Frederick Grassle describe mineral-spewing chimneys and newly discovered creatures living in warm-water oases around ocean-floor vents.

Hong Kong's Refugee Dilemma 709

The jam-packed British crown colony must now cope with last summer's flood of Asia's homeless: "boat people" from Vietnam as well as those fleeing from neighboring China. William S. Ellis and William Albert Allard document the human side of the situation.

COVER: *Dressed for the desert in Upper Volta, a Bella tribesman combines sunglasses and turban to shield against sun and sand.*
Photograph by George Gerster.

Scientists explore rifts in the seafloor where hot springs spew minerals and startling life exists in a

STRANGE WORLD WITHOUT SUN

ACROSS THE BOTTOM of the four oceans of the world runs the largest feature on the face of this planet, a mountain range and rift system some 40,000 miles long. Man has seen with his own eyes scarcely forty miles of this Mid-Oceanic Ridge.

But along those few miles in the past six years, scientists in tiny submarines such as *Alvin* have found, in those utterly dark nether depths of the sea, animals and mineral factories unlike any seen before.

In 1979, the latest in a series of expeditions went out into the Pacific to study spreading centers of the ocean floor. These are places where the thin, rigid plates that form the hard crust of our planet are pulling apart, separating as much as eight inches a year. In the cracks, molten magma wells up, meets cold seawater, and solidifies into a contorted landscape of black lava.

In such regions the scientists have been witnessing the all but believable. They have seen:

- Huge blood-red worms protruding from forests of white plasticlike tubes.

- Clams far larger than most shallow-water types, their meat scarlet with hemoglobin.

- Strange dandelionlike creatures moored by threads near fountains of warm water.

- Plumes of even hotter water — 350° C (650° F) or more — spewing black clouds of minerals from seafloor chimneys.

Since William Beebe's bathysphere dives in the 1930s and the descents of Jacques-Yves Cousteau, Jacques Piccard, and others in the 1950s and 1960s, the National Geographic Society has participated in and reported on many historic ocean explorations.

In May 1975, NATIONAL GEOGRAPHIC carried a full report on Project FAMOUS, man's first look at the Mid-Atlantic Ridge. In August 1976, we described dives into the yawning Cayman Trough in the Caribbean. In October 1977, the GEOGRAPHIC reported the astonishment of geologists who descended to the Galapagos Rift in the eastern Pacific and first discovered warm-water vents teeming with life. Rich in hydrogen sulfide and bacteria, these oases apparently attract larval organisms drifting in the currents.

This discovery set the stage for further dives in the Pacific. In 1978, French, U.S., and Mexican scientists using the French submersible *Cyana* explored the East Pacific Rise as 21° North, off the mouth of the Gulf of California. They found inactive vents and huge dead clams, similar to those discovered on the Galapagos Rift in 1977.

Then, in back-to-back expeditions beginning in January 1979 and continuing this fall, funded by the National Science Foundation and the Office of Naval Research, scientists returned to the Galapagos Rift and 21° N with *Alvin* and its support ships. On the pages that follow, two of the leaders report on expedition findings.

WE ARE PROUD that the National Geographic Society has played a part in these explorations. By a major research grant, as well as by providing cameras, film, and photographic experts and by operating color laboratories aboard the surface ships, the Society has markedly extended science's ability to see and record phenomena in the abyss.... — THE EDITOR

RETURN TO OASES OF THE DEEP

By ROBERT D. BALLARD and J. FREDERICK GRASSLE

BOTH WOODS HOLE OCEANOGRAPHIC INSTITUTION

THE SCENE a few feet outside *Alvin's* view port overwhelms. A riot of red-tipped worms, some of them 12 feet tall, grow around the organic-rich vents. Crabs and mussels are everywhere. These creatures that greeted geologists diving in 1977 now meet us, a group of biologists, chemists, and geologists returning to study the living oases of the Galapagos Rift.

The extraordinary worms have no eyes, no mouth, no gut, no anus. Laboratory dissections reveal they do have separate sexes, and most likely broadcast eggs and sperm into the water. Hemoglobin—red blood pigment—accounts for their bright color. Covering the solid spongy plume, more than 300,000 tiny tentacles arranged on flaps, or lamellae, absorb molecules of food and oxygen form the water. The blood carries this nourishment throughout the body.

The concentration of suspended food available at the vents is amazing. By one estimate, it is 300 to 500 times greater than just outside vent areas and four times greater than in productive surface waters.

Fountains of Life in the Abyss

WE LOOK directly into the heart of an active vent as *Alvin's* heat probe, at left, registers up to 13° C (55° F), much warmer than the usual deep-sea chill of 2° C. Yet heat is not the main lure for the flowerlike sea anemones, brown mussels, curling serpulid worms, and blind crabs gathered here, 2.5 kilometers below the surface. Sparse populations of similar animals survive the cold even at the sea's deepest point, 11 kilometers, existing on whatever organisms drift down from the sunlit surface.

Animals congregate at vents because of the enormous food supply based on bacteria. The bacteria, sulfur, and heat give the vent water its milky blue shimmer. Microbes exist everywhere in the sea, often in a state similar to suspended animation. Some types can metabolize hydrogen sulfide; when they find that nourishment here in vent water, they proliferate, providing food for clams, worms, and mussels.

We find that the mussels have a long larval stage. Such larvae, drifting great distances in ocean currents like plant seeds riding the wind, could start a community whenever a vent opens up. Then dead mussels, in turn, become food for scavengers, such as crabs.

The white brachyuran crabs—of a crustacean family not previously known—scramble into our fish-baited traps. In insulated containers kept at 2° C, dozens survive decompression on the hour-and-a-half ascent with *Alvin*. So we select this animal as the best living subject for studies on the relationship of temperature and pressure to metabolism, investigations pursued in a laboratory at the University of California at Santa Barbara.

When kept at sea-surface pressure of one atmosphere, the crabs did not live long. But those placed in a pressure vessel set at 250 atmospheres, the same as their home environment, behave normally and easily tolerate changes in temperature. The last survivor lived for more than six months.

New Ways to Study a New World

A JOURNEY to the deep sea is a little like going to the moon. We spend months preparing for an unknown realm but can stay only a few hours on the spot. And we have not just a new geology but also a complex, unfamiliar ecosystem to investigate.

Biologists are especially curious about respiration and growth rates at the vent, since elsewhere in the abyss metabolism slows down. Using *Alvin's* claw, we place mussels for 48 hours in a chambered respirometer to check oxygen uptake; others we leave in wire cages to test when we return this fall. The grenadier fish—common in the depths—may be attracted by *Alvin's* lights.

We look everywhere for new life forms. Microbiologist Holger W. Jannasch searches for bacteria in water samples, on rocks, and here on a mussel shell. Later, a scanning electron micrograph of the mussel's shell raises questions. The strings are stalks of bacteria. But what are the strange protuberances? Geologists thought they might be minute manganese nodules. Dr. Jannasch has found they are bacteria cells coated with manganese and iron.

Filtered vent water yields solid evidence that bacteria multiply rapidly within the vents by metabolizing hydrogen sulfide, carbon dioxide, and oxygen. Bacteria grow in mats and clumps in the subsurface spaces of porous rocks until the flowing water peels them off. The bacteria count is high, up to a million per cubic centimeter (less than a quarter teaspoonful). More than 200 different strains of bacteria are being kept alive at Woods Hole. The pink fish that we observed head down in vents may be feeding on bacterial masses.

FIELDS OF FOOT-LONG CLAMS, overrun by galatheid crabs, populate active vents along the rifts. A smooth-shelled individual from 21° N proves slightly larger than an eroded Galapagos specimen. Radiochemical dating at Yale University shows vent clams grow four centimeters a year, 500 times faster than a small deep-sea cousin that can live as long as a century. Galapagos clams have numerous large and yolky eggs, but we have not yet found how the clams disperse.

The meat inside is startlingly red, a rare sight in clams. Their hemoglobin has an unusually high affinity for oxygen, possibly an adaptation to periods of low oxygen.

A Marvelous Multitude

TO MARINE BIOLOGISTS, vent communities are as strange as a lost valley of prehistoric dinosaurs. The "dandelion," first seen suspended by filaments above the seafloor during the 1977 Galapagos expedition, proved to be a new siphonophore. Related to the Portuguese man-of-war, it consists of a gasbag for buoyancy, surrounded by hundreds of members with specific functions—some capture food, others ingest it, still others handle reproduction. Brought to the surface, the fragile animal started to fall apart, so we quickly put it in a fixative.

Another unusual animal, a small worm, forms a tube from minerals in the water, cementing itself near the chimneys at 21° N that spew solutions hotter than 350° C. This effluent cools so quickly on meeting the seawater that the worms don't actually live in the hottest water. Geologists dubbed them Pompeii worms, since they must survive a constant rain of metal precipitates. They turn out to be bristle worms, or polychaetes, which probably consume bacteria with feeding tentacles.

Among hundreds of specimens collected, we discover even more new species of whelks, barnacles, leeches, and a red-blooded bristle worm. While dissecting mussels, invertebrate zoologist Carl Berg finds the worm living in the mantle cavity.

On videotape from the CCD camera we can see such worms leaving the mussels we collect.

On board *Lulu*, geologist-author Ballard examines the largest tube worm brought to the surface; its body fills more than half of the 2.5-meter tube. Several juveniles had cemented themselves to this adult. We also find on such tubes a new variety of filter-feeding limpet, a living representative of Paleozoic fossils.

Lava Lakes and Frozen Pillars

THE SEAFLOOR near the vents gives us a big surprise. Geologists had believed that lava underwater always flowed slowly, forming bulbous pillows. Instead, we find, lava lakes fill depressions, much as on land. This means molten magma rushed up with such ferocity that cold seawater could not immediately harden it. The flow swirled and coiled before solidifying, a few hundred to a few thousand years ago, only yesterday in geologic time.

As it advanced across the cold seafloor, the lava capped water-filled cracks. This water heated and rose in a jet, hardening the lava it touched. After the lava lake drained, the hollow pillar—the mold of a water column—remained, with ledges like bathtub rings. Similar ledges line the lake edge three meters beyond.

On top of the pillar we see animals, perhaps tube worms. Sediments begin to collect, snowing down at the rate of five centimeters every

1,000 years, eventually blanketing the bottom as it moves away from the rift. Quiet for now, the ocean-floor crust will undergo intermittent rifting and eruption on a cycle of about 10,000 years. With time, the lava cracks, and water circulates down and up again, creating new vents. When cracks go deep enough, magma will again be released, and the cycle will repeat.

Our findings clear up major mysteries about the composition of ocean water. We once assumed that all its minerals had to come from river runoff. Yet the elements in the ocean were out of balance—not enough magnesium and too much manganese. Direct sampling of vent water proves that during circulation deep in the ocean crust, seawater drops off magnesium and picks up manganese. John Edmond, geochemical leader of the Galapagos II expedition, calculates that all the world's oceans circulate through the crust once every 10 million years.

Minerals Erupt at Hot Spots

LIKE A FACTORY at full throttle, a submarine chimney at 21° N belches hot mineral-laden water that rises through cold seawater pressing down at nearly two tons per square inch. As the solution mixes with the near-freezing water, it precipitates yellow, ocher, and reddish brown deposits of iron, copper, and zinc sulfides. When *Alvin* breaks off and retrieves a fragment, we learn its dull interior is sphalerite, a zinc sulfide, while its bright interior is chalcopyrite—fool's gold.

With *Alvin's* claw, we insert a temperature probe vertically into a "black smoker." The readout inside the sub spins off scale. Later, we determine that the water must be hotter than 350° C (650° F). But only one end of the plastic rod has melted. The far end is unaffected, showing that the solution cools instantly as it mixes with seawater.

As Mid-Oceanic Ridge exploration has shifted from relatively quiet spreading centers in the Atlantic to the Pacific's more active rifts, our anticipation has grown. Will we finally actually see molten lava erupting, and more exotic animals thriving, when we dive to the fastest spreading center known, off Easter Island on the East Pacific Rise?

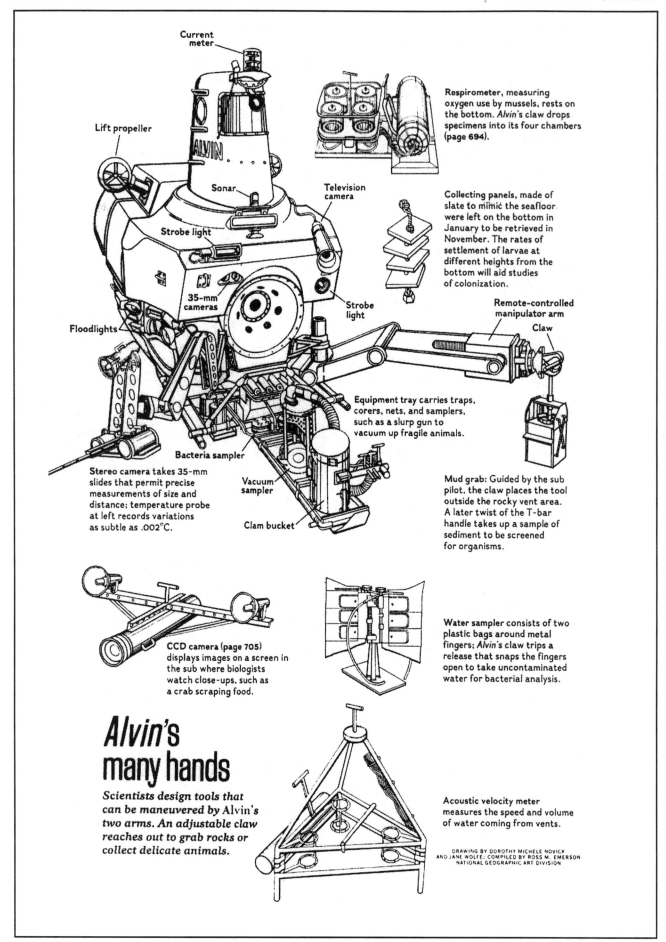

Current meter

Lift propeller

ALVIN

Sonar

Television camera

Strobe light

35-mm cameras

Floodlights

Stereo camera takes 35-mm slides that permit precise measurements of size and distance; temperature probe at left records variations as subtle as .002°C.

Bacteria sampler

Vacuum sampler

Clam bucket

Strobe light

Respirometer, measuring oxygen use by mussels, rests on the bottom. *Alvin*'s claw drops specimens into its four chambers (page **694**).

Collecting panels, made of slate to mimic the seafloor, were left on the bottom in January to be retrieved in November. The rates of settlement of larvae at different heights from the bottom will aid studies of colonization.

Remote-controlled manipulator arm

Claw

Equipment tray carries traps, corers, nets, and samplers, such as a slurp gun to vacuum up fragile animals.

Mud grab: Guided by the sub pilot, the claw places the tool outside the rocky vent area. A later twist of the T-bar handle takes up a sample of sediment to be screened for organisms.

CCD camera (page 705) displays images on a screen in the sub where biologists watch close-ups, such as a crab scraping food.

Alvin's many hands

Scientists design tools that can be maneuvered by Alvin's *two arms. An adjustable claw reaches out to grab rocks or collect delicate animals.*

Water sampler consists of two plastic bags around metal fingers; *Alvin*'s claw trips a release that snaps the fingers open to take uncontaminated water for bacterial analysis.

Acoustic velocity meter measures the speed and volume of water coming from vents.

DRAWING BY DOROTHY MICHELE NOVICK AND JANE WOLFE; COMPILED BY ROSS M. EMERSON NATIONAL GEOGRAPHIC ART DIVISION

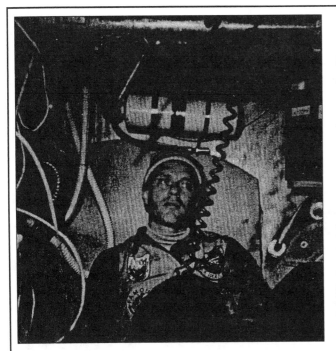

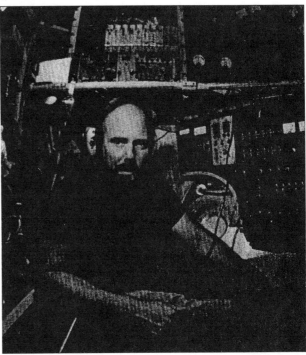

Seafloor explorers: Geologist Robert D. Ballard (left) has descended in Alvin to all the deep-sea spreading centers so far explored—the Mid-Atlantic Ridge, Cayman Trough, Galapagos Rift, and East Pacific Rise at 21° N—and described those adventures in the GEOGRAPHIC. Marine biologist J. Frederick Grassle (below) usually conducts ecological studies of diverse animal communities living in Atlantic sediments.

For the 1979 Galapagos expedition, other biological investigators were Carl J. Berg and Ruth D. Turner, Harvard University; James J. Childress, University of California at Santa Barbara, UCSB; Judith P. Grassle, Marine Biological Laboratory; Robert R. Hessler, Kenneth L. Smith, and George N. Somero, Scripps Institution of Oceanography, SIO; Holger W. Jannasch, Howard L. Sanders, and Albert J. Williams III, WHOI; David M. Karl, University of Hawaii; Richard A. Lutz and Donald C. Rhoads, Yale University; Jon H. Tuttle, University of Texas. Other geologists were Robin Holcomb and Tjeerd H. van Andel, Stanford University; Kathleen Crane, WHOI. Geochemists included John Edmond, MIT; John B. Corliss and Louis I. Gordon, Oregon State University; Michael L. Bender, University of Rhode Island. On the subsequent French-Mexican-U.S. geological expedition to 21° N, the principal scientists were Fred N.

Spiess and Kenneth C. Macdonald, co-leaders, and John A. Orcutt, all of SIO; Bruce P. Luyendyk, UCSB; William R. Normark, U.S. Geological Survey; Jean Francheteau and Thierry Juteau, Centre National pour L'Exploitation des Oceans; Arturo Carranza, Diego A. Cordoba, Victor Dias, Jose Guerrero, and Claude Rangin, Universidad Nacional Autonoma de Mexico.

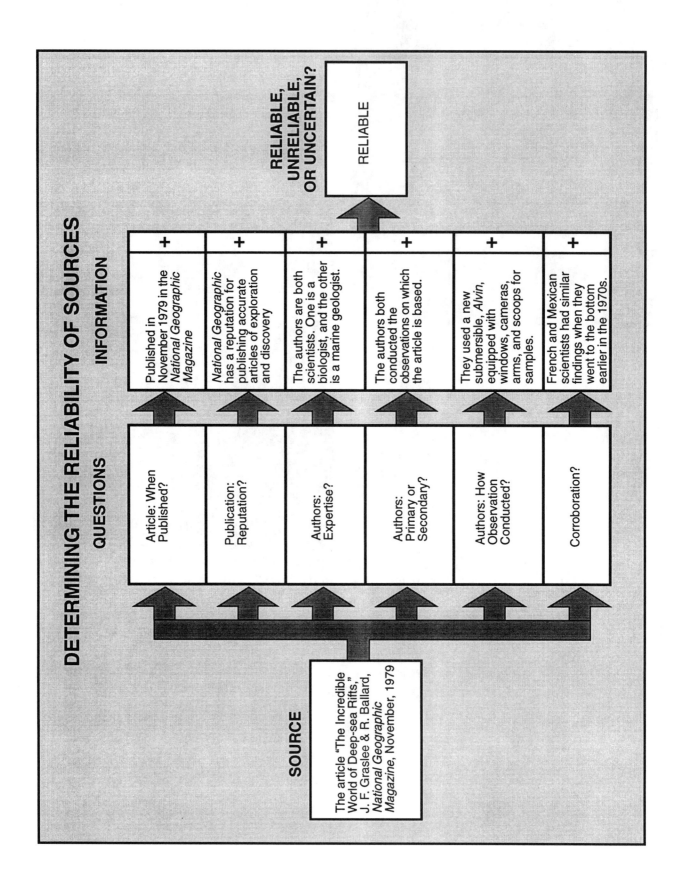

DETERMINING THE RELIABILITY OF SOURCES

RELIABLE, UNRELIABLE, OR UNCERTAIN?

RELIABLE

INFORMATION

+	Published in November 1979 in the *National Geographic Magazine*
+	*National Geographic* has a reputation for publishing accurate articles of exploration and discovery
+	The authors are both scientists. One is a biologist, and the other is a marine geologist.
+	The authors both conducted the observations on which the article is based.
+	They used a new submersible, *Alvin*, equipped with windows, cameras, arms, and scoops for samples.
+	French and Mexican scientists had similar findings when they went to the bottom earlier in the 1970s.

QUESTIONS

Article: When Published?

Publication: Reputation?

Authors: Expertise?

Authors: Primary or Secondary?

Authors: How Observation Conducted?

Corroboration?

SOURCE

The article "The Incredible World of Deep-sea Rifts," J. F. Graslee & R. Ballard, *National Geographic Magazine*, November, 1979

THE CANALS OF MARS

General Science, Astronomy   **Grades 6–12**

OBJECTIVES

CONTENT	THINKING SKILL/PROCESS
Students will learn about the surface of the planet Mars and will learn how astronomical observations are made.	Students will learn how to make judgments about the accuracy and reliability of observations based on the presence or absence of relevant factors.

METHODS AND MATERIALS

CONTENT	THINKING SKILL/PROCESS
Students read an observation report about the planet Mars made by an astronomer. They pool their background knowledge about instruments of observation in astronomy, in particular telescopes. They also read other reports about Mars.	A thinking map, a graphic organizer, and structured questioning emphasize factors that influence the reliability and accuracy of an observation. Collaborative learning enhances the thinking.

LESSON

INTRODUCTION TO CONTENT AND THINKING SKILL/PROCESS

- Have you ever heard an account of an event that you accepted as accurate but later found out that it wasn't? Jot down some of the details of that situation. Now tell your partner about it. **What went wrong with the account that made it incorrect?** Get three or four reports from students and write on the chalkboard their diagnoses of what went wrong with the reports. POSSIBLE ANSWERS: *It was based on rumors. It was a deliberate distortion for some ulterior motive. It was advertising to sell something. It was in a sensationalistic newspaper. The person who gave the account got information that was inaccurate from someone else and didn't know it. The person made a mistake in what he thought he saw because he was distracted.*

- Since we rely so much on information from others, it is important to make sure that we get accurate information. We make decisions about purchases based on information we get from others. We find out how to do things, as in sports, based on information we get from others. Think about all the different sources of information you use. Jot down a few of them. POSSIBLE ANSWERS: *newspapers, TV news, textbooks, teachers, salespeople, friends, dictionaries, telephone books, TV documentaries.*

- In this lesson, we will find out how to determine beforehand whether information we're getting is coming from a reliable source. We will examine a special kind of information—information that we and others get from observation.

- Science is one field in which we rely on observation to give us basic information. Whenever we conduct an experiment, we record our observations. Even complicated scientific theories are based on observation. What you read in your textbook about the planets is based on observations that people have made. We're going to look at some observations made by a scientist who drew some pretty startling conclusions. We're going to think about how to decide whether the observations are accurate.

THINKING ACTIVELY

- Read the following illustrated description written by someone who was interested in the planets and who decided to observe them to find out what they were like. This is his description of what he saw when he looked at the planet Mars.

> *15 July—I was amazed to see dark areas of blue-green that exactly typify the distant look of our own forests. Only a few months ago this area was pale yellow in color, suggesting the seasonal changing color of leaves. A projection stood out from the planet's surface, but it was clearly not a mountain peak, since the projection was not fixed in place, but suggested a cloud formation. The most startling observation of all was that of miles and miles of parallel lines, so geometrically regular that I am at a loss to show their ruled effect in my drawing. I'm certain that these bizarre parallel features are canals, laid down with as much precision as railway metals on Earth. Only intelligent life could have constructed such canals!*

- **What reaction do you have to this?** POSSIBLE ANSWERS: *I didn't know there was life on Mars. I wonder what the canals do. This must be science fiction.*

- Let's think about this. Sometimes, great discoveries sound fantastic. On the other hand, sometimes reports like this are wrong. How can we tell? It's clear that what this person is telling us about life on Mars is not something he observed directly. Rather, he is saying this because of his observation of canals on Mars, something he thinks only intelligent beings could construct. So let's think about this observation. How can we tell if it is accurate? Work together in groups of three or four to make a list of questions you would like to have answered to help you decide whether this report is accurate and reliable. List questions you would ask about the source and circumstances, as well as questions about the report. Use the diagram for questions about the reliability of sources for your list of questions. **Write them on the dotted lines on the left.** When the students have worked for about five minutes, ask them to report. Ask each group to mention one question, so that all the groups respond. List the questions on a large "class" diagram that you construct on the board or on a transparency. Then ask if students have any other questions that haven't been mentioned. Add those to the diagram.

**QUESTIONS ABOUT THE RELIABILITY
OF A SOURCE OF INFORMATION**

QUESTIONS   TYPES OF QUESTIONS

- We could try to answer these questions one by one, but perhaps there is a more organized way to do this. If we group questions, we can make a checklist of types of things we want to find out about the source. This will help us decide whether or not he or she is a reliable and accurate

From *The Search for Life on Mars: Evolution of an Idea,* by Henry S. F. Cooper, Jr. (Holt, Rinehart and Winston, 1976) observer. Group these questions together into a few basic categories, such as questions about the observer. Write these categories on the short lines on the top of the diagram and connect the appropriate questions to the dotted line that leads to it. Call for responses and write them on the class diagram. POSSIBLE ANSWERS: *The observer, how the observation was conducted (including the equipment and its condition), the report itself, corroboration.*

• Let's try to organize these types of questions and construct a thinking map that we can use to determine the accuracy of an observation. We can use the map as a checklist for making these kinds of judgments. Add relevant subcategories under each major heading. The list that students develop may resemble the thinking map shown at the right.

• Now let's go back to the report on the canals of Mars and try to answer some of these questions. Here's the title page of a book that the observer wrote about the canals of Mars and some additional biographical data about the observer. (Make available copies of pages 347–8). What information does this material provide that answers questions on your list? Record your results on this graphic organizer, "Determining the Reliability of Sources." Ask students to read some of the recorded answers to their questions. POSSIBLE ANSWERS: *The observer had the background and training to be a good astronomical observer. He was well qualified (professor at M.I.T.). He won many prizes, hence probably had a good reputation as an astronomer. He wrote Mars and Its Canals in 1906—before scientists had the sophisticated technology that we do now for observing planets. He used a large telescope for his time. He made many observations. He wrote this passage in his journal while he made the observation. He believed that canals were there before he looked through his telescope, perhaps creating a predisposition to believe this. There is no evidence of drinking or psychological disorder.*

• Try to reach agreement in your groups about whether you believe that this is an accurate and reliable observation. Explain why. In the box on the right side of your graphic organizer, write whether the observation report is reliable, unreliable, or uncertain.

• Use your textbooks and other sources for more information about the surface of the planet Mars. What would this information show about the credibility of Lowell's account? Is there any information that corroborates Percival Lowell's observations? Is any of this information more acceptable than Percival Lowell's report? Why? If you want to extend this activity, show students some of the Mariner photographs of Mars. These are contained in the book *MARS, As Viewed by Mariner 9*, published by NASA in 1976 and in other sources that are available at most

DETERMINING THE ACCURACY OF AN OBSERVATION

1. Which of the following features of the observer, observation, and report are present in this case?

Observer:
  Background?
  Qualifications?
  Usual reliability?
  Free of bias?
  State of mind?
  Physical ability to observe (eyesight, etc.)?
  Capacity to observe (proximity, direction, free of distraction)?
  Expectations/point of view?
  Vested interest in having audience believe the report?
Conduct of the observation:
  Frequency?
  Equipment?
    Strength or accuracy?
    Condition?
    How operated?
  Date and location?
  Replicated?
  Observation conditions?
Report:
  How soon after the observation?
  Details (drawings, photographs, graphs)?
  Language and findings expressed objectively?
Corroboration:
  By others?
  By me?

2. When you weigh these factors, how reliable would you judge the observation to be?

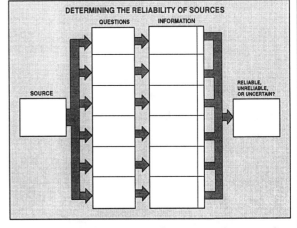

DETERMINING THE RELIABILITY OF SOURCES
QUESTIONS INFORMATION
SOURCE
RELIABLE, UNRELIABLE, OR UNCERTAIN?

libraries. Discuss with your students the differences in the technology and its reliability for the following circumstances: when a person observes Mars through a telescope on Earth, when photographs of Mars are taken through a telescope on Earth, and when a spacecraft near Mars sends back to Earth computer-enhanced pictures.

- **Prepare a report about what you think the surface of Mars is like. Illustrate this report. Explain why you think this is an accurate description. What unanswered questions do you have about the surface of the planet?** Discuss with students the planet's surface features that might have made Percival Lowell think he was seeing canals.

THINKING ABOUT THINKING

- **Let's put aside these questions about Mars now. Think about how you developed your checklist of things to consider in determining the reliability and accuracy of an observation report. What did you think about first, and how did you proceed so that you had a good checklist?** Answers should refer to listing the questions, categorizing them, and then transforming the categories into a checklist. Some students may also comment on the collaborative nature of the activity.

- **Is that a good way to develop a checklist? Can you think of other situations for which this strategy would be helpful?** Answers may refer to checklists for making purchases. *You could make a list of questions you would want answered to help you decide what to buy and a list of sources to consult to answer these questions.*

- **Is the checklist you developed for judging the reliability and accuracy of an observation report a helpful one to use for this purpose? Why?**

- **What advice would you give to another person about how to make sure that his or her observation reports are as accurate and reliable as possible?** Good advice includes using the checklist to plan how you are going to conduct and report on an observation.

APPLYING YOUR THINKING

Immediate Transfer

- **Suppose you were going to visit a nearby pond in order to make observations about the natural behavior of animals that lived in and around the pond. Plan your observation so that you will bring back the most accurate observation reports you can. Write out your plan.** Students should use the checklist and apply it to this specific observation.

- **Suppose a car hit the side of a truck. Several people saw the accident. Which of the following witnesses do you believe would give you the most accurate and reliable account? Which the least? Why?**

  A man getting into a car on the other side of the street

  A policewoman directing traffic at the next intersection

  The driver of the truck that was hit

  A three-year-old passenger in the back seat of the car that hit the truck

  A man looking out of his window from a third-story window on the other side of the street

Reinforcement Later

When the class is studying the following topic later in the school year, ask the given questions.

- We are now studying the effects of diet on health. We're going to look at various reports about diet and health in advertising, from health clinics, on milk cartons, etc. Apply your checklist to try to determine which of these is likely to be more reliable and accurate than others. What might you do to verify the information contained in these sources?

- We are studying the way of life of the Native Americans. Find four sources of information about the Native Americans. Use your checklist to decide which of these sources is likely to provide the most accurate and reliable information. Explain why. Compare these descriptions to your textbook and to a movie you've seen about the Old West. Describe two important things you've found out about the Native Americans in the Old West that you didn't know before and that you think are accurate. List some things that you think are not accurate. Explain.

- Think about a purchase you will be making in the next few months. What sources would give you the best information for making a good decision? Why?

ASSESSING STUDENT THINKING ABOUT RELIABLE SOURCES

Any of the transfer examples can serve as an assessment item. The example about the pond is particularly suited for this purpose. It can be used as pre- and post-tests to see whether the students have changed the way they make judgments about accuracy and reliability. Decide beforehand how many additional items from the checklist the student should mention in order to show improvement.

Here is an example of pre- and post-test responses that can be used as a paradigm for judging student improvement. In the pre-test, students were asked to plan an observation of animal behavior in and around a nearby pond. Student responses were predominantly like the following:

> I would plan to look at things.

> I would go there.

In the post-test given a few months after the *Canals of Mars* lesson, their responses included plans like the following, none of which were mentioned in the pre-test:

> I would bring binoculars.

> I would make sure I could get close enough to get a good view.

> It would have to be daylight.

> I would read up on what I was going to be observing beforehand so I knew what I was going to be looking at.

> I would bring a pad and pencil and write down what I saw when it was happening.

> I would bring someone along who would also take note of these things.

The post-test responses are more articulate than the pre-test responses. In this case, the scoring criteria for substantial improvement was that a student's plans must include mention of one or two factors from each of the major categories on the thinking map. These responses clearly show improved judgment about determining the reliability of an observation.

Sample Student Responses • The Canals of Mars

QUESTIONS ABOUT THE RELIABILITY OF A SOURCE OF INFORMATION

QUESTIONS	TYPES OF QUESTIONS		
	OBSERVATION		REPORT
	OBSERVER	CORROBORATION	
What is his background?	X		
What is his scientific reputation?	X		
For whom was the report written?			X
What kind of equipment did he use?	X		
Did he use the same equipment for all sightings?	X		
What was his state of mind? Was he clear-headed?	X		
Where was he when he made his observation?	X		
Did other accounts corroborate his report?		X	
In what form or publication did the report appear?			X
Was the report a translation or his own words?		X	
What were the weather conditions?	X		
In what year did he make the observation?	X		
When did he write the report?			X
Did he have normal sight?	X		
Was the equipment appropriately maintained?	X		
Was he typically trustworthy?	X		
What did he expect to see?	X		
Did he know how to use the equipment?	X		
How often did he observe it?	X		
Is the lens scratched?	X		
How long did he observe it?	X		
Did he believe in life on Mars prior to the observation?	X		
Did he make accurate observations of other planets?	X		
Was he drinking before he made the observations?	X		
Was a model made to verify how formations should look?	X		
Was he paid for this account? If so, by whom?	X		

MARS

AND ITS CANALS

BY

PERCIVAL LOWELL

DIRECTOR OF THE OBSERVATORY AT FLAGSTAFF, ARIZONA; NON-RESIDENT PROFESSOR
OF ASTRONOMY AT THE MASSACHUSETTS INSTITUTE OF TECHNOLOGY; FELLOW OF
THE AMERICAN ACADEMY OF ARTS AND SCIENCES; MEMBRE DE LA SOCIETE
ASTRONOMIQUE DE FRANCE; MEMBER OF THE ASTRONOMICAL AND
ASTROPHYSICAL SOCIETY OF AMERICA; MITGLIED DER ASTRO-
NOMISCHE GESELLSCHAFT; MEMBRE DE LA SOCIETE BELGE
D'ASTRONOMIE; HONORARY MEMBER OF THE SOCIEDAD
ASTRONOMICA DE MEXICO; JANSSEN MEDALIST OF
THE SOCIETE ASTRONOMIQUE DE FRANCE,
1904, FOR RESEARCHES ON MARS;
ETC., ETC.

ILLUSTRATED

New York

THE MACMILLAN COMPANY

LONDON: MACMILLAN & CO., LTD

1906

All rights reserved

THE CANALS OF MARS

Lowell's interest in Mars had begun late in the last century, when he became interested in reports of observations made in 1877 by Giovanni Schiaparelli, an Italian astronomer. Schiaparelli said he had seen faint lines on Mars, and he referred to them as *canali*. The popular British and American interpretation of the word *canali* was that it meant canals—which are, of course, man-made—rather than channels, which need not be. Nor did Schiaparelli make any attempt to clarify the interpretation; indeed, he once remarked; "I am very careful not to combat this suggestion, which contains nothing impossible"—a use of the double negative still favored by seekers after extraterrestrial life, particularly Carl Sagan. (It is, of course, a not ungrammatical use of the double negative.) Schiaparelli, whose eyesight was failing, continued to observe Mars on its close approaches to the Earth until about 1890. (Mars and the Earth pass each other in their orbits about every 2 years. At these times, they are said to be in opposition; exceptionally close approaches occur every 16 years.) Lowell, who had exceptionally good eyesight and was proud of it, took up the watch in 1894, when he set up an 18-inch telescope on a hill, which came to be called Mars Hill, outside Flagstaff, Arizona; this was the genesis of the Lowell Observatory, one of the first in this country to be situated in a remote spot for good visibility, and today a major astronomical institution. Then, as now, Mars watching had its difficulties. When the planet was low in the sky, so that the telescope's eyepiece was high off the ground, Lowell had to hang from a ladderlike scaffold that lined the observatory walls like the stall bars of a gymnasium. On a drawing board hooked to a convenient rung, Lowell made sketches of his observations. In color, Mars was a brilliant, splotchy orange red, though there were some darkish blue-green splotches as well. In 1659 the Dutch astronomer Christiaan Huygens—the first man to study Mars telescopically—had discovered white splotches at either pole, which he deduced were polar caps. The markings looked fuzzy and ill-defined, and they varied from time to time, the way a nearsighted man might view the image in a kaleidoscope; Mars had, in fact, the shifty splotchiness of a Rorschach inkblot, in which people are sometimes asked what they think they see.

During the next 20 years, Lowell concluded that Mars was laced with an elaborate webbing of canals, which, because of their extreme length, precision, and straightness, could have been created only by a highly advanced civilization. Lowell, who was as literate as Sagan (the poet Amy was his sister and his brother Abbott Lawrence was president of Harvard), recorded his observations in three very persuasive books. "Suggestive of a spider's web seen against the grass of a spring morning, a mesh of fine reticulated lines overspreads (the planet)," he wrote in *Mars and Its Canals*, which was published in 1906. "The chief difference between it and a spider's web is one of size, supplemented by greater complexity, but both are joys of geometric beauty. For the lines are of individually uniform width, of exceeding tenuity, and of great length. These are the Martian Canals."

From *The Search for Life on Mars: Evolution of an Idea*, by Henry S. F. Cooper, Jr. (Holt, Rinehart and Winston, 1976)

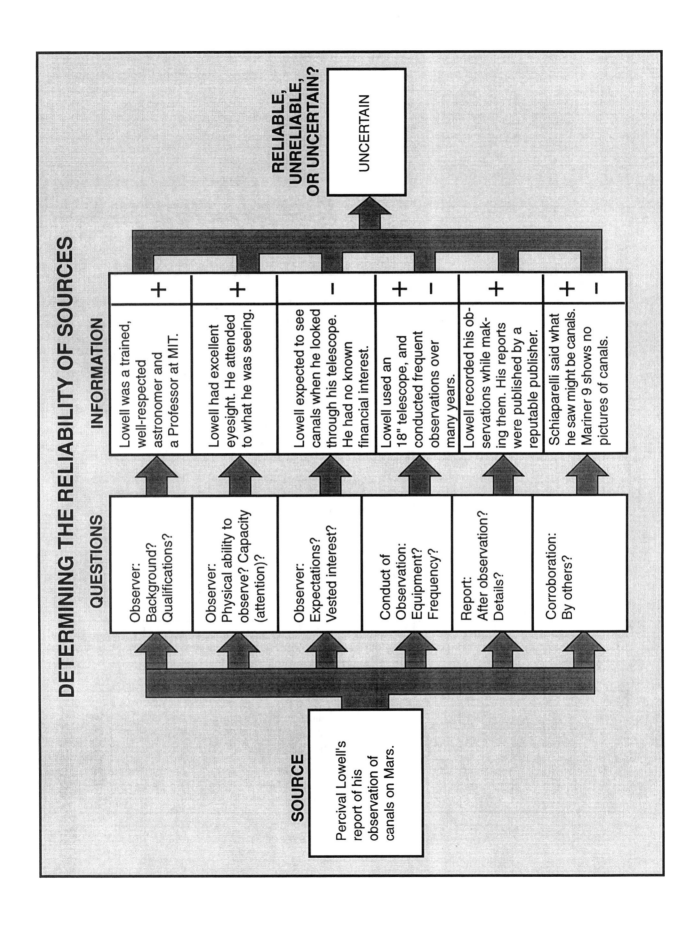

DETERMINING THE RELIABILITY OF SOURCES

RELIABLE, UNRELIABLE, OR UNCERTAIN?

UNCERTAIN

INFORMATION

+	Lowell was a trained, well-respected astronomer and a Professor at MIT.
+	Lowell had excellent eyesight. He attended to what he was seeing.
–	Lowell expected to see canals when he looked through his telescope. He had no known financial interest.
+ –	Lowell used an 18" telescope, and conducted frequent observations over many years.
+	Lowell recorded his observations while making them. His reports were published by a reputable publisher.
+ –	Schiaparelli said what he saw might be canals. Mariner 9 shows no pictures of canals.

QUESTIONS

- Observer: Background? Qualifications?
- Observer: Physical ability to observe? Capacity (attention)?
- Observer: Expectations? Vested interest?
- Conduct of Observation: Equipment? Frequency?
- Report: After observation? Details?
- Corroboration: By others?

SOURCE

Percival Lowell's report of his observation of canals on Mars.

DETERMINING THE RELIABILITY OF SOURCES LESSON CONTEXTS

The following examples have been suggested by classroom teachers as contexts to develop infused lessons. If a skill or process has been introduced in a previous infused lesson, these contexts may be used to reinforce it.

GRADE	SUBJECT	TOPIC	THINKING ISSUE
6–8	Science	Ozone layer depletion	What would determine the reliability of information about the degree to which the ozone layer is being depleted and the cause of that depletion? Refer to material published by Ralph Nader, Dixie Ray Lee, the petrochemical industry, etc.
6–8	Science	Animals	Many people have observed gorillas in the wild, but descriptions of the gorillas' behavior vary. Some describe these animals as ferocious, others as gentle and shy. How can you decide whether descriptions of such animals are accurate and reliable?
6–8	Science	Consumer biology	How would you determine the accuracy of advertising claims made about the effectiveness and gentleness of a shampoo?
6–8	Science	Consumer biology and safety	Evaluate sources like the following for reliability of technical information about automobile safety: "Car of Cars" by Frank Ford (*Children's Digest*, 1960); "The Auto Industry" by Tom Bench (*GMC Digest*, January, 1989); *Consumer Reports*. Explain why you rate them as you do.
6–8	Science	UFOs	What should you consider about an account by someone who claims to have seen a UFO in order to decide whether his or her story should be taken seriously?
6–8	Science	Tobacco and health	What is the best source of information about the implications of continuing to grow and sell tobacco? What role do a person's health concerns, economic interests and environmental protection awareness play in determining whether he or she is a reliable source of this information? What other factors are important to consider?
9–12	Biology	Evolution	Evaluate sources like the following for reliability of information about the origin of life on earth: biology textbook, the Bible, UFO newsletter, television special on PBS. Explain your evaluation for each.
9–12	Biology	Forensic evidence	Evaluate which source is most likely to be held by a forensic scientist to be the most reliable evidence of a crime: fingerprints, blood typing, DNA sample, footprints, fabric sample, etc. Explain your criteria for ranking each.
9–12	Biology	Consumer biology	Evaluate which source is most likely to be most reliable when seeking information about the nutritional value of a particular breakfast cereal: advertising claims, food labels, government reports, consumer guides such as *Consumer Reports*.
9–12	Biology	Observation and reporting	What would you consider in judging the accuracy of observations and data reported by a fellow student during an experiment (e.g., the study of phototropism and geotropism in plant biology)?
9–12	Biology	Environmental science	What would you consider to be the best source of information about the carrying capacity of a local forest for deer: a U.S. Forest Service report, a hunting newsletter, a bulletin from the local chapter of PIRG, or Greenpeace? Explain your answer.

DETERMINING THE RELIABILITY OF SOURCES LESSON CONTEXTS

GRADE	SUBJECT	TOPIC	THINKING ISSUE
9–12	Biology	Environmental science	What would you consider to be the most reliable source of information on the sources of sulfur and nitrous oxides in the environment: an automobile industry white paper, a government EPA report, a pamphlet put out by the local gas and electric company, an "infomercial" provided by an activist conservationist group, a story seen on the TV newsmagazine *60 Minutes*, an expose seen on the TV show *Hard Copy*. Why?
9–12	Biology	Data collection	Suppose you were collecting information about the variety of monera found in a local lake. What might you do to insure that the information you collect is reliable?
9–12	Chemistry	States of matter	What might you consider when determining the most reliable method for determining the state of matter for a substance: chemical formula, color, melting point, boiling point, density.
9–12	Chemistry	Predicting products of chemical reactions	What would you consider to be the most reliable method for predicting the products of an unknown chemical reaction: finding a similar reaction in the literature; analyzing the general equation of the reaction as combination, decomposition, replacement or combustion; looking it up in your textbook. Why?
9–12	Physics	Conservation of energy	An inventor claims that he has produced a perpetual motion machine. What would you have to know about the conservation of energy and momentum and the inventor in order to evaluate the reliability of his claim.
9–12	Physics	Electro-magnetism	An astronomer at Arecibo, the huge radiotelescope in Puerto Rico, announces that he has received responses to his radio message from the Great Cluster in Hercules. What would you have to know about electromagnetic waves and the scientist to assess the reliability of his announcement?

Chapter 14
Causal Explanation

Why is There a Need for More Skillful Causal Explanation?

When we try to prevent something we don't want (like a disease) or try to produce something we do want (like better crops or higher grades), we usually try to find the causes of these things. We can then try to alter causal conditions in order to block undesirable results or to foster desirable ones.

Knowing the causes of events or conditions can also enable us to predict them, although we may not be able to prevent them. Then we can try to avoid or minimize harmful effects. We know, for example, that certain conditions can cause severe weather. Predicting a storm based on knowledge of the presence of these conditions can give us time to find shelter.

The variety of contexts for causal explanation. Scientific investigation often involves trying to determine the causes of things. Knowing the cause of AIDS allows us to avoid conditions that lead to the spread of the disease. Understanding the causes of global warming allows for informed policies to prevent its serious consequences.

We also make causal judgments in other fields of study. When we try to explain what led to the Civil War, we are offering a causal explanation. When we attribute a specific motivation to a character or leader, we are offering a causal explanation of that person's behavior. When we explain why stock prices tumbled, we are offering a causal explanation of an economic trend. We also make causal explanations about everyday matters. When we try to figure out why the car didn't start, we are searching for a cause.

Problems regarding the way we think about causes. We can seldom find causes by direct inspection. Once the effect occurs, the cause is usually gone. In such situations, causal explanation is a matter of inference.

If we don't think carefully about what is causing something, we may jump to hasty conclusions that can create serious consequences. A hasty judgment about the cause of an illness may lead us to spend effort and money on the wrong treatment. A hasty judgment about a person's motive may cost that person his or her job or, in a murder case, his or her life.

There are two sources of hasty conclusions about causes. We often accept the first causal explanation that comes into our minds and become convinced that it is the correct one. Often, being anxious or fearing a situation or condition that has caused harm in the past, we become convinced that some current difficulty was brought about by the same cause. When the *Challenger* disaster occurred, many people believed immediately that it was sabotage. Like them, we sometimes don't consider other possibilities before making up our minds.

The second source of hasty judgments about causes is that, although we might think about other possibilities, we may not consider all of the relevant evidence to decide which explanation is the best one. The first bit of evidence we get for one possible cause sometimes leads us to accept it. Good detective story writers exploit this. It is all too easy to get us to think that a shady character committed the crime simply because someone saw him running from the scene.

The two basic problems with the way people usually think about causes are summarized in figure 14.1.

COMMON PROBLEMS WITH OUR THINKING ABOUT CAUSES

1. We consider only one possible cause and affirm it without thinking about other possibilities.

2. We take account of only a small sample of the relevant and available evidence in determining a cause.

Figure 14.1

If you find that you are limiting your causal judgment in either of these ways, you can learn to make more careful judgments about causes.

How Do We Engage in Skillful Causal Reasoning?

Tips for Determining Causes Skillfully. We can guard against selecting a cause hastily by considering many possible explanations of an event. Then it is easy to avoid affirming the first explanation that pops into our minds. Before we knew the cause of the *Challenger* disaster, many people speculated about a number of things that might have caused the tragedy. Although the explosion could have been caused by sabotage, it could also have been caused by ruptured "O" rings or a malfunctioning engine. Realizing that there are usually many possible causes reduces the tendency to jump to a quick conclusion about causation.

Thinking about what we can find out that would make one or another explanation plausible can help us avoid making a causal judgment on too little evidence. Often, it is helpful to list the kind of evidence that would be needed to justify selecting one possible explanation as the most reasonable one. Knowing what kind of evidence we need prompts us to search for information that we might otherwise overlook. For example, if my car doesn't start and my lights don't work, I may realize that I should not rule out the possibility that there is a broken wire, rather than a dead battery. This realization may then prompt me to check whether the wires are intact. When I find that they are, then a dead battery becomes a likelier cause.

Thinking carefully about what may be causing something can guide well-planned research in which we rule out various possibilities until the weight of evidence points to one causal explanation as the most plausible one. Conducting controlled experiments to gather necessary evidence is one of the more sophisticated techniques commonly used in scientific and medical research.

Skillful thinking about causes, therefore, involves four important matters:

- generating ideas about possible causes;
- considering what evidence would be necessary to show which is the probable cause;
- considering the evidence we have or gathering additional evidence that we need;
- making a judgment about the cause based on the evidence.

When we find that we have sufficient evidence to justify selecting one possible explanation, we are ready to affirm it. The investigation into the *Challenger* disaster is a case in point. Only after the investigating commission had sufficient evidence to rule out other possible explanations in favor of the rupture of protective rings did it advance this explanation as the most reasonable one.

We may attend to these four points and still find that we don't have sufficient evidence to make a causal judgment. Trying to determine what caused the extinction of the dinosaurs is this kind of causal issue. Nevertheless, even in this case, skillful thinking about what caused the extinction pays off. When we realize that we cannot determine a probable cause, we can defend that judgment by showing that a variety of possible causes are possible, yet insufficient evidence exists to pick one as the clear explanation.

Tools for determining causes skillfully. Considering these four factors (possible causes, necessary evidence to support causes, actual evidence, and judging whether a cause is warranted by the evidence) can help us make better causal judgments. The thinking map of causal explanation in figure 14.2 summarizes these key points. By asking and answering these questions carefully, we avoid the common problems with determining causes.

SKILLFUL CAUSAL EXPLANATION

1. What are possible causes of the event in question?

2. What could you find that would count for or against the likelihood of these possibilities?

3. What evidence do you already have, or have you gathered, that is relevant to determining what caused the event?

4. Which possibility is rendered most likely based on the evidence?

Figure 14.2

As we think about what caused something, we can use the graphic organizer in figure 14.3 to focus our attention on the evidence for a specific possibility. Using the diagram is particularly helpful when the evidence is complex. By recording the evidence on the diagram, we eliminate the necessity of holding it all in memory and can turn our attention to evaluating what it shows about the possible cause. The number of boxes can be expanded depending on specific needs.

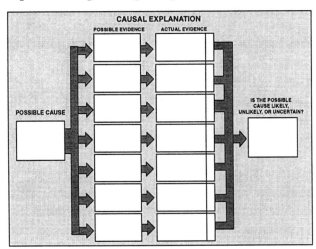

Figure 14.3

To use this graphic organizer, first brainstorm a list of possible causes of the event you are seeking to explain and record these on a piece of paper. You can then pick the ones that seem, initially, to be lively possibilities and explore them more fully using the graphic organizer. To do this, write one from your short list of lively possibilities in the "Possible Cause" box. Then brainstorm possible evidence that would support this as a likely cause. This evidence should be listed in the boxes under "Possible Evidence."

Then, using what you have projected as possible supporting evidence to guide you, try to determine whether such evidence exists. This may mean just looking around. (Do I see any footprints near the scene of the crime?) It could call for more extensive investigation (Have any witnesses come forward?) or even an experiment. (If I test his hands for powder burns, will they show up?) You might even remember something that is relevant. (Did I see him at the scene of the crime yesterday?)

When you find that there is such evidence, write it in the appropriate boxes under "Actual

Evidence." Each time you enter some evidence, you should indicate, with a plus or minus, whether it counts in favor of or against the possibility you are considering. Then, when you've done a satisfactory search for evidence, look at the weight of the evidence you've recorded and make a judgment about whether the possibility is likely, unlikely, or uncertain based on the evidence. Write this in the final box.

Causal chains. In some cases, we may wish to examine the remote causes of an event, as well as the immediate ones. We can then try to map a causal chain that leads to an event. In many cases, an important cause may be many steps back along the chain. If my car does not start, it may be because the spark plugs don't work well. The spark plugs may be faulty because the auto mechanic didn't install them properly and, during months of use, they have weakened.

Explaining past events in a causal chain can help us make better judgments about responsibility. We may find that people's past actions were involved in some of the causes that appear in the chain. If the automobile mechanic didn't install the spark plugs properly, the repair shop may be responsible for repairs resulting from their malfunction.

The graphic organizer in figure 14.4 can be used to record causal chains. It can be expanded as needed. The diagram can be used to depict a chain of possible causes before investigating whether this chain of events is likely to have occurred. It can also be used to record the results of such an investigation, i.e., what you have determined is the most reasonable chain of causes leading to the event in question.

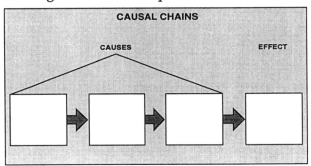

Figure 14.4

Multiple causal factors. Another consideration in determining causes lies in the fact that usually no single factor causes related effects.

Usually, a cluster of factors blend together to bring about effects. Lighting the match isn't, by itself, the cause of an explosion. Gas from a leak, in the right mixture with the oxygen in the air, is also necessary to create the explosion. The latter example is sometimes called a "standing condition," since it is constant, and the former example a "differential condition," since it is a new event that makes the difference.

Standing and differential conditions work together to bring about an effect. Each of them alone is not sufficient, but each seems necessary. Neither the gas nor the oxygen nor the lit match were sufficient in themselves to cause the explosion. Together, along with the other relevant standing conditions like the temperature, they are sufficient. Without one, no explosion occurs.

The graphic organizer in figure 14.5 can help us explore what factors blend together to bring about a specific event. In this case also, the graphic organizer can be used to record the results of speculation about a possible blend of causal factors leading to an effect or the most reasonable explanation supported by evidence.

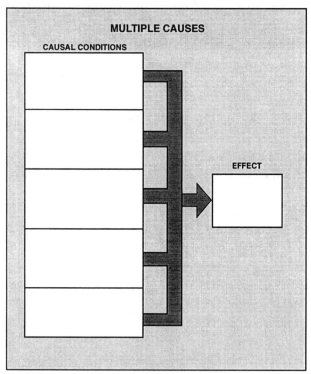

Figure 14.5

When should we think about causal chains and about multiple causes? When considering questions of human responsibility for events, we should extend our thinking beyond immediate causes and develop a causal chain that led up to an event. This can pinpoint places in the chain where human beings were involved. We've already discussed an example of this—the brakes failed on a car, and the basic responsibility lies in faulty work done earlier by an auto mechanic.

Similarly, when we're interested in finding out what caused some event in order to prevent a similar situation in the future, we may want to identify a variety of different factors that blended together to bring about an undesirable result. Perhaps some causes are more easily modified than others. Maybe it's better to fix the gas leak than to stop lighting matches. If we just focus on the differential condition, it may not occur to us that there may be an easier and safer way to correct the problem by altering problematic standing conditions.

Both in cases when you are trying to construct a causal chain and in cases when you are trying to determine multiple causes, you can use the same basic strategy outlined in the thinking map (figure 14.2) in order to make reasonable judgments. For causal chains, consider possibilities one-by-one in backwards order until you've constructed a plausible chain of causes. For multiple causal conditions, consider all the factors that may be present and together contribute causally to the effect, investigating evidence for each factor. In each of these cases, it is important to consider alternatives, rule out the implausible ones, and accept the plausible ones based on real evidence.

How Can We Teach Students Skillful Causal Explanation?

Asking students to identify already specified causes and effects, such as "identify the cause and the effect" activities in many instructional materials, is not yet engaging them in causal explanation. The students might be reading a story in which they are told that one effect of the main character's reassuring comments was to make her friends feel less anxious about a task they were about to perform. At the end of the chapter, a typical "cause and effect" question might be "Identify the cause of the group's feeling less anxious about their task." This is nothing more than a straightforward recall question that emphasizes identifying the reassuring comments of the main character as a cause.

There may, of course, be value in asking students to identify causes that are already specified. This does help to clarify the language of cause and effect and to make students more aware of the variety of causal relationships we experience in our lives. Such preliminary thinking about causes, however, is no substitute for causal explanation. Causal explanation involves trying to determine a cause when *we are aware of an effect and don't know what caused it.*

Even when students don't know what caused something, it is not enough to simply ask students what the cause is. "Why did the plague spread so rapidly in medieval Europe?" typically prompts quick answers, many of which don't involve much thought. Asking these sorts of "higher-order questions" doesn't teach students skillful causal reasoning.

Teaching causal explanation explicitly. Instruction in skillful causal explanation should help students internalize the pattern of thinking specified on the thinking map of causal explanation in figure 14.2. We want students to ask questions about possible causes and relevant evidence spontaneously when determining what caused something and to recognize whether they have answered those questions adequately.

The pattern of instruction in skillful causal explanation should follow the same structure we use in all infusion lessons.

- Introduce the skill explicitly.

- Guide students through an activity in which they use the skill by asking them about alternative explanations and evidence and by helping them to use the graphic organizer for causal explanation.

- Help them to reflect on what causal explanation involves and whether it is valuable to use the strategy on the thinking map.

- Provide them with plenty of opportunities to practice thinking in other contexts.

Initially, you will guide your students through this practice; eventually, they will guide themselves. The result will be that they will explain causes carefully and confidently.

Contexts in the Curriculum for Causal Explanation Lessons

The secondary school curriculum is replete with examples of causal explanation. The causes of wars, eclipses, population changes, miscommunication, and characters' actions described in stories are among the many applications throughout the curriculum.

Causal explanation is a type of thinking that is quite prevalent in the sciences. In fact, it plays a central role in the sciences; it is one of the key processes involved in what has been called "the scientific method."

Demanding that a causal hypothesis not be accepted unless it is supported by publicly observable evidence and unless competing hypotheses are likewise ruled out by publicly observable evidence is a fundamental tenet of science. Hence, causal explanation is one of those types of thinking that one would expect to find built into any comprehensive science curriculum.

In general, there are two types of contexts in which it is natural to develop causal explanation lessons in the sciences:

- *Causal explanations are important to learn in order to understand basic ideas and concepts in the curriculum.* In science, it is important for students to learn the causes of various phenomena—e.g., the weather, earthquakes, plant growth, and chemical equilibrium—in order to understand these phenomena better. Students can be guided to determine what the causes are in these examples by using skillful causal thinking. The more such lessons center on specific case studies, the more practice students will get in applying their knowledge in science to practical situations.

- *Causal explanations are important when either a controversy arises about the causes of important events studied in the curriculum or when no one yet knows what caused these events.* For example, lessons on the extinction of the dinosaurs, like the one we have included in this handbook, can help students gain a deeper understanding of dinosaurs and their worlds, as well as of scientific investigation. Clarifying what causes people to make choices can bring students insight about human motivation as well as an understanding of how to make reasonable judgments about why people do what they do.

By locating significant contexts like these, you have numerous opportunities to develop robust infusion lessons on causal explanation.

The menus we provide at the end of this chapter illustrate some of the specific opportunities that teachers have found in their science curriculum for the development of causal explanation lessons.

Model Lessons on Causal Explanation

We include three lessons that infuse instruction in skillful causal reasoning into content area teaching. The first is a middle school/high school science lesson on the extinction of the dinosaurs, the second a high-school biology/environmental sciences lesson on the causes of a catastrophic change in the health of a community, and the third a high school chemistry lesson on the diagnosis of a problem that arises in the manufacture of sulfuric acid. Each illustrates how you can focus students on the need for evidence in making causal judgments in ways that enhance the depth of their understanding in science.

In the first case, students explore the concept of the catastrophic extinction of a species, in the second the impact of pesticide pollution, and in the third the concept of chemical equilibrium. In addition, in the first, students also gain an understanding of what a scientific theory is and how it can be established, noting that there is not enough evidence to establish any of the theories that have been advanced about the extinction of the dinosaurs.

Each lesson illustrates the way structured questioning derived from the thinking map, together with a graphic organizer, can guide students through a process that leads to a well-founded critical judgment.

As you read each lesson, ask yourself the following questions:

- How does the graphic organizer facilitate a careful look at what the evidence shows about causes?
- What metacognitive strategies are employed to help students take charge of their own thinking?
- Are there alternative metacognitive strategies you could use?
- Are there other ways of reinforcing this kind of thinking besides the application examples mentioned in the lesson?

Tools for Designing Lessons on Causal Explanation

The thinking maps provide focus questions to be integrated into causal explanation lessons to guide student thinking. The thinking map below is used in each of the two lessons included.

The graphic organizer for causal explanation supplements and reinforces the sequenced questions on the thinking map. Graphic organizers for causal chains and for multiple factors show how many causes contribute to a specific effect.

The thinking maps and graphic organizers can guide you in designing the critical thinking activity in the lesson and can also serve as photocopy masters, transparency masters, or as models that can be enlarged and used as posters in the classroom. Reproduction rights are granted for use in single classrooms only.

SKILLFUL CAUSAL EXPLANATION

1. What are possible causes of the event in question?

2. What could you find that would count for or against the likelihood of these possibilities?

3. What evidence do you already have, or have you gathered, that is relevant to determining what caused the event?

4. Which possibility is rendered most likely based on the evidence?

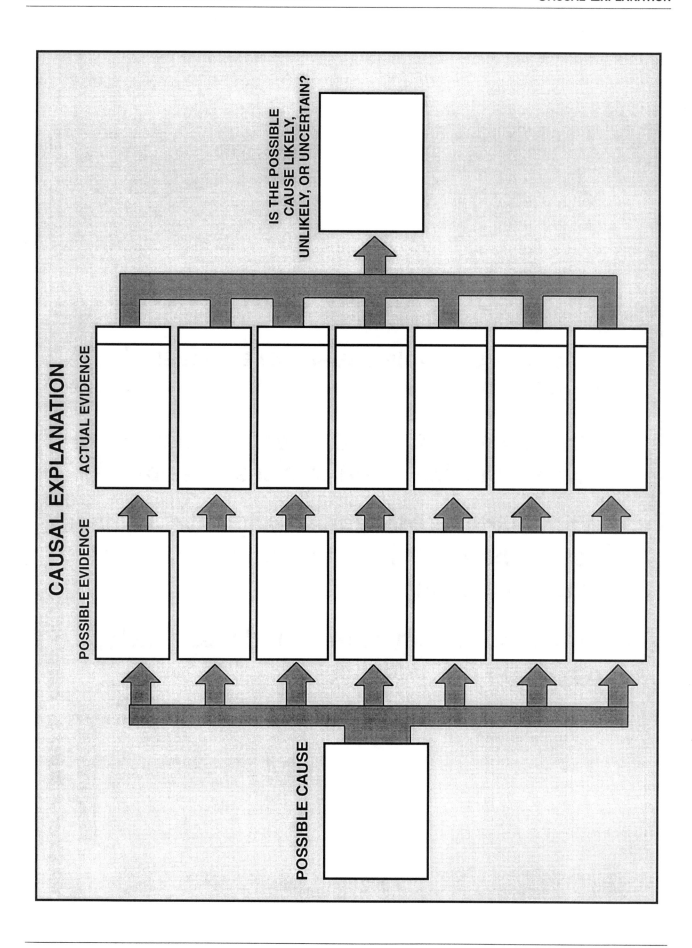

CAUSAL EXPLANATION

IS THE POSSIBLE CAUSE LIKELY, UNLIKELY, OR UNCERTAIN?

ACTUAL EVIDENCE

POSSIBLE EVIDENCE

POSSIBLE CAUSE

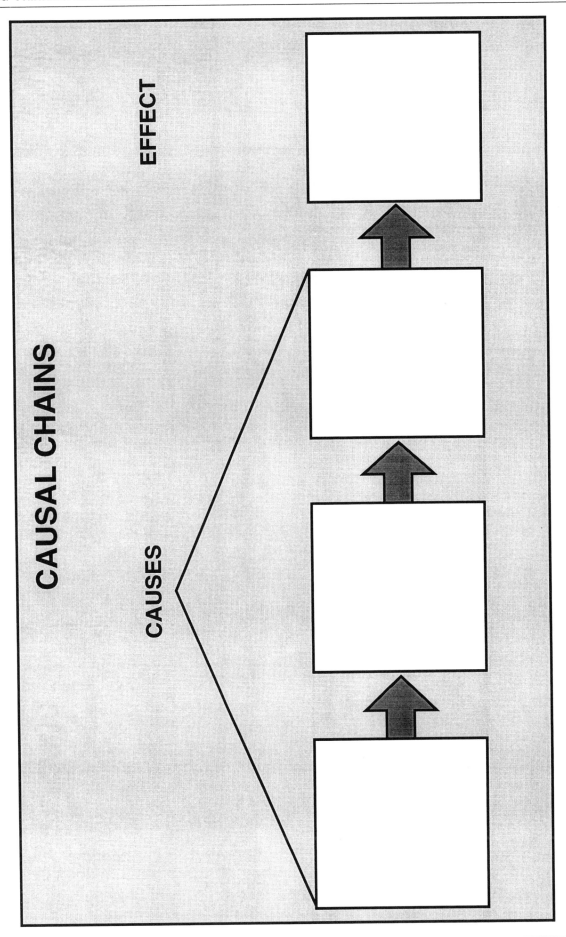

CAUSAL CHAINS

EFFECT

CAUSES

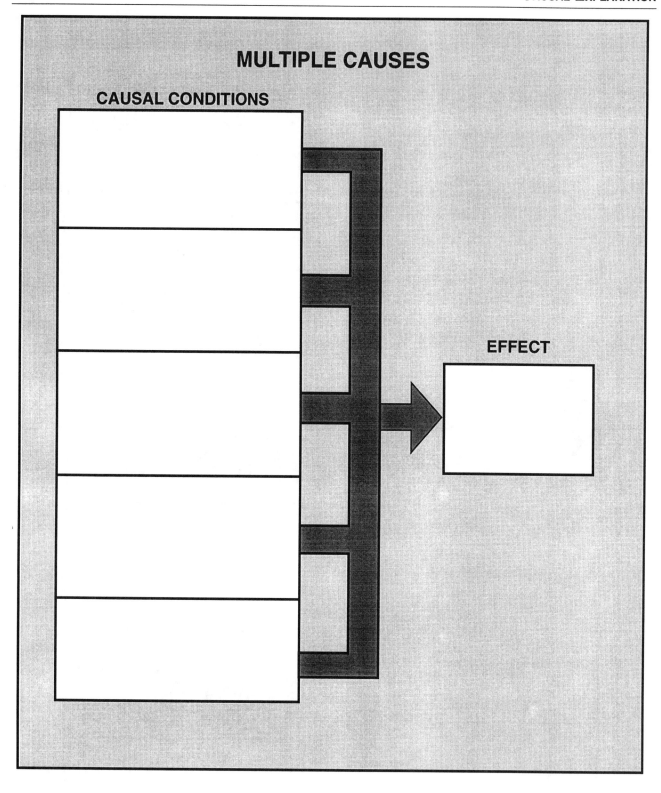

THE EXTINCTION OF THE DINOSAURS

General Science, Biology   **Grades 9–12**

OBJECTIVES

CONTENT

Students will learn different theories about the extinction of the dinosaurs. They will also learn about the different kinds of evidence (e.g., fossils, rare chemical deposits) that can provide us with information about what caused the extinction, and how to get this evidence. Finally, they will learn the nature of scientific theories.

THINKING SKILLS/PROCESS

Students will learn to develop alternative hypotheses and consider present evidence when trying to make a judgment about what caused something to happen.

METHODS AND MATERIALS

CONTENT

Students draw upon their prior knowledge about dinosaurs, fossils, and food chains from their texts or other sources. They collaborate to examine the content. Writing is used to elaborate details.

THINKING SKILLS/PROCESS

Structured questioning that follows the thinking map for the skill and a specialized graphic organizer are used to guide students through the thinking. Collaborative learning enhances their thinking.

LESSON

INTRODUCTION TO CONTENT AND THINKING SKILL/PROCESS

- When things that we don't like happen, we often try to find the cause. If we find the cause, we can sometimes fix it. If your TV picture is fuzzy and you know it's because the antenna is pointing in the wrong direction, you can fix it by moving the antenna. If that's not the cause, you may not be able to fix the picture until you find the real cause. Guessing won't help. Trying to find a cause requires some careful critical thinking. Can you think of a time when you guessed about a cause, got it wrong, and had to do more careful critical thinking to find the real cause? Write down an example of this and then discuss it with your partner. In your discussion, focus your attention on two questions: what did you do that led to your mistake, and what did you do to remedy this to find the real cause? STUDENT ANSWERS VARY BUT USUALLY INCLUDE IDEAS LIKE THE FOLLOWING: *I got an idea about what I thought caused the problem I was having trouble with my car and didn't think of anything else and the mechanic found that it was something else. My parakeet died, and I thought he was just old, but it turned out that I was feeding him the wrong food. I found that out because when I told my friend, who knows more about parakeets than I do, she told me that my bird wasn't really that old, so I thought that maybe it was something else. I thought about what it might be and wondered about the new food I had got. I found that the label said that this food was good to feed wild birds but not domesticated birds like parakeets.*

  As students list problems, make a list of "problems" in finding causes—for example, jumping to a conclusion about a cause without any evidence, deciding on the cause based on a small selection of inconclusive evidence, etc. Write the diagram (right) on the board

COMMON PROBLEMS WITH OUR THINKING ABOUT CAUSES
1. We consider only one possible cause and affirm it without thinking about other possibilities.
2. We take account of only a small sample of the relevant and available evidence in determining a cause.

as students report and ask them if they can fit into any categories the problems they have identified with how they have made judgments about causes before.

- Finding out what caused something is called "causal explanation." Doing this skillfully is a way to counter these defaults, as you each found out in this activity. Based on your reports, this involves thinking about possible causes and then deciding which is the likeliest cause based on evidence, after a thorough search for the evidence. Here is a thinking map for causal explanation that identifies these remedies and sequences them into a plan for skillful causal explanation. Do you think this expresses your ideas about what works in trying to make accurate judgments about causes? Most students agree that this makes their ideas about remedies clear and puts them together in a plan that would be better to follow than basing their causal explanations on the kinds of thinking that lead to misjudgments.

> **SKILLFUL CAUSAL EXPLANATION**
>
> 1. What are possible causes of the event in question?
>
> 2. What could you find that would count for or against the likelihood of these possibilities?
>
> 3. What evidence do you already have, or have you gathered, that is relevant to determining what caused the event?
>
> 4. Which possibility is rendered most likely based on the evidence?

- Scientists engage in causal explanation all the time. To cure a disease, scientists first try to find out what causes it. They develop a list of possible causes (called "hypotheses") and then conduct scientific research in order to gather evidence, after they plan the research to guide them in what they are looking for, to determine which of these hypotheses is shown to be the best explanation based on the evidence. In some cases, scientists have found the causes of diseases—for example, the cause of chicken pox—and have been able to develop ways to prevent the disease, like vaccination. In other cases, they are still trying to find out the causes—for example, in the case of cancer. When they find a cause, they may find a cure.

- One important technique that scientists use in conducting some of this research involves "controlling variables." For example, if you were trying to find out if one spot remover took stains out of fabric better than another, how would you design an experiment so that you could get evidence that showed this? Students usually respond that you would try the two spot removers on the same stains on the same fabrics but make sure that you were using the same amount, rubbing with the same force, etc. Identify this as "controlling variables," and discuss with students how making sure that the variables are controlled, when we can do this, is one important way of ruling out other possible causes and isolating one as the likely cause.

- Work with your partner to identify other instances in which scientists have faced mysteries about what caused certain things to happen, and have, after following the strategy you identified, found the causes of these things and have taken steps to prevent them from happening again based on this knowledge, or made them happen more often, when they are desirable. STUDENT ANSWERS VARY BUT SOMETIMES INCLUDE ITEMS LIKE THE FOLLOWING: *Scientists have investigated what makes crops grow well, and that has led to the development of fertilizers and other products that bring that about. In cases of plane crashes, investigators have found certain things responsible, like the rudder mechanism I read about, and that has led to changes that avoid such accidents. The causes of air pollution were investigated and the use of scrubbers in smokestacks resulted. Scientists investigated the connection between cigarette smoking and cancer and found evidence that one can cause the other. That has led to laws requiring warning labels on cigarettes.*

- Scientists are not interested solely in things that happen today. They are also interested in major changes that happened on Earth a long time ago. One thing that has puzzled scientists for a long time is what happened to the dinosaurs at the end of the Mesozoic Era. No one knows exactly why they became extinct, but scientists continue to ponder the cause.

• In this lesson, we're going to think about what might have caused the extinction of the dinosaurs and see what we can find out about this topic. But first, let's think about what we know about the dinosaurs themselves. Dinosaurs lived on Earth a very long time. Even though their numbers were large and many were huge animals, not one survives today. What other things do you know about the world of the dinosaurs? What aren't you sure about? Can you think of other things that you don't know but would like to find out? In conducting class discussion, use the graphic organizer below to help students organize their ideas. Give students copies (A larger version is at the end of the chapter.) and, after three or four minutes, ask for some of the items that they included on their diagrams, writing them on a similar diagram on the chalkboard. As they introduce ideas, help them with any technical terminology that is in use now. For example, if "Some ate plants only" is an item that they report, use the word "herbivorous." When you have filled the diagram with student responses, give them some background reading about dinosaurs, when they lived, what they ate, their variety, and the kinds of evidence people now look to about dinosaurs, e.g., fossils, etc. You can also show the students drawings and photographs of dinosaur bones that illustrate some of the commonly accepted facts about dinosaurs. (Do not give students any additional information at this time about extinction theories or evidence collected to support those theories.) Ask them whether these points answer some of their questions about dinosaurs, or whether they want to change anything on the diagram based on this information. Modify your master diagram accordingly. Leave this diagram on the board as a reference point to summarize the background knowledge the students will use in the activity.

Dinosaurs		
WHAT DO I KNOW?	WHAT DO I THINK I KNOW?	WHAT DO I NEED TO KNOW?

THINKING ACTIVELY

• Now, let's consider the fact that the dinosaurs disappeared from Earth some time ago. That's quite amazing, when you think about it. What could have caused this change? Before we try to decide about what caused the extinction of the dinosaurs, what should we think about first? Students usually respond that we should think about some possible causes. If students are not sure, refer them to the thinking map of causal explanation. **What are some different conditions that might have caused the dinosaurs to disappear from Earth?** Ask collaborative learning groups to brainstorm and list as many possibilities as they can identify. Give them a copy of the possible causes graphic organizer and ask them to write their ideas on the lines below "Possible Causes." Encourage different and unusual possibilities. Ask each group to mention one of their possibilities and write these on a master copy of the possible causes diagram on the chalkboard or on a transparency. Allow any group to add possibilities to their list that they have not included if they wish. Accept all responses; later, they will be subject to the test of evidence. POSSIBLE ANSWERS: *Disease, drastic depletion of the food supply (e.g., plants died), genetic changes that caused problems getting food, abduction by aliens, major changes in the climate, a catastrophic event (meteor impact or volcanic eruption that caused massive fires), a great flood, conflict with warm-blooded animals, limited agility, egg predators, evolution, fiercer animals killed them.*

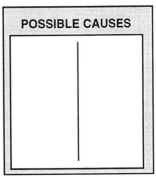

POSSIBLE CAUSES

• **Suppose you looked for clues about which of these possibilities was the likeliest explanation. What might you find today that could give you evidence for or against each possibility? How could you go about finding out these things?** Brainstorm some answers with the whole class. POSSIBLE ANSWERS: *We could find dinosaur remains in the ice and do medical tests for disease-causing viruses (disease). We could find dinosaur fossils in rocks and notice changes that cause problems, e.g., with teeth (genetic changes). We could find craters from large meteors (meteor impact) dating from* *the boundary of the Cretaceous and Tertiary Periods when the dinosaurs disappeared.* After this brainstorming session, give students the background material on fossil dating and collecting fossil and other relevant evidence about the Cambrian/Tertiary boundary period. Ask them to work together in their groups and "jigsaw" this reading material.

• **Imagine that, while looking for clues, you find the following evidence in various sedimentary rocks at the Cretaceous/Tertiary (K-T) boundary :**

  • **Lots of dinosaur tracks at one level, fewer at another, and then none.**

  • **Fossilized leaves and plants at one level and then very few.**

  • **More mammal tracks and bones in some levels than in others.**

  When the rocks are dated, you find that the plants and leaves coincide with a lot of dinosaur tracks. The layers with fewer plants came just before there were fewer and then no other dinosaur tracks. The number of mammal bones increased as the dinosaur tracks diminished.

• **What possible explanation does this evidence support?** Ask the students to discuss this in their groups and return with their responses. RESPONSE: *Drastic reduction in the food supply: the plants died, causing the plant-eaters to starve, which in turn caused the meat-eaters to starve.* **Scientists shouldn't accept an explanation until they have sufficient evidence to be sure of it. Otherwise, it is just a theory. Are the clues we just discussed sufficient evidence?** *No.* **Why?** POSSIBLE ANSWERS: *They are from only one location. Other explanations are still possible: Maybe only the dinosaurs in this one location didn't have enough food, but something else killed the dinosaurs living in other parts of the earth; or maybe something caused both the plants and the dinosaurs to die out at the same time.* **What other evidence would you need to be sure that this was the best causal explanation?** Draw a graphic organizer for causal explanation on the chalkboard or use a transparency or large poster. Write "Drastic depletion of the food supply" in the box for the possible cause. Then add what the students suggest to the boxes for "Possible Evidence." The graphic organizer should look like the sample at the right. POSSIBLE ANSWERS: *Similar but more extensive evidence from sites all over the world (decreasing fossils of plants as fossils of plant-eaters decrease), other animals that ate plants also disappearing from the fossil records, evidence about conditions that caused the plants to die, evidence of malnutrition in the bones of plant-eating animals that lived at the time.*

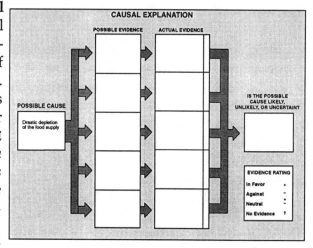

• **Work in your groups again. I'd like one of the groups to work on this possibility.** Ask for volunteers or assign this to one group. **The other groups should use the graphic organizer we have filled out so far as a model for what they are to do. They should pick one of the other possible causes and, as we did in the case of a drastic depletion of the food supply, make a list of possible evidence, using the graphic organizer for causal explanation.** After a few minutes, ask each group to report, displaying their diagram, or showing a transparency of it. Remind them that if other students in the class have ideas about additional possible evidence as a group reports, they should feel free to add it in the "Possible Evidence" column.

• **What actual evidence about the extinction of the dinosaurs can you get from your text or from other sources? What possible cause(s) does this evidence support? What possibilities does this evidence count against?** Each group should return to the graphic organizers in which they wrote down possible evidence that would support one of the alternative causal explanations. Look for evidence of the sort that you mentioned. Write the actual evidence you find for or against those possible explanations in the boxes marked "Actual Evidence" on your diagram. Rate the evidence by putting a plus next to it if it counts in favor of the possibility, a minus if it counts against, and a plus and a minus if it is neutral. Students usually find their textbooks very limited in actual evidence about the extinction of the dinosaurs. Provide additional sources from the school library, or ask the students to go to the library and find additional information. Each group should display their graphic organizer on a bulletin board or copy it on the chalkboard. Students usually find little evidence to support some theories (e.g., disease), but more to support others (drastic reduction in the food supply). They also realize that what killed the dinosaurs could have been a combination of factors. Nonetheless, they realize that even though some possibilities are better supported than others, not enough evidence exists to be sure that any of the possibilities is correct.

- When you are trying to find out what caused something, you will often realize that the cause of the event was, in turn, caused by something else. That may be important. For example, you may get to school late because you missed the bus. You may have missed the bus because you overslept. You may have overslept because your alarm didn't ring. Perhaps the alarm didn't ring because you didn't turn it on. All these events can be arranged in a *causal chain*. Use this diagram to arrange, in order, some of the events you think may have caused the extinction of the dinosaurs until you are satisfied that you have a plausible causal chain that reveals *the root cause* of their extinction. POSSIBLE ANSWERS: *The meat-eaters died because the plant-eaters died; the plant-eaters died because the plants died; the plants died because there was a cloud of dust around the earth; the dust was caused by a meteor hitting the earth and throwing up dust and debris into the atmosphere, or by massive volcanic eruptions that likewise caused a cloud of dust and debris.*

- Let's conclude this lesson by exploring two of the possibilities you have determined might have caused a drastic depletion in the food supply—massive volcanic activity and the impact of a meteor or other object from space like a comet. Does the evidence favor one theory?

- Let's first look together very carefully at one of these possibilities—volcanism, the destruction of dinosaurs as a result of prolonged global volcanic activity. What evidence would you expect to find if volcanos were the reason that dinosaurs became extinct? Use a new blank graphic organizer for causal explanation to record your thinking about what might prove this theory to be likely. After five minutes, ask students to report one of the items their group wrote on their graphic organizer, getting responses from as many groups as possible. Write these on a new transparency for massive volcanic activity 65 million years ago. TYPICAL STUDENT RESPONSES: *You would be able to locate a large number of volcanos that dated to 65 million years ago; you would find lava flows all over the world dating to 65 million years ago; you would find dinosaur fossils near the lava flows; you would find worldwide spread of soot and ash dating to 65 million years ago; you would find dinosaurs preserved in lava like people were at Pompeii; you would find evidence that other kinds of animals, not only dinosaurs, were killed at the same time; you could find evidence in modern volcanoes of the potential to produce a mass extinction; you would find evidence of damage due to acid rain resulting from all the sulfur pumped into the atmosphere; you would find evidence of tsunamis resulting from underwater volcanic explosions; you would find concentrations of iridium in or near ancient and modern volcanos; you would find shocked quartz in ancient volcanoes.*

- Now that you are focused on possible evidence, take a few minutes and read the passage in your handout titled The Volcanism Theory. See if the article brings to mind additional kinds of evidence you would expect to find. Add these to the list on your graphic organizer. Provide students with some resource material on volcanic activity like that included in this lesson. Do not yet provide them with material that gives evidence of this type of activity during the boundary of the Cretaceous and Tertiary periods. Allow several minutes and then ask for additions to the possible evidence list. Students usually add the following: *you would find evidence of a vast climatic cooling; you would find evidence of a magnetic field reversal happening at that time; traces of iridium and selenium dating from 65 million years ago would be scattered over the planet.*

• **Let's look at supporting evidence and objections to this theory and then we can evaluate just how valid this theory might be in explaining the extinction of the dinosaur.** Have students read material like that included in this lesson (e.g., from David M. Raup's book *The Nemesis Affair*, Officer and Page's *The Great Dinosaur Extinction Controversy*, and Evidence From the Volcanic Eruption Theory from *Great Unsolved Mysteries of Science*. **As you read the material, make a list of any facts or statements of evidence that would support or speak against the volcanism theory.** TYPICAL STUDENT RESPONSES: *Hot spots did exist when dinosaurs became extinct. Hot spots correlate to other mass extinctions. The Deccan Flats hot spot, which corresponds to the extinction of the dinosaurs, was active more than 500,000 years ago. Iridium-rich clay concentrations were found to be spread out over 500,000 years. Some hot spots are not associated with incidents of magnetic reversal. Tropical plants survived better than temperate plants. No evidence has been found of a global cold snap at the end of the Cretaceous Period.* **What does this evidence show about the likelihood of massive volcanic eruptions being the root cause of the extinction of the dinosaurs?** Students usually find that the evidence related to volcanism is inconclusive, although providing some support for this idea. When this activity is complete, you may wish to do the same thing for the impact theory.

THINKING ABOUT THINKING

• **Map out how you tried to figure out what caused the extinction of the dinosaurs. What did you think about first, next, etc.?** Ask students to work together in collaborative learning groups. Put their maps on the board. Try to reach consensus about what they did.

• **Compare the way you thought about the causes of the extinction of the dinosaurs to the way you ordinarily think about causes. Which do you think is the better way to try to find causes? Why?** Encourage students to reflect on what they thought caused some personal situation and how their understanding of what caused that situation could have been different.

• **Think about situations in which you wonder about causes, like what caused an illness or what is causing problems around the house. Plan what you might think about the next time one of these situations arises so that you can find the most likely cause.** Ask students to keep this plan in their notebooks, so that they can refer to it next time.

APPLYING THINKING

Immediate Transfer

• **We have studied other animals classified as endangered species. Make a list of these animals and pick one that you'd like to study further. Use your plan for causal explanation to determine what is causing this animal to be endangered. Based on this determination, make some suggestions as to what we might do to help these animals.** You can ask students to write a short analysis of the problem for the animal they choose, along with their suggested remedies. If you use a portfolio assessment plan, suggest to your students that they may wish to include these writings in their portfolios along with comments about the quality of the thinking.

• A number of things happen at school that many students and teachers are concerned about: noise in the cafeteria, a great many library books getting lost every year, etc. Select something that could be changed to make our school better. Try to find out the cause(s) of some of these situations, so that you can recommend some remedies.

Reinforcement Later

• Use the strategy for causal explanation to figure out the cause of the increase in the population of mammals after the Mesozoic Era.

• As we study early civilizations, we can find causes for changes in their way of life. When we studied ancient Egypt, we saw that the early Egyptians changed from being nomads to living in settlements along the river Nile. Try to figure out what caused this change and map out the causal chain that led to the development of the great civilization of the Egyptians.

WRITING/ART EXTENSION

Pick one of the possible causes of the extinction of the dinosaurs and write a story about how this could have happened. Draw some pictures to illustrate your story. Ask four or five students to show their pictures and explain their stories to the whole class.

THINKING SKILL EXTENSION: EVALUATING OTHER CAUSAL EXPLANATIONS

Ask students to use the school library and find books or articles in which people have discussed the extinction of the dinosaurs. Ask them to record the explanations that they find for extinction and the evidence offered to support them. Students should recognize that the different views about what happened to the dinosaurs are also scientific theories and that, for theories to be accepted, they must be well supported by evidence. They should report to the class on these different theories and discuss whether they are well supported and the evidence to support them.

THINKING SKILL EXTENSION: MULTIPLE CAUSES

Ask students to explore the possibility that multiple causes brought about the extinction of the dinosaurs. Provide them with copies of the blank "Multiple Causes" graphic organizer (at the end of the introduction) and ask them to list the causes that, based on their research, might have led to the extinction of the dinosaurs. A filled-in version is included at the end of this lesson.

ASSESSING STUDENT THINKING ABOUT CAUSES

To assess the skill of causal explanation, ask students to write an essay to answer any of the application questions or similar ones that you develop. They may critique someone else's causal explanation of some event or find a causal explanation for something that has happened (e.g., the plants in the classroom have all died; find out why). You can supplement their writing with interviewing. In all these assessment tasks, students should make their thinking explicit. Determine whether they are attending to each of the steps in the thinking map of causal explanation.

You can also use a portfolio assessment plan to gather information about your students' mastery of skillful causal explanation. For example, ask your students to continue to gather information about the extinction of the dinosaurs and to cull any additional evidence they find that supports or counts against the various hypotheses that they are exploring. Students may record their own progress in gathering such evidence (e.g., by using a research log). Their consideration of possible causes, possible evidence and actual evidence, and their judgments about the plausibility of the various possible causal explanations will reveal how well they have internalized the thinking involved in skillful causal explanation.

THE EXTINCTION OF THE DINOSAURS
SOURCE MATERIAL

DINOSAURS

About 600 million years ago, the diversity of life on earth changed dramatically; until then, there were prokaryotes, simple eukaryotes and only a few complex multicellular animals. But during the next 100 million years, the ancestors of all the phyla of the modern animal kingdom appeared. These included early members of the phylum chordata. Chordates include the classes Reptilia and Mammalia from which dinosaurs and humans evolved.

The geologic time scale deals in enormously long durations of time; an era is its longest unit of measurement. During the past 600 million years there have been three, the *Paleozoic* (570–248 mya), the *Mesozoic* (248–65 mya) and the *Cenozoic* (65 mya–present). It was during the Cambrian Period of the Paleozoic Era that modern animal phyla evolved. During the Mesozoic, the first dinosaurs and the first mammals appeared on earth. Eras are subdivided into periods. The Mesozoic Era is divided into three time periods, the *Triassic* (248–213 mya), the *Jurassic* (213–144 mya) and the *Cretaceous* (144–65 mya). The period of the Paleozoic just preceding the Mesozoic Era is the Permian; the period of the Cenozoic Era just following the Mesozoic is the Tertiary Period. Refer to Figure 1 below.

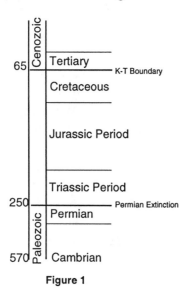

Figure 1

By 200 million years ago, during the Late Triassic Period, early mammals maintained a fragile existence as small shrewlike animals while some of their phyletic cousins, the dinosaurs, evolved into huge animals that dominated all other life forms.

Dinosaurs have been classified into two orders, *Saurischia* and *Ornithischia*, defined by the type of hip bone they have: Saurischians have a lizardlike pelvis, and ornithischians have a bird-like pelvis. Dinosaurs can also be classified as either small or large herbivores or carnivores.

Dinosaurs, exceedingly successful for about 150 million years, disappeared 65 million years ago at the end of the Cretaceous period. The reason for their extinction remains one of the great unsolved mysteries of our time.

DATING FOSSILS

Many theories explain why dinosaurs became extinct, and they all have one thing in common: They require correlating events happening in the Cretaceous environment with changes in dinosaur populations. In order to correlate geologic events with the extinction of the dinosaur, it is necessary to provide reasonably accurate dates for the age of rocks and the age of bones.

A number of methods are used to estimate the age of dinosaur fossils. They include using *radioactive isotopes* to date rocks that fossils are found in, creating a master diagram of rock layers called a *geologic column* to give relative ages of each layer, and using *index fossils* from species known to have existed for only a very short time.

Radioactive clocks use the decay of radioactive isotopes to measure time. Isotopes are elements that have too many or too few neutrons in their nucleus. For example, an *atom* of carbon has six protons and six neutrons in its nucleus and because it has an atomic mass of 12 is referred to as carbon-12. Most carbon found in nature is carbon-12, but a small percent of this element has 6 protons and 8 neutrons in the nucleus. These are *isotopes* of carbon and because they have an atomic mass of 14 they are referred to as carbon-14. The additional neutrons make the carbon 14 nucleus unstable, and occasionally a neutron in the carbon 14 nuclei disintegrates into a proton and an electron. An electron ejected from the nucleus of an isotope is called a Beta particle. After the electron is emitted from the carbon-14 nuclei, the isotope becomes a stable nitrogen atom with 7 protons and 7 neutrons. The process is called radioactive decay and can be measured with a Geiger counter. Ultimately, all the radioactive carbon-14 will decay to nitrogen-14 by beta particle emission. However, this takes a very long period of time. The time it takes one half of the carbon-14 isotopes to decay to nitrogen-14 atoms is called its half-life. The half-life of carbon-14 is 5700 years. This phenomenon can be used to date fossils. While alive, organic material absorbs carbon from its environment in food or carbon dioxide. As long as it lives, an organism like a plant or a dinosaur will have a fixed ratio of carbon-14 to carbon-12. However, once it dies, it no longer absorbs carbon from its environment and, therefore, the ratio decreases because of the constant decay of carbon-14 to nitrogen-14. If a fossil is found to have one half its original radioactivity, it is 5700 years old; if it is found to have one fourth its original radioactivity it is 11,400 years old. Carbon-14 dating can be used to date fossils from 200 to about 50,000 years ago. After about 4.5 half-lifes, the amount of carbon-14 left is too little to register on detection equipment.

A different isotope must therefore be used to date dinosaur fossils, which date from 65 million years ago to 210 million years ago. One isotope commonly used is potassium-40. It decays to argon-40 also by beta emission; however, it has a half-life of 1.3 million years. By measuring ratios of *isotopic* Potassium 40 to *atomic* Potassium 39, researchers can date fossils older than 50,000 years.

Radioisotope dating gives the *absolute* age of fossils and rocks. However, it is also possible to determine the *relative* age of bones and the rock layers they are found in. Because both geologic and evolutionary events occur in a sequential

manner, geologists and paleontologists have come to recognize that each strata of rock is older than the one above it (the law of *superposition*) and, therefore, the deeper the rock layer a fossil is found in, the older it is. In a particular area, rock layers are exposed and their sequence is recorded on a diagram called a *geologic column*. Refer to Figure 2. Then, the area is broad-

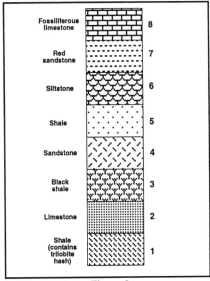

Figure 2

ened, and overlapping sequences are used to extend the column. Geologic columns are useful to paleontologists because they distinguish which fossils were deposited first and which animals coexisted. On the other hand, geologists can use the absolute age of particular fossils to date rock layers. The principle of superposition can be used in sequencing exposed rock strata, like those on the face of the Grand Canyon, as well as ordering sediment layers in core samples taken from the depths of the ocean floor.

Although many animals, like the trilobite, endured periods of time measured in eras, others had a short-lived existence measured in a million or two years. These short-lived species are very useful in dating other fossils. If they were not only short-lived but also wide-ranging, then they are referred to as *index fossils*. If an index fossil is found sharing the same strata as another fossil, then the age of the rock strata *and* the age of the other fossil can be pegged to a period within the short-lived existence of the index fossil. Because geologic time is measured in hundreds of millions of years, being able to date a rock layer and a fossil contained in it to within several million years is remarkable.

Physical Evidence: Iridium, Shocked Quartz, and Trapped Helium

Iridium is an element that does not occur in great concentrations in many places in the earth's crust. However, there are areas where it does. For example, in regions of volcanic activity, iridium is found in fine volcanic ash. Iridium deposits have also been found near some meteor impacts and also in fragments of meteors.

When you look at a grain of quartz under a microscope, the quartz appears to have smooth surfaces; however, after being subjected to high pressure and heat, the surface of quartz grains appear deformed and are known as "shocked quartz." Shocked quartz, for example, has been found at the sites of nuclear explosions.

During a volcanic eruption, helium gas gets trapped in lava and can be used to identify rock of volcanic origin.

SOURCE MATERIAL ON THE VOLCANISM THEORY

EXTINCTION

When the last animal of a species dies, no member of that species survives and the species has become extinct. This has not been a rare phenomenon during the long history of life on earth; in fact, we have inferred from the fossil record that 99% of all species that have ever lived have now perished. Extinction was always believed to be an ongoing process that resulted from gradual changes in climate. Species would flourish as long as their food supply was plentiful but would be slowly driven toward extinction as global or local climate patterns changed and threatened their diet and survival. It was thought that only rarely a more sudden event, for example a deadly bacterial infection, would wipe out a species. We see this today with the decline of the American chestnut tree as a result of a plant virus. The decline of a species and its ultimate extinction were therefore thought to result from very long-term and gradual causes. However, today, it is recognized by most paleontologists that in addition to the gradual destruction of species, catastrophic incidents of the simultaneous extinction of thousands of species have also occurred. These extinctions are known as *mass extinctions.* Understanding why dinosaurs disappeared 65 million years ago at the boundary between the Mesozoic and Cenozoic Eras starts with an explanation of mass extinctions.

Extinction events from *The Nemesis Affair: A Story of the Death of Dinosaurs and the Ways of Space* by David M. Raup. Copyright 1986 by David M. Raup. Reprinted by permission of W.W. Norton & Company, Inc.

"Extinctions are not uniformly distributed in geologic time. Some intervals, which we now call 'extinction events' or 'mass extinctions,' have many more than the normal number of species going extinct. These intervals were used by the early nineteenth-century geologists to label points in time that could be recognized worldwide. Names given to the periods of time between the extinction events are still with us.

"Thus, it is not surprising that the major extinctions were, and still are, at major boundaries in the geologic time scale. It is no coincidence that the dinosaurs went extinct at, or very near, what we know as the boundary between the Cretaceous and Tertiary periods (the K-T boundary). This is also the boundary between two large units: the Mesozoic and Cenozoic eras.

"The greatest mass extinction of all time was in the Permian period, some 250 million years ago, at or near the Permian-Triassic boundary, which is also the era boundary between the Paleozoic and Mesozoic. It has been estimated that this event eliminated as many as 96 percent of species living in the oceans at that time: a nearly complete destruction of all life."

THEORIES

The Volcanism Theory. During the late Cretaceous period, the world experienced an extraordinary outbreak of violent volcanic activity. It was brought about by a prolonged delay in the periodic reversal of earth's magnetic field during which "hot spots" are formed on the surface of the earth. The hot spots lie over regions of molten iron which, during a magnetic reversal, flow less freely, sort of get backed up, and, consequently, transfer huge quantities of heat upward. The upward heat and pressure produce an upward movement of magma, and volcanos result.

Lava, solid fragments, superheated steam, and immense clouds of volcanic ash and sulfur oxides were then ejected into the atmosphere on a global scale. During a period of half a million years, the accumulation of ash on the surface of the earth and on its oceans and the darkening of the sky led to acid rain, a cooling of the earth, the death of many, if not most, of its plants and, therefore, the death of dinosaur herbivores and ultimately dinosaur carnivores.

Evidence for and Objections to the Volcanic Eruption Theory

Hot Spots. From *Great Unsolved Mysteries of Science*

In Western India there was a large hot spot known as the Deccan Traps. It is of interest because it existed during the period of time when dinosaurs became extinct, some 65 million years ago. Other hot spots have been found dating to other mass extinctions during the history of the earth. Hot spots are of particular interest to those who prefer to blame mass extinctions on volcanic activity rather than the impact of an asteroid, meteor, or comet because where there are hot spots there are volcanos. Volcanos can produce the same environmental disasters—heat, ash, smoke—as meteor or asteroid impact.

Also supporting the volcanic theory is evidence that the element iridium, found in meteors and in areas of volcanic activity, is not confined to soil strata samples representing a brief span of time, but rather is spread down over more than a half a million years. Because hot spots remain active for periods of time of this magnitude, and celestial collisions are pretty much instantaneous events, it can be inferred that volcanic activity is a much more reasonable explanation for the events which led to the disappearance of dinosaurs and other mass extinctions.

Moreover, shocked quartz and mineral spherules are also found dispersed in strata over a similar period of time.

Also supporting the volcanic eruption theory is the fact that although the dinosaurs disappeared quickly in terms of the geologic time scale, they did not disappear suddenly. Rather than lasting a few years or decades, their extinction was a prolonged event lasting thousands of years. Proponents of the volcanic activity theory believe that a period of massive volcanic activity persisting over just such a time interval produced the changes in the environment necessary to cause the extinction of the dinosaurs.

The Volcanic Alternative From. David M. Raup, *The Nemesis Affair*

"The general volcanic interpretation of the K-T event implies a period of devastating volcanism, the like of which we have never seen—that is, if it is to cause the extinction of more than half the animal species on Earth. This seems incredible, but the human species has been around only a very short time and we have no real basis for saying that the level of volcanic activity we have experienced is typical.

"Is there other evidence of unusual volcanism about 65 million years ago? In fact, there is. An immense area in India is blanketed by thick basalts called the Deccan Traps. The Deccan volcanism went on for several million years, but nearly all estimates of the starting time fall close to 65 million years before present. A few other such basalt flows are known on other continents and of different ages, but because we have never witnessed this type of eruption, very little is known about environmental effects. It is not known whether the environmental effects would be local or global or whether they would involve fundamental properties of climate and atmospheric chemistry."

Ancient Eruption Abstract of article by Richard Monastersky in *Science News*, August 14, 1993.

Researchers found trapped helium inside old volcanic rock in the Deccan Traps in west India, which was a sign of ancient volcanic eruptions. The eruptions occurred between 65 and 68.5 million years ago and may have coincided with the extinction of the dinosaurs.

The Great Death

In his book *The Day of the Dinosaur*, Jon Man discusses the problems with oversimplifying the causes of dinosaur extinction. He focuses on the popular scientific assumption that a major contributing factor to the great death of the dinosaurs was their inability to survive the "cold snap" resulting from the blocking of the sun by the products of combustion spewing forth from worldwide fires ignited by either volcanic activity or meteor impact. "Though the dinosaurs might have been exterminated from many regions by cold, there is no apparent reason why they should not have survived in isolated pockets." Man agrees that dinosaurs were not equipped to survive prolonged cold; however, he believes that there were many areas near and at the equator that would have remained warm enough to support both herbivores and the carnivorous dinosaurs that fed upon them. Moreover, crocodiles, which should have perished, did not.

It appeared to Jon Man that the evidence linking the extinction of dinosaurs with a global drop in temperature caused by massive volcanic activity was far from conclusive.

Tambora and Krakatoa From Charles Officer and Jake Page, *The Great Dinosaur Extinction Controversy* (Pages 20-32, 61, 171-177). Copyright 1996 by Charles Officer and Jake Page. Reprinted by permission of Addison-Wesley Longman.

"The uninhabited island of Krakatoa, located between Java and Sumatra, was a chief navigational aid to sailors plying the Sunda Strait when suddenly, in 1883, it was lost to view. In August, on the 26th and 27th, eruptions nearly obliterated the island and set in motion a train of damage. More than 30,000 people were swept off Java and Sumatra by a sea wave called a tsunami, a low wall of water caused by the sudden vertical movement of the seafloor, often the result of earthquakes near deep sea trenches. These waves can propagate across the ocean, in the physicist's cool parlance, as "low-amplitude and long-wavelength disturbances," but as they approach the shallow waters near land, their amplitude increases. In other words, as a tsunami nears land, it rises up as a rushing wall of water some ten or more feet high. The one that followed after Krakatoa blew was caused either by ejected material plummeting into the sea or by collapse of the island structure itself.

"The Royal Society of London was able to catalogue more distant results as well, making this eruption the first geophysical event to be studied scientifically on a global scale. To begin with, it made one of the loudest sounds in history, being heard as far as 3,000 miles away on the island of Rodriguez in the Indian Ocean, where chief of Police James Wallis reported that 'several times during the night of the 26th-27th reports were heard coming from eastward, "like the distant roar of heavy guns.' Low-frequency sound waves circled the Earth as many as seven times and were picked up by barometric pressure gauges at stations around the globe.

"For a year after the eruption, many ships' logs noted floating pumice in the Indian Ocean. In March 1884, Captain Gray of the *Parthenope* found the central Indian Ocean strewn with pumice that was covered with barnacles, testimony to its long residence in the water.

"More lasting were the atmospheric effects of dust and sulfate aerosols that had been injected into the stratosphere. For three years in many parts of the world, the days were filled with a blue or green haze, with spectacular red glows just after sunset and just before dawn. In Poughkeepsie, New York, there was such 'an intense glow in the sky that fire engines were called in the morning' on November 27, 1884 more than a year after the eruption.

"Sixty-eight years before Krakatoa, in 1815, a monster volcano erupted. More than ten times as powerful as Krakatoa, it devastated Tambora on Sumbawa Island in the Indonesian archipelago. It was described by Sir Thomas Stamford Raffles, founder of Singapore at the time and was quoted in Officer's and Page's book:

"It began on the 5th day of April, and was most violent on the 11th and 12th, and did not entirely cease until July. The sound of the explosion was heard in Sumatra, at a distance of nine hundred and seventy geographic miles in a direct line, and at Ternate, in the opposite direction, at the distance of 720 miles.

"Out of a population of twelve thousand, only twenty-six individuals survived on the island. Violent whirlwinds carried up men, horses, cattle and whatever else came within their influence into the air, tore up the largest trees by the roots, and covered the whole sea with floating timber. ... On the side of Java, the ashes were carried to a distance of three hundred miles, and two hundred and seventeen toward Celebes, in sufficient quantity to darken the air. The floating cinders to the westward of Sumatra formed, on the 12th of April, a mass two feet thick and several miles in extent, through which ships with difficulty forced their way.

"The darkness occasioned in daytime by the ashes in Java was so profound that nothing equal to it was ever witnessed in the darkest night. ... Along the seacoast of Sumbawa, and the adjacent isles, the sea rose suddenly to the height of from two to twelve feet, a great wave rushing up the estuaries, and then suddenly subsiding. Although the wind at Bima was still during the whole time, the sea rolled in upon the shore and filled the lower parts of houses with water a foot deep.

"The area over which tremulous noises and other volcanic effects extended was one thousand English miles in circumference, including the whole of the Molluca Islands, Java, a considerable part of Celebes, Sumatra and Borneo."

"Dust and sulfate aerosols injected into the stratosphere by Tambora spread out globally, shielding incoming sunlight. In New England the following year, 1816, was called 'The Year Without a Summer.' ... The average temperature in June was seven degrees Fahrenheit below normal, crop failures were rife, and prices soared. Things were even more disastrous in Europe. Crop failures were universal, and food shortages led to local famines and anarchy.

"Even larger eruptions occurred in earlier times. In Sumatra some 75,000 years ago, the eruption of the Toba volcano left a collapse structure 30 by 60 miles in dimension. Ash was spewed over much of the Indian Ocean, leaving a layer on the bottom as much as four inches thick 1,300 miles away. Six hundred thousand years ago, in what is now Yellowstone National Park, an eruption spread ashes a few inches thick over most of the present Untied States west of the Mississippi. Both of these eruptions may well have wiped out most of their regions' flora and fauna as well as causing havoc on a global scale, but both regions recovered in a relatively short period."

The Deccan Volcanism and Selenium From *The Great Dinosaur Extinction Controversy*

"From the geologic record of the Deccan volcanism and from measurements of recent volcanic emissions, we have a clear picture of other environmental changes that occurred at K-T time. At its peak, K-T volcanism may have been a hundred times greater than what we experience today—that is, it would be something like a Mount St. Helen's (or bigger) happening every day of every year.

"During peak periods, chlorine emissions would have been 110 times those that are today estimated to cause an 8 percent reduction in the protective ozone layer. Increased ultraviolet radiation would have been present. Sulfur dioxide emission would have been fourteen times what arises from present-day fossil fuel emissions. That is a global average; acid rain for any given location or time could have substantially exceeded this value. Sulfur dioxide returning to the oceans as acid rain would have been sufficient to reduce the pH of surface waters from their normal 8.2 to 7.6 and maybe lower. Lowered pH means greater acidity, which reduces the availability of the carbonate ion needed in building (calcium-forming) plankton shells. Such a pH change would be long lasting—a particle of water has a lifetime of some fifty years in the 100-meter realm of surface waters.

"Volcanic aerosols, particularly sulfur dioxide, have a stratospheric residence of a year or so, and their effect from large volcanic eruptions is to cool the Earth's surface temperature. On the other hand, the associated carbon dioxide emission would lead to a greenhouse effect and global warming. ...

"...we do know that [vulcanism resulting from the building of the Rocky Mountains] lasted over a longer period of time, with eruptions at various locales that were, all together, probably of smaller total magnitude [than the Deccan Traps]. But rains brought volcanic dust and other airborne particles from this regional disturbance into the shallow seas of the Western Interior Seaway [now the Great Plains], where they collected on the bottom in succeeding sedimentary layers. ... The selenium concentrations in some of these shales have extremely high values—in the range of 10 to 20 parts per million.

"Now, when William Zoller analyzed airborne particles from Kilauea [Hawaiian volcano], the element that showed the greatest enhancement was selenium–an enhancement, in fact, of *40 million* ...

"Herbs and grasses grow with great facility in soils on the rock outcrops of the Great Plains. Many of these plants—particularly one called *Astragalus* by scientists and known otherwise as locoweed—accumulate selenium from the soils into their tissues; where it reaches levels of 500 to 1,000 ppm–levels that are highly toxic ...

"Seleniferous plants produce either acute or chronic poisoning in livestock. Acute poisoning usually arises from a single feeding on plants with high selenium concentrations of 500 ppm or more. In many cases, death follows within a few hours. ... Stockmen have learned to fence off areas of selenium-rich soils from their livestock and to remove locoweed either by uprooting it or spraying it with herbicides.

"Thus, the last active results of the titanic events of the K-T boundary dealt not with a bang but an agricultural program."

TYPICAL SOURCE MATERIAL ON THE IMPACT THEORY

THE THEORY

The impact theory states that the dinosaurs disappeared suddenly because of environmental and climate changes brought about suddenly as a result of an extraterrestrial object colliding with the earth. Different versions of this theory consider meteors, asteroids, and comets as possible sources of the catastrophic impact. The effects of an ancient cataclysm such as this are described in *National Geographic* by Rick Gore. Selections from his article "Extinctions" are reproduced below by permission of the National Geographic Society.

"The scenario is straight out of a science-fiction movie: Giant meteorite strikes earth, setting the planet afire. Volcanoes erupt, tsunamis crash into the continents. The sky grows dark for months, perhaps years. Unable to cope with the catastrophic changes in climate, countless species are wiped off the face of the planet.

"Many scientists believe that an object only 10 or so kilometers in diameter would initiate a chain of events that could ultimately cause a mass extinction. The impact explosion would have the energy of tens of thousands of hydrogen bombs. In an area hundreds of kilometers in diameter, the initial blast wave would produce enormous heat mixed with vaporized rock and water. Wildfires and pulverization of the earth's surface would follow. Simultaneously, a huge plume of dust would be ejected into the atmosphere. Devastating earthquakes would crack the earth's surface. As the weeks went by, a great mass of vaporized dust and water would circle the globe. Vegetation would become super dried in the hot winds of the blast wave. Wildfires would spread throughout the world, ignited by the surge of superheated material. During the ensuing years, the world would become dark, the skies covered in a thick mantle of sun-blocking dust. It would get colder. Plants would not photosynthesize efficiently, and many would die. Herbivores that depended on the plants would perish, and then carnivores that ate the herbivores would die. Plants and animals not suited to cold temperatures would perish. Vaporized nitrogen and sulfur oxides would produce years of highly acidic acid rain, further damaging the environment.

"Moreover, if the object dropped in the ocean, vaporized limestone (calcium carbonate) would produce huge quantities of carbon dioxide, initiating a greenhouse effect lasting for years. The climate would turn hot and animals and plants not suited to higher temperatures would die."

Evidence Relevant to the Impact Theory

Meteorites and Comets From *The Great Dinosaur Extinction Controversy*

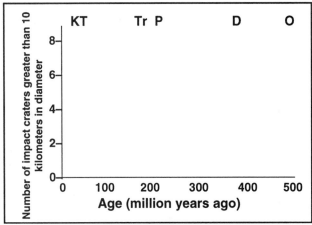

Table 1

"All it takes is a pair of binoculars to see that meteorites have made the Moon a pretty inhospitable place–and by implication the Earth, the major partner in this "two-planet" system. The Moon's surface, particularly its highlands, is pocked by craters great and small. Found therein are craters within craters, bespeaking devastating events about which astronomers argue. ... Clavius, for example, the largest crater on the visible side of the Moon, is 230 kilometers (about 150 miles) in diameter and is itself pockmarked by more than a half-dozen good-sized craters. ... On Earth, blessed as it is by large amounts of water, and far more activity, the results of such bombardment from without and volcanic activity from within have been largely wiped away....

"We know a great deal about these nomads [comets] in space. First of all, it is fair to say that comets are less important than they look. Mostly, they are what might be called special light effects.

"A comet is primarily a nucleus made up of small particles of interstellar dust along with frozen gases like methane and ammonia, and water. These cores, dubbed "dirty snowballs" by one astronomer, ran from maybe a hundred meters across to a few kilometers. As one of these snowballs nears the sun, its outer layers of ice melt, releasing some of its interstellar dust and gases to form the brilliant "head" or coma....

"A Dutch astronomer, Jan Hendrik Oort, proposed that they are in orbits, but vast ones, with their aphelia located in an enormous frozen cloud a light-year away from the Sun—a cloud that remains only hypothetical. The theory is that all comets we see...originate in Oort's cloud...

"A comet's nucleus might wreak a bit of havoc if it hit the Earth; but of the estimated 2,000 comets in the grip of the Sun's gravity, none come particularly close to Earth. Halley's nearest approach was 23 million kilometers in 1910.

"It would be injudicious to say that comets pose absolutely no threat to the Earth. But of far more concern are the ever-so-much-brawnier meteoroids, which we may assume are wayward visitors from the asteroid belt, which lies between the planets Mars and Jupiter, between 240 and 800 million kilometers from the Sun. This ring of rubble never managed to coalesce into a planet when the solar system came into its own. ...There are some 40,000 asteroids in the asteroid belt, and the vast proportion of them stay put out there beyond mars. Still, at least four of them—Adonis, Apollo, Geographos, and Hermes—with diameters of a kilometer or less, approach to within about a million kilometers of the Earth. ...

"In historical time, a huge event occurred that was almost certainly the result of some extraterrestrial visitation—but of what remains totally ambiguous. It occurred in the Tunguska River region of Siberia some 600 kilometers north of Vladivostok on June 30, 1908. That morning, passengers on the Trans-Siberian Railroad were stunned to see a meteor as bright as the Sun race across the sky from south to north and disappear beyond the horizon. Immediately, they felt a violent blast of air. Later, it was learned that the blast had been felt over a distance of 80 kilometers and flattened forests like so many matchsticks, in a radial pattern extending 30 kilometers from the center. It was a spectacular event, to be sure, but nothing like a major earthquake in destruction. ...

"Next to Meteor Crater in Arizona, the Ries Crater in southern Germany has been the most thoroughly studied impact site. The Ries Crater has a diameter of 24 kilometers and an age of 15 million years. ...

"The Montagnais Crater is a buried impact structure, as determined from seismic reflection profiling and drilling, off the coast of Nova

Scotia. It has a diameter of 45 kilometers and an age of 51 million years.

"Farther to the southeast, on the continental shelf off Virginia and New Jersey, Wylie Poag of the U.S. Geological Survey and colleagues have found, again by seismic reflection profiling and drilling, two additional impact structures in the sedimentary section. The Chesapeake Bay Crater is quite large, with a diameter of 85 kilometers; the adjacent Toms Canyon Crater has a diameter of 20 kilometers. Both are 35 million years old.

"Finally, there is the Red Wing Creek Crater in North Dakota. It is a buried structure of nine kilometers diameter with an age of 200 million years and is presumed to be of impact origin."

Louis Alvarez and Wendy Wolbach: the way it happened. The impact theory was born in the late '70s. Here, the theory is discussed in a *National Geographic* article written by Rick Gore after his conversation with Luis Alvarez, one of the scientists who first proposed it. "Extinctions" appeared in the June, 1989, *National Geographic*, Vol. 175 No. 6.

"The excitement began in 1978 when a team from the University of California and Lawrence Berkeley Laboratory found a large enrichment of the element iridium in a pencil-thin, 66-million-year-old layer of rock from Gubbio, Italy. This iridium-rich clay lay right at the boundary between the Cretaceous period, when there were dinosaurs, and the Tertiary, when there were none. (Scientists nickname this transition the K-T boundary.)

"Because iridium is rare on Earth but common in meteorites, the Berkeley scientists—Walter and Luis Alvarez, Frank Asaro, and Helen Michel—proposed that Earth had been hit by an asteroid ten kilometers (six miles) across. Wildly controversial at first, the proposal has since been backed up by abundant and convincing evidence from around the globe. Most scientists now concur that at least one great extraterrestrial object struck the planet around the time the dinosaurs died out.

"With Alvarez, I hike an Italian mountain road to inspect the Gubbio boundary clay. He digs out a chunk and hands it to me.

'You are holding debris from the impact,' he says. 'In the first days after Earth was hit, dust blanketed the entire world. It grew pitch-dark for one to three months. If the impact was on land, it probably got bitter cold. If it hit at sea, the water vapor could have created a greenhouse effect, making things hot. Hot nitric acid would have rained out of the atmosphere—a life-threatening rain that would have dissolved the shells of organisms.'

"That's not all. A surprising discovery by Wendy Wolbach, a graduate student at the University of Chicago, indicates that the world may have turned even nastier, as it did last summer at Yellowstone National Park.

"A red sun shines like the eye of an angry god through the pall of billowing smoke at Old Faithful. A rush of heat. A swirl of suffocating, sooty air.

"Suddenly, on the hillside behind the famous geyser, the gates of hell burst open, and a fire storm races down the slope. A million acres on fire. The worst conflagration to strike the vast Yellowstone ecosystem in history. An awesome, terrifying orgy of flame.

"Yet this holocaust is insignificant compared with what Wolbach believes happened that day 66 million years ago when Earth was hit. The entire world caught fire.

"Even as Yellowstone burns, Wolbach shows me her evidence in her Chicago office—scanning electron microscope pictures of soot particles embedded with the iridium layer from three widely separate sites—Denmark, Spain, and New Zealand. ...

'To get the amount of soot we find,' she says, 'as much as 90 percent of the world's forests must have burned.'

"Granted, the impact of a 10-kilometer body would be the equivalent of 10,000 times the power of all the world's nuclear weapons, but how could fire spread so disastrously across the globe?

'Even if it hit the ocean, the impact would have created a crater 300 kilometers across,' says Anders. 'A huge plume would have pushed the atmosphere aside. The fireball would have had a radius of several thousand kilometers. Winds of hundreds of kilometers an hour would have swept the planet for hours, drying trees like a giant hair dryer. Two-thousand-degree rock vapor would have spread rapidly. It would have condensed to white-hot grains that could have started additional fires.'

"In addition, lightning discharges like those in a volcanic eruption could have ignited wind-swept fires on all landmasses—fires that marched far faster than those at Yellowstone."

THE IMPACT THEORY: EVIDENCE AND OBJECTIONS

The evidence favoring the impact theory for the extinction of the dinosaurs is very similar to evidence cited in support of the volcanic activity theory. For example, the presence of iridium in rock formations dating back 65 million years is commonly given as strong evidence that an extraterrestrial object of substantial mass collided with the earth about that time. Iridium is not often found on the surface of the earth because it is a heavy element and during the formation of the earth nearly all of it sank to the earth's interior. However, meteors, asteroids, and other objects arriving from outer space are rich in iridium. Therefore, it should be no surprise that shortly after iridium deposits were found to occupy the same time frame as the disappearance of the dinosaurs, the impact theory was proposed.

In fact, rich iridium deposits were found all over the world at the K-T boundary. Moreover, the proportion of iridium found was identical to the proportion of iridium usually found in meteorites.

The discovery of mineral spherules sharing the K-T boundary clay with iridium is offered as further evidence in support of the impact theory. These spherules are similar to spherules routinely discovered in the vicinity of known meteorite impacts with the earth.

Shocked grains of quarts, also frequently found at meteorite impact sites, have also been uncovered in K-T boundary clay.

In the chapter "What Killed the Dinosaurs?" in *Great Unsolved Mysteries of Science*, the authors note that "a possible 'smoking gun' has also been identified. Some scientists believe that an impact occurred about 65 million years ago in the Caribbean Sea, near Mexico's Yucatan Peninsula." Recent discovery of high iridium content in K-T boundary clay on nearby islands appears to confirm this thesis.

Opponents of the impact theory are usually proponents of the volcanic activity theory. They are quick to point out that dinosaurs did not disappear as might be expected if their extinction resulted from a cataclysmic event as catastrophic as the collision of the earth with a substantial meteor or asteroid. They argue that dinosaurs took tens of thousands of years to disappear and therefore, the relatively slower processes of vulcanism are a more suitable explanation. To proponents of the volcanic activity theory, 10,000 years of gradual extinction is acceptable. They hold that the dinosaurs were plunging toward extinction anyway and would have disappeared with or without a celestial impact. "Supporters counter by saying that it is too much of a coincidence that an impact would occur at the exact point in time when the dinosaurs disappeared forever."

Nemesis, the Death Star From THE NEMESIS AFFAIR: A Story of the Death of Dinosaurs and The Ways of Space by David M. Raup. Copyright 1986 by David M. Raup. Reprinted by permission of W.W. Norton & Company, Inc.

"Astronomers have found that stars like our sun usually have a companion. The stars are known as a binary pair, and they orbit around their common center of mass extinctions. David Raup, a paleontologist, considers comets a likely suspect. Here, he proposes the following answer to the question: How did the meteors or asteroids that impacted on the earth 65 million years get here?

"Nemesis is one of several names given to our Sun's small companion star. This little star is now about two light-years away and moving away. But in another few million years, it will turn and head back toward the Earth. The inward trip will take another dozen or so million years, and before the orbit is complete, Nemesis will have passed close to the Sun. Close enough, in fact, to pass through the Oort Cloud, an envelope of billions of comets that go around the Sun in their own orbits beyond the outer planets. As Nemesis passes through the Oort Cloud, its own gravitational force will deflect some of the comet orbits in random ways. Most of them will be thrown out of the Solar System, but some will be sent in toward us. As a result, one or more of the errant comets will collide with Earth. And we know from the geological records of Earth history that such collisions can be devastating. One incident killed the dinosaurs, and another got the last of the crab-like creatures called trilobites. Many of the major biological crises of our past, the mass extinctions, were evidently caused by the environmental shock of what is known in the trade as "large-body impact." And because the Nemesis orbit has a fixed period of 26 million years, the biological catastrophes come every 26 million years. A great clock in the sky is controlling biological destinies on Earth.

Sample Student Responses • The Extinction of the Dinosaurs

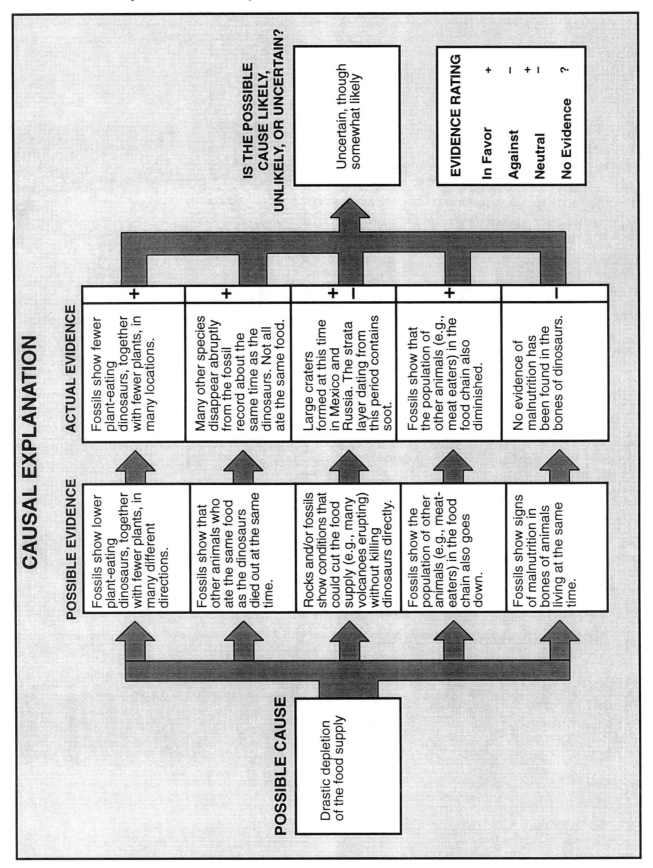

CAUSAL EXPLANATION

IS THE POSSIBLE CAUSE LIKELY, UNLIKELY, OR UNCERTAIN?

Uncertain, though somewhat likely

EVIDENCE RATING

In Favor	+
Against	-
Neutral	+ -
No Evidence	?

ACTUAL EVIDENCE

+ Fossils show fewer plant-eating dinosaurs, together with fewer plants, in many locations.

+ Many other species disappear abruptly from the fossil record about the same time as the dinosaurs. Not all ate the same food.

+ - Large craters formed at this time in Mexico and Russia. The strata layer dating from this period contains soot.

+ Fossils show that the population of other animals (e.g., meat eaters) in the food chain also diminished.

- No evidence of malnutrition has been found in the bones of dinosaurs.

POSSIBLE EVIDENCE

Fossils show lower plant-eating dinosaurs, together with fewer plants, in many different directions.

Fossils show that other animals who ate the same food as the dinosaurs died out at the same time.

Rocks and/or fossils show conditions that could cut the food supply (e.g., many volcanoes erupting) without killing dinosaurs directly.

Fossils show the population of other animals (e.g., meat-eaters) in the food chain also goes down.

Fossils show signs of malnutrition in bones of animals living at the same time.

POSSIBLE CAUSE

Drastic depletion of the food supply

Sample Student Responses • The Extinction of the Dinosaurs

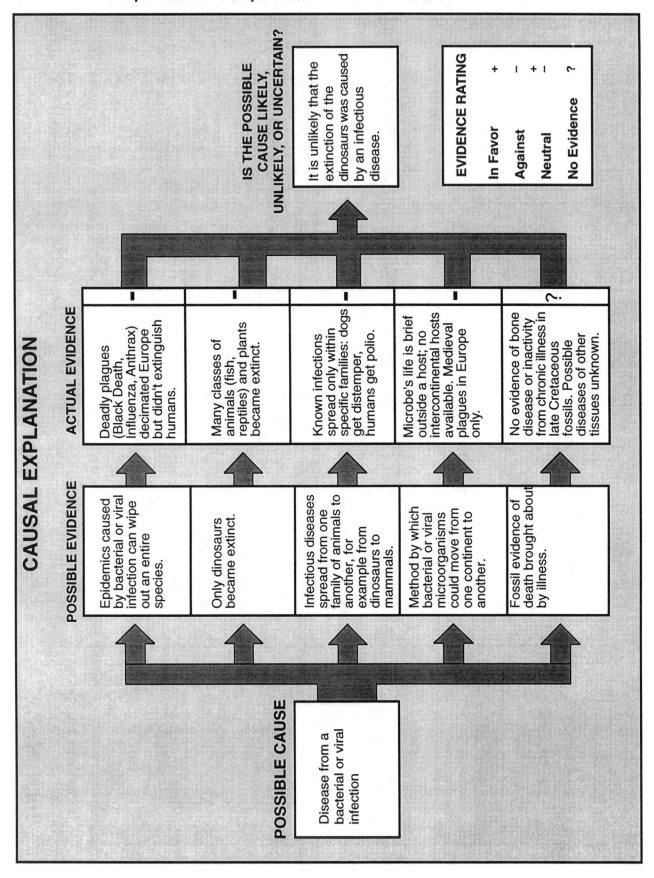

CAUSAL EXPLANATION

POSSIBLE CAUSE

Disease from a bacterial or viral infection

POSSIBLE EVIDENCE

- Epidemics caused by bacterial or viral infection can wipe out an entire species.
- Only dinosaurs became extinct.
- Infectious diseases spread from one family of animals to another, for example from dinosaurs to mammals.
- Method by which bacterial or viral microorganisms could move from one continent to another.
- Fossil evidence of death brought about by illness.

ACTUAL EVIDENCE

- Deadly plagues (Black Death, Influenza, Anthrax) decimated Europe but didn't extinguish humans.
- Many classes of animals (fish, reptiles) and plants became extinct.
- Known infections spread only within specific families: dogs get distemper, humans get polio.
- Microbe's life is brief outside a host; no intercontinental hosts available. Medieval plagues in Europe only.
- No evidence of bone disease or inactivity from chronic illness in late Cretaceous fossils. Possible diseases of other tissues unknown.

IS THE POSSIBLE CAUSE LIKELY, UNLIKELY, OR UNCERTAIN?

It is unlikely that the extinction of the dinosaurs was caused by an infectious disease.

EVIDENCE RATING

In Favor	+
Against	–
Neutral	+ –
No Evidence	?

Sample Student Responses • The Extinction of the Dinosaurs

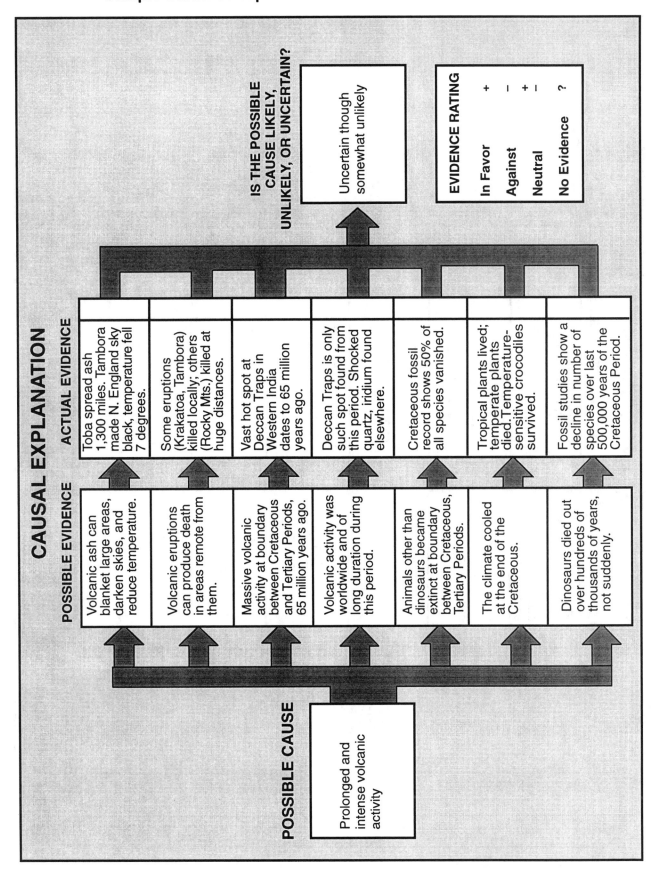

CAUSAL EXPLANATION

IS THE POSSIBLE CAUSE LIKELY, UNLIKELY, OR UNCERTAIN?

Uncertain though somewhat unlikely

EVIDENCE RATING

In Favor	+
Against	—
Neutral	+ —
No Evidence	?

ACTUAL EVIDENCE

- Toba spread ash 1,300 miles. Tambora made N. England sky black, temperature fell 7 degrees.
- Some eruptions (Krakatoa, Tambora) killed locally; others (Rocky Mts.) killed at huge distances.
- Vast hot spot at Deccan Traps in Western India dates to 65 million years ago.
- Deccan Traps is only such spot found from this period. Shocked quartz, iridium found elsewhere.
- Cretaceous fossil record shows 50% of all species vanished.
- Tropical plants lived; temperate plants died. Temperature-sensitive crocodiles survived.
- Fossil studies show a decline in number of species over last 500,000 years of the Cretaceous Period.

POSSIBLE EVIDENCE

- Volcanic ash can blanket large areas, darken skies, and reduce temperature.
- Volcanic eruptions can produce death in areas remote from them.
- Massive volcanic activity at boundary between Cretaceous and Tertiary Periods, 65 million years ago.
- Volcanic activity was worldwide and of long duration during this period.
- Animals other than dinosaurs became extinct at boundary between Cretaceous, Tertiary Periods.
- The climate cooled at the end of the Cretaceous.
- Dinosaurs died out over hundreds of thousands of years, not suddenly.

POSSIBLE CAUSE

Prolonged and intense volcanic activity

Sample Student Responses • The Extinction of the Dinosaurs

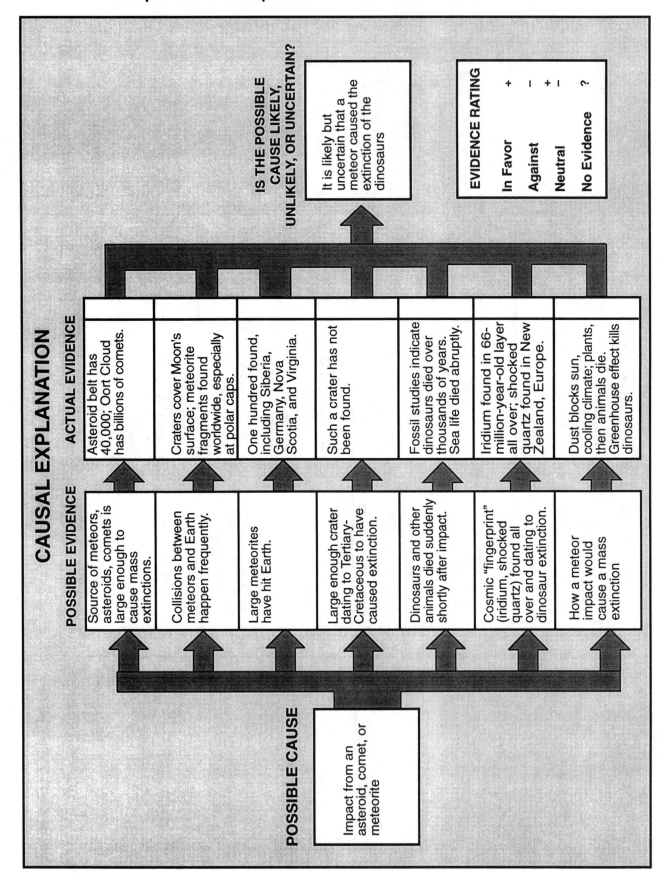

CAUSAL EXPLANATION

IS THE POSSIBLE CAUSE LIKELY, UNLIKELY, OR UNCERTAIN?

It is likely but uncertain that a meteor caused the extinction of the dinosaurs

EVIDENCE RATING

In Favor	+
Against	−
Neutral	+ −
No Evidence	?

ACTUAL EVIDENCE

Asteroid belt has 40,000; Oort Cloud has billions of comets.

Craters cover Moon's surface; meteorite fragments found worldwide, especially at polar caps.

One hundred found, including Siberia, Germany, Nova Scotia, and Virginia.

Such a crater has not been found.

Fossil studies indicate dinosaurs died over thousands of years. Sea life died abruptly.

Iridium found in 66-million-year-old layer all over; shocked quartz found in New Zealand, Europe.

Dust blocks sun, cooling climate; plants, then animals die. Greenhouse effect kills dinosaurs.

POSSIBLE EVIDENCE

Source of meteors, asteroids, comets is large enough to cause mass extinctions.

Collisions between meteors and Earth happen frequently.

Large meteorites have hit Earth.

Large enough crater dating to Tertiary-Cretaceous to have caused extinction.

Dinosaurs and other animals died suddenly shortly after impact.

Cosmic "fingerprint" (iridium, shocked quartz) found all over and dating to dinosaur extinction.

How a meteor impact would cause a mass extinction

POSSIBLE CAUSE

Impact from an asteroid, comet, or meteorite

Sample Student Responses • The Extinction of the Dinosaurs

CAUSAL CHAINS

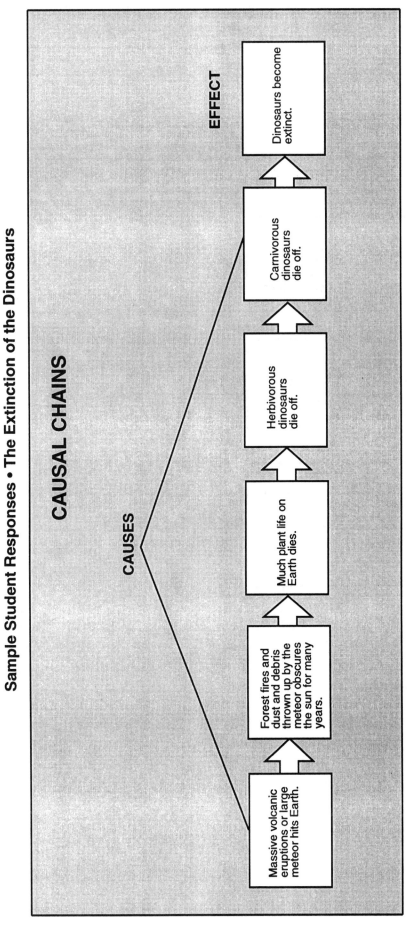

CAUSES

EFFECT

Massive volcanic eruptions or large meteor hits Earth.

Forest fires and dust and debris thrown up by the meteor obscures the sun for many years.

Much plant life on Earth dies.

Herbivorous dinosaurs die off.

Carnivorous dinosaurs die off.

Dinosaurs become extinct.

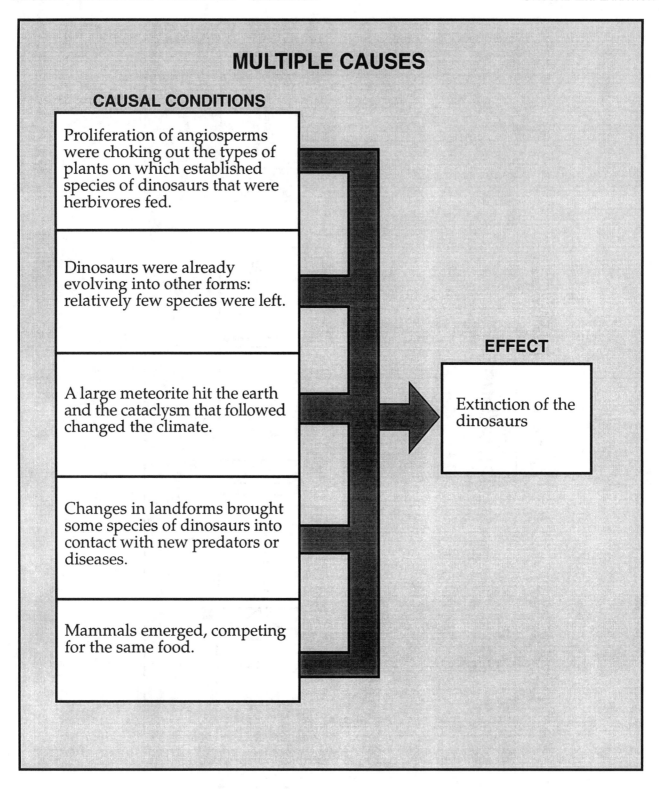

MULTIPLE CAUSES

CAUSAL CONDITIONS

Proliferation of angiosperms were choking out the types of plants on which established species of dinosaurs that were herbivores fed.

Dinosaurs were already evolving into other forms: relatively few species were left.

A large meteorite hit the earth and the cataclysm that followed changed the climate.

Changes in landforms brought some species of dinosaurs into contact with new predators or diseases.

Mammals emerged, competing for the same food.

EFFECT

Extinction of the dinosaurs

SULFURIC ACID

Chemistry   **Grades 10–12**

OBJECTIVES

CONTENT

Students will learn the sequence of chemical reactions that produce sulfuric acid and will attain an understanding of chemical equilibrium.

THINKING SKILL/PROCESS

Students will learn to identify and consider possible causes and assess their reasonableness based on evidence as a strategy for skillful causal explanation. They also learn to map causal chains.

METHODS AND MATERIALS

CONTENT

This lesson requires background material about chemical equations and about the behavior of sulfur and sulfur dioxide. Textual information and information on a graph will be used as the basis for the lesson.

THINKING SKILL/PROCESS

Students are guided by structured questioning to brainstorm possible causes and possible evidence in collaborative learning groups. A graphic organizer for causal explanation is also used. Student groups may report on transparencies of the graphic organizer or on large newsprint posters.

LESSON

INTRODUCTION TO CONTENT AND THINKING SKILL/PROCESS

• Last time I brought my car in for repair, I thought I knew why I was having trouble starting it. I thought it needed a tune-up, but I was wrong. It needed a whole new starter motor. If I had brought it in and asked for a tune-up, I would have paid for the tune-up, which I didn't need, and a new starter motor as well. Have you ever thought that you knew what caused something and found out later that you were mistaken? When you think you know what caused something, you are giving a causal explanation. Tell your partner about an example and why you were mistaken about the causal explanation.

• **How could you have avoided making these mistakes? Discuss this with your partner.** Write students' examples on the board with some suggested remedies. ANSWERS INCLUDE BUT ARE NOT LIMITED TO: *I could have investigated more so that I would have reasons for what I thought the cause was. I could have thought about what other things could be causing the problem and tried to figure out whether it was one of these things rather than what I thought at first: I often think of just one thing. I could have asked someone who knows more about it than I did. I could have been more careful and not jumped to a conclusion without making sure.*

• **Here's a plan for making careful judgments about what is causing something based on your ideas about how we can minimize our errors in making causal explanations. It makes use of two basic ideas: that we should think first about possible**

SKILLFUL CAUSAL EXPLANATION

1. What are possible causes of the event in question?

2. What could you find that would count for or against the likelihood of these possibilities?

3. What evidence do you already have, or have you gathered, that is relevant to determining what caused the event?

4. Which possibility is rendered most likely based on the evidence?

causes—called "hypotheses"—and then look for evidence that can be used to pick out the best causal explanation from among these possibilities. Does this sound like a good plan to use that might avoid some of the mistakes we commonly make about causes? Discuss this with the students. ANSWERS VARY.

- Making careful judgments about what is causing something is crucial in science. For example, before the 1950s, polio was a disease that crippled and killed many people every year. Scientists didn't know what caused it, so they were virtually helpless to try to prevent it. Through careful research, however, scientists acquired conclusive evidence that polio is caused by one of three viruses. Based on this research, they were able to develop a vaccine for the disease. As a result, today, polio has been nearly wiped out as a disease that threatens human well-being and life. Can you think of other examples of situations in which scientific research has led to the discovery of what is causing something we want to prevent? ANSWERS VARY.

- Scientists have also discovered the causes of things that interest us. For example, research into the use of chemical fertilizers has led to the improvement of crops of certain sorts. Can you think of examples in which scientific research has led to the discovery of what is causing something we want to perpetuate? ANSWERS VARY.

- Can you think of examples in which scientists do not know yet what has caused something because they don't yet have conclusive evidence? For example, we have many theories about why dinosaurs are extinct and evidence in favor of some of these, but no one of these theories has been universally embraced and scientific research about the extinction of the dinosaurs continues today. Can you think of similar unresolved questions about causes? ANSWERS VARY

- In this lesson, we're going to focus on a situation in which something is happening that we want to stop from happening. In order to figure out how to stop it, we must determine accurately what is causing the problem. If we make a mistaken judgment and then base our remedy on it, that may be costly, like my car repair would have been, and we may still not have solved the problem. To avoid this, we have to be careful by making an extensive inventory of possible causes, investigating each cause in terms of the evidence we discover, and determining whether we have evidence that is sufficient to establish one of the possible causes as the real cause. This time, the problem has to do with certain chemical processes that we have been studying, those connected with sulfur and oxygen.

THINKING ACTIVELY

- In order to think this problem through carefully, I'm going to ask you to put yourself in the place of a manager of a plant that produces sulfuric acid. Sulfuric acid has many uses in industry—for example, in etching metals and as an electrolyte in automobile batteries. Since this chemical rarely occurs naturally, it must be manufactured. In fact, it is produced in the largest quantity in the United States (69.45 billion pounds in 1980). What is the chemical formula for sulfuric acid? H_2SO_4

- In your plant, sulfuric acid is made by the contact process. This is a multistep series of reactions that starts with the burning of sulfur or sulfur compounds to produce sulfur dioxide. The next step, oxidation of the sulfur dioxide to form sulfur trioxide, is an equilibrium reaction with the following equation:

$$2\ SO_2\ (g) + O_2\ (g) \rightarrow 2\ SO_3\ (g) + 43\ kcal$$

The reaction requires a catalyst, vanadium oxide (V_2O_5), and the forward reaction is exothermic. The conditions in the chamber were adjusted for maximum production of sulfur trioxide. The chamber is routinely inspected and serviced once a year. Further steps in the process are not equilibrium reactions. The schematic diagram of the industrial chamber in the plant where the three reactions take place illustrates this.

STEP I

$$S (g) + O_2 (g) \rightarrow SO_2 (g)$$

STEP II

$$SO_2 (g) + O_2 (g) \underset{V_2O_5}{\rightleftharpoons} SO_3 (g) + 43 Kcal$$

STEP III

$$SO_3 (g) + H_2O (l) \rightarrow H_2SO_4 (l)$$

Schematic Diagram

- Today, you have found out about a problem at the plant. The production levels of sulfuric acid have, during the past week, begun to decrease. Your quality control staff have traced this problem to the equilibrium system (Step II). Something is causing the reduction in the quantity of sulfur trioxide produced. This, in turn, is affecting the production of sulfuric acid. Use this causal chain graphic organizer to indicate how far back in the causal chain the staff has traced the problem. Put a question mark above the box where the cause of the reduction in levels of sulfur trioxide is as yet unknown. Students write "Levels of sulfuric acid produced has decreased" in the box marked "effect," and then "Reduction in the quantity of sulfur trioxide" in the cause box immediately before the effect box. They put question marks above the boxes indicating causes of the reduction in levels of sulfur trioxide. The causal chain graphic organizer looks like this at this stage:

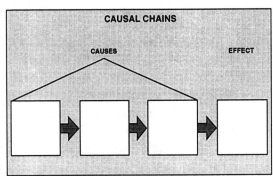

- Now please break into groups of four students each. Using your knowledge of the factors that affect equilibrium systems, list possible causal explanations for the decrease in production of sulfur trioxide. After a few minutes, ask the groups to share their possible causes. Write these on the chalkboard. Limit the responses to one a group

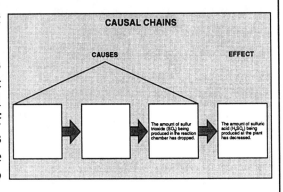

until every group has contributed to the list, then open the floor for any others that might not be on the list. ANSWERS INCLUDE: *The concentration of SO_2 and O_2 available for the reaction has decreased; the concentration of SO_3 remaining in the reaction chamber has increased; the temperature of the substances in the reaction chamber has increased; the pressure in the reaction chamber has decreased; the catalyst used is ineffective.* As groups report, ask how the possible cause they mention might have brought about a reduction in levels of sulfur trioxide. (For example, drawing out students on how an increase in the temperature or a decrease in the pressure in the reaction chamber might lead to lower levels of SO_3 being produced should bring out that these changes would favor the reverse reaction by shifting the equilibrium to the left in the equation. This is also why an increased concentration of SO_3 left in the chamber would inhibit the production of additional SO_3.)

• **I am now going to assign each group a different possible cause from your list to investigate. Use this causal explanation graphic organizer to develop a plan for searching for evidence that would support the idea that the possible cause you are working on is the real cause. Do this first by writing evidence you might find that would show that your possibility was likely in the boxes marked "Possible Evidence" and then develop a search plan for obtaining the evidence. Remember, you can get information from people who already have it as well as by direct research yourself. If you are not sure how you can obtain some of your needed information, circle it on your graphic organizer.** FOR EACH OF THE POSSIBLE CASES LISTED ABOVE TYPICAL STUDENT RESPONSES

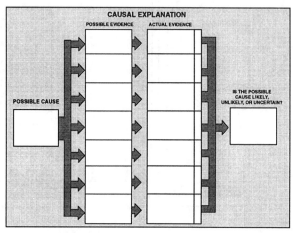

OFTEN INCLUDE: The concentration of SO_2 and O_2 available for the reaction has decreased: *cracks are detected in the input line, impurities are detected in the sulfur used to produce the SO_2; there has been less overall sulfur use at the plant during the period of decrease in levels of SO_3 while the production process hasn't changed; the standard instruments used to measure, monitor, and regulate input quantities of SO_2 and O_2 are old and outdated.* The concentration of SO_3 remaining in the reaction chamber has increased: *deposits of solid impurities have been noted on the walls of the output pipe, the pressure gauge on the output pipe shows higher than usual readings during the period of decrease in quantities of SO_3, the output regulation valve has stuck on occasion, the pump used to sustain the flow of SO_3 at a steady rate is old and its yearly maintenance was not conducted during the last two maintenance periods, the chamber reaction supervisor is new and inexperienced.* The catalyst used is ineffective: *some metallic screens marked V_2O_2 have had to be returned to the manufacturer because inspection revealed impurities in the V_2O_2, the log indicates that the V_2O_2 in the reaction chamber is overdue for replacement, cracks have been detected in the reaction chamber itself.*

• **Each group should now report briefly on their research plan and should ask the class for reactions to it and for help with any of your circled information. Then, regroup and, based on the reaction of the class, revise your research plan.** Students often help each other by identifying specific sources of information needed (e.g., a possible log book, purchasing records, etc.).

• **Now, we will look for real evidence, guided by your plans, and write that evidence in the boxes beside the relevant piece of possible evidence. If the evidence you actually find counts in favor of the possible cause you are investigating, put a plus in the column next to it; if it counts against, put a minus. Use this report on the equilibrium reaction issued by the quality control staff as well as any other information you've been given about the plant to initially gather evidence. When you are able to say whether your possibility is ruled out by the evidence, or is strongly supported, a member of your group should raise a hand.** When all groups indicate that they are ready to report, ask groups who have ruled out their possibilities to report first. Each time there is a report, ask the class if it concurs in the judgment. If so, strike the possible cause from the list on the board. If not, have the class discuss it and then vote about whether it should be eliminated. When the pruning of the list is completed, ask groups to report on those possibilities they think are supported by the evidence. Ask the class whether it concurs. If two or more possibilities remain that are judged equally likely, ask the class to regroup, have each group select one of these possibilities, and go through the process of gathering evidence again. When the class finally agrees on one of the possibilities as a likely cause, circle it on the board.

FACT-FINDING REPORT OF THE QUALITY CONTROL STAFF REGARDING THE EQUIPMENT AND MATERIALS USED TO PRODUCE SULFURIC ACID AT THE THREE-FORKS PLANT

1. Equipment
The reaction chamber is made of cast-iron and lined with acid-proof bricks. It was constructed 30 years ago. It is inspected every six months. The most recent inspection of three months ago found no changes in the chamber since its last inspection. No cracks, corrosion, or separation were found in the cast-iron, in the brick lining, or in the joints linking the chamber to the input and output tubes. Some deterioration of the brick pointing was found but was judged to be of minor significance.

The input and output temperature and pressure gauges were also the original equipment installed when the reaction chamber was constructed. Newer gauges are available, but the original gauges have not been replaced. The same is true of the input and output valves that automatically regulate the flow of gases into and out of the chamber according to the temperature and pressure. Every three months, the gauges and the valves are routinely inspected and the moving parts are serviced with lubricating oil. Service records indicate that the input-regulating equipment had its regular three-month inspection and service three months ago. This equipment was found to be in normal working order. Log records indicate that the output temperature and pressure gauges were found to be similarly in good working order and that they had also been routinely serviced at that time. No records indicate that the output-regulating valve was also serviced three months ago. One log record during the past three months indicates that the output valve was stuck in the closed position and had to be manually opened.

Both input and output pumps that bring the gases into the chamber and sustain the flow of the sulfur out of the chamber are also original equipment. Clear records indicate that the input pump has been routinely serviced up to and including its last inspection period and that it has been certified as in good working order. No records exist of an inspection or service of the output pump that occurred three months ago. The last service of that pump was 18 months ago.

2. Materials
The sulfur used to produce sulfur dioxide for the reaction is obtained from the Midland Chemical Company in Millville, Ohio. It is shipped by rail to the plant in sealed drums. It is quality-tested and certified by Midland before shipping, and it is routinely spot-checked at the plant for impurities. Midland sulfur has been used at the plant since it was constructed and remains in use today.

Receiving logs indicate that all drums received from Midland during the past two years have had Midland's certification sticker on them. Inspection logs at the plant indicate that the purity of the sulfur has remained constant during that period, except for a slight increase in arsenical pyrites but an increase well within the toleration limits for that contaminant.

The oxygen used is obtained from the air, and the water from deep wells on the plant property. No changes in chemical composition of the air used (e.g., due to pollutants) have been recorded in the routine inspections conducted every six months during the past two years. The water used in the reaction has come from the same source over the past two years. Inspection logs indicate a decrease in the trace calcium dissolved in the water three months ago in comparison to two years ago.

The vanadium catalyst has also been obtained from Midland. It was routinely replaced six years ago. Inspection indicates that the catalyst is routinely cleaned of dust deposits every three months and that the present catalyst was cleaned two months ago. The quantity of the vanadium catalyst remaining after the cleaning was 10% less than had been removed (probably due to normal loss during cleaning). The records indicate that the new catalyst was added at the time and that a mismeasurement increased the amount of the catalyst in the chamber by 5% in comparison to the amount removed before cleaning. Inspection of the residue from the latest cleaning also revealed trace amounts or arsenic in the contaminants that had not been noted before.

Logs also indicate that the same quantities of sulfur, oxygen, and water were used in the reaction during the past two years.

3. Personnel
The plant has had no personnel changes in the past three years. Both chamber reaction supervisors have been employed at least ten years at the plant, and their service records are excellent. The chamber reaction output supervisor is the employee who has been on the job the longest — 23 years.

4. Reaction Records
Besides indicating a decrease in the amount of sulfuric acid produced at the plant in the past two years, the reaction records entered every day indicate the following:

Normal average temperature and pressure readings on the input gauges (40-60 degrees C).
Increasingly higher than normal readings on the output temperature and pressure gauges (sometimes as high, for example, as 575 degrees C).
More than the usual deposits of arsenic on the inside of the output tube.

Respectfully Submitted   *Ralph Saunders*   Date *15 March 1997*

Ralph Saunders
For the Quality Control Team

- **Conditions like a decrease in the levels of sulfur trioxide cause results like a decrease in the levels of production of sulfuric acid at the plant, but you have found that such conditions also are caused by other factors that have changed in the process, such as an increase in the concentrations of sulfur trioxide in the chamber. But what has caused these other factors to change? Go back to your graphic organizers in your groups and see if any causes of the change we have identified as the likely cause of the decrease in levels of sulfur trioxide are indicated. If so, make note of these.** After a few minutes, ask for responses from the groups. ANSWERS INCLUDE: *Buildup of residue on the inside of the output pipe seems likely to have decreased the volume of SO_3 leaving the chamber, thereby increasing the pressure in the chamber, and hence the concentration of SO_3 in the chamber. This shifted the equilibrium to the left (in the equation), yielding less than expected amounts of SO_3 being produced. The defective pressure valve on the chamber contributed to our not detecting this problem. The residue buildup was probably caused by defective regulating equipment causing residue, which is normally cleansed out of the process at a later stage, to deposit on the walls of the output pipe.*

- **Now, complete your causal chain graphic indicating why the level of sulfur trioxide has been falling.** The remaining boxes should be completed with the possible cause that was judged likely, and what is judged to be the root cause of that situation.

• Circle the box on the causal chain graphic indicating which cause in the causal chain you think is the most reasonable to try to change in order to bring the levels of sulfur trioxide back up to acceptable levels. Write out a tentative plan for making the necessary changes. ANSWERS VARY.

THINKING ABOUT THINKING

• **Map out the strategy you used in deciding which cause(s) were best supported.** Ask the students to list/diagram what steps they went through in trying to determine the cause(s) of the depletion in levels of sulfuric acid produced at the plant. ANSWERS SHOULD INCLUDE: Generating possible causes, thinking about possible evidence, gathering and assessing actual evidence, and judging the cause(s). Display the verbal map and discuss whether this is a way of formulating their questions.

• **How was the way you thought about this issue different from the way you usually think about causes?** Ask students to map the way they usually think about causes and compare this with the thinking map just produced.

• **Do you think that this is a valuable way to think about what caused something? Why or why not? If not, how could it have been done better?** ANSWERS VARY. Many students say that this keeps them from making hasty judgments about causes, from acting just on a hunch, and from accepting what other people say about causes uncritically. They also say that it keeps their minds on the evidence that supports their ideas about causes. Some suggest that an alternative strategy might be to look at the evidence we have first, before thinking about possible causes, and then ask what possibilities the evidence suggests. Then, we can avoid a long brainstorming session for possible causes. Other students respond, however, that the brainstorming is valuable because we sometimes think of unusual possibilities that the initial evidence might not suggest, and these possibilities later turn out to be viable ones.

• **How did thinking about the possible causes of depletion of the levels of sulfuric acid and the evidence you would need to decide which was the best explanation help you to read the report on the equilibrium chamber?** ANSWERS TYPICALLY INCLUDE: *It made me look for specific pieces of evidence in the report. I formulated questions I needed to answer to make a judgment about my possible cause, and I looked through the report to try to find answers to these questions.*

• **Did thinking about what caused the problem help you understand what you have already studied about sulfur, sulfuric acid, and chemical equilibrium? How?** TYPICAL ANSWERS INCLUDE: *I understood the reactions better. I now have a much clearer idea about how chemical equilibrium systems work. I understand much more clearly how temperature and pressure can affect equilibrium reactions and how small changes can shift the equilibrium so that the desired balance does not occur. I now also have a much deeper understanding of the conditions under which sulfuric acid occurs and why this substance is not so prevalent in natural forms today on Earth. Before this lesson, I memorized the formulas; now, I understand them.*

APPLYING THINKING

Immediate Transfer

• **Select another important chemical process with which you are familiar. Plan how you would make a judgment about what might be causing this process to break down based on the strategy you mapped out for determining causes.** Ask students to work in pairs and report.

- While your physics instructor is demonstrating an experiment in which the decibel level of sound is being measured, the machine reporting the decibels of sound gives a number of incorrect readings. Use the strategy for causal explanation to develop a plan for finding out why.

- Reconsider the examples you described about causal judgments you made that were mistaken. Rethink one of them using the same strategy.

Reinforcement Later

Later in the school year when studying the following topics, ask the following:

- Write a scenario about chemical pollution in which the cause of the pollution is a mystery. Add details that will give clues about the causes, and then write a fictitious report about the pollution site in which these clues are hidden. I will ask you to bring these to class and give them to other students to figure out using the graphic organizers and thinking maps for causal explanation.

- Make a judgment about the cause(s) of acid rain in the northeastern states in the United States. Defend your judgment.

- Explain why some aspect of the operation of the school that you think could be improved is not operating efficiently. Suggest some possible remedies based on your causal diagnosis.

WRITING EXTENSION

Ask each student to write a report defending their choice of a causal explanation for the depletion of levels of sulfuric acid being produced at the plant. Have them include in the report the major competing explanations, why they were ruled out, and why the explanation chosen is the best explanation.

REINFORCEMENT OF OTHER THINKING SKILLS

Reinforce the skill of determining the reliability of sources of information. Beside each comment about actual evidence on the graphic organizer, students should indicate the source(s) they have for the information. Ask them to indicate a primary source by marking it with a (P) and a secondary source by marking it with an (S). Then ask them to explain why they think the source is reliable.

Reinforce the process of problem solving. Ask students to find the best way to solve the problem.

ASSESSING STUDENT THINKING ABOUT CAUSES

To assess this skill, ask students to write an essay to answer any of the application questions or any others like them that you develop. Ask them to make their thinking explicit. Determine whether they are attending to each of the steps in the verbal map for causal explanation.

FACT-FINDING REPORT OF THE QUALITY CONTROL STAFF REGARDING THE EQUIPMENT AND MATERIALS USED TO PRODUCE SULFURIC ACID AT THE THREE-FORKS PLANT

1. Equipment

The reaction chamber is made of cast-iron and lined with acid-proof bricks. It was constructed 30 years ago. It is inspected every six months. The most recent inspection of three months ago found no changes in the chamber since its last inspection. No cracks, corrosion, or separation were found in the cast-iron, in the brick lining, or in the joints linking the chamber to the input and output tubes. Some deterioration of the brick pointing was found but was judged to be of minor significance.

The input and output temperature and pressure gauges were also the original equipment installed when the reaction chamber was constructed. Newer gauges are available, but the original gauges have not been replaced. The same is true of the input and output valves that automatically regulate the flow of gases into and out of the chamber according to the temperature and pressure. Every three months, the gauges and the valves are routinely inspected and the moving parts are serviced with lubricating oil. Service records indicate that the input-regulating equipment had its regular three-month inspection and service three months ago. This equipment was found to be in normal working order. Log records indicate that the output temperature and pressure gauges were found to be similarly in good working order and that they had also been routinely serviced at that time. No records indicate that the output-regulating valve was also serviced three months ago. One log record during the past three months indicates that the output valve was stuck in the closed position and had to be manually opened.

Both input and output pumps that bring the gases into the chamber and sustain the flow of the sulfur trioxide out of the chamber are also original equipment. Clear records indicate that the input pump has been routinely serviced up to and including its last inspection period and that it has been certified as in good working order. No records exist of an inspection or service of the output pump that occurred three months ago. The last service of that pump was 18 months ago.

2. Materials

The sulfur used to produce sulfur dioxide for the reaction is obtained from the Midland Chemical Company in Millville, Ohio. It is shipped by rail to the plant in sealed drums. It is quality-tested and certified by Midland before shipping, and it is routinely spot-checked at the plant for impurities. Midland sulfur has been used at the plant since it was constructed and remains in use today.

Receiving logs indicate that all drums received from Midland during the past two years have had Midland's certification sticker on them. Inspection logs at the plant indicate that the purity of the sulfur has remained constant during that period, except for a slight increase in arsenical pyrites but an increase well within the toleration limits for that contaminant.

The oxygen used is obtained from the air, and the water from deep wells on the plant property. No changes in chemical composition of the air used (e.g., due to pollutants) have been recorded in the routine inspections conducted every six months during the past two years. The water used in the reaction has come from the same source over the past two years. Inspection logs indicate a decrease in the trace calcium dissolved in the water three months ago in comparison to two years ago.

The vanadium catalyst has also been obtained from Midland. It was routinely replaced six years ago. Inspection indicates that the catalyst is routinely cleaned of dust deposits every three months and that the present catalyst was cleaned two months ago. The quantity of the vanadium catalyst remaining after the cleaning was 10% less than had been removed (probably due to normal loss during cleaning). The records indicate that the new catalyst was added at the time and that a mismeasurement increased the amount of the catalyst in the chamber by 5% in comparison to the amount removed before cleaning. Inspection of the residue from the latest cleaning also revealed trace amounts or arsenic in the contaminants that had not been noted before.

Logs also indicate that the same quantities of sulfur, oxygen, and water were used in the reaction during the past two years.

3. Personnel
The plant has had no personnel changes in the past three years. Both chamber reaction supervisors have been employed at least ten years at the plant, and their service records are excellent. The chamber reaction output supervisor is the employee who has been on the job the longest—23 years.

4. Reaction Records
Besides indicating a decrease in the amount of sulfuric acid produced at the plant in the past two years, the reaction records entered every day indicate the following:

> Normal average temperature and pressure readings on the input gauges (40-60 degrees C).
> Increasingly higher than normal readings on the output temperature and pressure gauges (sometimes as high, for example, as 575 degrees C).
> More than the usual deposits of arsenic on the inside of the output tube.

Respectfully Submitted   *Ralph Saunders*   Date *15 March 1997*

Ralph Saunders
For the Quality Control Team

Sample Student Responses • Sulfuric Acid

CAUSAL EXPLANATION

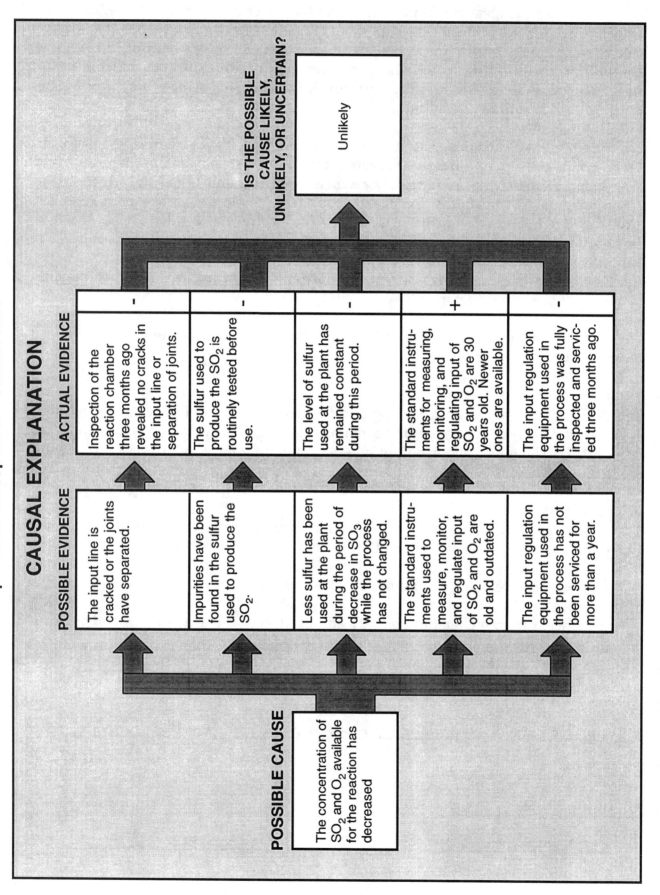

POSSIBLE CAUSE

The concentration of SO_2 and O_2 available for the reaction has decreased

POSSIBLE EVIDENCE

The input line is cracked or the joints have separated.

Impurities have been found in the sulfur used to produce the SO_2.

Less sulfur has been used at the plant during the period of decrease in SO_3 while the process has not changed.

The standard instruments used to measure, monitor, and regulate input of SO_2 and O_2 are old and outdated.

The input regulation equipment used in the process has not been serviced for more than a year.

ACTUAL EVIDENCE

Inspection of the reaction chamber three months ago revealed no cracks in the input line or separation of joints.

The sulfur used to produce the SO_2 is routinely tested before use.

The level of sulfur used at the plant has remained constant during this period.

The standard instruments for measuring, monitoring, and regulating input of SO_2 and O_2 are 30 years old. Newer ones are available.

The input regulation equipment used in the process was fully inspected and serviced three months ago.

IS THE POSSIBLE CAUSE LIKELY, UNLIKELY, OR UNCERTAIN?

Unlikely

Sample Student Responses • Sulfuric Acid

CAUSAL EXPLANATION

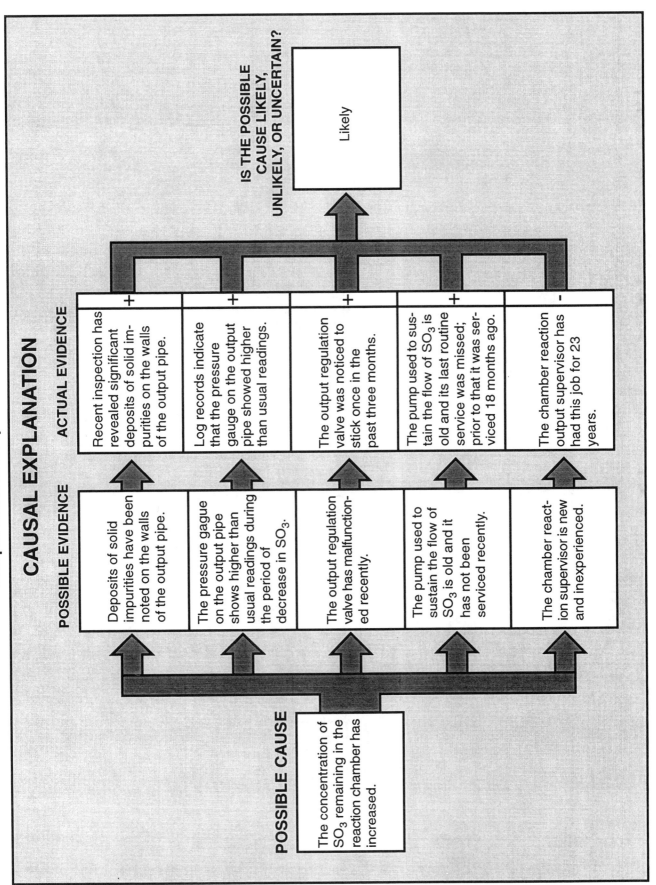

POSSIBLE CAUSE

The concentration of SO_3 remaining in the reaction chamber has increased.

POSSIBLE EVIDENCE

- Deposits of solid impurities have been noted on the walls of the output pipe.
- The pressure gague on the output pipe shows higher than usual readings during the period of decrease in SO_3.
- The output regulation valve has malfunctioned recently.
- The pump used to sustain the flow of SO_3 is old and it has not been serviced recently.
- The chamber reaction supervisor is new and inexperienced.

ACTUAL EVIDENCE

- (+) Recent inspection has revealed significant deposits of solid impurities on the walls of the output pipe.
- (+) Log records indicate that the pressure gauge on the output pipe showed higher than usual readings.
- (+) The output regulation valve was noticed to stick once in the past three months.
- (+) The pump used to sustain the flow of SO_3 is old and its last routine service was missed; prior to that it was serviced 18 months ago.
- (−) The chamber reaction output supervisor has had this job for 23 years.

IS THE POSSIBLE CAUSE LIKELY, UNLIKELY, OR UNCERTAIN?

Likely

Sample Student Responses • Sulfuric Acid • Causal Chain

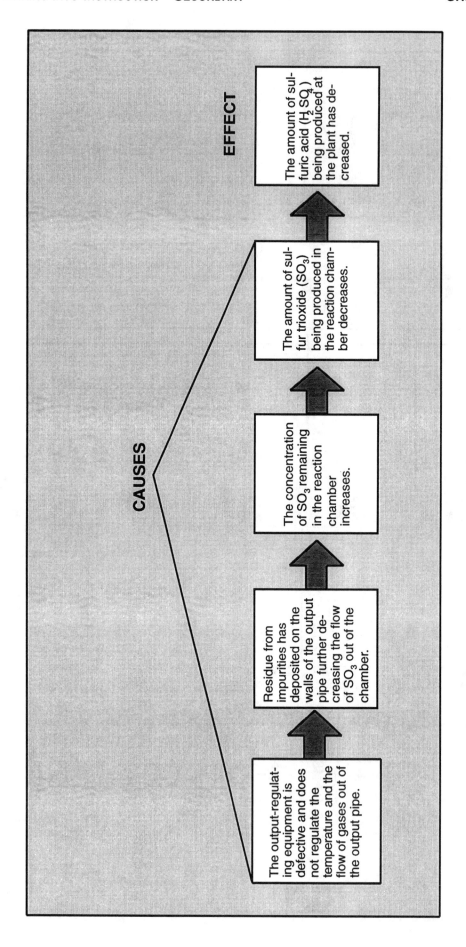

EFFECT

The amount of sulfuric acid (H_2SO_4) being produced at the plant has decreased.

The amount of sulfur trioxide (SO_3) being produced in the reaction chamber decreases.

CAUSES

The concentration of SO_3 remaining in the reaction chamber increases.

Residue from impurities has deposited on the walls of the output pipe further decreasing the flow of SO_3 out of the chamber.

The output-regulating equipment is defective and does not regulate the temperature and the flow of gases out of the output pipe.

Two Additional Diagrams That Have Been Used to Illustrate the Industrial Production of Sulfuric Acid.

These diagrams vary in the degree of complexity. They have been used in this lesson with more advanced students when the diagram in the lesson is below their level.

DIAGRAM 1

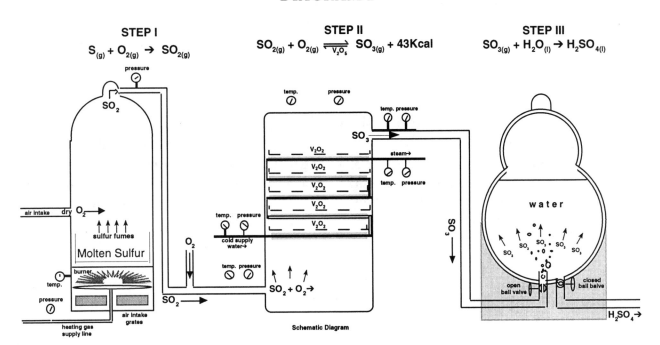

DIAGRAM 2

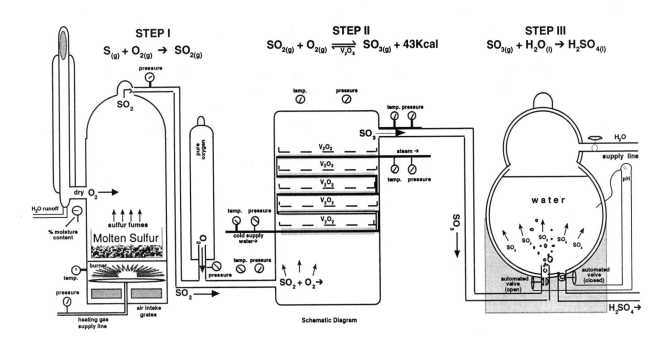

THE MYSTERY OF SILENT SPRING

Biology/Environmental Science   **Grades 9–12**

OBJECTIVES

CONTENT	THINKING SKILLS/PROCESS
Students identify effects of pollutants on the environment and the contributions of Rachel Carson during the formative years of the environmental movement.	Students will learn to develop alternative hypotheses and consider present evidence when trying to make a judgment about what caused something to happen.

METHODS AND MATERIALS

CONTENT	THINKING SKILLS/PROCESS
The lesson requires a copy of Chapter One of *Silent Spring* by Rachel Carson, and selections from an article, "Saving Mothers, Semmelweis and Childbed Fever" from *A History of Medicine*, Sutcliffe & Duin.	Structured questioning that follows the thinking map for the skill and a specialized graphic organizer are used to guide students through the thinking. Collaborative learning enhances their thinking.

LESSON

INTRODUCTION TO CONTENT AND THINKING SKILL/PROCESS

- When things happen that we don't understand or don't like, we often try to find out what is causing them. For example, when my car doesn't start on cold mornings, I want to find out why so that I can do something to prevent this from happening again. Or, when my tomato plants produce really large tomatoes, I want to find out if anything I did caused this (like how long I waited to plant the seedlings in my gardening, or my watering cycle) so that I can do it again. Can you think of times when you wanted to find out what caused something? Write down one or two in your notebooks. After a few minutes, ask for three or four examples from the students.

- Finding a cause is often a simple process; turn on a light and the monsters of your imagination disappear. But sometimes, when things are a little more complicated or unfamiliar, we think we know the cause and are really jumping to a conclusion without any evidence. Are any of your examples ones in which you thought you knew the cause but were mistaken? If not, can you think of any such additional examples? Write one or two in your notebook. Ask for two or three examples from the class after a few minutes. Students usually can produce a number of such examples. If they are having difficulty, tell them about a situation in which you thought you knew the cause of something but were mistaken.

- Let's think about these examples now. Can you think of anything you might have considered or thought about before you decided on the cause that might have led to a better sense of what really caused the effect you were concerned about—that might have avoided your mistake? How might you have done some more careful critical thinking in these cases? Write one or two ideas down in your notebooks. After a few minutes, ask for some ideas from the students. Write them on a transparency or on the chalkboard. STUDENT RESPONSES VARY BUT ARE TYPICALLY LIKE THE FOLLOWING: *Instead of jumping to a conclusion, I could have stopped and thought about what the cause might be and then eliminated the possibilities that didn't work. I should have made sure I had good reasons for thinking that what I thought was the cause was the right cause. I should have looked for evidence first.*

- **Here's a way to put your ideas together into a series of important questions to ask and answer well before you decide what caused something.** Draw this thinking map of skillful causal explanation on a transparency or on the chalkboard. **Notice that I have put in two questions about evidence: one about what evidence I might find that would show that the cause I am considering is the right cause, and one about what actual evidence I have found. That's how I sometimes make sure I have good reasons for thinking that a specific possibility is the right cause. I think about what I would need first and then look for it. Then, when one cause stands out and I've eliminated the rest, I feel pretty confident that I've** got the right cause. That's what I do when my car doesn't start, for example. Finding out the best explanation for what caused something is called "causal explanation."

> **SKILLFUL CAUSAL EXPLANATION**
>
> 1. What are possible causes of the event in question?
>
> 2. What could you find that would count for or against the likelihood of these possibilities?
>
> 3. What evidence do you already have, or have you gathered, that is relevant to determining what caused the event?
>
> 4. Which possibility is rendered most likely based on the evidence?

- **Finding the cause of complex or unfamiliar phenomena in science also can often require ingenious critical thinking. In medicine, for example, in order to find a cure for a disease, the first step is to isolate its cause. We're going to look at a classic example of a medical mystery and how one physician, Ignaz Semmelweis, solved it, to see if our ideas about skillful causal explanation coincide with what this physician did. The action takes place in Vienna, Austria, during the middle years of the 19th century. The place, the Vienna Hospital, is where Semmelweis worked as an assistant. The mystery to be solved is that at that hospital within a week or two of childbirth, many mothers were dying from a disease called "childbed" or "puerperal" fever. Childbed fever was particularly commonplace after hospital deliveries and was often referred to as "the terrible evil." It often produced fatal symptoms, including high temperature, pain, abscesses, peritonitis, septicaemia, delirium and heart failure. Read the passage in your source material handout titled** *Saving Mothers*, **which describes the conditions and attitudes of the time, and, in your groups, reconstruct a list of the possible causes of childbed fever considered at that time.** List student responses on the chalkboard or on a transparency under the heading "Possible Causes of Childbed Fever." Have each group contribute one or two responses. STUDENT RESPONSES: *It was caused by a "miasma"; it was caused by an infectious vapor in the atmosphere; it was caused by "putrid particles"; it was transmitted from cadavers by the dirty hands of interns; it was something carried by only women; it was "The curse of Eve"; it was caused by poverty; it was caused by medical school students.*

- **Semmelweis was also aware of these possibilities. But he focused on only one. Why did he rule out the others?** TYPICAL STUDENT RESPONSES: *He thought about all the possibilities but had evidence for only one.* Sometimes, they add that *some of the possible causes weren't logical.* Have students reread the selection and then give examples of the kind of thinking that might have led Semmelweis to disregard possible causes. STUDENT RESPONSES: *If only females got the disease, then why didn't females outside the hospital get the disease also. If the disease was found only in the hospital, then why didn't females in either hospital ward have an equal chance of surviving childbirth? If females got the disease, why didn't rich and poor females have an equal chance of getting the disease? If it was an infectious vapor, why wasn't it seen, smelled and equally present in all the hospital wards?* Students usually realize that Semmelweis discarded possible causes because he had evidence against them and didn't observe any evidence to support them.

- **Why did Semmelweis accept the idea that "putrid particles" were the cause of childbed fever?** FREQUENT STUDENT RESPONSES: *He discovered evidence pointing in that direction. Because he found no evidence supporting the other possible causes, he was able to eliminate them while at the same time, he thought he had good evidence to support the theory of putrid particles.*

- **Let's look carefully at how Semmelweis might have gone about finding evidence to support his hypothesis. Keep in mind that before you can look for evidence, you need to know what you are looking for. Galileo knew that if his theory were true, he should** *expect* **that when he dropped two objects from the leaning tower of Pisa, they would reach the ground simultaneously. Gregor Mendel similarly knew that if his theory were true, he should expect that a cross between two heterozygotes would produce a homozygote. In your groups, make a list of what Semmelweis might have expected to find as evidence in support of his hypothesis that putrid particles on the hands of interns were the cause of childbed fever.** TYPICAL STUDENT RESPONSES ABOUT WHAT SEMMELWEIS SHOULD HAVE EXPECTED: *that childbirths done at home were safer than those deliveries performed at the hospital; that mothers whose deliveries were performed in the ward by doctors who did autopsies were much more likely to get childbed fever; that deliveries performed by midwives at the hospital were much safer than deliveries performed by interns; that interns performing autopsies themselves might get the symptoms of childbed fever.*

- **How was Semmelweis able to prove his hypothesis?** STUDENT RESPONSES: *He looked for evidence that he expected to find; he compared mortality rates inside and outside the hospital; he compared the mortality rates for each of the wards; he compared mortality rates between intern and midwife deliveries; he looked for evidence that interns themselves were victims of childbed fever.* Tell students an epilogue that Semmelweis's discovery that putrid particles (the bacteria Mycoplasma haminas) in fact did cause puerperal fever and was transmitted on the dirty hands of interns who went directly from the autopsy room to the delivery ward was not immediately accepted by the medical community. In fact, it was not until 1879, some 30 years later, that Semmelweis was vindicated when the great French scientist Louis Pasteur, at a meeting at the Academy of Medicine in Paris, loudly defended Semmelweis's explanation that bacteria were the cause of most postoperative infection.

- **Does the thinking plan that Semmelweis seems to have followed look like the plan you developed for skillful causal explanation?** Students usually agree that the two are alike.

- **Bacteria, like mycoplasma haminas, are not the only causes of human suffering. Sometimes, we are our own worst enemy. We're now going to use causal explanation to solve another scientific "whatdunit,"** *The Mystery of Silent Spring.*

THINKING ACTIVELY

- **Rachel Carson was an aquatic biologist with the United States Fish and Wildlife Service. She had changed her college major from English to Biology after a biology professor rekindled her childhood fascination with nature. Her book** *Silent Spring,* **published in 1962, became hugely successful and equally as controversial. Moreover, Carson's book was to influence a president (John F. Kennedy), anger an industry, and act as a catalyst for worldwide change. Today, you are going to read a selection from** *Silent Spring* **and figure out why.**

- **Follow along with me as I read to you part of the first chapter of the book, titled "A Fable for Tomorrow."** Read the selection out loud.

From 'A Fable for Tomorrow'

There was once a town in the heart of America where all life seemed to live in harmony with its surroundings. The town lay in the midst of a checkerboard of prosperous farms, with fields of grain and hillsides of orchards where, in spring, white clouds of bloom drifted above the green fields. In autumn, oak and maple and birch set up a blaze of color that flamed and flickered across a backdrop of pines. Then foxes barked in the hills and deer silently crossed the fields, half hidden in the mist of the fall mornings.

Along the roads, laurel, viburnum and alder, great ferns and wildflowers delighted the traveler's eye through much of the year. Even in winter the roadsides were places of beauty, where countless birds came to feed on the berries and on the seed heads of the dried weeds rising above the snow. The countryside was, in fact, famous for the abundance and variety of its bird life, and when the flood of migrants was pouring through in spring and fall people traveled from great distances to observe them. Others came to fish the streams, which flowed clear and cold out of the hills and contained shady pools where trout lay. So it had been from the days many years ago when the first settlers raised their houses, sank their wells, and built their barns.

Then a strange blight crept over the area and everything began to change. Some evil spell had settled on the community: mysterious maladies swept the flocks of chickens; the cattle and sheep sickened and died. Everywhere was a shadow of death. The farmers spoke of much illness among their families. In the town the doctors had become more and more puzzled by new kinds of sickness appearing among their patients. There had been several sudden and unexplained deaths, not only among adults but even among children, who would be stricken suddenly while at play and die within a few hours.

There was a strange stillness. The birds, for example—where had they gone? Many people spoke of them, puzzled and disturbed. The feeding stations in the backyards were deserted. The few birds seen anywhere were moribund; they trembled violently and could not fly. It was a spring without voices. On the mornings that had once throbbed with the dawn chorus of robins, catbirds, doves, jays, wrens, and scores of other bird voices there was now no sound; only silence lay over the fields and woods and marsh.

On the farms the hens brooded, but no chicks hatched. The farmers complained that they were unable to raise any pigs—the litters were small and the young survived only a few days. The apple trees were coming into bloom but no bees droned among the blossoms, so there was no pollination and there would be no fruit.

The roadsides, once so attractive, were now lined with browned and withered vegetation as though swept by fire. These, too, were silent, deserted by all living things. Even the streams were now lifeless. Anglers no longer visited them, for all the fish had died.

• **Something bad happened here. Let's work toward solving the mystery. We are going to begin our investigation by carefully thinking about what we've read by asking two questions about each paragraph: what is it about and what is the point?** Meet in your groups, read the first para-

This paragraph is about	....and the point is
1	
2	
3	
4	
5	
6	

About-Point Content Reading Guide

graph, and then write a group statement on each line of this graphic organizer for each **paragraph in the selection.** Allow enough time for the groups to read and discuss the first paragraph and then have them report back. Write their responses on an About-Point diagram on the chalkboard and use them to create a class about-point statement for each paragraph. Students usually write that the first paragraph *is about a beautiful town in America;*

and the point is to make us want to keep it that way. They readily recognize that the second paragraph is *about the countryside and how natural and untouched it was;* they often write that the point is *to make us hope it continues to flourish.* The partially completed diagram with these statements should look like this:

This paragraph is about	....and the point is
1 a beautiful town in America	to make us want it to stay that way
2 how the countryside flourished for years	to make us hope it continues to flourish
3	
4	
5	
6	

About-Point Content Reading Guide

• **Now that we have carefully read this quote from the first chapter of Rachel Carson's book, what question immediately comes to your mind?** Students usually say that they want to know what happened to this town. **Why does this question come up immediately?** Typically, students say that the town was once prosperous and there has obviously been a change in its fortunes. **What kind of thinking is important to do well if you are to answer this question satisfactorily?** Causal explanation. **If we go back to the thinking map we used in thinking about Semmelweis and his attempt to explain causes, what should we avoid and what should we think about first in trying to uncover the causes of this change?** Typically, students answer that we should avoid making a hasty judgment about the cause without evidence; rather, we should think about possible causes and then try to get evidence that will show which of these possibilities is the most likely explanation. **In your groups, read the passage again and make a list of possible causes for the strange and mysterious events described in *Silent Spring*.** Allow ample time for groups to brainstorm a minimum of four possibilities. Then, have each group contribute one or two responses and list these on the board or on a transparency. TYPICAL STUDENT RESPONSES: *A nuclear accident spread radiation; the town was damaged by biological warfare; the town was poisoned by chemical warfare; the town was poisoned by industrial pollution; aliens attacked and made animals and people sick; the hole in the ozone layer got big and leaked too much radiation; a new virus infected all living things; evil spirits put a curse on the town; black magic or Satan cursed the town; polluted drinking water caused the problems; illness was caused by bacteria brought to the town by veterans from Desert Storm.*

• **As you have said, once we have a list of possible causes, we have to determine the best explanation from our list. How can we do that in an efficient way?** Most students respond quickly that they can follow the verbal map for skillful causal explanation again and consider, first, what evidence they might find that would point to one or another of the possible causes as likely. If necessary, remind them that, just as Semmelweis might have explored the possible causes for puerperal fever by evaluating evidence, so might they consider evidence that counts for or against their list of possible causes for Carson's *Silent Spring*.

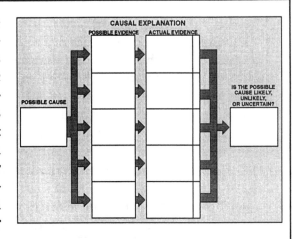

- Let's try this out by taking a closer look at one of the possible causes, "Radiation from a nuclear disaster." In order to help us organize our thoughts about evidence, we are going to use a graphic organizer that highlights evidence. We will determine whether radiation from a nuclear disaster is a likely, unlikely or uncertain cause of the silent spring, based on the evidence, by working through the graphic together. Display a transparency or create a graphic organizer for skillful causal explanation on the chalkboard. Write "Radiation from a nuclear disaster" in the "Possible Cause" box (refer to diagram on the right).

- Work in your groups again and think about the kind of evidence you would like to see that would point to radiation as being a likely cause of the silent spring. If radiation were the actual cause, what would you expect to find if you visited the town, for example? After four or five minutes, ask the groups to report by mentioning specific pieces of evidence they might find. Draw out these responses by asking extending questions like, "Why would that be evidence in favor of radiation?" Enter the possible evidence mentioned in the "Possible Evidence" column of the graphic organizer on your transparency or on the chalkboard. STUDENT RESPONSES INCLUDE BUT ARE NOT LIMITED TO: *Symptoms of radiation sickness in people and animals can be observed. Traces of radiation would be found in the town. A power plant, missile base, or submarine base is located nearby. A truck, train or airplane carrying nuclear waste crashed, derailed, or exploded in or near the town. A nuclear bomb exploded near the town.*

- Let us now use your list of possible evidence to research for actual evidence. If what we find includes what you have listed, what would that show about radiation as the basic cause of the devastating change described in *Silent Spring?* If some of this evidence is found, it would increase the likelihood that radiation was the cause. If a lot of what we have listed is found, it would make radiation likely. **Where might you find such evidence to support radiation as a likely cause of the tragedy?** Students usually respond immediately, "In the reading material." **Let's first reread the selection from "A small town in America" and see if we can find any actual evidence cited that matches the possible evidence we listed on the graphic organizer. We will start with "Symptoms of radiation sickness in people and animals." Read each paragraph carefully and see if you can discover any clues that would count for or against this possible piece of evidence. As you write this evidence in the actual evidence box next to "Symptoms of radiation sickness..." put a plus (+) next to it if it counts in favor of radiation as a cause and a minus (-) if it counts against.** STUDENT RESPONSES: *There is mention of "mysterious maladies" (+),"new kinds of sickness" (+), "sudden and unexplained deaths" (+), dead insects (+), lifeless streams (+), chicks that don't hatch (+), young pigs that died right after birth (+), sick sheep (+), "brown and withered vegetation" (+).*

- Now that we have evaluated one piece of evidence and have decided on whether it counts in favor of or against the possible cause, it is time to work in groups and complete the "actual evidence" boxes for each of the remaining pieces of "possible evidence." Assign or have each group select different evidence to search for and evaluate. Have them decide the nature of the actual evidence as they find it in the text and whether it counts for or against the possible cause. Allow enough time for each group to consider the evidence carefully and complete their assignment. Then have each group report back to the class. Record their responses on the class graphic organizer for "Radiation from a nuclear disaster."

- **The evidence we are looking at is limited to Rachel Carson's prose in the reading selection. Has she planted enough evidence in her story so that we can accept or reject radiation as an explanation for the afflictions that have fallen on this town? Is the possible cause likely, unlikely, or uncertain?** Although some evidence does support radiation as a possible cause, students usually reason that it is not compelling enough to convince them that the cause is "likely." They often reason that just because people, animals, and plants are sick, it doesn't necessarily follow that radiation caused the sickness. Students recognize that many other causes of sickness and disease are available— for example, chemical pollution and other "possible causes" under investigation. Students also focus on the passage "a strange light crept over the area and everything began to change," and many feel this is strong evidence that a nuclear explosion detonated near the town; however, most consider the passage vague and although it counts somewhat in favor of radiation, the text contains no explicit indication that this strange light is a nuclear explosion. The most important evidence students uncover in the text counting against radiation is absence of its mention from the narrative. Students are quick to point out that nowhere in the passage are the words radiation or radioactivity used. Likewise, no reference is made to a nuclear spill, a plant meltdown, or a problem at a nuclear facility. Students, therefore, find that, based on what is in the text, radiation from a nuclear disaster is an uncertain cause of Silent Spring.

- **When you investigate a possible cause like we just did and you find inconclusive evidence in one source, what might you do to probe further in the hope of resolving the uncertainty?** Look for more evidence somewhere else. **Where might you get additional evidence?** TYPICAL STUDENT ANSWERS INCLUDE: *Look in newspapers for reports of nuclear spills, etc. Look in magazines from the period. Consult the Atomic Energy Commission and ask them if they have any records of nuclear spills, etc. Find out whether the Army tested any nuclear bombs nearby.*

- **I'm going to ask you to investigate, in the same way, the other possible causes you've listed, and to expand your investigation to include such additional sources. The period in question is the late 1950s. Each group should select another possible cause for Silent Spring and use the graphic organizer for causal explanation to record your results. Begin by making a list of possible evidence you might find that would convince you that the possible cause in question was likely. Then, reread the selection from** *Silent Spring.* **Look for actual evidence in the reading selection that matches the possible evidence you list, fill in the appropriate columns, and mark whether each item counts in favor of (+) or against (-) the suggested causal explanation. Then, determine whether the possible cause is likely, unlikely or uncertain, based on the evidence. If additional evidence is needed, use the packet of resource material that includes newspapers, magazines, and various reports to try to find other actual evidence that counts in favor or against the possible cause. Record that evidence on the graphic organizer also.** Students usually find that, based on the selection from *Silent Spring,* many of the possible causes, including biological and chemical pollutants, are uncertain causes. Other than the strikingly unlikely causes such as Satan or evil spirits, students do not count out any of the possible explanations for the silent spring. In this case—where Rachel Carson's description is of a fictitious town—it is necessary to draw up similarly fictitious news clips like those in the resources at the end of the lesson. They should be varied and present students with challenging, if not conflicting, evidence that must be culled from the material. When students expand their research to include investigating source material like that included as resources in this lesson, they usually accumulate evidence that points to the extensive use of pesticides as the most likely possible causes. (An even more ambitious way to provide such resources is to insert such articles in facsimiles of newspapers, magazines, etc. so that the students have to search for them in a way that more closely represents what they would have to do if they were engaged in an authentic causal investigation. An alternative to the latter is to use a real case—i.e., the case of Love Canal—as the context for expanding the investigation of the causes of the silent spring in this town. Then real newspapers, etc. can provide authentic source material in which to search for real evidence to solve the mystery.) Whatever method is used, summarize the results of each group's evaluation by using the chalkboard or a transparency to list each possible cause and whether it was determined to be likely, unlikely or uncertain. Students nearly always have difficulty choosing one possible cause. They agree that a number of different causes could combine and account for the silent spring.

• **Now that you've made judgments about the likely causes of the silent spring in this American town, what are some possible things we might do to prevent such tragedies from happening again?** Brainstorm possible remedies with the students and list these on the chalkboard or on a transparency. STUDENT ANSWERS VARY BUT MAY INCLUDE: *Pass laws against the use of the pesticides used in this town. Educate people so that they don't abuse the environment. Try to develop pesticides that don't have this effect.* These responses can be used later to initiate a problem-solving lesson. Conclude this part of the lesson by discussing with the students the implications of a work such as *Silent Spring*, and discuss whether Rachel Carson's style in writing this book was legitimate for scientists, noting how such books can have an impact on public opinion and public policy in this country. Ask students whether Rachel Carson's love of nature and passion for protecting the environment has had an impact on their attitudes towards the environment.

THINKING ABOUT THINKING

• **After such complex thinking, it's always a good idea to stop to think about your thinking, especially when you are trying to follow a plan for your thinking. Let's review what plan we followed and how well we were able to do it.** Prompt your students to specify what plan they were following. When they indicate that it was the plan for causal explanation they developed earlier, write "Causal explanation" on a transparency or on the chalkboard. Most students indicate that it didn't seem difficult at all to follow the plan for causal explanation.

• **What was the sequence of questions you followed?** Students identify the questions on the thinking map and indicate that they started trying to come up with possible causes, thought about evidence they would need to show that these hypotheses were good ones, and then looked for this evidence, recording what they found and determining how likely the hypotheses were, based on the evidence.

• **Was it easy to follow these questions? If so, why?** ANSWERS INCLUDE: *You led us through the plan by asking us these questions. We used a graphic organizer that you asked us to work on from left to right and the graphic organizer incorporated the questions we put on the thinking map for causal explanation.*

• **In this a good way to determine what caused something?** Students often say that it is a good way to do this because you look for evidence that shows whether the possibilities are likely rather than just guess. They say that if there is good evidence they can be pretty sure the possibility that is supported by the evidence is the right one.

• **How could you be sure to follow this plan in the future?** TYPICAL STUDENT ANSWERS INCLUDE: *I could write the plan down and use it when I had to figure out what caused something. I could try to remember the questions from the thinking map. I could write down the graphic organizer when I had to find out what caused something and use it to guide me. I could work with someone else so that we could check each other.*

- It's a good idea for you to put this plan for causal explanation in your own words so that you develop your own plan for thinking through a complex problem like the mystery of the silent spring. Please do this in your notebooks.

APPLYING THINKING

Immediate Transfer

Shortly after the main activity in this lesson is completed and the students have developed a plan for causal explanation, ask them to work on the following causal explanation issues:

- During the past 25 years, a number of animals have appeared on the list of endangered species that the government keeps. Pick one of these animals and determine what caused it to become endangered using your plan for causal explanation.

- Bacteria have been living on the earth for close to four billion years. What is the best causal explanation for their survival?

- Causal explanation is used not only in science, but also in history. Pick one of the following historical events that you have studied and determine the best causal explanation of the event: the Industrial Revolution, the exploration of space, the production and use of metal tools, the "age of exploration."

Reinforcement Later

Later in the school year, ask the students to work on the following causal explanation issues:

- What are the main causes of acid rain? Explain why you think your explanation is the best explanation.

- Select a major character in one of the novels you are reading in English, and select an important turning point in the novel. Explain why the events that constitute this turning point occurred. Explain what evidence you have for your explanation.

- Select one of the following and explain why it occurred: an upswing in your performance in school, a downswing in your performance at school, your steady performance at school.

WRITING EXTENSION

Write a report on your findings about the silent spring. Explain which possibilities are unlikely and which are likely, citing your evidence in support of these judgments.

CONTENT EXTENSION

Differentiate the major types of insecticides used during the 1950s and determine their effects on living things in the environment in which they were used. Are any still in use today?

ASSESSING STUDENT THINKING ABOUT CAUSES

To assess this skill ask students to write an essay to answer any of the application questions or any others like them that you develop. Ask them to make their thinking explicit. Determine whether they are attending to each of the steps in the verbal map for causal explanation.

THE MYSTERY OF SILENT SPRING

I. SOURCE MATERIAL ON SEMMELWEIS AND CHILDBED FEVER

Saving Mothers

Childbirth today is usually a time of great joy; once labour is finished and the mother at last holds her new baby, there is little worry about her health. However, in the past, many mothers died within a week or two of giving birth from a disease called 'childbed,' or puerperal, fever: a high temperature, pain in the lower abdomen, swelling of the pelvic tissues, abscesses, peritonitis, septicaemia, delirium, and heart failure.

Childbed fever was particularly common in women who delivered in hospitals—and in the 19th century, these were almost always the poor, since the more well-to-do invariably gave birth at home.

One who refused to see the disease as the 'curse of Eve' was a Hungarian doctor, Ignaz Semmelweis. In 1847, as a 28-year-old assistant at the Vienna Lying-in Hospital, he noticed a striking difference between the incidence of deaths from childbed fever in the hospital's two obstetric wards: in Ward 1, 9.9 percent of newly delivered mothers died (and sometimes as many as 29.3 per cent); in Ward 2, it was 3.9 percent. Ward 1 became notorious, and patients would beg tearfully not to be placed in it.

The doctors in charge of the hospital, like others elsewhere in Europe, were certain that childbed fever was caused by a 'miasma'—an infectious vapour—present in the atmosphere. Semmelweis could not accept this, for surely the miasma would affect both wards equally. Other suggested causes were unsatisfactory.

Semmelweis found that Ward 2 was run almost exclusively by midwives, who were careful about cleanliness and that Ward 1 was the province of medical students, who entered directly from the dissecting rooms without washing their hands, wearing the same blood- and tissue-splattered coats in which they had performed autopsies. When one of Semmelweis's friends died of blood poisoning from a cut received while dissecting, having developed many of the same symptoms as the victims of Ward 1, he was sure of the cause of both. Something—Semmelweis called it 'putrid particles'—was being transferred from the cadavers of the dissecting rooms into a wound: the small cut on his friend's hand or the large 'wound' left behind after the placenta separates from the uterus following childbirth.

The young doctor must have been elated at his discovery, but he was also filled with guilt. He, too, had dissected many bodies, primarily to find out why so many of these poor women were dying. "Consequently, must I here make my confession that God only knows the numbers of women whom I have consigned prematurely to the grave."

Semmelweis acted quickly. Much to their surprise and anger, he insisted that the students wash their hands with a solution of lime chloride before entering Ward 1. The results were conclusive: Within one year, the mortality rate in the ward fell to just over 3 percent, and the following year, it further declined to 1.27 percent.

II. SOURCE MATERIAL ON RADIATION

Medical consequences of a Nuclear Conflict for the Human Population

From *Ecological and Demographic Consequences of a Nuclear War*, Yuri M. Svirezhev. ISBN 3-05-500193-1

"A great amount of toxic substances would enter the atmosphere as a result of nuclear explosions and the fires accompanying them. Calculations show that a tenfold increase in atmospheric contamination products can bring about a catastrophic growth of allergic diseases and diseases of the nervous system and organs of sense and respiration. Practically every person would become chronically ill."

In the wake of a nuclear exchange, "30 percent of the land surface in the middle latitudes" would receive a lethal dose of radiation. Peoples in areas that survived a lethal dose would still develop tumors, eye cataracts, gastro-intestinal disorders, sterility, and commonplace birth defects.

Radiation–As a Pollutant

From Environmental Science, Jane L. Person, ISBN 0-920008-41-0

"The nucleus [of an atom] contains small particles called neutrons and protons. The neutrons and protons are held together by large amounts of energy. However, the nuclei of some atoms are not stable. Energy and/or particles are released when these nuclei break down. This is called radiation.

"Radiation surrounds us. It is a part of our natural environment. Alpha, Beta, or Gamma radiation is ionizing radiation. This high-energy radiation can knock electrons from atoms. Some ionizing radiation enters Earth's atmosphere from outer space. We call this cosmic radiation. Solar radiation is mostly nonionizing, low-energy light waves.

"Radium is a naturally radioactive element. This means that radium is found in the earth's crust and its nucleus is so unstable that it breaks down without any outside force to start the process. The nucleus of a radium atom breaks down or 'decays' into a radon atom. Radiation is given off.

"Radiation is given off by trace amounts of uranium-238 ... the level of naturally occurring radiation is referred to as the background level.

"Man-made radiation sources include all of those activities in which man has created radioactive elements. Nuclear weapons testing resulted in areas where radioactivity levels were higher than natural levels.

"Manufacturers use radioactive substances to cure paint and coat nonstick fry pans. Other radioactive materials are used by water departments and oil companies to detect leaks.

Biological Effects of Radiation

From Environmental Science, Jane L. Person, ISBN 0-920008-41-0

"How does this radiation affect the human body? Most of our knowledge of the biological effects of radiation is the result of many laboratory experiments involving large numbers of plants and animals, but not humans. Humans who have been exposed to radiation because of medical treatments, nuclear accidents, or nuclear weapons have also been studied.

"The energy of radiation may cause chemical changes within the cell. The new chemicals may alter the structure of the cell or interfere with the cell's normal chemical reactions. The cell may no longer be able to carry out its normal functions. If these functions are vital to the life of the organism, death will result.

Clinical effects of doses below 1,000 millirem are not measurable by current technology. In general, we can say that the higher the dose, the more damaging will be the effects, and the quicker they will appear. The chart below shows the effects of specific doses of whole-body radiation given within a short period of time.

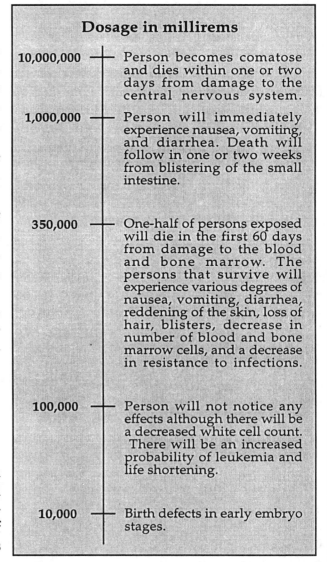

Dosage in millirems	
10,000,000	Person becomes comatose and dies within one or two days from damage to the central nervous system.
1,000,000	Person will immediately experience nausea, vomiting, and diarrhea. Death will follow in one or two weeks from blistering of the small intestine.
350,000	One-half of persons exposed will die in the first 60 days from damage to the blood and bone marrow. The persons that survive will experience various degrees of nausea, vomiting, diarrhea, reddening of the skin, loss of hair, blisters, decrease in number of blood and bone marrow cells, and a decrease in resistance to infections.
100,000	Person will not notice any effects although there will be a decreased white cell count. There will be an increased probability of leukemia and life shortening.
10,000	Birth defects in early embryo stages.

Radiation Effects on Mammals and Birds

From *Ecological and Demographic Consequences of a Nuclear War*, Yuri M. Svirezhev. ISBN 3-05-500193-1

Irradiation doses lethal for some mammalian and bird species. $LD_{50/30}$ denotes a dose that brings about the death of 50% of irradiated population in the course of 30 days.

Small rodents	300
Goat	350
Goldfinch	600
House sparrow	625
Pigeon	920

Incident at Windscale

Early nuclear reactor designs used graphite (carbon) as a structural material for the reactor core —the area where controlled nuclear fission produces heat. Although initially felt to be safe, graphite cores in some reactors caught fire and burned out of control. One of these was the Windscale Plutonium Reactor in England. There, in 1957, graphite materials used to manufacture its core became exposed to high heat and air. The resulting fire released radioactive contamination over a hundred-square-mile area.

Three Mile Island–Five Years Later

From Environmental Science, Jane L. Person, ISBN 0-920008-41-0

Harrisburg, September (1985)—The state Health Department said in a report today that it has found no evidence of increased cancer among area residents due to the 1979 accident at Three Mile Island but that long-range monitoring will be continued."

More than five years after the accident, the department of health conducted a study to determine if the number of new cancer cases and the number of deaths due to cancer was larger than expected. Why had they waited this long? Cancer caused by radiation usually takes a long time to develop. It may take 10 to 20 years before cancer, caused by radiation, is diagnosed. Even leukemia, a cancer which has a relatively short period of development, usually isn't detected until five years after exposure to radiation.

As of 1985, no evidence suggests that the accident at TMI has caused an increase in the cancer death rate. The question, 'Did the accident at Three Mile Island cause an increase in cancer?' can be answered only thus: "not yet."

Chernobyl Disaster Killed Many, But Threatens Even More

Nuclear disasters such as the explosion at Chernobyl, Russia, release lethal amounts of radioactivity into the atmosphere. It is estimated that those very close to the accident received more than 300,000 millirems of radioactivity, as evidenced by the high fatality rate in the town proper. Chernobylians began showing up for treatment within hours of the tragedy, exhibiting cysts, blisters, bleeding sores and a multiplicity of gastrointestinal acute disorders including bloody vomiting, severe cramps, extreme nausea, and horrible pain. Years after the accident, the high incidence of leukemia and cancer are constant reminders of the importance of nuclear power plant safety.

III. SOURCE MATERIAL ON INSECTICIDES

Insecticides

Two types of insecticides were used during the 1950's to control the spread of insects, the chlorinated hydrocarbons such as DDT and organic phosphates such as malathion and parathion. Both categories are described in *Silent Spring* by Rachel Carson.

Chlorinated hydrocarbons: Chlorinated hydrocarbons are known to affect the liver, an organ which cleanses the body of toxins and poisons. The importance of a healthy liver was described in *Silent Spring*:

"Without a normally functioning liver the body would be disarmed –defenseless against the great variety of poisons that continually invade it Some of these are normal by-products of metabolism, which the liver swiftly and efficiently makes harmless by withdrawing their nitrogen. But poisons that have no normal place in the body may also be detoxified ... In similar ways the liver deals with the majority of the toxic materials to which we are exposed. ... A liver damaged by pesticides is not only incapable of protecting us from poisons, the whole wide range of its activities may be interfered with ...

"In connection with the nearly universal use of insecticides that are liver poisons, it is interesting to note the sharp rise in hepatitis that began during the 1950's and is continuing a fluctuating climb. Cirrhosis also is said to be increasing."

Chlorinated hydrocarbons also affect the nervous system:

"This has been made clear by an infinite number of experiments on animals and by observations on human subjects as well. As for DDT, the first of the new organic insecticides to be widely used, its action is primarily on the central nervous system of man; the cerebellum and the higher motor cortex are thought to be the areas chiefly affected. Abnormal sensations of prickling, burning, or itching, as well as tremors or even convulsions may follow exposure to appreciable amounts, according to a standard textbook of toxicology."

Insecticides such as DDT can also be absorbed through the skin. In the following selection Rachel Carson describes what happens when researchers allowed their skin to be exposed to DDT:

"Our first knowledge of the symptoms of acute poisoning by DDT was furnished by several British investigators, who deliberately exposed themselves in order to learn the consequences. Two scientists at the British Royal Navy Physiological Laboratory invited absorption of DDT through the skin by direct contact with walls covered with a water-soluble paint containing 2 percent DDT, overlaid with a thin film of oil. The direct effect on the nervous system is apparent in their eloquent description of their symptoms: "The tiredness, heaviness, and aching of limbs were very real things, and the mental state was also most distressing ... [there was] extreme irritability ... great distaste for work of any sort ... a feeling of mental incompetence in tackling the simplest mental task. The joint pains were quite violent at times.

"Another British experimenter who applied DDT in acetone solution to his skin reported heaviness and aching of limbs, muscular weakness, and 'spasms of extreme nervous tension.'

Organic Phosphates: The other class of insecticides used during the 1950s, organic phosphates, *are* also dangerous because they destroy enzymes that are necessary for the proper functioning of the nervous system. Chemicals called neurotransmitters are required to transfer nervous impulses from neuron to neuron. For example, in order to move your finger, a signal must pass from your brain to the muscles of your hand. In order to do so, the "signal" must transfer from the end of one nervous cell (neuron) to the beginning of the next and ultimately the signal must pass from the end of the last neuron into muscle tissue. In order to accomplish this neurotransmitters are necessary. However, when the nervous impulse is completed, the neurotransmitter must be destroyed; otherwise, impulses would continue to fire and your finger would move without your even willing it to. An enzyme, acetylcholinesterase, is used to breakdown acetylcholine after it has done its job. Acetylcholine is the neurotransmitter that passes impulses between the brain and the muscular system. If the acetylcholine is not destroyed after a muscle moves, then "tremors, muscular spasms, convulsion, and death quickly result." Rachel Carson explains what happens when the nervous system comes in contact with organic phosphates:

"This contingency has been provided for by the body. A protective enzyme called cholinesterase is at hand to destroy the transmitting chemical once it is no longer needed. By this means a precise balance is struck and the body never builds up a dangerous amount of acetylcholine. But on contact with the organic phosphorus insecticides, the protective enzyme is destroyed, and as the quantity of the enzyme is reduced that of the transmitting chemical builds up. In this effect, the organic phosphorus compounds resemble the alkaloid poison muscarine found in a poisonous mushroom ...

"Repeated exposures may lower the cholinesterase level until an individual reaches the brink of acute poisoning, a brink over which he may be pushed by a very small additional exposure. ... A chemist, thinking to learn by the most direct possible means the dose acutely toxic to human beings, swallowed a minute amount, equivalent to about .00424 ounce. Paralysis followed so instantaneously that he could not reach the antidotes he had prepared at hand, and so he died. Parathion is now said to be a favorite instrument of suicide in Finland."

During 1982, the Pesticide Subcommittee of the Committee on Environmental Improvement sponsored "a symposium addressing environmental and health issues concerning pesticides." The symposium was organized to address environmental issues which were raised twenty years earlier by Rachel Carson in *Silent Spring*. The symposium was able to concur with many of Rachel Carson's findings. For example they were also able to report that symptoms of organic phosphate insecticide poisoning "include numbness, muscular weakness, tremors, convulsions, paralysis, depression and a sense of mental incompetence."

Both chlorinated hydrocarbons and organic phosphates have been shown to interfere with enzymes necessary for the production of ATP from ADP. ATP is a molecule necessary for energy production and transport in living cells. Without it, or without enough of it, the molecular machinery of the cell winds down. Moreover, infertility may result from a lack of ATP in the fertilized egg.

Cancer and congenital deformities may also develop. DDT and several other insecticides were shown in 1960 to cause tumors in laboratory animals.

High levels of DDT have been found in bird and mammal eggs. In *Silent Spring Revisited* the authors describe avian pesticide research done at the Patuxent Wildlife Research Center.:

"Patuxent scientists were aware of the reproductive problems faced by large predatory birds and were reasonably certain that DDT was involved. Ospreys and bald eagles were severely contaminated by pesticides and had massive reproduction problems. Experiments were begun with the new breeding colony of kestrels (small falcons) in an attempt to elucidate the problem. Before the results of these experiments were available, a British scientist reported that eggshell weight in European birds of prey had decreased since the introduction of modern pesticides. The work was soon repeated in the United States by a group at the University of Wisconsin. Thus, shell thinning was shown by correlative data to be a major mechanism of pesticide impact on wildlife populations.

Shell thinning was soon demonstrated in captive kestrels. The birds, fed a combination of dieldrin and DDT, had reduced shell thickness and the attendant breakage and disappearance of eggs seen earlier in free-living raptors (birds of prey)."

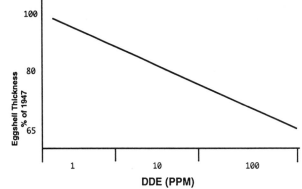

Graph depicting the relationship of DDE (a breakdown product of DDT) and eggshell thickness in brown pelicans from *Return to Silent Spring.*

Ground water contamination: Pesticides used before 1970 posed little threat to drinking water because they were not soluble in water. However, more recent pesticides are water-soluble and easily absorbed into the ground. Modern insecticides are formulated this way in order to minimize their accumulation on the surface where animals and plants can readily ingest and absorb them. In other words by making insecticides less dangerous to wildlife, they have become a greater threat to our potable water supply. However, according to the authors of *Silent Spring Revisited*, the threat is not too alarming:

"Groundwater contamination from pesticide use does not necessarily mean that drinking water supplies will be affected unless the ground water is the supply source. Privately owned shallow wells in unconfined aquifers are probably at greater risk than large municipal wells in deeper, confined aquifers.

"[However, ground water] contamination by pesticides involves some risk (pesticides by definition create this risk) and probably will have to be evaluated carefully in view of this risk potential. Pesticides account for 20% of the total agricultural chemical market, and some 3.7×10^5 metric tons of chemicals involving approximately 600 active ingredients are currently applied to agricultural land."

- Effects on Fish:

Pesticide Concentration	Percentage of Fertile Eggs	Percentage of Normal Larvae	Percentage of Normal Juveniles
0.32	86.9	13.8	6.9
0.18	97.7	23.6	16.7
0.10	99.2	100.0	99.0
0.05	96.9	100.0	90.2
Control	95.4	100.0	100.0

"Problems caused by pesticides," *Environmental Science*

Effects of Pesticide on Early Life Stages of Fathead Minnows

In her textbook *Environmental Science*, Jane L. Person discusses "superbugs" and bio accumulation, two additional problems caused by pesticides:

"When the synthetic pesticides first appeared, doctors had visions of no more diseases spread by pests and farmers imagined a day when no more pests would harm their crops. Unfortunately, neither vision came to pass ...

"In 1962 the book *Silent Spring* was published. Rachel Carson wrote the book so that the public would be aware of the problems associated with the use of pesticides. The book caused quite an uproar. Many people credit the publication of *Silent Spring* as the event that stimulated an evaluation of the risks of using pesticides. Today we are much more aware of some of the 'problems' caused by pesticides.

"Superbugs: People expected the pesticides to eliminate all pests. That didn't happen. Even before it was approved for public use in 1947, some house flies survived when sprayed with DDT. Why were some flies surviving when others died? Changes occur in the DNA of some flies that enable them to produce enzymes which detoxify or make the pesticide harmless. These genetic changes make the insect immune or resistant to the pesticide.

"Each time a pesticide is used, more insects become resistant to it. These resistant insects breed and produce offspring that are resistant to the pesticide ... Most pests develop resistance to a chemical within five years, but sometimes it takes only one season.

"Today more than 400 species of insects and mites are resistant to pesticides. This number has doubled since 1965.

"Bioaccumulation: Only very small amounts of the pesticides that are chlorinated hydrocarbons will dissolve in water, but they easily dissolve in fat. This was first thought to be a major advantage because the pesticides would not be easily washed away by rain.

"When pesticides like DDT are carried into lakes and streams, much of the chemicals deposited in the soil at the bottom of the water or absorbed by organisms living in the water. Cell membranes are made out of protein and fat. The chlorinated hydrocarbons are easily absorbed through the cell membranes of the aquatic organisms.

"Most organisms do not have the enzymes needed to break down the pesticides ... The pesticides begin to accumulate in the organisms. As the pesticide moves up the food chain, it becomes more and more concentrated. This is called bioaccumlation."

In the late '50s a study of water pollution at California's Clear lake indicated that fish-eating birds contained 50 times the concentration of pesticides as did the fish they fed on. What happened as a result is described in *Environmental Science*:

"The populations of some species of fish-eating birds, such as ospreys and eagles, began to decline. In Connecticut one colony of ospreys declined from 200 breeding pairs in 1939 to 12 pairs in 1965. Scientists frequently found broken eggs and eggs that did not hatch. The egg yolks in the unhatched eggs contained high levels of DDT."

Pesticides can also cause human illness or poisoning. Pesticide poisoning has affected large numbers of people throughout the world, as summarized in *Environmental Science*:

"In 1972 the World Health Organization estimated that 500,000 people were victims of pesticide poisoning, and approximately 5,000 people died. During 1977 more than 1500 cases of pesticide illness were reported in the state of California. [California uses more pesticides than any other state.] Almost half of these illnesses involved farm workers, most of the workers were pickers. No one knows how many cases of pesticide poisoning are not reported. Some feel that the number of unreported cases is much higher than the number of reported cases ...

"Most agricultural pesticides are applied by airplanes. A study of aerial spray pilots (crop dusters) conducted by the FAA (Federal Aviation Agency) showed most of the pilots had mild to moderate symptoms of pesticide poisoning. Another study showed that spray plane pilots had a high number of auto accidents on their way home from work.

"Some workers have died from pesticide poisoning. Others suffer severe nervous system disorders ... At least two pesticides (endrin and parathion) have killed dozens of people and made hundreds of others ill when flour was improperly stored near leaking containers of pesticides."

IV. SOURCE MATERIAL ON AIR POLLUTION

The Effects of Air Pollution

On people: "On December 4, 1952, a large mass of cold air began to spread across Britain. Late in the evening and early the next morning, fog began to form. At first it was just another London Fog. But when the city awoke, tons of smoke from millions of coal stoves poured into the cold, foggy air. Power plants added more smoke as people began to use more electricity. As people went to work, their cars and buses added more pollutants to the cold, morning fog. Throughout the day, the coal furnaces of the factories and industries added smoke.

The fog became contaminated by the smoke. This mixture of smoke and fog later became known as smog. It was difficult to see across the street. For several days, it was necessary to use the car's headlights to drive. Even street lights were needed.

Within twelve hours, the first people began to die. By the time the winds came and the smog was carried out to sea, many people were ill. The death rate climbed during the next few weeks, resulting in some four thousand deaths.

Donora is a small town on the Monongahela River in Pennsylvania. It lies in a valley surrounded by hills and had a steel mill, a plant that made sulfuric acid, and a zinc-reducing plant. The major source of energy for these industries was coal. In October, 1948, after three days of smog-filled air, many people complained of throat irritation, hoarseness, coughs, shortness of breath, and nausea. One third of the 14,000 living in Donora became ill and seventeen people died. The normal daily death rate is two."

On animals: "During the 1952 London disaster, a cattle show was in progress. Three hundred fifty cattle were being exhibited. Five of these died, 52 became seriously ill, and 9 had to be destroyed. The sheep and swine showed no effects. The horses in London showed no effects. Cases of bronchitis and pneumonia increased among some animals at the zoo."

On plants: "Plants filter the air. They frequently show the effects of pollution before animals do. ... When landscape designers select plants for areas with heavy traffic patterns, they frequently choose the ginkgo tree. The ginkgo is more resistant to pollution than many other species.

Sulfur dioxide enters the pores in the leaf along with carbon dioxide. As the cells die, damage can be seen as dried, whitened areas on the leaf. Even if no damage is seen, sulfur dioxide interferes with growth and decreases the yield of some plants.

Nitrogen dioxide and ozone, both present in photochemical smog, cause leaf damage and retard the growth of some plants."

Death of the Mountain

In 1897, the New Jersey Zinc Company was formed by the joining of several smaller companies. A site was needed for a new smelting plant. At the smelting plant, the ore is heated to remove the minerals. The smelter must be located near the ore supply and be near a good source of fuel.

A site was chosen along the Lehigh River in Pennsylvania. It was named Palmerton for Stephen S. Palmer, the president of the company. ... The Blue Mountain, just south of Palmerton, is no longer the beautiful landscape that Stephen Palmer knew. It is covered with the sun-bleached skeletons of dead trees. Nearly 1,200 acres of the mountain are severely eroded and support little or no plant growth ...

For many years, the mountain and the town have been exposed to air pollution and industrial waste from the zinc plant. Zinc oxide, cadmium, lead and other air pollutants have been deposited in the soil. Laboratory studies show that the high level of zinc in the soil prevents the germination of seeds by inhibiting root growth.

A recent study determined the level of heavy metals in white-tailed deer shot by hunters in the area near the smelters. ... Concentrations of cadmium, lead and zinc were higher in deer killed near the smelters. Older deer had higher levels of cadmium in the kidneys that the younger deer."

Disaster in Bhopal

"It came on the evening wind that drifted through the shantytowns. ... The lucky ones, alerted by the suffocating odor, escaped. Thousands did not. Some perished in their sleep. Others awoke, dizzy and nauseated, their eyes on fires and their lungs filling with fluid until they could no longer breathe, dying from exposure to a chemical few had heard of in perhaps history's worst industrial accident."

The site of the world's largest industrial accident was the Union Carbide Pesticide Plant in Bhopal, India. On December 3, 1984, water accidentally entered a tank containing methyl isocyanate (MIC), a chemical used in making pesticides. The chemical reaction caused the pressure in the underground storage tank to increase. The safety system failed, and the very toxic MIC gas filled the air.

Some of the MIC broke down into hydrogen cyanide, a poison used in the Nazi gas chambers in World War II. Unfortunately, no one realized this, and a cyanide antidote was not used. Government figures show that 2,500 people died and another 200,000 people were affected. People are still dying because of lung or heart problems caused by the gas. Some complain of chest pains, breathlessness, and pains in their muscles. Those with lung damage cannot get enough oxygen for hard physical labor.

V. NEWSPAPER ARTICLES COLLECTED IN THE AFTERMATH OF SILENT SPRING

Outbreak of hepatitis puzzles physicians

John Flynn has been practicing medicine in this area for more than 30 years. He has treated his share of mumps, measles, and chicken pox and occasionally he diagnoses hepatitis, a disease of the liver. But Dr. Flynn is puzzled. During the past few months he has diagnosed hepatitis in 17 people, four in the same family. When he arrived at the Smith farm last week for a house call, he was puzzled to find John Smith, his wife, and his two children exhibiting hepatitis symptoms. Dr. Flynn has contacted the Center for Disease Control in Atlanta, which is sending an epidemiologist.

County suicide rate explodes

The county had seven suicides last month. "That is more than we have seen in seven years," said county coroner Lou Phips. "Maybe it's just a coincidence, but something's definitely wrong here."

Local artist losing his 'touch'

Renowned landscape artist Paul Jones, best known for his stylized paintings of corn and wheat fields, was taken suddenly ill last week. "My right arm went numb, I started shaking, I dropped my brush ... I barely had enough energy to reach the barn and call for help," he related as he convalesced at the hospital. Doctors are not sure why his hand is numb or why he starts shaking. They suspect a neurological problem and have sent for a neurologist.

Town clerk quits job, loses pension

Stephanie Collins, just weeks before official retirement, has suddenly quit her job. "I just don't want to work anymore ... I feel like I'm jumping out of my skin. I am making too many stupid mistakes," she explained. What makes this situation particularly bizarre is the fact that Collins stands to lose half of the pension she would have gotten had she retired three weeks from now. When reminded of this, she said in a weary voice, "I just can't handle the job anymore."

Help Wanted advertisement

Help wanted: experienced crop dusters. Apply Ace Crop Dusting, County Airport.

'Hercules' described by CAD at 4H club celebration

The County Agriculture Department introduced farmers and other interested parties to a new strain of fly they have nicknamed "Hercules." Members of the 4H club proudly displayed trapped specimens of Hercules—so named because they survived this year's annual insecticide "fogging"—during their annual end-of-summer barbecue.

New woes at the county zoo

Only last month, the county zoo sadly announced the death of "Sabo," its prize gorilla, from what appeared to be "depression." He just lost interest in food and showed no enthusiasm for anyone or anything. To add to the zoo's troubles, yesterday, the zoo, long a sanctuary for the region's birds, announced its eagle population had fallen on hard times. According to ornithologist Todd Graham, "For a number of years, the variety and number of birds nesting in our sanctuary has declined. But this year, we are faced with a tragedy." It appears that only a handful of eagle chicks were born. Graham blames the problem on eagle eggshells which, he says, are lately too thin and fragile to survive.

Clean air study gives 'thumbs up'

State measurements of airborne toxic pollutants reveal the county is "within established norms." The report went on to praise local townships for effectively controlling the amount of soot and other airborne products of combustion released into the atmosphere.

Problems at Golden Pond

John Brown suspected that something was wrong out at Golden Pond. For years, he has been happily fishing there. Recently, he can't catch so much as a minnow. Local wildlife management officials ran a test on the pond and found contamination. The identity of the pollutant has not been found yet; however, powerboats will be kept off the lake for the time being.

Sample Student Responses • The Mystery of Silent Spring

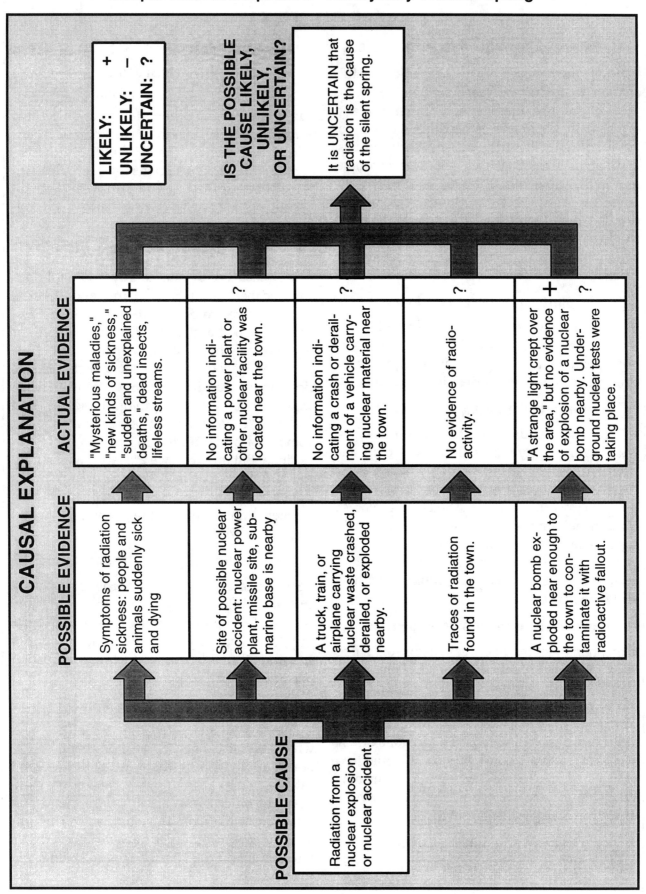

CAUSAL EXPLANATION

LIKELY: +
UNLIKELY: –
UNCERTAIN: ?

IS THE POSSIBLE CAUSE LIKELY, UNLIKELY, OR UNCERTAIN?

It is UNCERTAIN that radiation is the cause of the silent spring.

ACTUAL EVIDENCE

+ "Mysterious maladies," "new kinds of sickness," "sudden and unexplained deaths," "dead insects, lifeless streams.

? No information indicating a power plant or other nuclear facility was located near the town.

? No information indicating a crash or derailment of a vehicle carrying nuclear material near the town.

? No evidence of radioactivity.

+ ? "A strange light crept over the area," but no evidence of explosion of a nuclear bomb nearby. Underground nuclear tests were taking place.

POSSIBLE EVIDENCE

Symptoms of radiation sickness: people and animals suddenly sick and dying

Site of possible nuclear accident: nuclear power plant, missile site, submarine base is nearby

A truck, train, or airplane carrying nuclear waste crashed, derailed, or exploded nearby.

Traces of radiation found in the town.

A nuclear bomb exploded near enough to the town to contaminate it with radioactive fallout.

POSSIBLE CAUSE

Radiation from a nuclear explosion or nuclear accident.

Sample Student Responses • The Mystery of Silent Spring

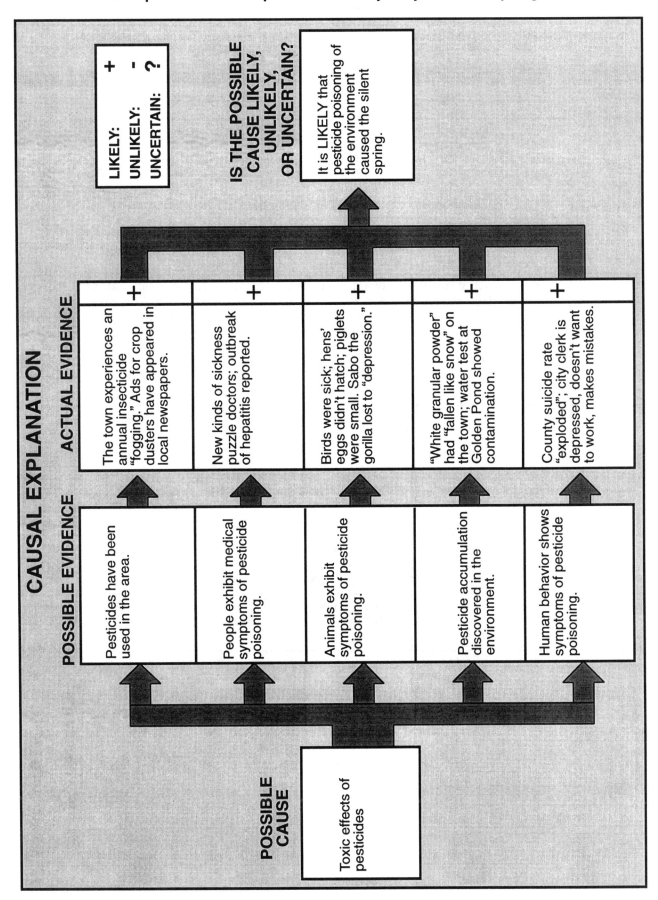

CAUSAL EXPLANATION

LIKELY: +
UNLIKELY: −
UNCERTAIN: ?

IS THE POSSIBLE CAUSE LIKELY, UNLIKELY, OR UNCERTAIN?

It is LIKELY that pesticide poisoning of the environment caused the silent spring.

ACTUAL EVIDENCE

+ The town experiences an annual insecticide "fogging." Ads for crop dusters have appeared in local newspapers.

+ New kinds of sickness puzzle doctors; outbreak of hepatitis reported.

+ Birds were sick; hens' eggs didn't hatch; piglets were small. Sabo the gorilla lost to "depression."

+ "White granular powder" had "fallen like snow" on the town; water test at Golden Pond showed contamination.

+ County suicide rate "exploded"; city clerk is depressed, doesn't want to work, makes mistakes.

POSSIBLE EVIDENCE

Pesticides have been used in the area.

People exhibit medical symptoms of pesticide poisoning.

Animals exhibit symptoms of pesticide poisoning.

Pesticide accumulation discovered in the environment.

Human behavior shows symptoms of pesticide poisoning.

POSSIBLE CAUSE

Toxic effects of pesticides

CAUSAL EXPLANATION LESSON CONTEXTS

The following examples have been suggested by classroom teachers as contexts to develop infused lessons. If a skill or process has been introduced in a previous infused lesson, these contexts may be used to reinforce it.

GRADE	SUBJECT	TOPIC	THINKING ISSUE
6–8	Science	Origin of the earth	What are explanations for the origin of the earth? How might you decide on the best explanation?
6–8	Science	Environmental science	What are possible explanations for the disappearance of species from a lake? How might you determine which explanation is the best one?
6–8	Science	Plants	What are possible explanations for the movement of plants toward light and away from the ground? How might you test to determine which possibility is most likely?
6–8	Science	Energy flow in food chains	What are the possible ecological relationships between producers and consumers in a food chain? How might you determine the best arrangement of producers and consumers in a food chain consisting of a zebra, grass, and a lion?
6–8	Science	Astronomy	What are the possible explanations for the movement of the planets and stars? How might you determine which is the best explanation?
6–8	Science	Meteorology	What are explanations for changes in the weather? What needs to be known to make the best forecast? (Consider changes in termperature, air pressure, and the amount of moisture in the atmosphere.)
6–8	Science	Human biology	What are possible explanations for people growing bigger and living longer today than they did 2000 years ago? What is the most likely reason(s)?
6–8	Science	Natural selection	What are explanations for the development of the large brain in primates and especially in humans? What evidence might you find to support the best explanation?
6–8	Science	Evolution	What are explanations for the survivability of bacteria on Earth? They have been around for 3.8 billion years and are still going strong.
9–12	Biology	Protein synthesis	What are the possible explanations for cellular production of an enzyme that does not work? (Consider mutations in DNA, mistakes in transcription of DNA, mistakes in transition of RNA, mistakes in secondary, tertiary and quaternary polypeptide structure.)

CAUSAL EXPLANATION LESSON CONTEXTS

GRADE	SUBJECT	TOPIC	THINKING ISSUE
9–12	Biology	Mendelian genetics	What are the possible genotypic explanations for the phenotype Brown Eyes/Brown Hair? How might you test for the correct genotype?
9–12	Biology	Fossilization	What are the possible explanations for the forming and preservation of fossils during millions of years? (Consider the effects of weathering, minerals, cover, etc.)
9–12	Biology	Extinction	What are possible explanations for the extinction of the dinosaurs? What evidence would you expect to find to help you determine the most likely explanation?
9–12	Biology	Human biology: immune system	Once it was determined that HIV attacked the human immune system, what were possible immune system targets for the virus? What might researchers have done to narrow the list of possibilities to T cells?
9–12	Biology	Origin of life on Earth	What are possible explanations for the emergence of life on Earth? What evidence might you find to help you reach the most likely explanation? (Consider special creation, spontaneous generation, evolution from simple forms via natural selection, and exobiology.)
9–12	Chemistry	Combined gas law	What are explanations for an increase in the pressure of a gas? How might you determine which explanation is the most likely?
9–12	Chemistry	Periodic table	What are explanations for the reactivity of different groups within the periodic table? Why do some groups bond more readily than others?
9–12	Chemistry	Properties of water	What are possible explanations for the expansion of water at 4 degrees centigrade?
9–12	Chemistry	Atomic theory	What are explanations for the composition of matter? (Consider theories of Democritus, Dalton, Thomson, Rutherford and Bohr.) How might you test which theory is best?
9–12	Physics	harmonics	What are possible explanations for the collapse of the Tacoma Narrows bridge? What evidence would you expect to find to indicate the best cause?
9–12	Physics	Nature of light	What are possible explanatons for the nature of light? (Consider wave and particle theories.) How could you demonstrate each quality?

CHAPTER 15
PREDICTION

Why is Skillful Prediction Needed?

Whenever we expect something to happen, we are making a prediction. If we like what we expect to happen, we may take action that will enable us to experience it. If we do not like it, we try to avoid it. I may expect that it will rain this afternoon. If my prediction is right, bringing an umbrella will prevent me from getting wet.

Types of Predictions

What will the weather be like next week? What will the economy be like next year? Will I keep my job? Will global warming occur? Will global warming happen in my lifetime? These are all direct questions about trends or general conditions in the future. To answer them well, we have to make predictions.

We also make predictions when we are concerned about the effects of an event that is occurring now or that we think might occur. For example, if we hear of an approaching hurricane, we may prepare for the effects of wind or water.

In other situations, we might be wondering about the results of something that might occur but hasn't yet. We might, for example, be considering specific options in making a decision and want to predict the consequences of adopting one or another. I may want to take a trip to a ski resort on a vacation. I need to consider certain things: How much will it cost? What will the food be like? Does the terrain allow the kind of skiing I enjoy? Predicting likely outcomes minimizes my chances of being disappointed and allows me to choose more carefully. In this case, skillful prediction is an important subskill of decision making.

The three types of predictions include

- predicting trends or general conditions;
- predicting the effects of a particular event;
- predicting the consequences of options.

Problems with the Way We Make Predictions

The most common problem with predicting is that sometimes we do not take time to think about what might happen in the future. We do not ask questions like, "What might it be like tomorrow?" or "What might happen as a result of what's going on today?" If we do not ask such questions, we may experience situations that we do not expect and may not like.

Even when we do consider what might happen in the future, we may develop unrealistic expectations that result in unnecessarily, costly disappointments. Often, we let our hopes or fears lead us to anticipate opportunities, rewards, or difficulties that we do not have a good reason to expect. I may go to a movie theater early because I'm worried that I won't be able to get a ticket. If I have no good reason to think the movie is popular, I may unnecessarily cut my meal short and rush to the theater to find that few other people are interested in the movie. We take a risk in making hasty predictions that are no more than guesses about what is going to happen.

We must also be aware of another more subtle problem with the way we make predictions. Sometimes, we have reasons for thinking that something will happen but do not attend to significant information that can change our predictions. I may rush to meet a friend who has said that he will join me at a local restaurant at 6 P.M. However, I may forget that my friend has never been on time in the past. Not taking account of all the available relevant information that has bearing on our predictions is another problem with the way we develop expectations about the future.

It is important to make sure that our predictions are well-founded. Skillful prediction involves thinking carefully about what we expect and how likely it is to happen.

The three main problems in prediction are summarized in figure 15.1 (on the next page).

COMMON PROBLEMS WITH THE WAY WE MAKE PREDICTIONS
1. We don't raise questions about what might happen as the result of a particular circumstance.
2. We don't consider how likely our predictions are.
3. We don't take into account all the relevant evidence in determining how likely our predictions are.

Figure 15.1

How Do We Make Skillful Predictions?

Tips on making predictions skillfully. Because predictions are always made before what is predicted happens, they are always inferences. As such, predictions require good support to be well-founded. Only then are we justified in feeling confident that our predictions are likely to happen. Basing predictions on good support is the crux of skillful prediction.

The same pattern of skillful prediction is helpful in each of the three types of prediction identified earlier. To make predictions skillfully, it is important to think first about what *might* happen. Future occurrences could be trends (like a recession next year), results we might expect of some natural occurrence (like the increased risk of skin cancer because of the thinning of the ozone layer), or consequences of some action we are considering (like an enjoyable evening at the movies). These occurrences are not yet predictions because they are tentative. But it is important to be aware of these possible outcomes in order to determine whether we should affirm them as predictions. In effect, they serve as hypotheses that we are considering about the future.

Predicted occurrences may or may not happen. People sometimes mistakenly believe that *possible* occurrences are *likely* occurrences without any real basis for doing so. What are good reasons for believing that an occurrence is likely to happen?

Generally, we have to look for available evidence in order to determine whether a prediction is likely. This search usually involves information about similar events in the past. If many trees have been blown down when hurricanes have hit coastal towns, it is likely that trees will be blown down in a coastal town in the path of an impending storm of the same magnitude.

Sometimes, we have more exact information about likelihood. If, in 80 percent of the cases in the past when conditions in the atmosphere have been as they are now, it has rained within a few hours, then we have evidence that now there is an 80 percent chance of rain. Frequencies of past occurrences provide the type of evidence many people rely on to make quantitative predictions of likelihood. Statistical evidence provides strong support for the likelihood that an event will occur.

This evidence, however, may not be enough to make the outcome likely. We should also think about evidence that might count against our predictions. If there is some important difference between the event you are considering and what you are comparing it to in the past, that discrepancy may provide evidence against the likelihood of a consequence. Although most hurricanes cause trees to fall, storms with winds under 65 miles per hour seldom uproot trees. If an approaching storm has winds under 65 miles per hour, then we have evidence against the likelihood that trees will be blown down.

Making skillful predictions about trends or general conditions, rather than specific events like a hurricane, operates the same way. Will a current economic recession continue? What will the economy be like next year? If factors are present that are frequently correlated with an economic trend continuing, like increasingly high unemployment and a decline in business productivity, then we can predict the economic consequences with some confidence.

Sometimes, frequencies are summarized in general statements. I might be planning a trip to London in September. When I find out that Septembers are rainy in London, I can plan to bring appropriate clothing. The information that Septembers are rainy in London is based on the frequency of past occurrences of rain in London in September. That frequency makes it a pretty good bet that it will rain again this September.

We must recognize, however, that predicting trends is, in general, more tenuous than predicting the results of an event like a hurricane. A broad and complex range of factors can affect things like economic trends. Gathering information about them is often difficult.

When we consider all of the relevant information that counts for and against the likelihood of something we predict, we should weigh the evidence in order to judge whether the prediction is likely, unlikely, or uncertain. If, based on the evidence, it is likely, then the prediction is well-founded and we can explain why by referring to the evidence that supports it.

If we can't get evidence about these past frequencies, we must consider another way to get good reasons for thinking that an occurrence is likely. We may get information from a reliable source who we believe has consulted the primary evidence about such situations. We judge whether a weather reporter who predicts rain is reliable by knowing of his or her "track record." How often have this person's predictions been validated? Asking about evidence of past effectiveness is another good way to decide whether a prediction that someone makes is well-founded.

A Thinking Map and Graphic Organizer for Making Skillful Predictions

How can these ideas help us to make better predictions? The thinking map in figure 15.2 contains a sequence of important questions to ask in making skillful predictions. We can also use a graphic organizer to supplement the think-

ing map and to write down information in response to these questions.

The graphic organizer in figure 15.3 illustrates the important points in skillful prediction highlighted on the thinking map.

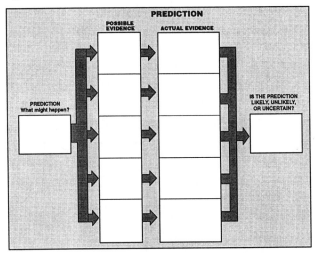

Figure 15.3

In using this graphic organizer, write what you need to know to confirm the likelihood of your prediction in the "Possible Evidence" column before writing available information in the "Actual Evidence" column. Then, when you've accumulated the available evidence in the "Actual Evidence" column, you can compare it to what you put in the "Possible Evidence" column to determine whether you have enough evidence to judge whether or not the prediction is likely.

Multiple effects. Sometimes, it isn't just one possible future occurrence that concerns us, but many. For example, if you were a public official in a town on the east coast of the United States and were concerned about the hurricane bearing down on the town, you would want to make an accurate judgment about the range of effects of the hurricane so that you could be adequately prepared. Trees blown down might be one of these effects, and you would want to make sure that everyone knew of this danger. If you expected significant coastal flooding, you would certainly plan to evacuate people from their homes in coastal regions. The flooding is a function of the tides as well as the storm, so in assessing the likelihood of such flooding, this additional information will be necessary. And, of course, depending on the strength of the

---

SKILLFUL PREDICTION

1. What might happen?

2. What evidence might you get that would indicate that this prediction is likely?

3. What evidence is available that is relevant to whether the prediction is likely?

4. Based on the evidence, is the prediction likely, unlikely, or uncertain?

Figure 15.2

---

winds, you could expect lots of debris being blown about, some potentially life-threatening. Similarly, possible power outages, pollution of the water supply, blockage of roads, and looting may also be consequences you will want to attend to if there is a likelihood of their occurrence. Taking the time to remind yourself that these things might happen is an important first step in skillful prediction.

The simple but important graphic organizer in figure 15.4 encourages a systematic approach to listing the varied potential effects.

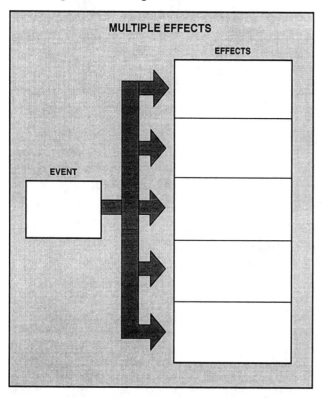

Figure 15.4

In listing predicted consequences under "Effects," we should recognize that these, too, have the status of *possible* consequences until we have sufficient evidence to judge them to be likely. Developing such a list is taking the first step in ascertaining what is likely to happen. Each consequence listed should, therefore, be subjected to a search for supporting evidence of likelihood before we affirm that it will happen.

Predicting the consequences of options skillfully. Decision making provides a specialized context in which it is important to predict a range of consequences well. We want to determine the positive and negative consequences of different courses of action we are considering—their risks and benefits. The thinking maps for prediction (figure 15.2) and for decision making (chapter 3, figure 3.2) are modified and blended to produce the thinking map in figure 15.5, which stresses predicting options' consequences.

The "Multiple Effects" diagram in figure 15.4 is modified to produce the graphic organizer in figure 15.6 for considering the likelihood of consequences predicted as the outcomes of op-

---

PREDICTING THE CONSEQUENCES OF OPTIONS

1. What consequences might result from a specific option?

2. Does each consequence

   a. count for or against the decision?

   b. rank as important?

3. How likely is the consequence?

   a. Is there evidence that counts for or against the likelihood of the consequence?

   b. Based on all the evidence, is the consequence likely, unlikely, or is its likelihood uncertain?

4. Is the decision advisable in light of the significance and likelihood of the consequences?

Figure 15.5

---

tions. This prediction strategy can supplement the decision-making strategy provided in Chapter 3.

Notice that steps (1) and (3) in the thinking map (figure 15.5) for predicting the consequences of options (and the "consequences" and "evidence" columns in the graphic organizer, figure 15.6) relate to prediction but prompt us to assess the likelihood of consequences somewhat differently from the way likelihood is assessed using the standard prediction strategy. In the standard prediction strategy, we take time to consider what evidence would show that the predicted outcome is likely before we search for actual evidence. In this form of predicting the

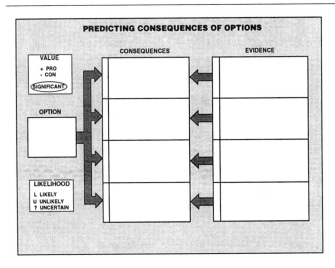

Figure 15.6

consequences of options, one dispenses with considering possible evidence that may need to be researched and attends directly to the actual evidence that is available. We can then more efficiently blend our judgments about the likelihood and significance of the consequences in order to determine whether the option is an advisable one.

How Should We Teach Students to Make Skillful Predictions?

Just asking students to make predictions will not teach them strategies for skillful prediction. For example, students are often asked to predict how a story they are reading will end. They are then asked to finish the story to see if they predicted the right ending.

To teach skillful prediction, students should, in addition, be asked to attend carefully to what is in the story in order to make a *well-founded* prediction that can be backed up with reasons. Teaching students to attend to evidence in an organized way when they make predictions is lacking in activities in which they are merely asked to tell what they think will happen.

Teaching Skillful Prediction Explicitly

Both the thinking maps and the graphic organizers are valuable tools for constructing infusion lessons on prediction. They can guide students in their thinking as they make predictions so that they get in the habit of asking the right questions and confirming that their answers support their judgments.

Following the guidelines for infusion lessons used in this handbook, such lessons should introduce students to skillful prediction, involve them in an activity in which they engage in skillful prediction, and provide them with opportunities to both reflect on how they can make skillful predictions and how they can apply this way of thinking to a variety of important predictions. Given these lessons, students will come to internalize the thinking process involved in skillful prediction.

Contexts in the Curriculum for Lessons on Skillful Prediction

There are a myriad of natural contexts in the secondary school curriculum into which instruction in skillful prediction can be infused. Predicting the weather, earthquakes, how characters in a story or novel will behave, or how historical characters will react to important events are common examples of such contexts, as are the predictions of broader outcomes in science, like global warming, and in history, like an economic depression. The curriculum contexts in which decisions are studied also provide opportunities for lessons in which predicting consequences skillfully, as an important ingredient in good decision making, can be emphasized.

Prediction, like causal explanation, is an especially important type of thinking in the natural sciences. Like causal explanation, prediction is one of the key thinking processes used in what has been called "the scientific method." One of the main ways of testing hypotheses in science is to predict observable events that would happen in certain circumstances if the hypothesis were correct and then to test whether these predictions occur. Furthermore, scientific knowledge provides support for many predictions that we make that helps us to gain benefits and avoid risks. For example, predicting, based on what we know about hurricanes, that it is likely that a hurricane heading for a coastal town will cause severe damage when it hits can provide warning for the townspeople in time enough to get out of harm's way. Similarly, predicting that there will be an eclipse of the sun at a certain time can allow us to prepare for observations of the sun that we might not otherwise be able to make. Hence, we can expect that any science curriculum will contain numerous opportunities for prediction as well as activities (e.g., lab activities) involving prediction.

In general, three basic curriculum contexts provide opportunities for prediction lessons to teach students skillful prediction while enhancing content learning in science:

- *The outcomes of events are important to us.* How will the hull of the submarine react when the water pressure increases as it goes deeper into the ocean? What can we expect to happen to the crops when the temperature falls below freezing? What can we expect to occur if the Earth is subject to global warming? How well does the body operate in space? What will happen if the rain forests are destroyed? These, and many other questions about outcomes of certain natural conditions, provide robust contexts for infusion lessons on predicting in standard science curricula in secondary school.

- *The consequences of decisions studied in the curriculum are important to us.* Predicting the likelihood of consequences of choices is the main emphasis in prediction lessons. Public policy decisions often provide such contexts. For example, fluoridation of the water supply in various communities has been debated and, in some cases, adopted, because of projected beneficial consequences for children. Ask students how they would predict the outcome of this debate in their community. The banning of DDT, the use of nuclear power, and adopting electricity as a primary form of energy for human use are examples of such contexts. Risky and heroic decisions make good contexts for prediction lessons and can accentuate qualities of character that students should become aware of. For example, Marie Curie's work on radium is an example that all students study. What could she have predicted the consequences to be?

- *Students have to make certain predictions themselves as part of their learning.* Students' lab work provides an excellent context for les-

sons on prediction, for example. In these contexts, students discover scientific principles by using a process of hypothesizing, predicting results if the hypothesis is true, testing these predictions, observing the results, and using this information to judge the viability of the hypothesis. Contexts in which students design experiments, rather than simply following procedures outlined in lab manuals, provide richer contexts for infusion lessons.

A menu of suggested contexts for infusion lessons on prediction can be found at the end of the chapter.

Model Lessons on Skillful Prediction

We include two prediction lesson in this handbook. The first is a middle-secondary school lesson on the population explosion. The second is a high school physics lesson in which predictions are made about the forces created by a falling body and their effect on human beings. In this lesson, students have to predict the effect of adding certain kinds of loops to the tracks in the famous roller coaster Desperado, in Las Vegas.

The Population Explosion is a model for lesson on prediction in which there is a complexity of relevant evidence. *Desperado* demonstrates how prediction can be handled using the more complex strategy for predicting the consequences of options. This lesson helps students develop a habit of skillful prediction by guiding them to think about what might happen as the result of some event or circumstance and then to weigh the evidence about the likelihood of these possible results. As you examine these lessons, ask yourself the following questions:

- How are students introduced to the need for skillful prediction?

- How is the need for evidence to support predictions emphasized?

- What content objectives are enhanced by the predictions the students make?

- What additional contexts are there for reinforcing skillful prediction after the lesson has been completed?

Tools for Designing Prediction Lessons

The thinking maps that follow provide questions to guide student thinking in prediction lessons. These maps can also be used to help students reflect on how to engage in skillful prediction.

The graphic organizers that follow each thinking map are included for your use in prediction lessons. They supplement and reinforce the guidance in critical thinking provided by the sequenced questions derived from the thinking map.

The thinking maps and graphic organizers can guide you in designing the critical thinking activity in the lesson and can also serve as photocopy masters, transparency masters, or as models that can be enlarged and used as posters in the classroom. Reproduction rights are granted for use in a single classroom.

SKILLFUL PREDICTION

1. **What might happen?**

2. **What evidence might you get that would indicate that this prediction is likely?**

3. **What evidence is available that is relevant to whether the prediction is likely?**

4. **Based on the evidence, is the prediction likely, unlikely, or uncertain?**

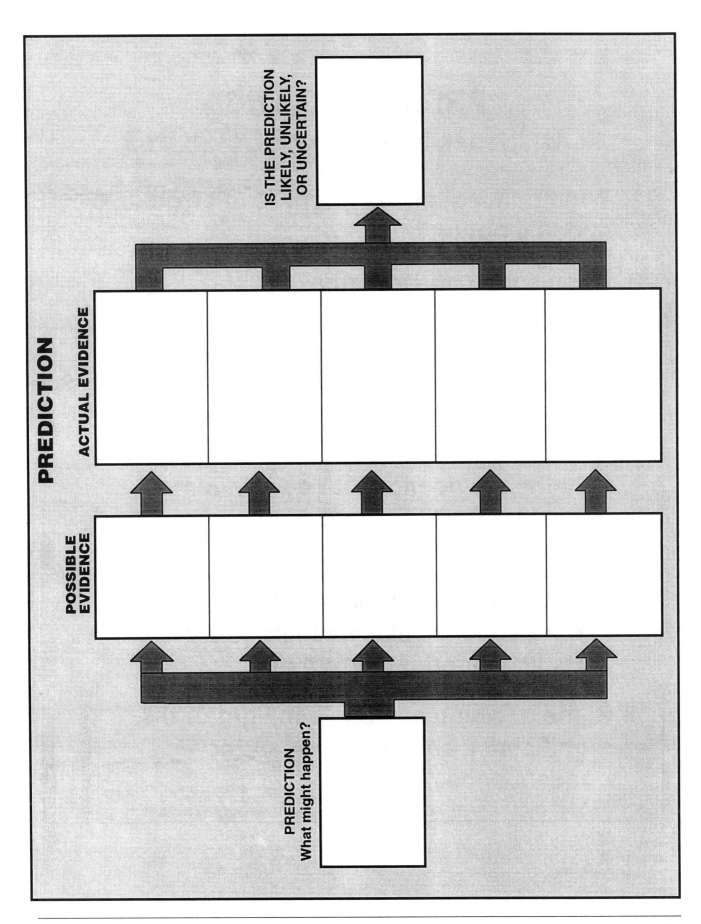

PREDICTION

IS THE PREDICTION LIKELY, UNLIKELY, OR UNCERTAIN?

ACTUAL EVIDENCE

POSSIBLE EVIDENCE

PREDICTION
What might happen?

PREDICTING THE CONSEQUENCES OF OPTIONS

1. **What consequences might result from a specific option?**

2. **Does each consequence**

   a. **count for or against the decision?**

   b. **rank as important?**

3. **How likely is the consequence?**

   a. **Is there evidence that counts for or against the likelihood of the consequence?**

   b. **Based on all the evidence, is the consequence likely, unlikely, or is its likelihood uncertain?**

4. **Is the decision advisable in light of the significance and likelihood of the consequences?**

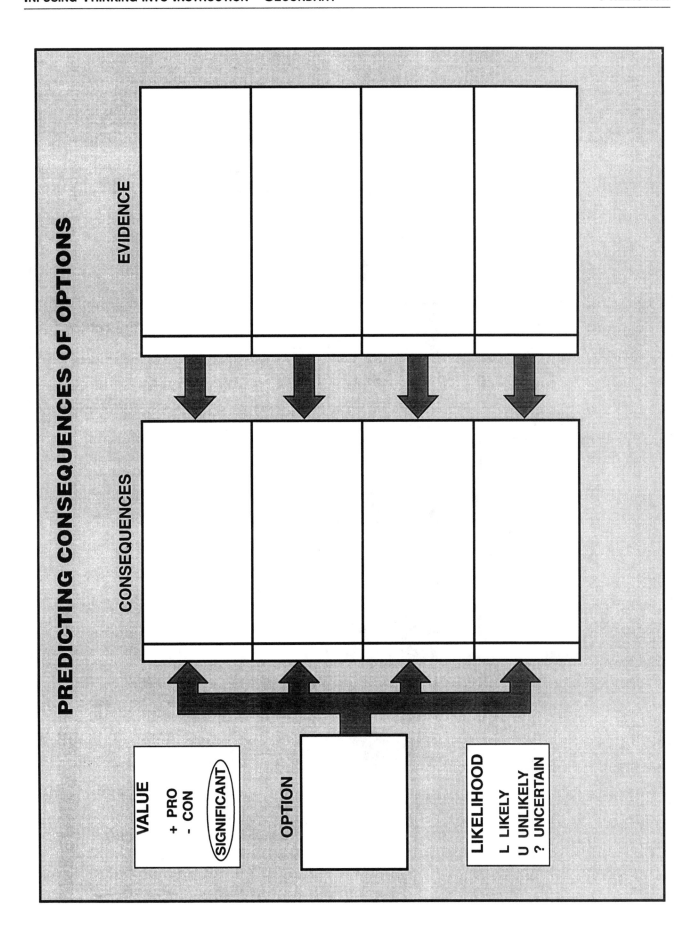

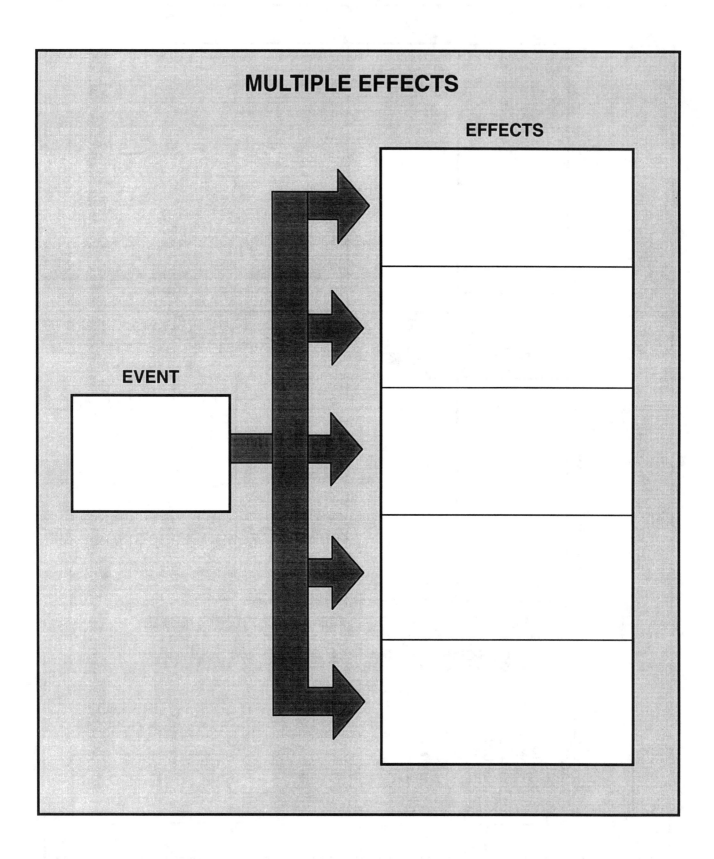

MULTIPLE EFFECTS

EFFECTS

EVENT

THE POPULATION EXPLOSION

General Science, Biology, Environmental Science   **Grades 8–12**

OBJECTIVES

CONTENT

Students will learn about the factors that contribute to population size and the needs of growing populations. They will learn and use the concepts of exponential population growth, the carrying capacity of an environment, zero population growth, and a population crash. Students will apply this knowledge to human population growth, as projected for the 21st Century.

THINKING SKILL/PROCESS

Students will develop skill at predicting by considering evidence for the likelihood of their predicted consequences.

METHODS AND MATERIALS

CONTENT

Students will examine data about population growth and construct various graphs to map and project population growth. A resource packet on population experiments is used. Graph paper is needed. Students work in partnership teams.

THINKING SKILL/PROCESS

Students' predictions are guided by structured questioning and the use of a graphic organizer that highlights key points in skillful prediction, especially the need for supporting evidence of the likelihood of the prediction. Students work together in collaborative learning groups on this activity.

LESSON

INTRODUCTION TO CONTENT AND THINKING SKILL/PROCESS

- Can you think about a time when you thought that something was going to happen but it didn't? Write down what you thought would happen and then what actually happened in your notebooks. Why do you think what you expected didn't happen? Discuss this with your partner and then write why you think what you expected didn't happen in your notebooks. After a few minutes, ask for three or four examples and list your students' responses on the chalkboard or on newsprint. **Now discuss with your partner whether there was anything you could have found out beforehand that would have made you realize that it might not happen. Make a list of things you might have found out in your notebooks.**

- When you think that something will happen before it happens, you are making a *prediction*. Predictions are always about the future. We can't know the future the way we know what's happening now, but we can find things out now that indicate whether what we predict is *likely, unlikely, or uncertain*. Those things that you could find out now that help you decide how likely a prediction is count as *evidence* for or against the prediction. When you think about how likely a prediction is and base your judgment on evidence, you are practicing *critical thinking*. Go back to what you have written in your notebooks and attach the following labels to the different items:
  Prediction
  What happened

Evidence you might have got at the time that would have indicated that your prediction probably wouldn't happen.

If you got that evidence, how would you have rated your prediction: likely, unlikely, or uncertain? Write this in your notebooks also with an arrow pointing to the prediction. Ask for a few examples from the class. Each time a student explains whether his prediction remains likely, unlikely or uncertain, ask why.

- **Why do you think that predicting accurately is very important?** STUDENT ANSWERS VARY BUT SOMETIMES INCLUDE: *You might make plans based on your prediction and then be disappointed. You might get hurt if you predict that something will be safe and it isn't. You could end up doing things that you don't have to do. For example, if I think it is going to be cold and wear my coat but it turns out to be warm, I really didn't have to wear my coat.* **What should you do to make sure your prediction is as accurate as possible?** STUDENT ANSWERS SOMETIMES INCLUDE: *You should try to find out in advance if anything might cause your prediction to be mistaken. You should look for evidence that your prediction may or may not turn out the way you think it will.*

- **Let's put together some of these ideas about how to make skillful predictions. We'll make a thinking map that can guide us in making more accurate predictions. The thinking map contains questions to ask and to answer carefully as you are making a prediction. The map helps you think critically about your predictions. If, as you use this strategy, you find that your prediction is not accurate, you should then modify it according to the evidence you have.**

> **SKILLFUL PREDICTION**
>
> 1. What might happen?
> 2. What evidence might you get that would indicate that this prediction is likely?
> 3. What evidence is available that is relevant to whether the prediction is likely?
> 4. Based on the evidence, is the prediction likely, unlikely, or uncertain?

- **Let's use these ideas to think about a really simple prediction. If you are going outside and it's sunny, you may predict that it is not going to rain and decide not to take your raincoat. When you see that it's sunny, that is your evidence for thinking that it will not rain. Is that a good reason? Why?** *No. Sometimes when it's sunny it gets cloudy and rains.*

- **If you hear on the radio that thunderstorms will move into your area before you get back from school, how might that influence whether or not you take a raincoat? Why?** Students often respond that the report would change their minds and that they might decide to wear their raincoats. They sometimes say that they remember times when they got wet in an unexpected storm. **Whose prediction is more reliable: yours or the weatherperson's? What reasons might you have for reconsidering your prediction of sunny weather?** POSSIBLE ANSWERS: *Weather reports are based on records and observations with equipment that we don't commonly have. The weather service bases its predictions on conditions in an area larger than we can see. All of this usually provides evidence that is better than just noticing that it is sunny.*

- **What information might help the weather service to predict precisely** *when* **it is going to rain?** POSSIBLE ANSWERS: *The differences in barometric pressure on either side of a front that may influence its intensity or speed, records of the speed of approaching storms based on past ones.*

THINKING ACTIVELY

- Among the many things that we predict, change is one of the most frequent concerns of ours. The weather is one good example, but there are many others. Science comes in when studying how things change so that future changes are predictable and we can prepare for them. You have been reading in your textbook about how populations change in relation to their food supply. For example, when there is a shortage of food, the population of animals that feed on that food tends to shrink; when it is abundant, population growth is evident. Suppose that the food supply is constant, however. **Will the population grow at a constant rate?** Students are often quick to realize that other factors may play a role in determining the size of a population. They may mention disease or migration of predators as other factors affecting population size.

- **Let's consider a specific case of human population growth—the growth in the size of the population of a major city. For example, Las Vegas is a city on the move. Its population is growing as rapidly as any other city in this country. It is now a little over a million. In 1972, the population of Las Vegas was approximately 500,000. Most of the increase was due to people moving there from other places in the United States. Work with your partner and construct a graph that will project the population growth of Las Vegas at the same rate over the next 25 years.** After a few minutes, ask for student responses and for student displays of their graphs. Most students project a doubling of the population of Las Vegas during the next 25 years and use either a bar graph or a line graph to illustrate this. **Who might be interested in these projections?** STUDENT ANSWERS VARY BUT TYPICALLY INCLUDE: *City planners who have to plan for adequate roads and parking spaces, real estate developers who think about investing in property and building homes, casino owners, food chains that have to plan for the location of new stores in the area; public utilities that need to meet future demands for water, sewer, and power.*

- **If you were one of these people, would you rely on these projections? Why or why not?** Some students will say that they would rely on these projections; others recognize that these projections assume that factors that might change the population growth rate are constant and that many other things could change the population growth rate. Students are not sure how reliable these projections are. When this response is given, ask the students what factors they can identify that might change the growth rate either up or down. Ask the students to work in pairs again and to brainstorm at least five such factors. Ask for reports, accepting only one item from each group until most of the groups have reported, then solicit any additions from any of the groups that wish to add items to the list. Write these on a transparency or on the chalkboard. POSSIBLE ANSWERS: *Land for expanding the city may not be available, the death rate might go down because of advances in medicine, there might be an epidemic, the government might declare casino gambling illegal, there might not be enough water in the desert around Las Vegas for all those people, people might feel pretty comfortable in Las Vegas, make a lot of money and decide to have larger families; property taxes could go down and people from other areas might relocate to Las Vegas to build homes.*

- **In order to gain more confidence in these projections, therefore, what would you have to find out?** Students usually respond quickly that they would have to find out that these things were not going to happen. **Realistically, we can't find out now that these things are definitely not going to happen because they are in the future. Think about our earlier discussion of prediction. What might be available to us now that can help us with these predictions?** Students usually respond that we could get evidence now that indicates the likeliness of these things. If they are having trouble responding, refer them to the thinking map that was developed earlier about skillful prediction. When students recognize that they might get evidence that would help here, ask them what they would do about their predictions of population size if it seems to indicate that some of these inhibiting factors are probably going to occur or that some of the accelerating factors are probably going to occur. Students usually respond that we should then adjust our predictions downward or upward accordingly. In the same spirit, students also usually acknowledge that if the evidence indicates that the factors that affect growth rate will probably remain as they have been over the past 25 years, we should keep the projections the way they are now. You can sum this up for the class by saying something like: Overall, when we get evidence, we can weigh the factors that we think will push the population of Las Vegas upward beyond our projections against those that we think will push the population downward and come out with a result. When we do this, we can have more confidence that our predictions of population growth in Las Vegas are reliable.

- **Let's shift our focus now and do the very same thing with the overall population of human beings on Earth. For this activity, we will develop some graphs of human population growth and use a special graphic organizer that represents the thinking strategy for skillful prediction we have been discussing. This will enable us to make some judgments about the likely population growth trends of humans in the future. First, however, let's think about the main factors that contribute to changes in size of the human population. They are clearly not connected with the migration of people from one place to another, as they are in Las Vegas. What do you think the main determiners of the growth in human population are?** Students are usually very quick to recognize that the number of people born and the number of people that die are the main determiners of population change. **Let's consider just these factors for a moment. If the birth rate—the number of people born during any given period—equals the death rate—the number of people who die during the same period—what effect will this have on the size of the human population?** Most students state that this condition will keep the population stable because when the birth rate equals the death rate, zero population growth (ZPG) is achieved. (In high school classrooms, explain to students that it will not grow or shrink during this period. Explain to students that when this happens, we say that although it *appears* that the population will immediately stop growing, nevertheless it will take many generations of continued ZPG before the world population would in fact stabilize. Ask students why the population will continue to grow for many years even after the birth and death rates have become equal. The reason has to do with the age structure of the world population: more than 50% of its people are under the age of 25 and their offspring will contribute to a growing population for many generations.)

- **The human population has not shown this kind of stability; rather, it has grown. Many scientists estimate that our species has been around for about 200,000 years. But during 99% of that time—all but 2,000 years—the human population has grown with astonishing slowness. Although the pace of population growth did increase during this period, it changed so sluggishly that a graph of population growth during this period looks quite gradual. It is estimated that the worldwide human population 12,000 years ago (10,000 B.C.) was, perhaps, 50 million people. Five thousand years later, it had climbed to 140 million. By A.D. 1, almost 2,000 years ago, worldwide population had grown only to about 250 million. In your groups, use these statistics to draw a graph of world population versus time for human population growth between 12,000 and 2,000 years ago.** Students have little difficulty with drawing a graph of human population growth during this period. The figure below shows a typical stu-

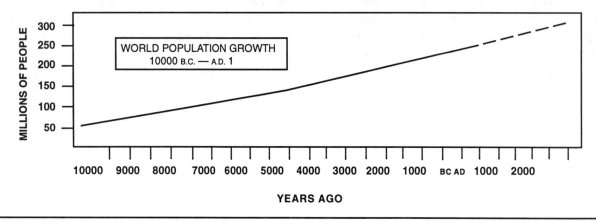

dent response. It is important that students realize that population growth during this period wasn't perfectly linear, but nonetheless increased at nearly the same rate each generation.

• **If the rate of worldwide population had continued to increase slowly, what would you expect the population of the world to be in the year 2000? Use your graphs to estimate what it might have been by** *extrapolating*—**using information you have to predict what will or would have happened.** Most students recognize that it would be about 300 million people. However, this is not the way the human population has grown during the past 2,000 years. There are about 6,000 million (6 billion) humans on earth today. To better understand what this means in human terms, consider these statistics: It took about 5,000 years for the world population to double from 125 million to 250 million in A.D. 1. However, by 1650, the world population had doubled again, having reached 500 million. But this time, it had taken only 1,600 years or so to double. *Within 200 years*, world population had doubled once again to reach 1 billion, in 1850. Between the middle of the 19th century and 1930, world population doubled again and reached 2 billion, this time doubling in only 80 years. Forty-five years later, by 1975, world population further doubled and reached 4 billion people. **What can you say about the rate of population growth during the past 2,000 years as compared with the rate of population growth during the preceding 10,000 years?** Students immediately realize that the world population during the past 2,000 years is not only increasing much faster than it did during the previous 10,000 years, but is also increasing at a faster and faster rate.

• **Let's draw a graph of population growth during the past 2,000 years. Use the data about human population growth we just discussed. Be prepared to compare it with the previous graph your group has plotted.** Although students at times have difficulty finding an appropriate range for each axis, ultimately their graphs resemble the typical "J" graph for population growth like the one below. **What you have drawn is called a "J" graph, and it represents**

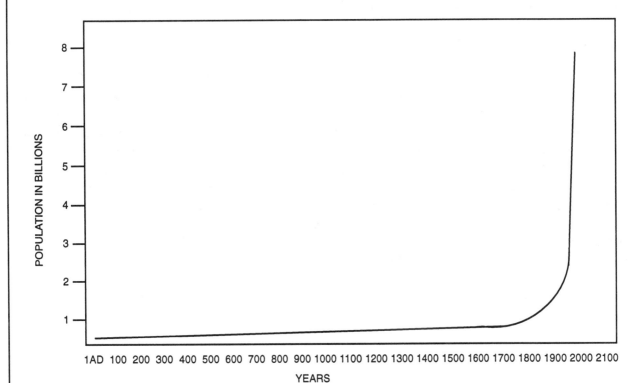

a kind of growth that is called "exponential growth." That means that it takes shorter and shorter periods of time for the population to double. Compare the two graphs that your groups have drawn. What can you say about them? TYPICAL STUDENT RESPONSES: *The first graph showed slow population growth over a long period of time while the second graph showed faster and faster population growth over shorter and shorter periods of time.*

- **Here is a graph showing population growth on earth from 10000 B.C. to A.D. 2100. It is a combination of the two graphs you have already drawn.** Show students an overhead transparency or distribute copies of the graph shown below (and reproduced at the end of this lesson) to each group.

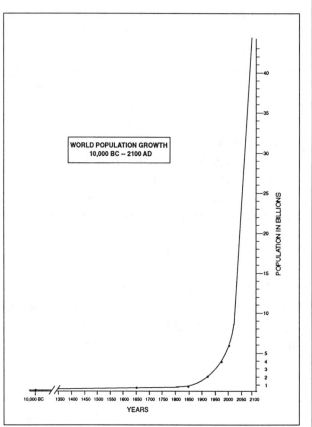

- **Now let's figure out what this all means and do some thinking about it. Based on this graph, what is the size of the human population now?** *About 6 billion people.* **About how long will it take the population to double and reach 12 billion people?** Although the scale is tight, students usually predict that the world population will double in about 30 to 40 years.

- **Let's extrapolate again; if the rate of population growth increases in the manner indicated by this graph, what will the population of people on Earth become as we approach the end of the next century?** About 40-50 billion. That represents a lot of people, nearly 10 times the number of people on Earth now. ... This is what people mean by the "population explosion." This was not something that people were concerned with in the 1700s and 1800s, but you can understand from this graph why we are concerned about it now. If the human population growth continued at this rate for 100,000 years, can you imagine how many people there would be? At this rate, it looks like human beings will fill up all the space on the planet sometime soon. Think about this for a few seconds (pause). **How many of you think this is likely?** Most students say that this couldn't happen. **Why do you think that this would not be a very good prediction? Work with your partner again and think about what would probably prevent such a huge population explosion. Make a list of four factors that might influence how much the human population grows.** After a few minutes, ask for responses from the teams, allowing only one item to be mentioned per team. Write these on the chalkboard or on a transparency under the heading "Factors That Might Influence Human Population Growth." STUDENT ANSWERS USUALLY INCLUDE: *We might run out of food. There wouldn't be enough air to breathe. We might fight with each other and kill each other over scarcer and scarcer resources. We might run out of material to fight diseases with. We might poison the atmosphere with more and more carbon dioxide, carbon monoxide, and other things that would kill a lot of people. The Earth might get hotter and the heat would kill us. A nuclear war or reactor disaster might kill millions of people.*

- **If such a huge population explosion that would fill all the space on the earth is prevented by these factors, what do you think might happen instead?** Let's look at some information that

we have about other living things to see if we can get some ideas about what might happen to humans. Divide the class into three groups of students and have them do a "jigsaw" by reading three different selections from the resources material at the end of this lesson and then reporting on the situation to the class. Give the first group of students the selection on the growth and subsequent leveling off and stability of a population when it reaches the carrying capacity of the environment and zero population growth results. Ask them to graph the situation described. Give the second group of students the selection describing experiments with the growth of bacteria and ask them to graph the population growth and subsequent decline. Explain to students that when a population expands rapidly like this and then dies off, population experts say that the population has "crashed." Finally, assign the selection on the fly experiment in which overshooting of the carrying capacity was induced, leading to undershooting, and then to overshooting and undershooting in a recurring pattern. Introduce them to these

terms. Ask each group to report and to exhibit the graphs that represent the population growth in each situation.

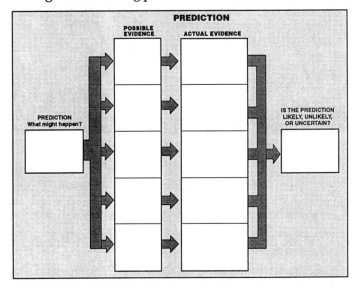

- **Let's now try to predict which trend is the most likely scenario for human population growth—the prediction that at a certain point our population will crash, the optimistic prediction that we will reach the carrying capacity of our environment and attain a stable zero population growth, or the population oscillation represented in the third graph. Use the graphic organizer for prediction (right). Each group should write the scenario they studied, applied to the human population, in the box on the left for what might happen. For example, the group that studied the experiment with bacteria should write, "The human population will crash,"** and the other groups should write similar predictions that match the situations they have studied in the first box on their diagram. In the boxes under possible evidence, write what information you might get that would tend to support the likelihood of this prediction. For example, the first group might write in one of the boxes under "Possible Evidence" that the food supply for humans is finite, limited, and does not grow. After your group has written in the boxes a number of possible facts that would make your prediction likely, we will pause in this lesson to give you a chance to gather information to determine whether what you have written in the possible evidence boxes is in fact the case. If you find that it is, write that in the actual evidence box next to it and put a plus in the column next to the information, indicating that the evidence you have found counts in favor of the likelihood of your prediction. Use your textbooks, other resources in class, and the school library to research the evidence. Record what you find in your graphic organizer. Then, your group will reconvene and share what they have found. After all of the actual evidence is recorded, your group should rate the predicted trend as likely, unlikely, or uncertain.** Student responses vary, but many students find that there is potential for increasing the food supply through natural (e.g., ocean farming) as well as artificial means (e.g., genetic engineering), but they recognize that the planet does have a limited capacity for producing food. In addition, they may find that food production has increased so that, though hunger and famine still exist on the planet, there is a general trend toward decreasing world hunger and increasing the availability of food to those who need it. They may also find that techniques like fish farming and genetic engineering have increased the supplies of some foods. On the other hand, many students become concerned that pollution of the environment is having an adverse effect on some foods, making them less available, if not inedible (e.g., fish from heavily polluted lakes). They may also note a trend toward greater and greater pollution of the environment in ways that directly affect human beings (e.g., air pollution and an increase in lung cancer and emphysema). Advances in medicine sometimes suggest to some students that such trends may be counterbalanced by our ability to medically remedy problems that humans face as the result of the increased toxicity of the environment. In addition, some students note that the trend toward better and better ways of travelling in space provides some evidence in favor of a stabilization of the population on Earth as more and more people may become space travellers and colonizers of other planets. Finally, birth control methods and attempts by some governments to limit family size are noted as a possible trend that also counts in favor of stabilization of the population. In general, students conclude that though there is a great deal of uncertainty with regard to human population growth on Earth, a population crash is somewhat less likely than zero population growth but that the oscillation model is probably the most likely. Completed graphic organizers of typical student responses for each of these cases follow this lesson.

THINKING ABOUT THINKING

- **What did we do in this activity that is different from the usual ways you make predictions?** POSSIBLE ANSWERS: *Usually, I don't think about whether my predictions are accurate, I just make them. I rarely ask whether there is any evidence for my predictions when I make them. I never compare my reasons for my predictions with the information I think I would need to make a good prediction.*

- **In your own words, construct a thinking map of skillful prediction that you can use in the future to be sure that your predictions are as accurate as possible.** Discuss with your students how the methods they sketch in their thinking maps produce predictions that aren't just guesses. Ask the students to write their thinking maps in their notebooks for future reference. If there are differences between what the students produce, ask them to discuss these differences.

- **When would you use this strategy for prediction? In what circumstances might you not use it? Discuss this with a partner.** Ask students for some of their ideas. POSSIBLE ANSWERS: *I would use it when something dangerous might happen, when what I am predicting is important to me, and when I have the time to get evidence. Only rarely would I not use it: when there is no time to get the information I need, and when the prediction is not important; however, if there was no time to get a lot of information and the prediction was important, I would do the best I could.*

- **How would using this process affect your confidence in your predictions? How would using this process affect your confidence in your decisions?** Usually, students say that it makes them feel more confident in both their predictions and the decisions they base on these predictions because they've looked for evidence and can back up their predictions with reasons.

APPLYING THINKING

Immediate Transfer

Shortly after this activity on human population growth, ask the students to do more activities in which they use the strategy for skillful prediction. Here are three possible follow-up activities:

- **Scientists are warning us about the possibility of "global warming"—an increase in the temperature of Earth due to industrial gases acting as a blanket in the atmosphere. They project that the planet's average temperature could increase by approximately 5 degrees over** the next 50 years. Use skillful prediction to determine the likely impact of this increase in temperature on how we live.

- Choose one of your school's athletic teams. Use skillful prediction to determine the likelihood that your team will have a good record this year.

- Use skillful prediction to determine how long it is likely to take you to prepare for your next test so that you can do well.

Reinforcement Later

Later in the school year, come back to skillful prediction and ask students to review the strategy and use it again in activities like the following:

- Apply skillful prediction to decide whether a currently endangered species, such as the Florida panther, is likely to become extinct.

- Apply skillful prediction to a recent historical event that you have studied or to some current news event to determine what its consequences are likely to be.

- We're going to study weather prediction. We often hear a weather report calling for a 50% or a 70% chance of rain, for example. Determine what evidence the weather service uses to make predictions of this sort. What other kinds of predictions can be made with this degree of specificity? Gather relevant information about graduating seniors in your high school and make a numerical prediction concerning some important outcome, for example how many will go to college. The weather predictions are based on statistical data about how often, in similar circumstances in the past, it has rained. If it has rained in 7 out of 10 occasions on which the weather conditions have been like they are now, then there is a 70% chance of rain, all other things being equal. Similar statistical predictions have been made about earthquakes, chances of getting into auto accidents, and about chances of contracting certain diseases.

CONTEXT EXTENSION

Ask students to use the school library to bring back information about actual population planning strategies that have been used by countries to adjust the birth rate.

ASSESSING STUDENT THINKING ABOUT PREDICTIONS

Any of the transfer examples in this lesson can serve as assessment items for determining the extent to which students are thinking about their predictions skillfully. Make sure information is available for students to use in supporting their predictions. Determine whether students are thinking about possible evidence, making use of actual evidence, and considering how likely the predictions are based on the evidence used.

A variation on this approach is to ask students to critique predictions made by others. Give them examples of predictions and ask them to determine whether these predictions are well-founded.

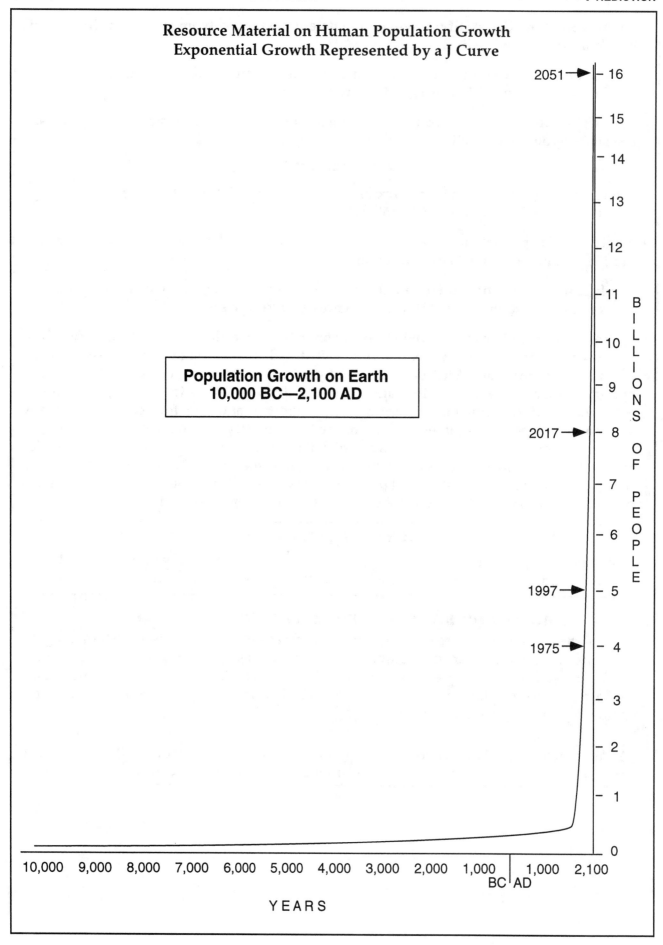

Resource Material on Human Population Growth
Exponential Growth Represented by a J Curve

**Population Growth on Earth
10,000 BC—2,100 AD**

2051 → 16

15

14

13

12

B 11
I
L
10
L
I 9
O
N 8 ← 2017
S
7
O
F 6

P 5 ← 1997
E
O 4 ← 1975
P
L 3
E
2

1

0

YEARS

RESOURCES ON THE BEHAVIOR OF POPULATIONS

1. Population Stability and Zero Population Growth

When the birth and death rates in a population remain constant and the birth rate is double that of the death rate, the population grows exponentially, from 2 to 4, from 4 to 8, from 8 to 16, etc., provided no other factors such as an emigration boom influence its size. This is called "exponential growth." Such growth can be drawn on a graph such as the one featured previously representing the growth of the human population. The population curve on this graph is usually in the form of a "J" curve.

Exponential population growth will not go on forever. Usually, factors in the environment limit the growth of a population, such as the amount of food and nourishment needed by the individuals in the population that the environment provides. As a population's size increases, the same resources must be shared by more and more individuals. The birth rate may go down because of the decreasing supply of resources, and the death rate may go up. When the birth and death rates are in balance and the resources remain constant, the population becomes stabilized at some constant size. When the population has stabilized, the environment reaches what is called its "carrying capacity."

Typically, when a small population grows, the growth is slow at first. As numbers increase, its speed picks up. As it approaches carrying capacity, it tends to slow, then level off once the carrying capacity is reached. When this occurs, the population has what is called zero population growth. The births and deaths are balanced so that the population size does not change, even though new members of the population are constantly being born into it and older members are dying.

2. Population Instability and Population Crashes

Nature is not always kind to populations. Sometimes, resources required by a population become scarce because of natural disasters. Sometimes, we destroy or damage natural resources by pollution. Hence, the carrying capacity of an environment could fluctuate considerably and, in particular, decrease so much that a flourishing population might find itself declining rather than growing. (This will occur, typically, when the death rate in a population exceeds the birth rate.)

An extreme case of a population decline is called a population *crash*. An experiment in which this is illustrated is one in which a single bacterium is put in a culture flask with a complete supply of needed nutrients. In half an hour, the bacterium undergoes division into two bacteria. After another half an hour, each of these divides, thereby increasing the population size to four bacteria. If no cells die and this repeats, the number will double every thirty minutes. Thus far, this population exhibits typical exponential population growth. After 9 1/2 hours, the population will be greater than 500,000 bacteria and, as we might expect, in just another half an hour there will be 1,000,000.

In reality, about 25 percent of the bacteria in the flask die between doublings. How does this affect the growth rate of the population? Well, as you might expect, it slows down the actual growth of the population. Instead of doubling every half-hour, it now doubles every two hours. But this does not mean that the explosion of a million bacteria will not occur. Rather, it will certainly occur, but it will take 30 hours instead of 10 hours. The growth rate of the population is still exponential and a J-shaped curve shows up when it is plotted on a graph.

So why aren't these bacteria populating the whole Earth? The reason is simple. In this experiment, the bacteria were initially provided with nutrients needed for growth and reproduction. But there was a finite amount of such nutrients. As soon as the population reached the carrying capacity, it began to level off. Almost at once, the population then went into a rapid decline and completely died off. Not a single bacterium remained. This was because the nutrients were not replenished and when carrying capacity was reached, it lasted for just a moment. Because the amount of food for these organisms immediately declined because it was being used by the vast numbers of bacteria that had entered the population. Soon, the food was used up. With no more nourishment remaining, the death rate increased dramatically and the population died off.

The experimenters tried the experiment again but this time replenished the nutrients for the bacteria. What do you think happened? This had no effect on the population. It still crashed after its initial exponential growth. What happened clearly indicated that other factors limit population size. As the bacteria functioned, they, like other living things, produced waste products. These waste products are not harmful in small quantities, but with a huge population and a confined space, they became toxic and killed themselves.

3. Population Variations due to Overshooting and Undershooting the Carrying Capacity of the Environment

Thus far, it has been in only extreme, rare, and highly controlled situations (like laboratory experiments) that whole populations have crashed. The more usual pattern is one of growth and decline as the carrying capacity of an environment fluctuates because of variations in natural conditions that lead to temporary shortages, like local food shortages, or because of disease, like the spread of the Black Plague in Europe in the Middle Ages, killing more than 75 million people.

An experiment conducted in controlled laboratory conditions illustrates the growth and decline type of population change pattern in a pure form. A scientist experimented with a population of flies (in this case sheep blowflies) by feeding their larvae limited amounts of beef liver while at the same time feeding the adults unlimited amounts of sugar and water. When the adult population density was high, a correspondingly high number of eggs were laid by the flies. The resulting larvae devoured all the food before completing their development that would have led them to adulthood, and they all died. Adults also died through the natural processes they were always subject to. Hence, the adult population got much smaller because no

new adults entered the group to take the place of the flies that died of natural causes. Consequently, fewer eggs were consequently. This meant that there would be fewer larvae and less competition for food. Hence, some larvae were able to mature into adults. The population size continued to fall because the new adults were not yet mature enough to lay eggs. When the new adults matured, the adult population again increased exponentially—again and again.

An example of the same type of situation has been reported from one of the Pribilof Islands off the coast of Alaska. In 1910, four male and 22 female reindeer were brought to the island. In 30 years, the population had increased to 2,000. This exceeded the carrying capacity of the environment because there was not enough food for so many reindeer to eat and survive. The plants eaten by the reindeer were devoured and almost disappeared. As a result of the diminished resources, the population of the herd had fallen in 1950 to eight. Fortunately, with fewer reindeer and less competition for food, the food supply gradually replenished itself and the population is on the upswing again.

This story of a natural overpopulation, then decline, adjustment, and now an increase again shows us how fragile a relationship is maintained between a population and the environment in which it lives. Had a severe rainfall shortage coincided with the depletion of the natural food of the reindeer, it might have killed off this food supply forever. Then, there would be little hope that the reindeer could come back relying on nature alone. They would be in the same position as the bacteria in the previous experiment whose population crashed completely. This also points up how our awareness of these fragile relationships and how they work can lead us to intervene where nature fails. As we have done for numerous endangered species, we are able to step in and provide food where needed. But this also points up an important message for us as humans in thinking about our own population. Who will step in to help if we find ourselves in the position of the bacteria, or the unfortunate reindeer, whose food supply no longer exists?

Sample Student Responses • Graphs of Population Growth

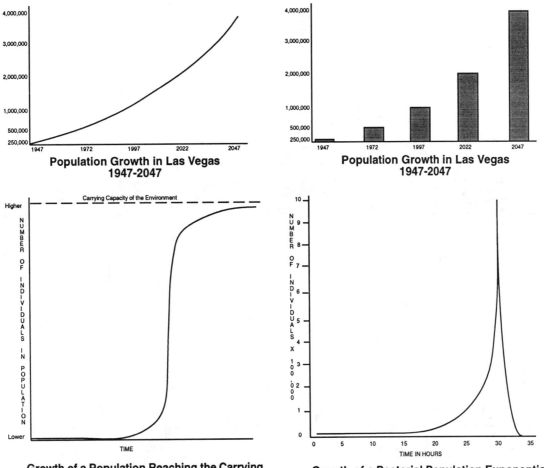

Population Growth in Las Vegas
1947-2047

Population Growth in Las Vegas
1947-2047

Growth of a Population Reaching the Carrying Capacity of the Environment, then Stabilizing. Exponential Growth, then Zero Population Growth. Represented by an S Curve

Growth of a Bacterial Population Exponential Growth, Then Population Crashes

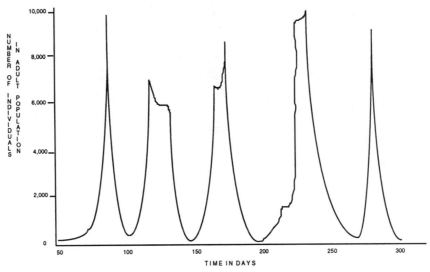

Growth of a Blowfly Population Exponential Growth,
Then Overshooting–Undershooting Cycles

Sample Student Responses • The Population Explosion

PREDICTION

POSSIBLE EVIDENCE   **ACTUAL EVIDENCE**

IS THE PREDICTION LIKELY, UNLIKELY, OR UNCERTAIN?

It is uncertain that the human population will crash; the present growth trend makes this unlikely, though it is a risk.

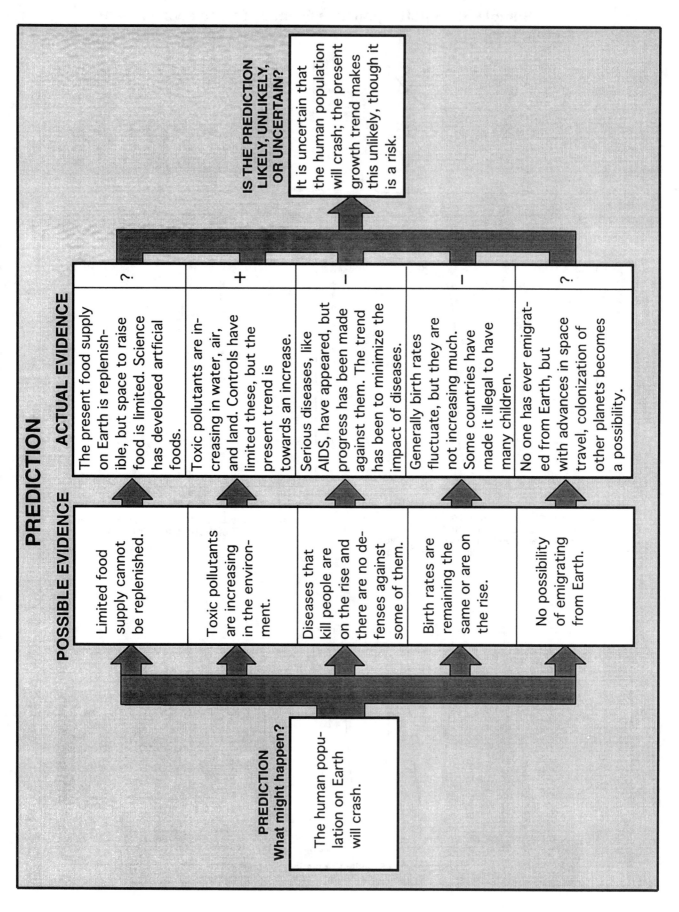

POSSIBLE EVIDENCE	ACTUAL EVIDENCE	
Limited food supply cannot be replenished.	The present food supply on Earth is replenishible, but space to raise food is limited. Science has developed artificial foods.	?
Toxic pollutants are increasing in the environment.	Toxic pollutants are increasing in water, air, and land. Controls have limited these, but the present trend is towards an increase.	+
Diseases that kill people are on the rise and there are no defenses against some of them.	Serious diseases, like AIDS, have appeared, but progress has been made against them. The trend has been to minimize the impact of diseases.	–
Birth rates are remaining the same or are on the rise.	Generally birth rates fluctuate, but they are not increasing much. Some countries have made it illegal to have many children.	–
No possibility of emigrating from Earth.	No one has ever emigrated from Earth, but with advances in space travel, colonization of other planets becomes a possibility.	?

PREDICTION
What might happen?

The human population on Earth will crash.

Sample Student Responses • The Population Explosion

PREDICTION

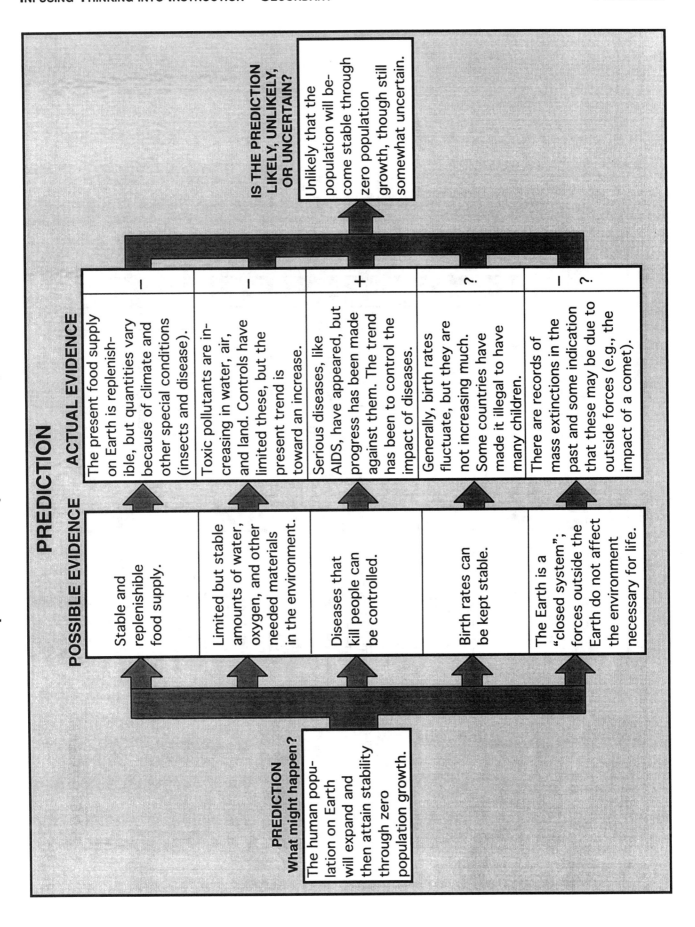

PREDICTION
What might happen?

The human population on Earth will expand and then attain stability through zero population growth.

POSSIBLE EVIDENCE

- Stable and replenishible food supply.

- Limited but stable amounts of water, oxygen, and other needed materials in the environment.

- Diseases that kill people can be controlled.

- Birth rates can be kept stable.

- The Earth is a "closed system"; forces outside the Earth do not affect the environment necessary for life.

ACTUAL EVIDENCE

The present food supply on Earth is replenishible, but quantities vary because of climate and other special conditions (insects and disease).	–
Toxic pollutants are increasing in water, air, and land. Controls have limited these, but the present trend is toward an increase.	–
Serious diseases, like AIDS, have appeared, but progress has been made against them. The trend has been to control the impact of diseases.	+
Generally, birth rates fluctuate, but they are not increasing much. Some countries have made it illegal to have many children.	?
There are records of mass extinctions in the past and some indication that these may be due to outside forces (e.g., the impact of a comet).	– ?

IS THE PREDICTION LIKELY, UNLIKELY, OR UNCERTAIN?

Unlikely that the population will become stable through zero population growth, though still somewhat uncertain.

Sample Student Responses • The Population Explosion

PREDICTION

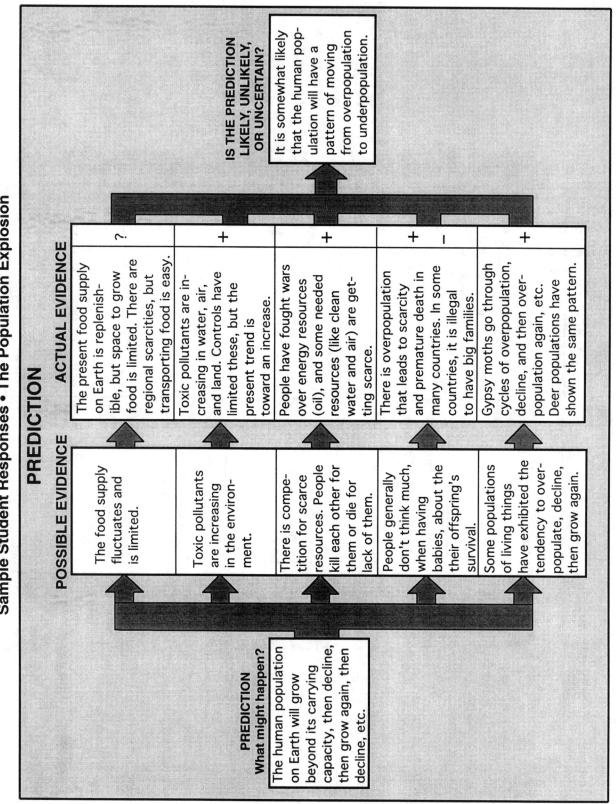

POSSIBLE EVIDENCE

ACTUAL EVIDENCE

IS THE PREDICTION LIKELY, UNLIKELY, OR UNCERTAIN?

It is somewhat likely that the human population on Earth will have a pattern of moving from overpopulation to underpopulation.

POSSIBLE EVIDENCE	ACTUAL EVIDENCE	
The food supply fluctuates and is limited.	The present food supply on Earth is replenishible, but space to grow food is limited. There are regional scarcities, but transporting food is easy.	?
Toxic pollutants are increasing in the environment.	Toxic pollutants are increasing in water, air, and land. Controls have limited these, but the present trend is toward an increase.	+
There is competition for scarce resources. People kill each other or die for lack of them.	People have fought wars over energy resources (oil), and some needed resources (like clean water and air) are getting scarce.	+
People generally don't think much, when having babies, about their offspring's survival.	There is overpopulation that leads to scarcity and premature death in many countries. In some countries, it is illegal to have big families.	+ −
Some populations of living things have exhibited the tendency to overpopulate, decline, then grow again.	Gypsy moths go through cycles of overpopulation, decline, and then overpopulation again, etc. Deer populations have shown the same pattern.	+

PREDICTION
What might happen?

The human population on Earth will grow beyond its carrying capacity, then decline, then grow again, then decline, etc.

DESPERADO

Physics   **Grade 11–12**

OBJECTIVES

CONTENT
Students will learn how to make calculations of energy, centripetal force and gravity for moving objects.

THINKING SKILL/PROCESS
Students will develop skill at predicting and evaluating consequences by weighing their significance and considering evidence of their likelihood.

METHODS AND MATERIALS

CONTENT
Students will view segments of the video *World's Greatest Roller Coaster Thrills*, read segments of *Six Flags Great Adventure Physics*, and examine diagrams of roller coaster design.

THINKING SKILL/PROCESS
Predicting consequences is guided by structured questioning and a graphic organizer that highlights key points in skillful prediction. Collaborative learning promotes discussion of options, consequences, and evidence.

LESSON

INTRODUCTION TO CONTENT AND THINKING SKILL

- One summer day in 1992 a young woman and her fiancé boarded a free-fall ride at a theme park in New England. Free-fall rides pull a car straight up more than 150 feet and then drop it down a rail that curves horizontally as it approaches the ground. During a few brief seconds, the riders fall free and experience weightlessness. This type of ride requires two major safety restraints: one safety bar over the shoulders and the other across the waist. The shoulder restraints are passive. The waist restraints, however, like automobile seat belts, uncoil and need to be locked in a clasp on the opposite side of the rider's body. The ride will not start unless the seat belt for every passenger is fully uncoiled from its winding spool. As an additional precaution, ride operators are required to visually check each occupant's safety restraints before the car is allowed to ascend. On this day, as witnesses later noted, the young lady was seen holding the uncoiled waist belt that had not yet been secured in the clasp, and then, as the car slowly ascended, waving her arms frenziedly as she and her boyfriend struggled to secure her seat belt. They failed. Twenty seconds later she lay dead on the ground. She had fallen out of the car as it dropped. The ensuing investigation confirmed that, although the girl's safety belt had been completely pulled from its rewind spool, she had been unable to secure it to its clasp. Because she had held it close to its dock the attendant thought it was secured when he inspected the car visually. Because the belt was completely unspooled, the "fail-safe" interlock located in the rewind spool engaged and released the car for ascent.

- Designing safety into potentially dangerous structures such as amusement park rides requires identifying nearly all imaginable risks and then developing ways to avoid or manage them. Identifying risks involves being able to *predict consequences*. The tragedy on this free-fall ride might have been avoided had enough questions been raised beforehand about human behavior on the ride and the shortcomings of a "foolproof" restraint system.

- **Prediction is important in making decisions about safety because good design judgments depend on consequences that we can predict, some of them, at times, ruinous and deadly. In your groups, think about several disasters that you have heard or read about that might have been avoided if designers had been able to predict them in advance.** Give students time to identify several disasters and ask them to write a list in each group. STUDENT RESPONSES INCLUDE: *fires at night clubs, airplanes that crash, the sinking of the Titanic, the collapse of the Tacoma Narrows bridge, the death of three astronauts aboard Apollo 1, the Challenger tragedy, the explosion of the Hindenburg.* Construct a list of these disasters on the chalkboard or on a transparency based on reports from the groups. (Ask each group to mention only one or two from their list.)

- **In your groups, determine, for each accident, what risk was overlooked and what might have been done to avoid it if we had been able to predict it in advance.** Give students time to evaluate their accidents. Have them report back and list their responses on the chalkboard or transparency next to the disaster. STUDENT RESPONSES INCLUDE: *overcrowding and inadequate egress from nightclubs; faulty cargo bay latches on DC-10s; inadequate life boats and transverse bulkheads that did not go far enough above the Titanic's waterline; extremely difficult emergency egress through the space module hatch; changes in booster rocket "O" rings in cold temperatures; safeguards against ignition of hydrogen gas during docking in an electrical storm.*

- **Our predictions about consequences shouldn't be just guesses, hopes, or fears, but** *judgments that we make based on evidence.* **Only when we make judgments based on evidence can we trust our predictions and make realistic assessments about risks and dangers that justify us in taking steps to change things so that these risks and dangers can be avoided. Pick one of the disasters on our class list and discuss in your groups what evidence might have been obtained beforehand that might have indicated how much risk was involved and that might have pointed the way to doing something that could have prevented the disaster.** STUDENT ANSWERS VARY BUT SOMETIMES INCLUDE: Nightclub fires: *Fire drills in night clubs might have indicated how long it would take for a crowd of a certain size to leave. If that was longer than an average fire takes to spread, then we would have evidence of a risk in the future when the night club is just as crowded.* Difficulty in opening the space module door and getting through it: *The designers might have tested how quickly a person could open the hatch and get away in comparison to how quickly an oxygen-fueled fire would spread in an enclosed space of the same time. This would provide evidence that indicated that if a fire started, the crew probably would not make it out.*

- **We've been practicing in class a strategy for skillful decision making that involves consideration of our options, their consequences, how important specific consequences are, and then choosing the best option. If we look at these examples in this manner, what we have been thinking about is how to tell whether the nightclub exit doors, the bulkheads on the Titanic, etc. are good options by skillfully predicting the risks of that design. Here is a thinking map for skillfully predicting the consequences of options based on evidence that can be a helpful guide in doing this kind of thinking:**

PREDICTING THE CONSEQUENCES OF OPTIONS
1. What consequences might result from a specific decision?

2. Does each consequence
   a. count for or against the decision?
   b. rank as important?

3. How likely is the consequence?
   a. Is there evidence that counts for or against the likelihood of the consequence?
   d. Based on all the evidence, is the consequence likely, unlikely, or is its likelihood uncertain?

4. Is the decision advisable in light of the significance and likelihood of the consequences?

- **Granted that hindsight in thinking about these disasters is always "20/20," it should be evident to you that disaster often follows from inadequate risk analyses; only after an exhaustive review of all contingencies is completed and evaluated can safety be reasonably anticipated. We've been studying the physics of motion—speed, acceleration, gravitational force, the conservation of energy, and** the mechanics of angular motion. Today, we will be applying what we know about motion to a real situation that will allow us to practice the strategy for skillful prediction we just introduced. We are going to be redesigning a roller coaster that presently exists in the United States. Roller coasters are potentially dangerous amusement park rides. Their designers strive, however, to make them perfectly safe while they give people the maximum thrill. Although roller coaster design has evolved from the beautifully scaffolded "woodies" to the looping alloy "steelie" monsters of today, the physical forces that act to produce the illusion of danger, thrills, and safe design have remained astonishingly the same. We will start by viewing segments of a tape which shows different world class roller coasters in operation. (Alternatively, distribute diagrams of roller coasters.) Then, when we turn to the one we will be working on—called "Desperado" in Las Vegas—you will have seen some alternative design options. We will focus our attention on predicting the possible risks and benefits of two of these specific design options.

THINKING ACTIVELY

- **Let's look more closely at roller coasters and determine what design elements produce an exciting ride. In your groups, make a list of the things that make roller coasters exciting.** Show pictures of different roller coasters and, if possible, show segments of or, if time permits, the entire video tape, *World's Greatest Roller Coaster Thrills (60 minutes, Goldhill Video, Thousand Oaks, California).* Solicit responses from the whole class and list these on the chalkboard. SAMPLE RESPONSES: *Height of coaster, number of loops, spirals, boomerangs, corkscrews, tunnels, speed of coaster, illusion of danger, kind of car (standing up, sitting down, hanging), number of g's at bottom and top, location (e.g., on the side of a mountain, on top of a pier over the ocean).*

- **Now let's look carefully at Desperado.** Show film segment or transparency of the diagram of Desperado and then ask: What's not there that you have seen in other roller-coasters? Solicit responses from the whole class and list them on the chalkboard. SAMPLE RESPONSES: *There are no loops. There are no spirals. There are no boomerangs. There are no corkscrews. The cars are just plain sit-down cars. The location is on a flat, featureless desert, not a place with an exciting view.*

- **What might the owner (Wild Bill's Casino Resort) do to make Desperado more exciting? What are some of his options?** Students usually recognize that although it is very tall and fast, Desperado would be a whole lot more exciting if it had some of the "death-defying" design elements they already identified in other coasters. Solicit answers from the whole class and write them down on the chalkboard. SAMPLE ANSWERS: *Add a loop. Add a spiral. Add a boomerang. Add a corkscrew. Replace sit-down cars with stand-up cars. Replace sit-down cars with hanging cars.* Most students recognize that it would not be feasible to move the roller coaster to a location with more breathtaking scenery, like the Grand Canyon.

- **Now let's try and narrow the field of possibilities somewhat. Meet in your groups and assume that you are the redesign team for the Desperado project. This is your first meeting and you have convened to review this list of possibilities and narrow it down to those which are feasible and practical. You will need to consider each possibility in terms of how much of the original coaster will have to be changed and, relative to each other, which changes are likely to cost the least.** Allow students ample time to evaluate each design element. SAMPLE RESPONSES: *Changing the kind of car would be particularly difficult and expensive; hanging cars probably would need entirely new rail configurations; stand-up cars would require rebuilding nearly all track underpasses. Boomerangs reverse the direction of a car 180 degrees, which would mean having to build a second boomerang in addition to tearing out and realigning a lot of track and support towers.*

Spirals would require building high support towers. Loops or corkscrews might be easiest and least expensive because they would occupy only a short distance of horizontal track Students usually narrow the possibilities to adding a loop or a corkscrew.

- **Let's see if we can retrofit a loop into Desperado to make it a more exciting ride. What initial information do we need to know in order to be able to predict the consequences of using different types of loops? Assume that friction is zero and that the height and position of the major "hills" are too expensive to alter.** SAMPLE RESPONSES: *Location of the proposed loop; size of the loop; shape of the loop; kinetic energy of the rider as he enters the bottom of the loop; speed of the car entering the loop; g's on the rider, layout of the existing coaster, type of car.*

- **Let's take a closer look at Desperado.** Pass out copies of the Desperado diagram included at the end of the lesson. **Read the descriptive material about Desperado and in your groups examine the diagram of its layout and elevations. Notice that a proposed location for the retrofit has already been chosen between hill number 2 and hill number 3. In your groups, discuss why.** Students usually recognize that a car needs a certain amount of kinetic energy to get up and around the loop and therefore needs to be placed in a valley after a hill with adequate height; students also realize that if the lead hill is too large, then the forces acting on the rider within the loop will be unsafe. SAMPLE RESPONSES: *Cars would be going too fast (80 mph) at the bottom of the first hill; cars would not have enough energy after hill number 3 to make it around the loop; a tunnel would have to be removed at the bottom of the first hill; the bottom of the first hill doesn't have enough space for a loop.* Solicit responses from each group. Use a transparency of the diagram to clarify their ideas.

- **Read the source material titled *History of the Roller Coaster* and then in your groups discuss the possible shape of the loop you are designing.** Allow time for reading and discussion. Then discuss the differences between circular and clothoid loops as described in the reading. Students usually realize that increased gravitational force (g's) at the bottom and reduced g's at the top of the loop produce a thrilling sense of danger but that too many g's can kill or hurt you, and too few can be equally as dangerous if you are at the top. Ask students to extract the information they need about the effects of decreased or increased gravitational forces on people from the text so that it will be available for use in predicting what will happen to people who might be riding in a car in the different loops they will be considering. Ask: **What should we do to predict which design option—circular or clothoid—is best, given where we will put the loop and its size?** Students usually respond that by calculating the magnitude of the forces at the top and bottom of each loop, they can choose the best and safest one.

- **Before we predict the consequences of using each kind of loop, let's review what we know about the relationship between potential and kinetic energy, angular motion, centripetal force, and centripetal acceleration that will be helpful in making these predictions.** At this point, review the *Forces and Vertical Circles* source material and make sure that students understand and can apply the formulas listed as *Useful Formulas*.

- **Now let's think about one of these options, the circular loop. In order to help in predicting the consequences of using a circular loop, we're going to use a graphic organizer for skillful prediction of the consequences of options. It has places to write down the consequences we initially predict, the evidence we have found for and against the likelihood of those consequences, and how those conse-**

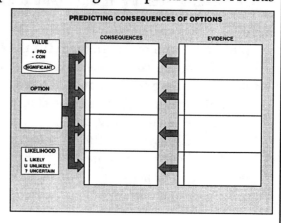

quences affect the feasability of the option we are considering. To start, in your groups make a list of all the consequences you can think of that might result from building a circular loop 40 meters in diameter between hills 2 and 3. Write these in the consequence boxes of the graphic organizer after you've written "Use a circular loop" in the "Option" box. Have each group report back with one or two responses. Enter these on a transparency of the graphic organizer or on a drawing of it on the chalkboard or on a flip chart. POSSIBLE ANSWERS: *riders would pass out at the bottom of the loop; riders would fall out at the top of the loop; the ride would make you sick; a circular loop would cost a lot to build; the loop would be unsafe; riding around the loop would be very exciting.*

- Now let's think about whether using a circular loop in Desperado is feasible, given our design constraints (e.g., 40 meters in diameter), by determining the likelihood of these consequences and rating them in favor or against the option. In order to do this, first put a plus or a minus in the column to the left of the consequence to indicate whether it is a pro or a con in relation to the option. Also, circle the consequences that you think are especially important so that they can be given more weight in weighing the pros against the cons. Point out the *Design Constraints* listed on page 458. Ask students to report and enter their judgments on the graphic organizer. If students disagree, ask a few students on each side to explain why they rate the consequences as they do and then ask for a show of hands from the class, indicating the majority and minority views (if there is a minority view) on the graphic organizer.

- Let's look at the first consequence, "Riders would pass out at the bottom of the loop." Work in your groups to determine whether you have enough relevant information to judge the likelihood of this prediction. Students usually respond that they read in the source material that people passed out on the circular loop of the flip flap roller coaster at Coney Island. They also read that people pass out at force factors more than 4.5. Write, "If the force factor is more than 4.5, then riders will pass out" in the evidence box on the same line as the consequence.

- What additional information do you need to provide evidence that it is likely that a rider would pass out on the 40m circular loop proposed for Desperado? Students usually respond that the force factor at the bottom of the loop could be calculated and that if it were more than 4.5, then riders would be likely to pass out. Have students work together in their groups to calculate the force factor at the bottom of the loop. Assist them where necessary and then ask the groups to report. If there is a difference of opinion, ask the students who differ to explain how they got their result and discuss the calculations with the class. ANSWER: *Force Factor at Bottom is 6.3.* Enter this answer in the Evidence box adjacent to the consequence and the calculations that support it in the "Calculation" box next to it.

- When you look at the evidence for passing out at the bottom of the loop, is it likely, unlikely, or uncertain that a force factor of 6.3 would cause a rider to pass out? Students usually respond that it is likely that at least some riders will pass out. Place an "L" above the arrow that connects the evidence box with the consequences box.

- Let's finish the rest of the consequences. In your groups, select one consequence and determine what relevant information you have related to it and whether you can get additional information that provides evidence for or against it. If calculations are necessary to derive the evidence, do them and add them and the relevant evidence to the appropriate boxes. If students seem uncertain about the procedure, repeat it with another consequence. Give students plenty of time to do the necessary calculations and to complete the evidence and calculation boxes. Then, have each group report their results and enter their responses on the class graphic organizer. As information is entered as evidence relevant to each consequence, be sure to ask whether the evidence makes the projected consequence likely, unlikely, or uncertain. Write an "L" for likely, a "U" for unlikely or a "?" for uncertain over the arrow connecting each evidence box with the relevant consequence. Calculations reveal that the force factor at the top of the loop is .51, not enough to keep people in their seats, and that the range in force factors is broad as the car transverses the loop. Students recognize that this supports the prediction that riders will fall out of the car at the top of the loop (unless restrained) and that they may get nauseated as the car goes around the loop. At the same time, they recognize that such a ride will, in all likelihood, give the riders a great thrill.

- **Is the circular loop a feasible option for redesigning Desperado?** Ask students to discuss this question in their groups. Most students agree that the risk of harm on a circular loop and the probable high cost of construction and retrofitting restraint devices counts heavily against using a circular loop, despite the high degree of excitement that is also likely.

- **Evaluate the option of using a clothoid loop. Do this assignment as homework.** The results can be reviewed at the chalkboard or on a transparency the next day. Be sure to explain how a clothoid loop differs from a circular loop. Generally, students agree that a clothoid loop is a much more viable option than a circular loop, especially because the force factors are within the boundaries beyond which riders can fall out at the top or pass out at the bottom. However, they also recognize that no final decision about adding a clothoid loop should be made by the designers of Desperado until a careful cost analysis is conducted.

THINKING ABOUT THINKING

- **What did we do in this activity that may be different from the usual ways we make decisions?** Students usually respond by saying that they often don't think about consequences and rarely, if ever, take the time to figure out how likely the consequences are. Ask students to recall the questions they asked and answered as they thought about which type of loop would be better for Desperado. Compare these to the questions on the thinking map for predicting the consequences of options.

- **When would you use this way of thinking? In what circumstances might you not use it?** Some students say that this way of thinking should be used for important decisions when we have the time to think about things. Other students agree that we should use it for important decisions but say that even we don't have enough time to gather all of the information needed; we should always do as much as we can to think about these decisions in this way.

- **Is it valuable to take the importance and likelihood of consequences into account in making decisions? Explain.** Students usually respond that it is valuable because one can tell how much danger is involved in doing certain things and also how likely it is that your decision will have good consequences.

- **How would using this process affect your confidence in your decisions?** Students usually respond that they would feel very confident since they would have a better idea ahead of time about what was going to happen if they made certain decisions.

- **What insight did evaluating consequences give us about the way roller coasters work that we might have missed if we had not examined them this way?** ANSWERS VARY BUT USUALLY INCLUDE: *When I thought about the evidence for the negative consequences a rider would face, I realized how we can calculate exactly how fast a moving car will go on a roller coaster and how much force will be exerted. It made me realize how what we've been studying in physics really helps me understand how things work in the world.*

APPLYING THINKING

Immediate Transfer

- Calculate friction on Desperado and then introduce it into your calculations for the radii of the circular and clothoid loops. Have students then predict the consequences of friction on force factors at the top, side, and bottom of each loop. Students can then adjust the dimensions of each loop to compensate for losses in energy due to friction.

- Identify examples of decisions you are making that require careful prediction of consequences. Use this prediction strategy to think through one of them. Students can do this assignment as homework and share their results the next day.

Reinforcement Later

- For a given loop, predict the consequences of changes in the height and slope of the leading hill on the force factors inside the loop.

- Many people have expressed concern about the impact of global warming on our lives in the 21st Century. Use the strategy for predicting embedded in the way you have predicted the consequences of options to make judgments about the impact of an increase of 2 degrees in the average temperature at sea level. Do the same for 5 degrees. Make all your calculations explicit.

- In American History, you have studied Martin Luther King's decision whether to proceed past the bridge in Selma, Alabama, until the Supreme Court ruled on the constitutionality of the march to Montgomery. Think through this decision using this strategy for predicting consequences. Do you think he make the right decision? Why?

WRITING EXTENSION

Write a report including illustrative diagrams and all of your calculations to Arrow Dynamics about the relative merits of a circular and clothoid loop of 40 meters located between the second and third hills.

ASSESSING STUDENT PREDICTIONS

To assess how skillfully students predict the likelihood of consequences and evaluate their effect in a decision-making context, ask them to write an essay on a controversial topic calling for a decision. Ask them to make their thinking explicit. Determine whether each step in the verbal map for predicting consequences (options, consequences, evidence, likelihood, and importance) was taken into account in making their decision of the best course of action.

REDESIGNING DESPERADO
SOURCE MATERIAL ON THE HISTORY, DESIGN, AND PHYSICS OF ROLLER COASTERS

HISTORY OF THE ROLLER COASTER

As long ago as 1650, Russians were building 70-foot-high timber ice slides for public amusement. Two hundred years later, the French switched from ice to closely space rollers and used a track for the ramp. The name "roller coaster" has stuck. Wheels on the cars replaced rollers on the track by 1875. In 1884, the first oval coaster, The Serpentine Railway, was built. It reached a top speed of 15 mph. By the end of the 19th century, "scenic railways" could be found throughout America. Picturesque scenes were painted on the walls of lighted tunnels through which the coasters passed.

Early attempts to build and successfully market a circular looping ride met with failure, with one notable exception, Coney Island's Flip-Flap, which produced more than 6 g's entering the loop; few rode it, but many paid full admission to see the hazardous effects of the ride on the few who risked experiencing it. Severe neck injuries, cervical trauma, and blackout often resulted.

The first figure 8 roller coasters heralded what has become known as the "Golden Age of Roller Coasters." These wooden giants curved back and over themselves. By 1929, America had nearly 1,500 of them. By far the most famous coaster ever built, and considered by all to be the standard against which the illusion of roller coaster danger is measured, is the Cyclone, built in 1927 at Astroland in Coney Island. Cyclone is still in operation today.

The Great Depression of the 1930s hit the amusement industry particularly hard; and although a number of coasters were built after World War II, fewer than 200 wooden giants could be found by 1960.

With Disneyland and the introduction of the steel roller coaster ("steelie"), theme parks created a new demand for faster, taller, and longer coasters. The resulting boom in coaster business was led by Arrow Dynamics, which designed Disney's first steel coaster, The Matterhorn Bobsled ride. Its innovative design and construction using tubular steel track and nylon wheels has become the archetype for coasters produced since. Among Arrow's innovations were development of the corkscrew and the teardrop-shaped clothoid loop.

Desperado

Located outside of Las Vegas at Wild Bill's Casino Resort, Desperado (1994) is the tallest (209 ft) roller coaster in North America and has the longest drop (225 ft). At 5,900 feet long, it is second only to The Beast (7,400 ft), in length. Reaching speeds in excess of 80 mph at the bottom of its first hill, Desperado is the fastest roller coaster in the world and requires only 2 minutes and 43 seconds to transit its mile-plus length. Desperado cost $10 million to build; is made of tubular steel; and its second hill, at 155 feet, is still taller than the tallest of wood coasters. However, Desperado lacks inversion, vertical loops, corkscrews, and boomerangs.

FORCES AND VERTICAL CIRCLES

According to Newton's first law, an object moves at a constant speed in a straight line unless an unbalanced force acts on it. To make an object move in a circle, an unbalanced force directed toward the center of the circle must be applied. The sum of the forces applying this net, center-directed (centripetal) force is fixed.

$$\sum F = F_c = mv^2/r$$

When a person on an amusement park ride is moving in a vertical loop, two forces are involved in supplying this centripetal force—the seat and gravity. Gravity comes in the form of the rider's weight.

Taking a closer look at the top, bottom, and side forces acting on a rider moving along a vertical loop should help you evaluate looping roller coaster problems.

When the rider is at the bottom of the loop, the force of the seat on the rider and the gravitational force on the rider are in opposite directions with the seat force aimed toward the center of the loop.

$$\sum F = mv^2/r$$
$$F_{SEAT} - mg = mv^2/r$$

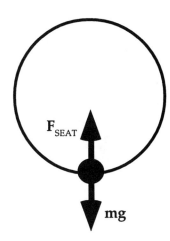

to determine the centripetal force acting toward the center of the circle.

$$\sum F = mv^2/r$$
$$F_{SEAT} + mg = mv^2/r$$

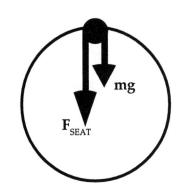

When the rider is at the side of the loop, only the seat force that is directed toward the center can provide the centripetal force needed for curved motion. The gravity force is pulling straight down and is not a consideration in determining the center-directed centripetal force.

$$\sum F = mv^2/r$$
$$F_{SEAT} = mv^2/r$$

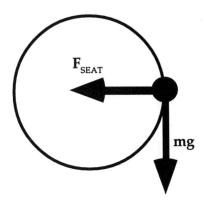

When the rider is at the top of the loop, the seat force and gravity work together to create the curved motion. Therefore, they must be added

If the rider is moving fast enough, the seat has to exert a large force to keep the rider in the loop and the rider *feels* right-side-up even though he may be upside down.

Often, the ratio of the seat force, F_{SEAT}, and the rider's weight, mg, are used to describe the effects of circular motion on the rider. The greater the force factor, the heavier the apparent weight experienced by the rider.

force factor = F_{SEAT}/mg

The first looping coasters were circular. However, in order for the coaster to be going fast enough at the top to stay pinned to the track, the speed at the bottom had to produce forces more than 6 times normal weight (force factor = 6).

The clothoid or teardrop-shaped loop combines a large radius arc at the bottom (entry) part of the loop with a much smaller radius arc at the top. With its large radius at the bottom of the arc becoming a small arc at the top, forces at the bottom of the arc are minimized while the centripetal force needed at the top is large enough so that the force on the person can be maintained at more than his weight, while the forces on the rider as he is entering the loop at high speed remain low.

DESPERADO: A CLOSER LOOK

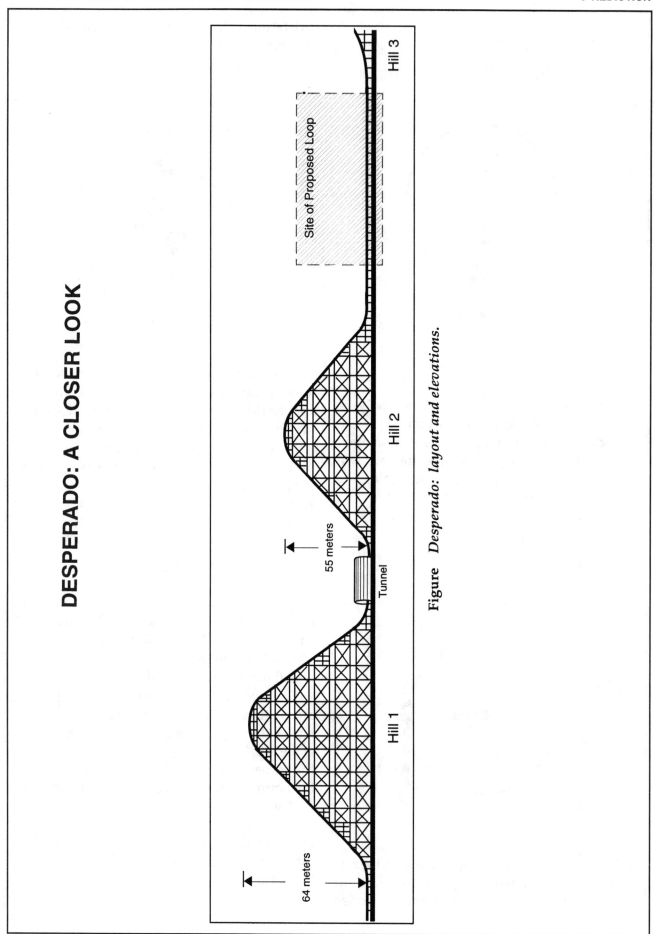

Figure *Desperado: layout and elevations.*

Note that the clothoid loop has two parts: the long shallow arc leading into the clothoid and the closed loop, which includes the small radius arc at the top.

When computing centripetal force, force factors, and g's at points along a clothoid loop, the entering radius of the shallow arc is R and the radius at the top of the loop is r. Refer to the layout diagram of Desperado (left) to locate the loop and determine height measurements.

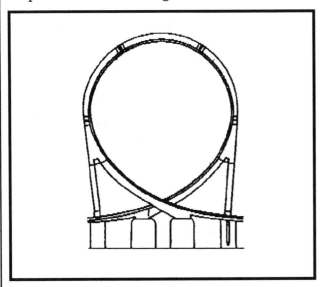

Fig. Typical clothoid loop.

Useful Formulas: (Six Flags Physics)

Acceleration in g's

$$gs = a/9.8 \text{m/s}^2$$

Centripetal Acceleration

$$a_c = v^2/r = r\pi^2 r/T^2$$

Centripetal Force

$$F_c = ma = mv^2/r = m4\pi^2 r/T^2$$

Gravitational Potential Energy

$$PE = mgh$$

Kinetic Energy

$$KE = 1/2mv^2$$

Circular Tangential Speed

$$v = 2\pi r/T$$

Newton's Second Law

$$f = ma$$

Force Factor

$$\text{force factor} = \text{applied force/wt}$$

Design Constraints

- **Acceleration:** People are uncomfortable at accelerations of more than 3.5 times gravity and begin to pass out at about 4.5 to 6 times gravity. On the other hand, people can feel equally uncomfortable experiencing accelerations approaching zero, which leave them with a sense of "falling out." Moreover, at accelerations approaching 6 times gravity, structural design becomes exponentially more expensive and coaster safety becomes increasingly more difficult to guarantee. Cases of vascular medical emergency have been caused by excessive roller coaster acceleration. Similarly, whenever accelerations approach zero g's, the likelihood of riders' falling and the cost of rider restraint systems increases quickly.

- **Size of circular loop:** Because of space restrictions within Desperado, a coaster loop is restricted to a radius of 20 meters (40m dia).

- **Size of clothoid loop:** Space restrictions between the base of the second hill and the clothoid loop restrict the radius of the shallow arc to 45 meters and the radius of the enclosed loop to 9 meters.

- **Height of leading hill:** 55 meters; not feasible to alter the height of the existing structure.

- **A frictionless system:** To simplify calculations, assume that Desperado operates in a friction-free environment.

- **Rider's weight:** 50 kg

- **Velocity at the top of the leading hill:** just enough to push over.

REDESIGNING DESPERADO
Teaching Notes: solving for force factors
on circular loops

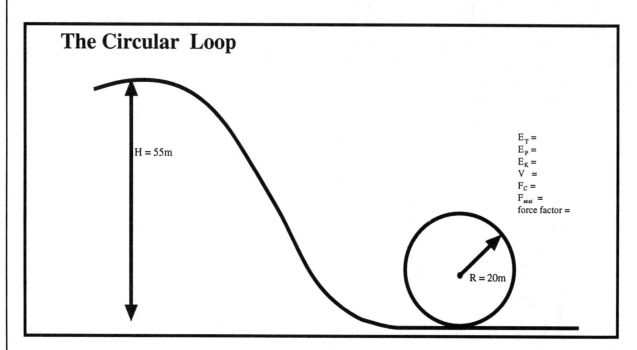

The Circular Loop

H = 55m

R = 20m

$E_T =$
$E_P =$
$E_K =$
$V =$
$F_C =$
$F_{seat} =$
force factor =

Find total energy, E_T

Total energy of the system is the potential energy of the rider at the top of the leading hill.

$E_T = E_P = mgh$

$E_T = 50kg(9.8N/kg)55m = 26,950J$

Find force factor at the bottom of the loop.

$E_T = E_P + E_K$

At the bottom of the loop, h = 0, therefore

$E_P = 0$ and $E_K = E_T = 26,950J$

To find velocity of the rider, solve the equation for kinetic energy for v.

$E_K = 1/2mv^2$; $v = \sqrt{E_K/m} = 32.8m/s$

Find centripetal force at the bottom.

$F_C = mv^2/r = 50kg(32.8m/s)^2/20m = 2690N$

Find total forces acting on the rider at the bottom.

$F_C = \Sigma Forces = F_{SEAT} - mg = 2690N$

Find seat force acting on rider.

$F_{SEAT} = F_C + mg = 2690N + 50kg(9.8N/kg) = 3180N$

Find force factor at the bottom of the loop.

force factor = appl'd force/wt = F_{SEAT}/wt = F_{SEAT}/mg

force factor = 3180N/50kg(9.8N/kg) = 6.5

Find force factor at the top of the loop.

Calculate the force factor at the top of the loop in the same manner. Same total energy of the system; subtract potential energy at height of 40 meters to determine kinetic energy of rider at top of loop; solve kinetic energy equation for velocity at top; use v to determine centripetal force at top; determine seat force and force factor at top.

$E_T = 26,900J$: $E_P = mgh = 50kg(9.8N/kg)(40m) = 19,600J$;

$E_K = E_T - E_P = 7,350J = 1/2mv^2$; $v = 17.1 m/s$;

$F_C = mv^2/r = 50kg(17.1m/s)^2/20m = 740N$

$F_C = \Sigma Forces = F_{SEAT} + mg$; $F_{SEAT} = F_C - mg$

$F_{SEAT} = 740N - 490N = 250N$;

force factor = F_{SEAT}/mg = 250N/490N = .51

Find force factor at the side of the loop.

When the rider is at the side of the loop, weight is not a factor in calculating forces which provide the centripetal force needed for curved motion. Therefore, the seat force is equal to the centripetal force acting on the rider. Calculate centripetal force by first finding potential energy, kinetic energy, and velocity halfway up the loop, and then solving for centripetal force.

$E_T = 26,900J$; $E_P = mgh = 50kg(9.8N/kg)(20m) = 9800J$;

$E_K = E_T - E_P = 17,100J = 1/2mv^2$; $v = 26.2m/s$;

$F_C = mv^2/r = 50kg(26.2m/s)^2/20m = 1716N$;

$F_{SEAT} = F_C = 1716N$;

force factor = F_{SEAT}/mg = 1716N/490N = 3.5

REDESIGNING DESPERADO
Teaching Notes: solving for force factors
on clothoid loops

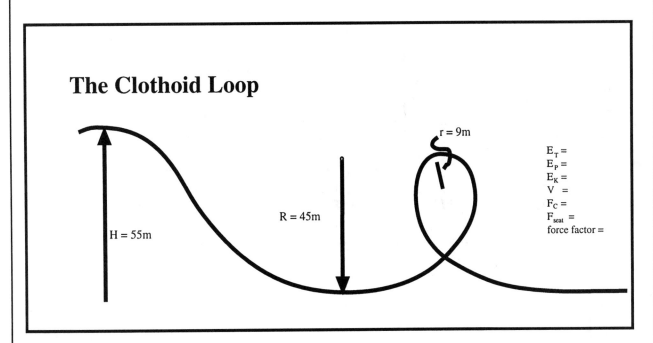

The Clothoid Loop

H = 55m

R = 45m

r = 9m

$E_T =$
$E_P =$
$E_K =$
V =
$F_C =$
$F_{seat} =$
force factor =

Find the total energy, E_T

The clothoid loop has a constantly changing radius of curvature. This means the force at the bottom can be minimized while the force at the top can be maximized by just changing the radius. At the top of the lead hill

$E_K = 0$

$E_T = E_P = mgh = 50kg(9.8N/kg)(55m) = 26,950J$

Find force factor at the bottom of the loop.

Forces entering the bottom of loop are calculated using the 45 meter radius, R;

$E_T = E_P + E_K$; $E_K = E_T - E_P = 26,950 - 0 = 26,950J$;

To find velocity of the rider solve the equation for kinetic energy for v.

$E_K = 1/2mv^2$; $v = \sqrt{EK/m} = 32.8m/s^2$

Find centripetal force at the bottom.

$F_C = mv^2/r = 50kg(32.8m/s)^2/45m = 1195N$

Find total forces acting on the rider at bottom.

$F_C = \Sigma Forces = F_{SEAT} - mg = 1195N$

Find seat force acting on rider.

$F_{SEAT} = F_C + mg = 1195N + 50kg(9.8N/kg) = 1685N$

Find the force factor at the bottom of the loop.

force factor = $F_{SEAT}/mg = 1685N/490N = 3.4$

Find the Force Factor at the top of the loop.

As the rider approaches the top of the loop, he enters a circle with a much smaller radius; it therefore takes a greater force to stay in a circle. The calculation, upside down at the top of the clothoid, is the same used as at the top of the circular loop. Note that the top of the clothoid

is the same height as the top of the circular loop and that the speed of the rider at any given height is the same; it is the amount of force necessary to turn the circle that increases on a clothoid.

$E_T = 26,900J$; $E_P = mgh = 50kg(9.8N/kg)(40m) = 19,600j$

$E_K = E_T - EP = 7,350J = 1/2mv^2$; $v = 17.1$ m/s;

$F_C = mv_2/r = 50kg(17.1m/s)^2/9m = 1625N$

$F_C = \Sigma Forces = F_{SEAT} + mg$; $F_{SEAT} = F_C - mg$;

$F_{SEAT} = 1625N - 490N = 1135N$

force factor = $F_{SEAT}/mg = 1135N/490N = 2.3$

Find the force factor at the side of the loop.

Usually the top 130 degrees of the loop is circular with radius r; the clothoid of regularly changing radius makes up the ascending and descending parts of loop. The basis for estimating the force factor at the side of the clothoid is:

1) When the rider is perpendicular to the ground the loop is already very tight with a radius closer to r than to R, so we take the radius to be 17 meters.

2) The height at the side of the loop is the height of the loop minus the radius r, or 31 meters.

Then we can calculate the centripetal force just as we did for the circular loop.

$E_T = 26,900J$; $E_P = mgh = 50kg(9.8N/kg)(40m-9m) = 15190J$;

$E_K = E_T - E_P = 11710J = 1/2mv^2$; $v = 21.6m/s$;

$F_C = mv^2/r = 50kg(21.6m/s)^2/17m = 1372N$;

$F_{SEAT} = F_C = 1372N$;

force factor = $F_{SEAT}/mg = 1372N/490N = 2.8$

Sample Student Responses • Desperado

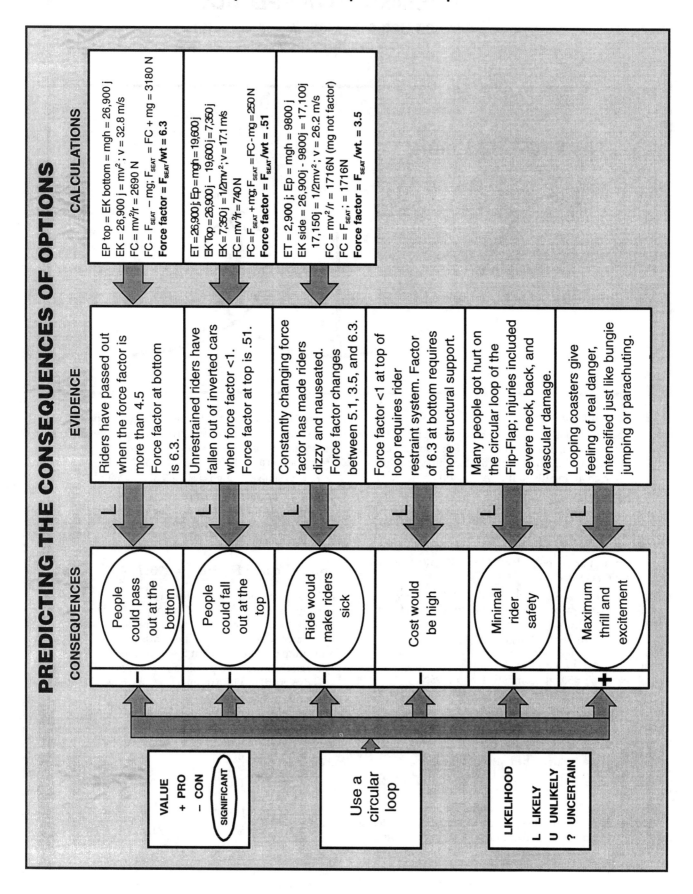

PREDICTING THE CONSEQUENCES OF OPTIONS

CALCULATIONS

EP top = EK bottom = mgh = 26,900 j
EK = 26,900 j = mv²; v = 32.8 m/s
FC = mv²/r = 2690 N
FC = F$_{SEAT}$ − mg; F$_{SEAT}$ = FC + mg = 3180 N
Force factor = F$_{SEAT}$/wt = 6.3

ET =26,900j; Ep=mgh=19,600j
EKTop=26,900j − 19,600j=7,350j
EK=7,350j = 1/2mv²;v=17.1 m/s
FC=mv²/r=740N
FC=F$_{SEAT}$ +mg; F$_{SEAT}$ = FC−mg =250N
Force factor = F$_{SEAT}$/wt = .51

ET = 2,900 j; Ep = mgh = 9800 j
EK side = 26,900 j - 9800j = 17,100j
17,150j = 1/2mv²; v = 26.2 m/s
FC = mv²/r = 1716N (mg not factor)
FC = F$_{SEAT}$ = 1716N
Force factor = F$_{SEAT}$/wt. = 3.5

EVIDENCE

Riders have passed out when the force factor is more than 4.5
Force factor at bottom is 6.3.

Unrestrained riders have fallen out of inverted cars when force factor <1.
Force factor at top is .51.

Constantly changing force factor has made riders dizzy and nauseated.
Force factor changes between 5.1, 3.5, and 6.3.

Force factor <1 at top of loop requires rider restraint system. Factor of 6.3 at bottom requires more structural support.

Many people got hurt on the circular loop of the Flip-Flap; injuries included severe neck, back, and vascular damage.

Looping coasters give feeling of real danger, intensified just like bungie jumping or parachuting.

CONSEQUENCES

People could pass out at the bottom —

People could fall out at the top —

Ride would make riders sick —

Cost would be high —

Minimal rider safety —

Maximum thrill and excitement +

VALUE
+ PRO
− CON
SIGNIFICANT

Use a circular loop

LIKELIHOOD
L LIKELY
U UNLIKELY
? UNCERTAIN

Sample Student Responses • Desperado

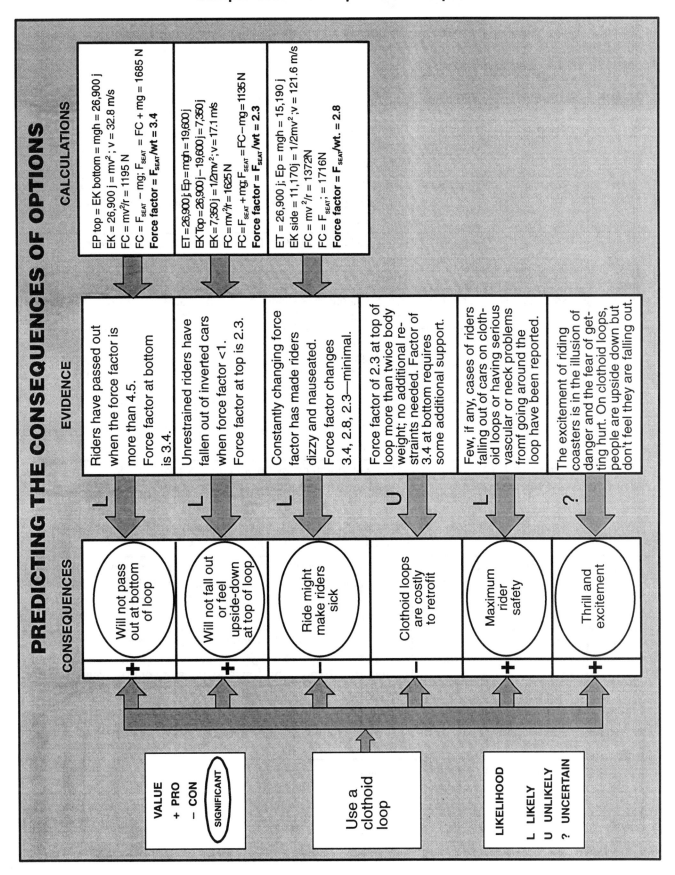

PREDICTING THE CONSEQUENCES OF OPTIONS

CALCULATIONS

EP top = EK bottom = mgh = 26,900 j
EK = 26,900 j = mv²; v = 32.8 m/s
FC = mv²/r = 1195 N
FC = F_SEAT − mg; F_SEAT = FC + mg = 1685 N
Force factor = F_SEAT/wt = 3.4

ET = 26,900 j; Ep = mgh = 19,600 j
EKTop = 26,900 j−19,600 j = 7,350 j
EK = 7,350 j = 1/2mv²; v = 17.1 m/s
FC = mv²/r = 1625 N
FC = F_SEAT + mg; F_SEAT = FC−mg = 1135 N
Force factor = F_SEAT/wt = 2.3

ET = 26,900 j; Ep = mgh = 15,190 j
EK side = 11,170 j = 1/2mv²; v = 121.6 m/s
FC = mv²/r = 1372N
FC = F_SEAT; = 1716N
Force factor = F_SEAT/wt. = 2.8

EVIDENCE

Riders have passed out when the force factor is more than 4.5.
Force factor at bottom is 3.4.

Unrestrained riders have fallen out of inverted cars when force factor <1.
Force factor at top is 2.3.

Constantly changing force factor has made riders dizzy and nauseated.
Force factor changes 3.4, 2.8, 2.3—minimal.

Force factor of 2.3 at top of loop more than twice body weight; no additional restraints needed. Factor of 3.4 at bottom requires some additional support.

Few, if any, cases of riders falling out of cars on clothoid loops or having serious vascular or neck problems fromf going around the loop have been reported.

The excitement of riding coasters is in the illusion of danger and the fear of getting hurt. On clothoid loops, people are upside down but don't feel they are falling out.

L   L   L   U   L   ?

CONSEQUENCES

(Will not pass out at bottom of loop) **+**

(Will not fall out or feel upside-down at top of loop) **+**

(Ride might make riders sick) **−**

(Clothoid loops are costly to retrofit) **−**

(Maximum rider safety) **+**

(Thrill and excitement) **+**

VALUE
+ PRO
− CON
(SIGNIFICANT)

Use a clothoid loop

LIKELIHOOD
L LIKELY
U UNLIKELY
? UNCERTAIN

PREDICTION LESSON CONTEXTS

The following examples have been suggested by classroom teachers as contexts to develop infused lessons. If a skill or process has been introduced in a previous infused lesson, these contexts may be used to reinforce it.

GRADE	SUBJECT	TOPIC	THINKING ISSUE
6–8	Science	Hostile environment	Predict whether blue green algae can survive in desert conditions. Explain what adaptations might be useful to ensure survival.
6–8	Science	Inherited traits	Use a Punnett square to predict the chances that a blue-eyed mother and a brown-eyed father will have a brown-eyed child. Explain your prediction.
6–8	Science	Space survival	Predict what it would be like to live on board a space station in orbit around the earth. Explain your reasons for the predictions.
6–8	Science	Ecology	What is likely to result in the biosphere from the destruction of rain forests? Explain why you predict these results.
6–8	Science	Environment	Use information about population growth, energy use, global warming, and waste disposal to predict what a day in your community might be like in 2020. Explain how this information leads to your predictions.
6–8	Science	Ballistics	Predict which projectile will land further from the point of firing—one fired at an angle of 30, 45, 90, or 125 degrees. Explain your answer.
6–8	Science	Astronomy	What would you predict would be the consequences of a meteor, one mile in diameter, hitting the earth? What are your reasons for thinking your results would occur?
6–8	Science	Weather	What would you predict would happen to Miami if a Force 5 hurricane hit it? Explain your reasons.
6–8	Science	Ocean-ography	Predict what it would be like to live on board a habitat at the bottom of the ocean for an extended period of time. Explain what precautions you would take and preparations you would be likely to make.
9–12	Biology	Population growth	Predict the outcome of mixing two different species of paramecium (P. aurelia & P. caudatum) in a culture with limited food. Explain the consequences of your predictions.
9–12	Biology	Adaptive advantage	Predict adaptive advantages that nocturnal animals might evolve; find examples in nature to validate your predictions. (Consider animals such as the owl, the bat, and deep ocean fishes.)
9–12	Biology	Population biology	Predict the consequences for a regional state forest of making hunting illegal. (Consider positive and negative consequences and explain which is most important.)
9–12	Biology	Environ-mental science	What are the consequences of using DDT to kill mosquitoes causing malaria in a suburban community? Consider how the pesticides affect flora and fauna in the ecosystem

PREDICTION LESSON CONTEXTS

GRADE	SUBJECT	TOPIC	THINKING ISSUE
9–12	Biology	Greenhouse effect	Predict the consequences of the greenhouse effect on the ecosystem. (Consider temperature effects in the atmosphere, in the oceans, on rainforests, and on arable land.)
9–12	Biology	Photosyn-thesis	Predict the effects of growing plants in different frequencies of light (i.e., green, red, white). Explain your reasoning and relate your explanations to the colors of various plants.
9–12	Biology	Homeostasis	Predict the human homeostatic responses to increases and decreases in room temperature. (Consider sweating, vasoconstriction, shivering, goosebumps, vasodilation, behaviors like adjusting clothes and leaving the room). Explain why you predict what you do.
9–12	Biology	Hormones: adrenaline	Predict the consequences, in terms of physiological responses, of fear. (Consider the effects of adrenaline on pulse, blood pressure, pain receptors, circulation, etc.). Explain why you predict what you do.
9–12	Chemistry	Periodic table	Predict the chemical reactivity of a molecule of ethane, methane, and acetlyene when given the magnitude of their bond dissociation energy. Explain your prediction.
9–12	Chemistry	Bonding	Given two atoms, predict whether they will form covalent, ionic, or hydrogen bonds. Illustrate your predictions with electron dot structures and, where appropriate, discuss in terms of valence electrons.
9–12	Physics	Resultants and equilibrants	Predict the resultant of two or more vectors and describe its equilibrant. Describe an example from your experience to illustrate how you determined the equilibrant.
9–12	Physics	Simple machines & mechanical advantage	Predict which of the simple machines—wheel and axle, wedge, lever, jackscrew, pulley, or inclined plane—would have the greatest ideal mechanical advantage and the most utility in raising a 2-ton safe 10 feet.
9–12	Physics	Motion	Use principles of motion and force to predict how cars are likely to move in a variety of driving conditions.
9–12	Physics	Trajectories	Plot exactly what trajectory a missile of a given weight fired at a 45-degree angle from a given location would have and where it would land. Explain your reasoning. Plot variations in the angle of launch and force to get it to the same target.
9–12	Physics	Trajectories	In order to be able to go into orbit around the planet Mars, a rocket of a given weight would have to be fired with the right thrust at the right angle from Earth at the right time of year. Determine the average weight of a Saturn rocket and its rate of weight loss as it burns fuel, and plan a rocket launch that would result in the final stage of the rocket entering into orbit around Mars. Explain why you make the predictions you do.

CHAPTER 16
GENERALIZATION

Why Is Skillful Generalization Necessary?

We often get to know someone better by learning more about that individual. I may find out that a friend is planning a trip to Hawaii because she is interested in birds, and Hawaii is noted for exotic birds. This knowledge may enrich our relationship. If I, too, have an interest in birds, I know that she is someone with whom I can share this interest.

As important as learning such new things about my friend may be, what I learn is limited. It informs me only about an individual. If I want to know if I can also share this interest with other individuals, what I have just learned doesn't help much. I have to learn about other people's interests individually.

Sometimes, though, we expand what we know by learning something about *types* of things. Roses smell sweet. Antelope run fast. Diamonds are hard. These are generalizations that apply°to *all* things of a certain sort (roses, antelope, or diamonds), not just to one individual. While we learn things about specific individuals, we can also learn generalizations that inform us about all individuals of a certain type without having to investigate each of these individuals separately.

The Importance of Generalizations in Our Thinking

Using generalizations makes our thinking much more efficient. I know that roses smell sweet. On the basis of this general knowledge, the next time that I want to buy sweet-smelling flowers, I can decide quickly that I will buy roses. I do not have to sniff around until I find an individual flower that smells sweet. Accepting the general statement licenses me to be confident that *any new* individual rose I get will smell sweet.

Our knowledge of general statements can help us avoid danger, too. I know that certain kinds of jellyfish have a poisonous sting. If I recognize the distinctive crown of a Portuguese man-of-war jellyfish floating on the surf, I can be confident of its harmful nature because of my general knowledge of this kind of jellyfish.

Common Problems about the Way We Generalize

Because of their efficiency, generalizations run through much of our thinking. We learn many generalizations from others but also develop them ourselves.

Our tendency to generalize, however, can lead ourselves and others astray. It is risky to accept uncritically all generalizations that occur to us or that we hear from others. Not all generalizations are well-founded. If the large, juicy grapes in a grocery store advertisement look wonderful, I may think that all the grapes at the supermarket are large and juicy. Though they may be, often we find that the quality of what was advertised isn't as good as it appeared in the ad. It is easy to make hasty generalizations.

Stereotyping and various forms of bias and prejudice also revolve around faulty generalizations. On my first visit to a school, I may find that there is a group of students in the corridors being rowdy. Because of this, I might say to myself, "The students in this school are rowdy and poorly behaved." This is a generalization that could affect my attitudes about this school. Indeed, this could be a case of stereotyping. I might carry this belief with me and think of anyone who came from that school as rowdy and poorly behaved.

This generalization may well turn out to be inaccurate. Such a small sample is not adequate support for this generalization. The supporting evidence is too narrow. These could be the only rowdy students in the school. Are they representative of the rest of the students? If they are not representative, then the sample is too narrow. If we are going to use generalizations in our thinking, we should make sure that they are well-founded.

Generalizing in either of these circumstances (based on a small sample or based on a narrow sample that does not represent the whole) is often called the fallacy of "hasty generalization." Figure 16.1 contains a list of these problems in generalizing.

COMMON PROBLEMS WITH THE WAY WE GENERALIZE

1. We often generalize based on a small number of individuals.

2. We often generalize based on a sample of individuals without knowing whether or not it is representative of the whole group.

Figure 16.1

What Does Skillful Generalization Involve?

Usually, generalizations are supported by studying a number of individuals (a "sample") that fall under the generalization. This is why generalizing is often characterized as "inductive" reasoning. We start with knowledge about individuals and infer general conclusions.

A Strategy for Determining Whether a Generalization is Well-founded

Since most generalizations are about all things of a certain type, like *all* roses or *all* Portuguese man-of-war jellyfish, we can't hope to support most of them completely. Still, the better the sample we study, the more likely the generalization will be correct.

Before we generalize, we should make sure we gather information about a number of individuals. How large a number is needed to support a generalization? This depends on the selection procedure used. What is really important is that we should make sure that these individuals are *representative* of the whole group. This involves not just the size but also the selection of the group of individuals that we use as a basis for the generalization.

Using a nonrepresentative, or skewed, sample is one of the most common mistakes in generalization. If I think that my state favors a certain political candidate because everyone in my neighborhood does, I am making this mistake. Maybe my neighborhood has some special reason for preferring this candidate and people in other communities do not.

Choosing a large number of individuals *at random* from the total population is one way to minimize mistakes from misleading samples.

Doing this effectively, however, requires that we sample a relatively large number. If, on the other hand, we pick individuals *at random from representative groupings* (e.g., some from each of the different areas of the city), we may not need such a large number. If we do not use either method, the sample may be biased.

Whichever method of selection we choose, before we can support a generalization sufficiently, we should have reasons for thinking that our sample is representative of the whole group. Attending to the sample size and its selection allows us to judge the strength of support it provides for the generalization. This is what makes generalizing skillful.

The thinking map for generalization in figure 16.2 prompts reflection on the adequacy of a sample for well-founded generalization.

SKILLFUL GENERALIZATION

1. What generalization is suggested?

2. What sample is needed to support that generalization?

3. Is the sample being used large enough?

4. Is the sample being used like the whole group?

5 Is the generalization well supported by the sample?

6. If not, what additional information is needed to support the generalization?

Figure 16.2

A Graphic Organizer for Skillful Generalization

The graphic organizer in figure 16.3 provides spaces for writing a description of the sample used to support a generalization and for assessing the characteristics of the sample.

How Can We Teach Students to Generalize Skillfully?

Teaching this skill should involve having students work with examples of specific generalizations. They should be prompted to think about the sample that the generalization is based on and to determine its size and representativeness. On the basis of this determination, they

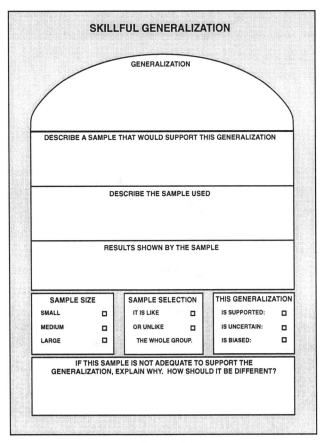

SKILLFUL GENERALIZATION

GENERALIZATION

DESCRIBE A SAMPLE THAT WOULD SUPPORT THIS GENERALIZATION

DESCRIBE THE SAMPLE USED

RESULTS SHOWN BY THE SAMPLE

SAMPLE SIZE		SAMPLE SELECTION		THIS GENERALIZATION	
SMALL	☐	IT IS LIKE	☐	IS SUPPORTED:	☐
MEDIUM	☐	OR UNLIKE	☐	IS UNCERTAIN:	☐
LARGE	☐	THE WHOLE GROUP.	☐	IS BIASED:	☐

IF THIS SAMPLE IS NOT ADEQUATE TO SUPPORT THE
GENERALIZATION, EXPLAIN WHY. HOW SHOULD IT BE DIFFERENT?

Figure 16.3

should be given an opportunity to reflect about the support the sample provides for the generalization. The goal of your teaching should be to help students raise these questions and gather information to answer them when they consider accepting a generalization.

In situations in which generalizations are not founded on an adequate sample, students can use the descriptions that they wrote of what an adequate sample would be like as a basis for an investigation. This may lead to additional support that may make the generalization well-founded. For example, students can develop a polling strategy that they think creates an adequate sample. On the basis of this sample, they can legitimately generalize about students' attitudes regarding recycling.

Constructing Lessons to Teach Students Skillful Generalization

Using the graphic organizer for skillful generalization will help your students evaluate gen-

eralizations. In addition, though, you should use the thinking map to make the strategy explicit. You should also provide opportunities for students to reflect on the process and to use it in different contexts. When you combine the thinking map and the graphic organizer in the infusion lesson format, you can teach students skillful generalization.

Contexts in the Curriculum for Lessons on Skillful Generalization

Generalizations appear in almost every subject students learn. Science instruction relies heavily on generalizations: Hot air rises, hummingbirds migrate in the winter, and acid is corrosive to metals. Generalizations about social groups and societies appear in social studies (e.g., functioning communities feature a division of labor, industrialization raises the standard of living). Students make generalizations about people as they read stories, poems, and novels in language arts. In art, generalizations students learn about materials guide their use (clay can be shaped by the pressure of our fingers but retains its shape if left undisturbed). That bats, baseballs, footballs, and basketballs behave in certain ways is the basis for decisions that students continually make in playing sports.

Generalization plays a central role in science. The methodology in science is utilized not only to support judgments about the causes of specific events in the world; its power lies in its goal of discovering and establishing the laws of nature by which our world works. These laws of nature are general in character. Hence, in both the substantive science curriculum and the laboratory work that students do in physics, biology, and chemistry, contexts abound for lessons on skillful generalization.

The curriculum features two contexts in which students can become engaged with generalizations in science:

- *Generalizations others have made are presented directly to students.* For example, students learn about the properties of certain materials, e.g., metals. They also learn the general laws that govern the behavior of these materials (e.g., the gas laws).

• *Students are asked to make their own generalizations based on information that is provided or that they gather.* For example, many hands-on science activities involve students in situations in which they are asked to draw general conclusions from their own observations. In many science curricula, students are also asked to design studies that will involve data collection sufficient to support generalizations. In lessons on generalization, one main focus should be on the type of sample needed. Planning a research project in this way is a particularly useful way to help students develop skill at generalization.

A menu of suggested contexts for infusion lessons on generalization can be found at the end of the chapter.

Model Lesson on Skillful Generalization

The model lesson on generalization is a biology lesson in which students make generalizations and then critique them based on the need for a good sample. In this lesson, students themselves develop the standards for a good sample.

As you read this lesson, consider the following focus questions:

• How can this lesson be a vehicle for explicit instruction in scientific methodology?

• What other examples can reinforce instruction in skillful generalization?

Tools for Designing Lessons on Generalization

The thinking map and graphic organizer for skillful generalization guide students in their evaluation of a suggested generalization.

The thinking maps and graphic organizers can guide you in designing the critical thinking activity in the lesson and can also serve as photocopy masters, transparency masters, or as models that can be enlarged and used as posters in the classroom. Reproduction rights are granted for use in single classrooms only.

SKILLFUL GENERALIZATION

1. What generalization is suggested?

2. What sample is needed to support that generalization?

3. Is the sample being used large enough?

4. Is the sample being used like the whole group?

5 Is the generalization well supported by the sample?

6. If not, what additional information is needed to support the generalization?

SKILLFUL GENERALIZATION

GENERALIZATION

DESCRIBE A SAMPLE THAT WOULD SUPPORT THIS GENERALIZATION

DESCRIBE THE SAMPLE USED

RESULTS SHOWN BY THE SAMPLE

SAMPLE SIZE		SAMPLE SELECTION		THIS GENERALIZATION	
SMALL	☐	IT IS LIKE	☐	IS SUPPORTED:	☐
MEDIUM	☐	OR UNLIKE	☐	IS UNCERTAIN:	☐
LARGE	☐	THE WHOLE GROUP.		IS BIASED:	☐

IF THIS SAMPLE IS NOT ADEQUATE TO SUPPORT THE
GENERALIZATION, EXPLAIN WHY. HOW SHOULD IT BE DIFFERENT?

ALL MICE AROUND THE WORLD?

General Science, Biology   **Grades 6–12**

OBJECTIVES

CONTENT

Students will learn about the behavior of mice by observing two studies of this animal.

THINKING SKILL/PROCESS

Students will learn to base generalizations on an adequate number of representative samples.

METHODS AND MATERIALS

CONTENT

Students will observe two separate feedings of mice. Students will also design their own field studies to supplement the ones they read about.

THINKING SKILL/PROCESS

An explicit thinking map, graphic organizers, and structured questioning emphasize a thinking strategy for well-founded generalization. Collaborative learning enhances the thinking.

LESSON

INTRODUCTION TO CONTENT AND THINKING SKILL/PROCESS

- When I was shopping at a market, I noticed a sign above the grapes: *These Grapes Are Delicious: Sample a Few Before You Buy Them.* **So I tasted one that was on the top. It was sweet and seedless. I decided that all the grapes would be tasty, so I bought some. Do you think that was a good way to decide that they all were tasty?** Some students say that this is a good way to decide, but most say that one grape isn't enough ("Maybe you tasted the one good grape.") and that one from the top is from only one place ("Maybe the grocer put the sweet ones on the top."). **If you think this isn't a good way, work with a partner and figure out a better way to decide.** After a few minutes, ask for two or three of the plans. POSSIBLE ANSWERS INCLUDE: *Try about five grapes. Pick them from different places in the bin. Make sure you try some from underneath and from the back. Don't eat them all from one bunch; make sure you try grapes from many bunches.*

- **What you've been doing is planning how to make a good** *generalization.* **A generalization is a statement that begins with an "all" and says that something is true of all things of a certain sort. "All the grapes in the bin are tasty" is the generalization we've been discussing. With your partner, list three more generalizations that you think you know.** Write six to ten of these on the board as the groups respond. POSSIBLE ANSWERS: *All the students in this school are nice. All the people in this country are Americans. All heavy objects sink in water. All candy is sweet.*

- **When deciding whether a generalization is one you should accept, you usually can't look at every individual. It would probably not be possible to find out about every person in this country before you could say whether they were all Americans. Usually, we base our generalizations on a smaller** *sample,* **like the grape that I tasted. In order to generalize about all of the things in the group, like all of the grapes, you suggested the kind of sample needed. Let's review what you said about the kind of grape sample required.** ANSWERS: *You need more grapes and from different bunches and locations so that you get a better idea that the ones you tasted were like all of them.*

- What ideas about samples should we think about whenever we need to make a generalization? POSSIBLE ANSWERS: *A sample should not be too small, and it should be like the whole group, or from many parts of the whole group.* To be sure that our generalizations are based on a good sample, let's use these ideas for a thinking map of skillful generalization.

- We make generalizations in science based on studying a sample of things. For example, if we've studied turtles and list things that we learned, some of you might write that turtles have shells, that they travel slowly, and that they lay eggs. These are generalizations. They are about all turtles. We're going to conduct some research on another type of animal in order to learn about its behavior. The animal is the mouse.

> ## SKILLFUL GENERALIZATION
>
> 1. What generalization is suggested?
> 2. What sample is needed to support that generalization?
> 3. Is the sample being used large enough?
> 4. Is the sample being used like the whole group?
> 5. Is the generalization well supported by the sample?
> 6. If not, what additional information is needed to support the generalization?

THINKING ACTIVELY

- One important thing that we can learn about animals is their behavior. When we study animal behavior, we can learn important things about that animal—what it eats, how it spends its time, how it relates to its young, etc. We know that some animals—fish, for example—bear offspring but don't care for them. In fact, some fish have been known to eat their offspring. Other animals, however, take care of their young. Lions, for example, protect their young from predators. We have learned these things about animals by observing their behavior and by generalizing, so that when we meet up with an individual animal of a certain sort, even though that particular animal was not studied, we nevertheless know what to expect from it. When near a group of young lions, for example, we can be sure that we are in some danger from the mother because we know that lions protect their young and attack anyone who seems to threaten them. Can you think of some things you already know about animal behavior? Mention a few of these. STUDENT ANSWERS VARY.

- Let's now think about the behavior of the mouse. What do you already know about mice? What do you think you know but are not sure of? What don't you know but would like to find out? Work with your partner and make a list of things that fall into one or another of these three categories on the KWL graphic organizer (right). ANSWERS VARY.

WHAT DO I KNOW?	WHAT DO I THINK I KNOW?	WHAT DO I NEED TO KNOW?

- One way to try to answer some of your questions is to study these animals, to observe their behavior. We can't study the real behavior of some animals in the classroom, for example the elephant or the lion, but we can study the behavior of others. Animals like gerbils and mice are often kept in science classrooms so that we can study them. We're going to study the behavior of mice to see what we can learn about them. To do this, I've set up some equipment in the front of the room in which you can all observe how the mice I'll be showing you behave in certain situations. We'll be looking at two groups of mice—white mice and gray mice—to see whether they behave any differently. We will be observing two things about the mice: how they behave when they are mixed together and how they behave when they eat. I'd like you to record what you observe on this observation sheet, and then draw conclusions about the behavior of mice from your observations. Put a large cage with two compartments on a table in the front of the room. Have four white mice in one compartment and four gray mice in the other. Then do two things. First, open the barrier between the mice and let mice from one side go into the other side. Ask the students to write down what they observe the white mice doing and what they observe the gray mice doing. Then ask the students to write down what they can conclude about the social behavior of white and gray mice together in a group. Ask students to volunteer to report their findings. ANSWERS INCLUDE: *The white mice behaved exactly as the gray mice did: they sniffed each other on some occasions, sniffed the cage and ignored each other on other occasions, etc. The white and gray mice didn't behave any differently.* Conclusion: *White and gray mice behave the same towards each other in groups.* Second, put food in the cage and ask students to record what they observe about the eating behavior of the mice. (Make sure the white mice haven't been fed in a while, and the gray mice have been fed amply just before class. Do not tell this to the students yet.) Ask the students what conclusion they draw from this behavior. ANSWERS INCLUDE: *The white mice quickly ran to the food and each mouse devoured the food until it was gone. The gray mice ignored the food and nibbled on only a few morsels in a casual way.* Conclusion: *White mice attack and devour food; gray mice eat more casually.*

• **What is the difference between your observations and your conclusions?** Ask for volunteers to respond from the class. If students have difficulty responding, remind them of the discussion of the other animals earlier. ANSWERS INCLUDE: *The observations were about a sample of mice and our conclusions were generalizations about all mice in the world.*

• **Let's use this graphic organizer for generalization to record the sample, the generalizations, and to assess whether the sample was a good one to base the generalizations on. Use one graphic organizer for each generalization.** Ask the students to write the generalizations at the top of the graphic organizer and to write the sample used in the appropriate box.

• **One way to assess whether the sample was a good one is to test the generalizations by trying the same experiment with different mice—with a different sample.** This time, duplicate the experiment but use both white and gray mice that have been normally fed. When you ask students for the results, they should acknowledge that the results for social contact remain the same but that there is no difference in eating behavior between the white and gray mice this time. Explain the difference in the way the mice were fed in the first experiment when students report their findings.

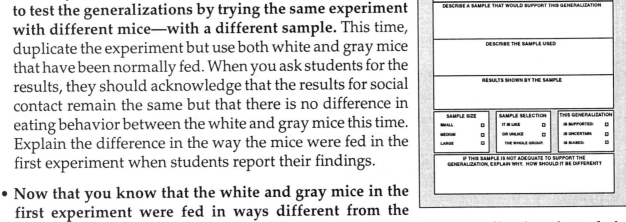

• **Now that you know that the white and gray mice in the first experiment were fed in ways different from the normal, how would you assess them as a sample to be used for a generalization about their eating behavior?** *They are not a good sample; they are a biased sample.* **Record this on your graphic organizer and explain what would have been a better sample to use and why.** ANSWERS INCLUDE: *It would have been better to use a sample in which there were no other differences between the white and gray mice and in which we knew that the mice used were not special in any way other than their color. If we knew that they were selected randomly, or that they were not different from other white and gray mice, we would be in a better position to say that the generalization was right.*

- **Are there any other generalizations about eating behavior now suggested by the samples?** Ask students to work in their groups again. ANSWERS INCLUDE: *Hungry mice devour their food. Mice that have eaten recently don't eat much even if food is present.* **Are the combined samples from the two experiments good ones to base these generalizations on? If so, why? If not, explain what you might do to make sure your sample is a good one upon which to base the generalization.** Students usually exercise more caution at this stage and indicate that they would be uncertain even though they knew that this sample was like the overall population of mice. They usually respond to this by saying that they should increase the sample size to many mice—more than 100, some say—and/or that they should use a sample of mice from around the world. However, many students think that this latter condition is not necessary if we know that the mice in the sample are anatomically like mice anywhere. This would make the sample representative, they say. When students have completed their reports, they can be given the opportunity to continue the experiment by using the sample they think would support their generalization better.

THINKING ABOUT THINKING

- **What do we call the kind of thinking we just did?** *Generalization.*

- **What questions did we ask as we did this kind of thinking?** Students should mention the questions on the thinking map for skillful generalization. If they are having trouble, point to some of the questions and ask if each was considered. Students identify the following questions: *What is the generalization that is being considered? What is the sample? Is the sample large enough? Is it like the whole group (representative of the whole group)?*

- **Is it a good idea to support generalizations in this way? Why or why not?** POSSIBLE ANSWERS: *If you can't test all of the things the generalization is about, then you have to use a sample. The generalization could be wrong if the sample is too small or unlike the group as a whole. If we make sure that the sample is a good one, we will probably make a good generalization.*

- **In the lesson, we did not think about the kind of sample that would be needed to support the generalization until after we had decided that the sample used was not a good one. Is this the best time to think about what kind of sample would be a good one? If so ,why? If not, why not?** Most students respond that it is better to think about this question before taking the sample so that we can avoid making mistakes in relying on too small or nonrepresentative a sample, as we did in the first experiment with the mice.

- **What advice would you give a friend about making good generalizations? Would you advise your friend to use the diagram we used? Why? If you think the diagram is a good idea, how would you advise your friend to use it?** ANSWERS VARY.

APPLYING YOUR THINKING

Immediate Transfer

- **Pick one of the questions you identified in the "I need to know" column of the chart you developed at the beginning of this lesson, develop a generalization about mice that you would like to test, and design an experiment that will provide you with a good basis for affirming this generalization.**

- **In trying to find out how bees communicate so well that when one bee returns from a food source, other bees can fly directly to it even though it is quite a distance away, the German biologist Karl von Frisch conducted a series of observations of bee behavior for 25 years. He** observed the bees in a special glass-walled hive that he constructed and placed feeding stations stocked with sweetened water at various distances from the hive. He concluded that bees communicate when food is near by performing what he called a "round dance" and when food is some distance by performing what he called a "waggle dance." Von Frisch published his work in a book called "The Dance Language and Orientation of Bees" published in 1967. Consult this work and assess whether his generalizations about bee communication were well-founded.

- School elections come up every year. Suppose you had to predict for whom the majority of students in the school would vote but couldn't ask every student. Design a way of getting information that would allow you to generalize from your sample.

Reinforcement Later

- When we were studying the planets in the solar system, we learned where they were, how large they were, etc. What generalizations can you make about planets from your study of our solar system? Why do you think these are good generalizations?

- Draw a picture of a Native American family as they lived in the 1800s. Compare your drawings with others in the class. What are the objects and customs pictured by students in your class? Now, compare your drawings with pictures of five different tribes from different parts of the country in the 1800s. Were your drawings good samples of what Native Americans were like? Why or why not? If your drawings were not good samples, what additional information would you need and how could you get it?

- List some school rules. What generalizations can you make about the purposes of these rules? Explain why you think these are good generalizations.

- Stereotyping is a form of generalization in which a group of people (or animals) is represented by an individual who displays qualities that are manifested only by some members of the group yet are taken to be manifested by all members of the group. For example, Native Americans are often stereotyped in western movies as aggressive if not vicious, yet we know that many Native American tribes were peaceful. Identify stereotypes you find in TV commercials and explain why these are stereotypes and not good generalized representations of the groups they stand for.

ASSESSING STUDENT THINKING ABOUT GENERALIZATION

Ask students to select one of the generalizations they thought was true at the beginning of the lesson. They should write or discuss with a partner whether the generalization was based on a good sample. If not, what would one do to get a good sample? As they try to solve the problems, confirm that in their writing or discussions the students are raising the questions on the thinking map.

Sample Student Responses • Mice Around the World

SKILLFUL GENERALIZATION

GENERALIZATION

White mice devour their food; gray mice nibble at it and eat very little.

DESCRIBE A SAMPLE THAT WOULD SUPPORT THIS GENERALIZATION

Four white mice and four gray mice were
given the same amount and type of food.

RESULTS SHOWN BY THE SAMPLE

The four white mice devoured their food.
The four gray mice sniffed and nibbled at
their food but ate very little of it.

SAMPLE SIZE		SAMPLE SELECTION		THIS GENERALIZATION	
SMALL	☒	IT IS LIKE	☐	IS SUPPORTED:	☐
MEDIUM	☐	OR UNLIKE	☒	IS UNCERTAIN:	☐
LARGE	☐	THE WHOLE GROUP.		IS BIASED:	☒

IF THIS SAMPLE IS NOT ADEQUATE TO SUPPORT THE GENERALIZATION, EXPLAIN WHY. HOW SHOULD IT BE DIFFERENT?

This sample is not adequate because it is very small and the white mice have been starved, while the gray mice were adequately fed before the experiment. The sample would be better if it were considerably bigger, both gray and white mice were fed the same for some days before the experiment, we made sure that the mice used were typical white and gray mice, and the type of food was varied.

Sample Student Responses • Mice Around the World

SKILLFUL GENERALIZATION

GENERALIZATION

Hungry mice devour their food; mice that
have just eaten sniff food, but ignore it.

DESCRIBE A SAMPLE THAT WOULD SUPPORT THIS GENERALIZATION

Four hungry white mice and twelve recently fed
mice, four white and eight gray, were given the
same amount and type of food.

RESULTS SHOWN BY THE SAMPLE

The four hungry mice devoured their food.
The twelve recently fed mice sniffed at the
food, but ate very little of it.

SAMPLE SIZE		SAMPLE SELECTION		THIS GENERALIZATION	
SMALL	☒	IT IS LIKE	☒	IS SUPPORTED:	☐
MEDIUM	☐	OR UNLIKE	☐	IS UNCERTAIN:	☒
LARGE	☐	THE WHOLE GROUP.		IS BIASED:	☐

IF THIS SAMPLE IS NOT ADEQUATE TO SUPPORT THE GENERALIZATION, EXPLAIN WHY. HOW SHOULD IT BE DIFFERENT?

This sample is not adequate because it is very small although it seems to be
representative of the whole group the generalization is about (the hungry mice are
like other hungry mice and the mice that have just eaten like others that have just
eaten). The sample would be better if it were considerably bigger, both gray and
white mice were in the first group, and the type of food was varied.

GENERALIZATION LESSON CONTEXTS

The following examples have been suggested by classroom teachers as contexts to develop infused lessons. If a skill or process has been introduced in a previous infused lesson, these contexts may be used to reinforce it.

GRADE	SUBJECT	TOPIC	THINKING ISSUE
6–8	Science	Movement of astronomical objects	Is the generalization that all planets, moons, and asteroids revolve a well-supported generalization? Explain. Do all planets, moons, and asteroids rotate on an axis? Is that a well-supported generalization? Explain.
6–8	Science	Galaxies	Many "stars" seen at night are really galaxies containing millions of stars. What would be wrong with the generalization that all of the "stars" we see at night are really galaxies if it was based on seeing pictures of a dozen galaxies that look like stars at night?
6–8	Science	Animals	Suppose that you were observing the behavior of humpback whales so that you could learn something about their food and migration patterns. Sketch out how you would conduct this observation so that you would have enough data to make some generalizations.
6–8	Science	Reproduction	Fish may produce millions of eggs; mammals often produce one per cycle. What generalization might you draw from that and how might you confirm it?
9–12	Biology	Medication	Drug companies often gather research information regarding the effectiveness of drugs on male subjects only to eliminate any possible effects that the hormonal fluctuations of women might have on the data. Is this practice appropriate to support using the drug?
9–12	Biology	Animal behavior	Suppose you observed baboons in the Ambroseli National Park in Kenya for three years and found that the older males you observed were more aggressive than the younger. What generalization does this suggest? What else, if anything, would you want to know before you accepted this generalization?
9–12	Biology	Animal cells	If you notice that the specimen of animal cells that you observe through a microscope has a cell membrane, cytoplasm, a nuclear membrane, mitochondria, and vacuoles, what would you need to do to generalize whether all animal cells have these features?

CHAPTER 17
REASONING BY ANALOGY

Why is There a Need for Skillful Reasoning by Analogy?

We sometimes raise questions about familiar things that we can't answer. For example, I may question how computer programs work, but I don't know enough about them to understand the technical language involved in answering this question. I could take the time to find out directly by taking a computer course. Sometimes, however, an analogy helps us learn more quickly. Someone may suggest that using a computer program is like using a code. If I know the codes, I can use this knowledge to understand how computer programs work.

When we note that two different kinds of things are alike in various ways, we draw an analogy between them. If, based on that analogy, we think what we know about one thing is also true of the other, we are *reasoning by analogy*.

Uses of Reasoning by Analogy

People appeal to analogies to justify drawing various conclusions about one thing based on its similarity to another. For example, many draw an analogy between the president of the United States and the captain of a ship. This comparison creates an image of leadership and authority. Most of us learned about atoms by using the analogy of the solar system. This comparison creates the image of powerful forces binding the atom together into a single system of particles revolving around a nucleus. More recently, the Persian Gulf War was compared to World War II in that one country was invaded by another.

A special case of reasoning by analogy may arise if we have a question about something and finding an answer directly would be risky. Can a bridge withstand extreme stress? How do humans react if they eat only a certain diet? If these issues can't be determined by conducting research directly with the bridge or with humans because of the risks involved, research using something similar may help. Engineers may simulate the stresses on a model bridge. Researchers may try the diet on an animal.

Other Uses of Analogies That Do Not Involve Reasoning by Analogy

Recognizing that two things are analogous can serve other purposes besides allowing us to form conclusions about one thing based on our knowledge of another. Analogy is the basis for similes, metaphors, and personifications. When Shakespeare said, "My love is like a red, red, rose," he meant to highlight certain characteristics that his love exhibits that are also manifested in a beautiful rose. The speaker is not drawing or affirming a conclusion about his love based on the analogy. Rather, the analogy communicates the author's perception that his beloved is lovely, soft, and fragile. In using analogous objects metaphorically, we highlight the characteristics that the two objects have in common; in analogical reasoning, we infer something new about one of the objects based on the fact that they have characteristics in common.

Problems with the Way We Use Analogical Reasoning

The problem with reasoning from simple analogies, like comparing guinea pigs to human beings, the president to the captain of a ship, or the Persian Gulf War to World War II, is that sometimes they mislead us into drawing incorrect conclusions. The analogy between the president and the captain of a ship may mislead one to expect that the president is the sole authority in this country. This is a mistake: The Congress and the Judiciary put constraints on presidential power.

When we don't think critically about using analogies to draw such conclusions, the mere fact that the two items compared are alike may be enough for us. Yet often, as in the case of the power of the president, the similarities are not strong enough to support analogical reasoning.

Comparing the Persian Gulf War to World War II in order to justify our fighting against Iraq presents a different problem. These wars were, indeed, similar in significant ways. There was an important difference between them, however. In World War II, the United States was attacked by the Japanese. In the Persian Gulf War, Kuwait was attacked by Iraq. This difference is crucial and makes the analogy a poor one to use in supporting the intervention of the United States. This is, of course, not to say that there aren't better analogies or reasons.

Figure 17.1 contains a summary of the two main causes of faulty analogical reasoning.

COMMON PROBLEMS WITH REASONING BY ANALOGY

1. We note that two things are alike but don't ask whether they are enough alike to support the conclusion we wish to draw.

2. We note that two things are alike but don't ask whether there are differences between them that weigh against drawing the conclusion we wish to draw.

Figure 17.1

Analogical reasoning in which these questions are not asked is tenuous and incomplete.

What Does Skillful Reasoning by Analogy Involve?

Skillful reasoning by analogy involves initially searching for things that are similar to what we are trying to understand. Then, we must decide whether any of these similar things yields good analogies. Even if someone else suggests an analogy, we still have to determine whether it is a good analogy for our purposes.

To decide whether we can use a suggested analogy, *we must first determine how the two things are alike and whether their similarities are significant*. The likenesses may be too superficial to extend what we know about one to the other. I may wonder which of the following is a good analogy to help students understand how blood carries nutrients to various parts of the body: a conveyor belt, water in a pipe, or the postal service. The postal service is similar in that its members carry things around, but the similarities run no deeper. Water in a pipe seems to be more analogous to the blood in the body. Both are liquid, move in tubular enclosed vessels, and carry substances that are dissolved. This image is clearly better for helping students un-

derstand how the blood nourishes the body.

When we ascertain that two things are significantly alike, we usually describe them as analogous. *Determining whether there are any significant differences between analogous objects is the next important step in skillful reasoning by analogy.* The similarities between humans and guinea pigs are not superficial. Similarities in body functioning are significant. But we should not draw conclusions about humans from studies of guinea pigs on the basis of similarities alone. There may be basic differences between guinea pigs and humans that weigh against drawing these conclusions. Not every difference is important, however. But some may be. For example, we may find that guinea pigs are genetically different from humans in their ability to fight certain diseases. However similar they are, this difference would be quite significant in determining whether the analogy supports drawing conclusions about the risk of disease for humans because certain substances cause disease in guinea pigs.

We can take two tacts in trying to find out whether significant differences between the things compared make drawing a specific conclusion tenuous. First, we may list as many of the differences as we can discover and then ask whether any are significant. When we compare guinea pigs and humans, we find many differences. None of the differences, however, may be known to block in guinea pigs the effects of an experimental drug being developed for humans. In that case, the conclusion that if the drug is effective in guinea pigs, it will work in humans is supported by the analogy because differences that would interfere are not known to exist.

The second strategy is to identify what factors would block the effectiveness of the treatment and find out whether humans and guinea pigs function alike regarding that factor. If we find no such differences between guinea pigs and humans, we can reasonably extend to humans similar effects to those found in guinea pigs.

Both strategies for evaluating differences should be used in a careful application of reasoning by analogy. This will assure thoroughness in this type of thinking and counter our tendency to ignore or trivialize differences.

A Thinking Map and Graphic Organizer for Skillful Reasoning by Analogy

The thinking map in figure 17.2 can be used to guide us through skillful reasoning by analogy.

REASONING BY ANALOGY

1. What things are similar to the object or idea that you are trying to understand?

2. Which are similar in significant ways?

3. What do you know about these things that you don't know about the thing you are trying to understand?

4. Are there any differences between the two that could affect whether what you are trying to understand has these features?

5. What can you conclude regarding what you are trying to understand based on this analogy?

Figure 17.2

Note how it begins by having students generate analogies and then has students use the analogies to support conclusions. The corresponding graphic organizer is shown in figure 17.3.

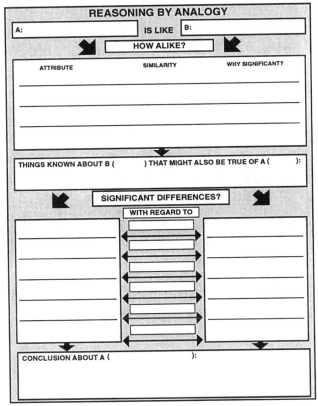

Figure 17.3

In using the graphic organizer in figure 17.3, we should, of course, make explicit the basis for the analogy by writing the significant similarities at the top. Then, we make explicit potential conclusions that might be drawn from this analogy by indicating information we have about the second item that we think might be true of the first. This can guide us in searching for significant differences. The two strategies for finding significant differences will make this a thorough search. We should write any significant differences we find in the spaces provided for differences. Finally, we reaffirm the tentative conclusion in the box at the bottom if no significant differences are apparent. If significant differences are present, we should write "None" in the box at the bottom.

How Forceful is Reasoning by Analogy?

We may develop a line of analogical reasoning fairly quickly by using information that we already have about one thing to suggest answers to questions about another. We should be cautious about reasoning by analogy, even if it is conducted skillfully. At most, the conclusions we draw using this kind of reasoning are supported only by indirect evidence. They are at best well-supported suggestions that should never substitute for getting information needed to answer our questions more directly.

How Can We Teach Skillful Reasoning by Analogy?

In teaching reasoning by analogy, it is not enough to ask students to draw analogies between something they are studying and something they already know. This stops short of *reasoning* by analogy. They should also be asked to note why the two items are analogous, what conclusions or insights they might draw about one from knowledge of the other, and whether there are significant differences between the two things making such inferences problematic.

Repeated practice using this thinking strategy with a variety of analogies will reinforce the use of this skill. Students can develop their own analogies or critique analogies offered by others. Make sure that you are explicit about the strategy by using the thinking map in figure 17.2 and graphic organizer in figure 17.3.

Contexts in the Curriculum for Lessons on Reasoning by Analogy

Using analogies to extend knowledge is useful in almost every area of the curriculum. Because of analogies between mice and humans, the way mice react to certain diets often enables scientists to draw conclusions about how it is likely that humans will react. Analogies between geometrical figures, between parents and children in literary works, between wars, and between linguistic forms in different languages are only a few analogies that appear regularly in standard secondary school curricula. Each of these provides a potential context for an infusion lesson on reasoning by analogy.

Reasoning by analogy plays a significant role in science. In most science contexts, it is a substitute for the use of more direct methods of gathering information, when the latter is impractical or physically impossible. For example, because of the impracticality of testing potentially harmful drugs on humans, they are often tested on other living things. Because of the indirectness of information obtained when reasoning by analogy, the conclusions are often much more tentative than if they were obtained directly. Nonetheless, reasoning by analogy remains the sole route to certain sorts of advancements in our understanding of how the world works, however tentative they may be. For example, our understanding of what the other planets in the solar system are like is derived mainly from how analogous they are to the earth, taking account of significant differences. Until we have more direct access to them, either via space probes or in person, analogical reasoning may be the best we can do to advance our understanding.

The curriculum lends itself to two main contexts for lessons in reasoning by analogy:

- *Explicit analogies are used to help students understand important concepts or ideas.* For example, the atom is often compared to the solar system, electricity to water flowing through a pipe, and the human body to a machine. When analogies are used directly in science curricula, they provide us with an opportunity for a rich infusion lesson.

- *Students are asked to construct analogies, simulations, or models to help them understand important concepts and ideas.* For example, students construct models to understand the structure of molecules. Models of RNA and DNA that students construct are often useful in helping students learn more about these essential molecules of life. Students are also sometimes guided to draw analogies between the way certain mechanisms operate in the human body and devices used in industry.

A menu of suggested contexts for infusion lessons on reasoning by analogy can be found at the end of the chapter.

Model Lessons on Reasoning by Analogy

Included are two lessons on reasoning by analogy included in this booklet. One is a biology lesson in which an analogy is drawn between a living cell and a factory. The second is a physics lesson in which sound waves are recognized as analogous to water waves and students determine what they can learn about sound waves from this analogy and their firsthand observation of water waves. The following questions can be helpful:

- How is reasoning by analogy different from comparing and contrasting two things?

- What are different ways of introducing the thinking skill in the lesson introduction?

- What additional examples can you think of for reinforcing this skill?

Tools for Designing Lessons on Reasoning by Analogy

We include the thinking map for reasoning by analogy and additional graphic organizer to assist in teaching skillful reasoning by analogy.

The thinking map and graphic organizer can guide you in designing the critical-thinking activity in the lesson and can also serve as photocopy or transparency masters or as posters in the classroom. Reproduction rights are granted for use in a single classroom only.

REASONING BY ANALOGY

1. **What things are similar to the object or idea that you are trying to understand?**

2. **Which things are similar in significant ways?**

3. **What do you know about these things that you don't know about the thing you are trying to understand?**

4. **Are there any differences between the two things that could affect whether what you are trying to understand has these features?**

5. **What can you conclude regarding what you are trying to understand based on this analogy?**

REASONING BY ANALOGY

A: _____ IS LIKE B: _____

HOW ALIKE?

ATTRIBUTE	SIMILARITY	WHY SIGNIFICANT?

THINGS KNOWN ABOUT B (   ) THAT MIGHT ALSO BE TRUE OF A (   ):

SIGNIFICANT DIFFERENCES?

WITH REGARD TO

CONCLUSION ABOUT A (   ):

THE CELLULAR FACTORY

General Science, Biology, Anatomy, Physiology   **Grades 8–12**

OBJECTIVES

CONTENT

Students will review the structure and function of the parts of a typical animal cell. This lesson emphasizes the organization and function of organelles as components of the cellular apparatus which synthesize protein.

THINKING SKILL/PROCESS

Students will learn how to draw a conclusion based on an analogy by determining whether there are significant similarities and differences between the two things compared.

METHODS AND MATERIALS

CONTENT

Reading from the textbook about parts of a cell and their function in protein synthesis is combined with the use of a special graphic organizer to identify retained background knowledge and lack thereof. Electron micrographs showing the structure of cellular components and transparencies of idealized drawings of organelles would be useful.

THINKING SKILL/PROCESS

Structured questioning, a graphic organizer, and a thinking map guide students through reasoning by analogy. Students will also work in collaborative learning groups.

LESSON

INTRODUCTION TO CONTENT AND THINKING SKILL/PROCESS

- At times, we struggle to learn something new that becomes astonishingly clear after we draw an analogy between it and something old and familiar. Frequently, teachers rely on this tactic to clarify a new concept. For example, analogies are often used to explain the behavior of animals. Noted author and physician Lewis Thomas in *The Lives of a Cell (1974),* draws an analogy between the behavior of ants and the behavior of humans. Read or have students read the following quote from "On Societies as Organisms" in *The Lives of a Cell.*

  "Ants are so much like human beings as to be an embarrassment. They farm fungi, raise aphids as livestock, launch armies into wars, use chemical sprays to alarm and confuse enemies, capture slaves. The families of weaver ants engage in child labor, holding their larvae like shuttles to spin out the thread that sews the leaves together for their fungus gardens. They exchange information ceaselessly. They do everything but watch television."

- This comparison can lead to the conclusion that ants, like humans, live in a cooperative but, at times, practical society where they work together for mutual survival. When you draw a conclusion like this, you are inferring that because so many aspects of human and ant behavior seem similar, then they might be similar in other and broader ways. When you draw a conclusion like this, you are *reasoning by analogy.* Many authors use analogies to dramatize their ideas. For example, Ayn Rand, in her book *Atlas Shrugged,* repeatedly returns to the theme that just as the world would fall if Atlas shrugged his shoulders, so might the world economy collapse if Big Business shirked its responsibility. Think of some analogies that you have used, heard, or read. In your groups, write them down by completing sentences like this:
  _____ is like _____

- Sometimes, analogies can be misleading. Valid and useful conclusions can be drawn from analogies only if the differences between the things being compared in no way compromise the appropriateness of their similarities. For example, it would be possible to compare a human being with a brick. Both occupy space, have mass, are composed of atoms, etc. But how useful would this analogy be in describing the fundamental attributes of a living organism such as man? In that context, the differences between a brick and a man overwhelm any similarities and render the analogy useless. Here's a thinking map of reasoning by analogy that emphasizes important questions we ask to avoid drawing conclusions from analogies not well-supported.

REASONING BY ANALOGY
1. **What things are similar to the object or idea that you are trying to understand?**
2. **Which things are similar in significant ways?**
3. **What do you know about these things that you don't know about the thing you are trying to understand?**
4. **Are there any differences between the two things that could affect whether what you are trying to understand has these features?**
5. **What can you conclude regarding what you are trying to understand based on this analogy?**

• In our biology class, we've begun to study the biology of the cell. The cell is the basic unit of all living things. What are its basic components? Ask the class, and as students identify the components of a cell, write them on the chalkboard. ANSWERS INCLUDE: *The nucleus, cytoplasm, cell membrane, and organelles.* Studying these components and how they function is biology at the microscopic level. The basic question we will try to answer is how cellular components are organized to produce the major product of cells—proteins.

• It is not possible to observe the form and function of many cellular components by using the light microscope in class. When it's not easy to observe something directly, a situation is created in which we might want to use more indirect methods. *Models* are often used to explain difficult or hard to observe phenomena in nature. For example, representing the functions of the brain by employing a computer model or comparing the organization of atomic particles to the organization of the solar system are frequently cited examples of using a model of something familiar to draw conclusions about something obscure and less known. When we use models like this, we are engaging in reasoning by analogy. Reasoning by analogy is especially useful here because we already know certain things about cells and can build on them to form analogies; referring to something analogous to the cell and its metabolic processes that we know better may be very helpful in giving us an understanding of what we cannot directly see. Think about things that are analogous to cells. In your groups, make a list using the form: A cell is like _____.

Have groups report back and list student responses on the blackboard or on a transparency. TYPICAL STUDENT RESPONSES: *A cell is like a computer chip; a cell is like an army; a cell is like an automobile factory; a cell is like a building block; a cell is like a brick; a cell is like an egg.* I'm going to select one of these analogies and we will see what we can learn from it in class. When we've finished, your group can select another and see what you can learn from that analogy. The one I'd like us to work with in class is this: A cell is like an automobile factory. We will evaluate this analogy and see what we can learn from it by using the thinking strategy for reasoning by analogy we just uncovered.

THINKING ACTIVELY

• What are some similarities between an automobile factory and a cell that lead us to think that they are analogous? Break into your groups again and use the graphic organizer for reasoning by analogy to first make a list of similarities between an automobile factory and a cell. Think about how an automobile factory works and list those characteristics that may also be used

to describe how a cell works. After a few minutes, ask each group to contribute examples of similarities to a list you are keeping on the board or on a transparency. Initially, limit each group to contributing only one similarity so that many groups can participate. RESPONSES OFTEN INCLUDE: *Both need an energy source. The primary function of both is to make a product. Both need to export their product. Both assemble a product from parts. Both have a center of operations. Both are protected from the environment by an enclosure.* **Add any similarities you don't have to your group's list.**

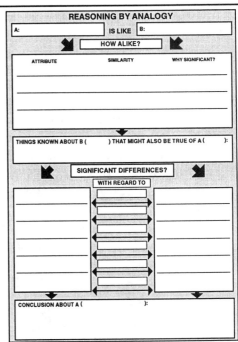

- **If an analogy is to be useful, it must be possible to find significant similarities between the two things being compared. In your groups, discuss each similarity in terms of its importance and why it is or is not significant.** Ask students to write their ideas by completing the "Why Significant?" column on their graphic organizer and then ask the groups to share their responses by reporting back to the class. As these reports are given, prompt extended responses by asking each reporting group questions of elaboration. Then, ask the class to respond if they don't agree with these assessments and facilitate a short open-class discussion of the issue, culminating in a "sense of the class" vote on whether the similarity is significant. RESPONSES INCLUDE: *"Both assemble a product from parts" is very significant because neither a cell nor a factory can assemble a product without having an inventory of parts. "Both have a center of operations" is very significant because the cell and the factory could not operate without instructions from a managing authority. "Both export their products" is very important because without being able to distribute products, both would fail to meet demand for their product. "Both are protected from a hostile environment by an enclosure" is very important because the manufacturing process is sensitive to changes in temperature, humidity, etc. "Both need an energy source" is very important because without energy, the process of production would stop.* Students usually agree that they find enough significant similarities between an automobile factory and a cell to make this a good analogy.

- **Now, let's think about things we know about an automobile factory that we don't yet know about cells but that, based on this analogy, we might be able to say are likely to be true of cells as well. Work together in your groups and write these in the box marked "Things known about B (a factory) that might also be true of A (a cell)."** RESPONSES USUALLY INCLUDE: *An automobile factory imports raw materials such as steel, rubber and glass from external sources to make automobiles. An automobile factory makes cars sequentially on an assembly line. A factory uses machines and tools to make an automobile.*

- **Before we can feel confident that we can draw these conclusions about cells, what must we do?** Students usually show that they understand the need to eliminate any differences "that make a difference" between factories and cells before they can affirm that these truths about factories are, in all likelihood, true of cells. *Are* **there differences between a factory and a cell that might make it difficult or invalid to learn about the organization of the cell from the operation of a factory? Work with your groups again and include differences that you note between automobile factories and cells.** STUDENT RESPONSES OFTEN INCLUDE: *A cell is living; a factory is an inorganic industrial plant. Cells can spontaneously reproduce more cells; a new factory must be planned and designed. Cells have a predictable lifespan; a factory exists as long as it is kept up. A cell makes thousands of different proteins; a particular car factory makes a few different models.*

- **Now, think about whether these differences are significant and write your responses on the graphic organizer as well.** Typically, students find that major differences between the cell and the automobile factory are ones of product and not process. They evaluate such differences, therefore, as not compromising the validity of the analogy. For example, students usually agree that the fact that a cell produces thousands of proteins but that an automobile plant produces far fewer models of cars does not significantly invalidate the analogy because it has to do with quantity of products and not with the process whereby these products are produced.

- **Based on the similarities and differences you've listed for the automobile factory and the cell, what conclusions can you draw about using the automobile factory as a model for the cellular factory? What new insights about cells do you now have that derive from what you know about automobile factories that you didn't initially know about cells?** Have students meet in their groups to discuss an overall conclusion. STUDENT RESPONSES TYPICALLY INCLUDE: *Proteins are probably made from some kind of overall plan. Raw materials necessary to build proteins must, in all likelihood, be imported into the cell. Complex proteins are probably somehow assembled in a sequential process from these raw materials. In all likelihood, a cell must use something analogous to machinery and tools to manufacture proteins.*

THINKING ABOUT THINKING

- **When you were evaluating whether or not the analogy between a cell and an automobile factory was useful and valid, you followed the thinking map for reasoning by analogy. Let's review how this played itself out in this particular lesson. What were you thinking as you went through the process of selecting an analogy and drawing insight about cells from it? What did you think about first, next, etc.?** TYPICAL STUDENT RESPONSES: We *thought of all the ways that a cell and an automobile factory are similar. We removed those that were not relevant to how a cell and a factory produce things. We then looked for things that we knew about the manufacture of automobiles that might be applied to the manufacture of proteins in the cell. We then considered differences between the cell and the factory and determined whether or not these differences would have a bearing on the validity of the analogy with regard to the manufacturing process. We determined that they were not significant and concluded that the analogy that the cell is like an automobile factory was a good analogy. We then went back to the new ideas we introduced that might be true of cells as well as factories and affirmed that it was likely that these were features of the operation of cells in manufacturing proteins.*

- **Did following this verbal map help you in your reasoning? If so, how? Is this way to engage in reasoning by analogy any better than the way you used to do it?** STUDENT ANSWERS INCLUDE: *The verbal map reminded me to think broadly and consider many similarities. It also reminded me to make sure I think about the possibility of significant differences as well. When I've used analogies in the past, I haven't thought about differences. It also focused my attention on the fact that I should reach a specific conclusion about the thing that I knew the least about based on the analogy.*

- **How did the process of using the graphic organizer help you to practice reasoning by analogy?** Students usually express their satisfaction that the graphic organizer helps them to focus on the problem at hand, helps them with evaluating one item at a time, gives them a format to follow, allows them to write down their thinking rather than keep it in their heads, and saves time.

APPLYING THINKING

Immediate Transfer

Shortly after the lesson, engage the students in the following activities:

- **Explore one of the other analogies for a cell that you came up with earlier. What insights about cells can you draw from this analogy? Why?**

- Draw an analogy between one of the parts of the human body in the following list and an object you are familiar with: the heart, the pituitary gland, the eye, the knee, the larynx, the central nervous system. Use reasoning by analogy to conclude how the organ works.

- Draw an effective analogy between some current event and some other event you have studied in history. What insights does this give you about the current event?

Reinforcement Later

- Use reasoning by analogy to develop some ideas about how a complex biological process that you are studying operates.

- Use reasoning by analogy to develop some ideas about how volcanic activity operates.

- Develop an analogy between a problem in your community and a national problem that this country has had to face. Based on the analogy, come up with possible solutions.

CONTENT EXTENSION

In previous class work, we studied a number of chemical processes and principles related to organic molecules: chemical bonding, the structure of carbohydrates, lipids, proteins and nucleic acids (including DNA transcription), mRNA translation, and the molecular mechanisms of protein synthesis. This information may help us to understand in greater depth how the manufacture of proteins might follow a plan, how raw materials might get selected and imported into a cell, what kind of sequential process might go into the production of proteins, and what, in cells, might be analogous to machinery and tools in a factory. Let's review what we know about these processes and principles first, being sure to keep these basic questions in mind. Work in groups of four to conduct this review. Each group will work on one of these topics. Use this graphic organizer (left) to help you write down your level of understanding of the mechanism you are reviewing. When the groups are ready, ask for reports. Ask a representative of each group to write five items from each column on a diagram of this graphic organizer on the chalkboard. When each group's record has been recorded, open the class for questions. If students dispute about what is in the "I know" column, move it into the "I think I know column." Discuss with the class how they might get answers to the questions listed in the "I need to know" column. Tell the class that they should raise their hands as they get ideas that help them to understand the aspects of the operation of cells that have been derived from the automobile factory analogy. Record these ideas on the chalkboard, and allow time for discussion of them as they are mentioned. Elaborate the mechanisms involved yourself for the class as appropriate. STUDENT ANSWERS INCLUDE: *The mRNA in a cell could act as the cellular factory's blueprint. Cellular tools and machines might be the organelles—for example, the ribosomes, endoplasmic reticulum, etc. Raw materials imported from the outside environment could include glucose, minerals, oxygen, water, etc.*

Chemical Processes Important In Biology		
I KNOW	I THINK I KNOW	I NEED TO KNOW

ASSESSING STUDENT REASONING BY ANALOGY

To assess students' use of reasoning by analogy, any of the transfer examples in this lesson can be used as prompts. Ask students to make explicit their thinking. In your assessment of their thinking, make sure you attend to whether they are raising the questions identified in the verbal map for reasoning by analogy.

Sample Student Responses • The Cellular Factory

REASONING BY ANALOGY

A: A LIVING CELL	IS LIKE	B: AN AUTOMOBILE FACTORY

HOW ALIKE?

SIMILARITY	SIGNIFICANT? WHY?
Both utilize an energy source.	Very important; assembling products depends on energy.
Both make a product from parts.	Very important; assembling products depends on parts.
Both have a center of operations.	Very important; management of production is essential.
Both export their products.	Very important; without delivery products cannot be used.
Both are protected by an enclosure.	Very important; enclosures protect the production process.

THINGS KNOWN ABOUT B (Automobile Factories) THAT MIGHT ALSO BE TRUE OF A (Cells)
An automobile factory imports raw materials. An automobile factory manufactures cars in a sequential process on an assembly line. Automobiles are manufactured from an explicit set of plans. A factory uses machines and tools to assemble an automobile.

HOW DIFFERENT?

DIFFERENCE	SIGNIFICANT? WHY?
A cell is living; the factory is an inorganic industrial plant.	Not important; process is important and not the physical nature of the facility.
Cells can spontaneously reproduce more cells; a new factory must be planned and designed.	Not important; the manufacturing process is not affected by how its enclosure is produced.
Cells have a predictable lifespan; a factory exists as long as it is maintained.	Not important; cells live long enough to meet product demand.
Any cell manufactures thousands of different proteins; a car factory manufactures only ten or so different car models.	Not important; the manufacturing process isn't related to the number of different products.
Millions of other cells produce similar products. Automobile plants are relatively few.	Not important; the manufacturing process isn't affected by the number of producers.

CONCLUSION ABOUT A (Cells):
In all likelihood, proteins are made from some kind of overall plan. Raw materials necessary to build proteins must be imported into the cell. Complex proteins are somehow made in a sequential process from these raw materials. A cell probably uses something analogous to machinery and tools to manufacture proteins.

SOUND WAVES

Physical Science, Physics   **Grades 9–12**

OBJECTIVES

CONTENT

Students will learn the universal characteristics of wave motion and apply this understanding to sound waves.

THINKING SKILL/PROCESS

Students will learn what an analogy is, how to specify the ways in which analogous things are similar, and how to determine whether ideas suggested by the analogy are accurate by searching for important differences.

METHODS AND MATERIALS

CONTENT

Students will observe a wave demonstration using a ripple tank/overhead projector apparatus. Students will read *Ripple While You Work* from the *Conceptual Physics Lab Manual* or similar wave experiment guide.

THINKING SKILL/PROCESS

Structured questioning, the use of a graphic organizer, and a thinking map guide students through reasoning by analogy. Students' thinking is also enhanced by working in collaborative learning groups.

LESSON

INTRODUCTION TO CONTENT AND THINKING SKILL/PROCESS

- Sometimes, it is difficult to explain a phenomenon in nature because it is too obscure or too remote to be easily observed. For example, gravity cannot be directly observed. Teachers, however, often use the model of a rock being whipped in a circle to communicate the effects of gravity on the angular motion of the moon. Being able to compare events and phenomena that cannot be observed to models that are easily seen and understood is called *teaching by analogy*. Attributing characteristics that are easily identified in one analog to another analog in which such identification cannot be carried out so easily is called *reasoning by analogy*.

- It is important to consider just how alike and unalike two things are when using reasoning by analogy. If we don't, we might find that although the things we are comparing are alike in some respects, they are also unlike in other respects, and these differences are quite important. Because of the differences, we may not be able to infer the hidden properties of one from those we know that the other has. For example, would we be able to learn anything about plexiglass (a transparent plastic sheet) by comparing the properties of a glass window to the properties of a plexiglass window? In your groups, make a list of characteristics that are common to both a glass and a plexiglass window. STUDENT RESPONSES: *Both transmit light; both are practically colorless; both come in thin sheets; both transmit heat (poor insulators); both are waterproof; neither is biodegradable.*

- Glass and plastic have a lot in common, but it might be a mistake to consider glass a replacement for plastic and vice versa without knowing their differences. Now, let's make a list of the differences between glass and plastic: STUDENT RESPONSES: *Glass is brittle and plastic is flexible. Glass doesn't melt at low temperatures like plastic does. Glass transmits much more light. Glass does not age and remains clear; plastic discolors over time. Glass is difficult to scratch; plastic scratches easily. Glass is inert and will not react to a stringent chemical like an acid or a detergent; plastic reacts to many such chemicals. Glass is made from inorganic silicone; plastic is made from organic molecules (petrochemicals).*

- **It is obvious that although glass and plastic have many properties in common, they have many significant differences. Do you think that you could predict, with confidence, how plastic would react in a novel application, if you knew how glass reacted in a similar situation?** Students usually comment that it would be difficult to predict how glass or plastic would react in a new situation because they have found so many seemingly significant differences in the properties of the two substances.

- **Let's make explicit what is involved in skillful reasoning by analogy so that we can follow a plan when we engage in it. Here is a thinking map that can guide us:** Write the thinking map on the chalkboard or on a large piece of newsprint and post it on the wall or chalkboard.

- **Today, we are going to learn about how sound travels by studying sound waves. Sound waves, however, are waves that we cannot see, although we can tell certain things about them by our experience of how they behave, such as that they**

> **REASONING BY ANALOGY**
>
> 1. **What things are similar to the object or idea that you are trying to understand?**
>
> 2. **Which things are similar in significant ways?**
>
> 3. **What do you know about these things that you don't know about the thing you are trying to understand?**
>
> 4. **Are there any differences between the two things that could affect whether what you are trying to understand has these features?**
>
> 5. **What can you conclude regarding what you are trying to understand based on this analogy?**

can produce an echo if they bounce off a hard smooth object. Can we learn more about sound waves by using reasoning by analogy? What could we compare sound waves to that might serve as the basis for such reasoning? STUDENT ANSWERS VARY BUT USUALLY INCLUDE: *Light, water waves, electricity, lightning, hot and/or cold ocean currents.* **I'm going to pick one of these that seems plausible as an analogy for sound waves—water waves. We will consider whether we can use characteristics of water waves, which we will produce in an apparatus called a "ripple tank," where such waves are easily seen, to act as a model for sound waves. If so, then we can use reasoning by analogy to disclose things about sound waves that are not easy, or at times, possible to see.**

THINKING ACTIVELY

- **The first step in evaluating the validity of using the analogy** *A sound wave is like a water wave* **to draw conclusions about water waves is to make a list of all the significant characteristics and properties that both kinds of waves have in common.** Ask students to meet in groups and write "A sound wave is like a water wave" in the appropriate boxes, and to make a list of similarities between water waves and sound waves in the "How Alike" box on the graphic organizer:

After three or four minutes, have students report back, and then use a transparency of the graphic organizer for reasoning by analogy, or the chalkboard, to make a class list of those characteristics that sound and water have in common. TYPICAL STUDENT RESPONSES INCLUDE: *Both travel through a physical medium (waves travel through water,*

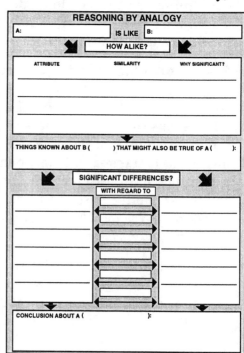

sound travels through air and water). Both have velocity. Both can be produced in different sizes. Both can travel over long distances. Sound radiates from its origin in all directions, and water waves do the same (e.g., a rock thrown into a pond); sound and water waves will bounce off hard objects (e.g., an echo, a riptide).

- **Are there other characteristics of waves that might also be characteristics of water and sound waves? I am going to demonstrate additional wave characteristics to you so that you might have an opportunity to decide if they can be added to your "How Alike" list based on what you know about sound waves.** Show the ripple tank apparatus and demonstrate how it works. Then, demonstrate the different wave patterns and other characteristics as described in the "Ripple While You Work" exercise or another typical lab guide.

- **After I demonstrate each water wave characteristic, meet in your groups and determine if a sound wave has this characteristic based on your experience with sound or by conducting a similar experiment with sound. If it does, add it to your "How Alike" list. For example, observe what happens when I poke my index finger into the ripple tank one time.** Usually students correctly observe a single circular wave to radiate from the point of impact outward in all directions. If students don't recognize that this means that water waves move at the same speed in all directions, ask them what this suggests about the speed that water waves travel away from the point of origin. Then ask, "Does sound move in all directions at the same speed?" Students usually realize that sounds like church bells can be heard equally in all directions. Write on the class transparency that water and sound waves move in all directions at the same speed.

- **Let's look at another characteristic of water waves that you have not considered. Observe what happens when I poke my finger in the water many times.** Strike water with your finger slowly at first and then quickly after. Ask students to comment on the effects of each experiment. Then ask: "What do you call this characteristic of water waves?" Most groups recognize this phenomenon as frequency and relate frequency to the pitch of sound. Using the ripple tank, continue to show water wave properties to the class that they can also identify in sound waves based on either their experience or a simple experiment. Include new similarities on the class transparency or on the chalkboard as the class reports them.

- **Now that you have compiled your list of observable similarities between water and sound waves, let's decide if these similarities are significant. In your groups, review your list one item at a time and note whether that characteristic is a significant similarity and why.** Allow students enough time to discuss and record their responses and then have them report back to the class explaining why the similarities are significant when they identify them as such. Enter their responses on the class transparency or the chalkboard. Then ask, **"Are the similarities between water and sound waves significant in general?"** Students usually find that they are.

- **Let's return now to the ripple tank. I'm going to demonstrate a few other characteristics of water waves.** Demonstrate positive and negative interference, reflection and refraction among other characteristics. **What can you say about these?** Students often note that they can't tell from their experience whether sound waves have these characteristics. **Does the fact that you can't tell whether sound has these characteristics as well mean that sound doesn't have these characteristics?** Students usually recognize that this doesn't follow. Rather, they just can't tell. Now, let's consider using reasoning by analogy. If, in fact, water and sound waves have many properties in common, perhaps they also have resonance, interference, diffraction and refraction as common characteristics. Let's record these properties in the box titled "Things Known About Water That Might Also Be True of Sound Waves."

- **Before we can infer that sound has these four properties, what should we make sure of?** Students usually recognize that they have to determine any significant differences between water and sound waves. If they don't recognize this, call their attention to the questions on the thinking map of reasoning by analogy that you have on the wall or on the chalkboard. You might also want to ask students why it is important to consider any significant differences between water waves and sound waves. Students usually have little difficulty in reasoning that significant differences might cause differences in the properties they are considering.

- **In your groups, discuss and then make a list of differences between water and sound waves.** Allow students adequate time to think about the differences, compile a list, and report back to the class. Then enter their responses on the class transparency or the chalkboard. STUDENT RESPONSES USUALLY INCLUDE: *Water waves travel only in water; sound waves travel in the air, water and in other mediums. Water waves are visible; sound waves are invisible. Water waves can travel over very long distances in the ocean; sound waves dissipate very quickly in the air.*

- **In your groups, evaluate the significance of each of these differences and state why you think so on your group's graphic.** Have students report back and enter their responses on the class graphic. Usually, students find few, if any, significant differences between water and sound waves.

- **In your groups, discuss and then draw a conclusion about sound waves based on the strength of the analogy as you evaluate it.** TYPICAL STUDENT RESPONSE: *Like water waves, sound waves have, in all likelihood, the properties of resonance, interference, refraction and diffraction.*

- **Reasoning by analogy like this is, at best, a way to draw reasonable conclusions about phenomena that are difficult to observe directly. Often, reasoning by analogy can lead us to investigate further to confirm directly by experiment whether what we are inferring from our analogy actually has these properties. Work in your groups and consider one of the properties of sound waves that you concluded they likely have. Design an experiment with sound that can provide you with confirmation that sound waves do, indeed, have this property.** STUDENT ANSWERS VARY.

THINKING ABOUT THINKING

- **What do we call the kind of thinking we just did?** STUDENT ANSWER: *Reasoning by analogy.*

- **How did you do that kind of thinking? What did you think about first, second, etc.?** POSSIBLE ANSWERS: *I thought about how two things, water waves and sound waves, were alike, and then I thought about whether they were so different that we could not infer anything about sound waves that we did not know from features of water waves that we observed.*

- **Is there anything you have to be careful about when you learn new things by analogy?** POSSIBLE ANSWERS: *Yes. The analogy could make me think that everything true of one thing in the analogy is true of the other, and that could be misleading. I should always look for important differences that may show that my conclusion is a poor one.*

- **What questions would a thinking map for reasoning by analogy include? Put these in your own words and keep the thinking map you construct so that you can use it again when you are doing reasoning by analogy.** Student responses should suggest the ideas on the thinking map for reasoning by analogy.

APPLYING THINKING

Immediate Transfer

- You're learning about different kinds of animals in science. One type of animal has a shell. What are some animals that have shells? POSSIBLE ANSWERS: *Clams, snails, turtles, crabs, lobsters, mussels, and oysters.* People sometimes say that the shells of such animals are analogous to houses. Using the questions on your thinking map for reasoning by analogy, evaluate whether this is a good analogy. What can you learn from the analogy?

- You've learned in history about the periods in American history when a great many immigrants came to live and work in the United States. How is this analogous to a new student entering your school? From this analogy, what can you learn about how these people felt when they came to the United States and how other people felt about them?

- When scientists test a new drug for people, they usually test it first on mice. If the drug works on mice, they sometimes conclude that the drug is good for people. What is the analogy? Use your thinking map for reasoning by analogy to decide whether or not the scientists' conclusion is a good one.

Reinforcement Later

Later in the school year when you cover the following topics, ask students to draw analogies and explain what they can learn from them.

- What analogies can be drawn between industrial pollution and things that happen in the school or at home? What can you learn from these analogies?

- What analogies can be drawn between reading stories and watching TV? What can you learn from these analogies?

ASSESSING STUDENT REASONING BY ANALOGY

Any of the application examples can serve as assessment items to demonstrate whether students are reasoning by analogy skillfully. Remember, it is not enough to ask them to find analogies. They should also be able to formulate new ideas about one thing suggested by its analogy to another. In addition, they should be able to support or show reason for rejecting claims based on analogies by citing similarities or differences that verify or count against these ideas. Ask the students to make their thinking explicit to show how well they are considering whether two things are really analogous and whether they can support their judgment that the analogy does or does not provide them with accurate ideas. Use the thinking map as a checklist to make sure they are following all of the steps in using this skill.

Sample Student Responses • Waves

REASONING BY ANALOGY

| **A:** A SOUND WAVE | **IS LIKE** | **B:** A WATER WAVE |

 HOW ALIKE?

SIMILARITY	**SIGNIFICANT? WHY?**
Both travel through a medium.	Very. Without a medium, neither could propagate.
Generated by putting pressure on medium.	Very. Both are caused by the same type of event.
Both have velocity.	Very. Motion is a basic feature of waves.
Both come in different sizes (amplitude).	Very. Without amplitude waves would be flat & not waves.
Both move away from their point of origin (dispersal) and radiate in all directions.	Very. Both respond to energy input the same way.
Both bounce off hard objects (reflection).	Very. Without, both would transfer all energy on impact.
Both move the same speed in all directions.	Very. This makes both predictable.
Both obey s = frequency x wavelength.	Very. Shows how predictable they both behave.
Both make longitudal waves.	Very. Similar waves means similar behavior.
Both make transverse waves.	Very. Similar waves means similar behavior.
Size of wave is proportional to energy input.	Very. Propagation of both predictable.

THINGS KNOWN ABOUT B (A WATER WAVE) **THAT MIGHT ALSO BE TRUE OF A** (A SOUND WAVE):
A water wave demonstrates positive and negative interference, resonance, refraction, and diffraction.

 HOW DIFFERENT?

DIFFERENCE	**SIGNIFICANT? WHY?**
Water waves travel only in a liquid; sound waves travel in gases and solids as well.	Not Very; both liquids and gases are fluid, and their fluidity determines how they generate waves, even though they are different states of matter.
Water waves are visible; sound waves are invisible.	Not Very; visibility relates to how we perceive these waves and is a function of the medium in which they travel.
Water waves can travel for hundreds of miles; sound waves travel close to their origin.	Not Very; how far they travel relates to the size of the wave, not its wave-like qualities.
Sound waves affect our sense of hearing; water waves do not.	Not Very; this difference also relates to our perception and is a function of the medium.

CONCLUSION ABOUT A (SOUND WAVES):
Even though water waves and sound waves occur in media which exhibit some different properties, it is likely that just as water waves demonstrate resonance, positive and negative intereference, refraction, and diffraction, sound waves do also.

REASONING BY ANALOGY LESSON CONTEXTS

The following examples have been suggested by classroom teachers as contexts to develop infused lessons. If a skill or process has been introduced in a previous infused lesson, these contexts may be used to reinforce it.

GRADE	SUBJECT	TOPIC	THINKING ISSUE
6–8	Science	Cancer	Use an analogy to explain how cancer grows and spreads.
6–8	Science	Cells	How is cell respiration like burning fuel? Is this a good analogy to help us understand important features of cells in the body?
6–8	Science	Cells	Use information about cell functioning to describe how one cell can be analogous to a whole living organism.
6–8	Science	Electricity	How is electricity in a wire analogous to water flowing in a pipe? What can you learn about electricity from this analogy?
6–8	Science	Acids and bases	Use the analogy of baking soda and stomach acid to suggest ways to minimize the damage of acid rain on a pond. What might limit your plan?
6–8	Science	Endangered species	Select a species that became extinct. Find out what scientists believe may have caused that species to die out. Which endangered species today may be threatened for the same reason? How might its extinction be prevented?
6–8	Science	Human eye	Is a camera a good analogy for the human eye?
6–8	Science	Nervous system	Is the telephone system a good analogy for the human nervous system?
9–12	Physics/ general science	The Dragons of Eden	Carl Sagan used a one-year calendar to represent the history of the universe. Review the comparison of the calendar (January 1 to December 31) to cosmic time (the Big Bang to the present). What insights does the calendar comparison give about the relative length of man's history on Earth?
9–12	Biology	Populations	Use reasoning by analogy to assess warnings about the population explosion based on studying what happens to a population of fruit flies in a closed system with a finite amount of food.
9–12	Biology	Cells and tissues	Use information about the specialized function of a cell to explain the function of the tissue it is part of (e.g., red blood cells and blood, neuron and nerve, fibroblast and lung tissue).
9–12	Biology	Evolution of life	Use information about the environmental abiotic factors that are believed to have been requisite for the evolution of life on Earth to suggest ways that life may have evolved on Mars.
9–12	Biology	Bones and muscles	How is the mechanical advantage gained by using a lever to lift a weight like the mechanical advantage gained by the evolved location of the insertion and origin of muscles in bones? Is this analogy strong enough to draw conclusions from? What conclusions can you draw?

REASONING BY ANALOGY LESSON CONTEXTS

GRADE	SUBJECT	TOPIC	THINKING ISSUE
9–12	Biology	The central nervous system	How is the central nervous system like a computer? What conclusions about the central nervous system does this analogy support?
9–12	Biology	Animal classification	How is the Linnaean system of classifying animals like the classification system used by the post office to deliver mail?
9–12	Biology	Enzymes	Is a lock and key a good analogy for the enzyme-substrate complex?
9–12	Biology	Genetics	Is heredity in peas, Drosophila, and bacteria sufficiently analogous to heredity in other animals and humans so that principles of heredity gained from research on them would be true of humans?
9–12	Biology	Research and use of medication	Drug companies often gather research information regarding the effectiveness of drugs on male subjects only. Is this research practice appropriate to support using the drug for women?
9–12	Chemistry	Polymers	What can be learned about the chemistry of polymers by characterizing the reactivity of their monomers? (Consider amino acids/proteins; nucleic acid/DNA/RNA; fatty acids and glycerol/lipids; and monosaccharides/polysaccharides.)
9–12	Chemistry	Molecular structure	Construct molecular models that show how acids and bases can neutralize each other.
9–12	Physics	Nuclear forces	How is the electrical attraction between oppositely charged ions similar to the magnetic attraction of the opposite poles of a magnet? Can this provide an analogy on the basis of which we can learn something new about nuclear forces?
9–12	Physics	Nuclear forces	Make an analogy between the dynamics of springing a mousetrap with a ping-pong ball balanced on it in the middle of 1,000 similarly cocked mousetraps. Note the behavior of neutrons and atoms and how the rate of fission changes over time.
9–12	Earth Science	Earthquakes	How does the analogy between the earth's masses and plates help us to understand how earthquakes occur? What other analogies have people employed to help us understand earthquakes? Are these analogies better or worse than plate tectonics? Why?

CHAPTER 18
CONDITIONAL REASONING

Why Is It Necessary to Engage in Skillful Conditional Reasoning?

Whenever we draw a conclusion based on information in the form of "if … then" (a "conditional" judgment), we are engaging in *conditional reasoning*. For example, suppose that you know that if a person you are rooting for in the tennis match wins *this* game, she wins the championship. When she wins this game, you have every right to celebrate; you draw the conclusion that she's won the championship.

Someone who does not put these ideas together may not realize that this player has won the championship even if he finds out that she has won this game. Conditional reasoning is one way that we combine information to extend our knowledge.

Conditional reasoning pervades our thinking. It is involved in both our everyday activities and our professional work. In fact, we engage in conditional reasoning with such frequency that we often are not even aware that we are doing it. I know that if I buy something, I own it. I know that if I drive more than 55 mph on the highway, I'm breaking the law.

We carry with us a multitude of information about conditions under which a variety of things happen and use this information to advance our knowledge every day of our lives.

Conditional reasoning is a form of deductive reasoning. Conditional reasoning is different from causal explanation, prediction, generalization, and reasoning by analogy in which *we use evidence to support the likelihood of a conclusion*. Reasoning based on evidence is often called "inductive" reasoning. The fingerprints of the alleged murderer on the murder weapon and his footprints near the body strongly support the hypothesis that he was the murderer, but his guilt isn't *proven* with 100% certainty.

In contrast, when you conclude that the person you are rooting for has won the tennis championship, the reasoning you are engaged in is an example of *deductive* reasoning. The conclusion that you draw from the two pieces of information isn't just made highly likely by that information. *It follows from it with 100% certainty.* If the information is accurate, it cannot turn out that she did not win the championship.

The caveat in the last sentence is important. Like reasoning based on evidence, the information we start with in deductive reasoning has to be accurate if we are to have confidence in the accuracy of the conclusions we draw. If it is inaccurate, there is no guarantee that the conclusion is accurate.

An environmental scientist may know that if there are certain chemicals in a community's water supply, the water isn't safe to drink. There is some indication that these chemicals may be present, but he isn't sure. He realizes that he must investigate further. Until he does, he cannot affirm the conclusion that the water is not safe to drink based on conditional reasoning. But if he finds that the information about the chemicals *is* accurate, you'd better not drink the water.

Conditional reasoning is only one type of deductive reasoning. Other types involve reasoning with statements containing "all" and "some" ("categorical" statements) and, especially, drawing conclusions by combining two such statements ("syllogisms"). Still other types of deductive reasoning involve combining statements containing "and," "or," and "not" to draw conclusions ("propositional reasoning").

If I know that all mammals are warm-blooded and that all warm-blooded creatures have hearts, and I draw the conclusion that all mammals have hearts, I am engaging in categorical or syllogistic reasoning. If I reason that I must have left my car keys at the restaurant because I know that either I left them there or in the bookstore, and I've found out that they aren't at the bookstore, I'm engaging in propositional reasoning.

The study of arguments in which categorical statements are used is called "syllogistic" logic. The study of arguments in which "and," "or," and "not" statements are used is called "propositional" logic. Propositional logic also is the branch of logic in which conditional arguments are studied.

Problems we encounter when we engage in conditional reasoning. Sometimes, we make mistakes in conditional reasoning.

Suppose I am the environmental scientist who is investigating the water supply of a community for traces of certain chemicals. I am conducting this investigation because of rumors that a chemical plant had illegally dumped this chemical nearby. I am quite concerned because I know that if these chemicals are in the water, the water is unsafe to drink.

As I am conducting this investigation, a colleague informs me that she has just completed her investigation of the same water supply and determined that it was, indeed, unsafe to drink. I would be making a mistake to conclude that it was polluted with the same chemicals that I was investigating and then to recommend action against the chemical company. A water supply can become polluted in many ways.

Likewise, if I found out that the water supply *does not* contain any traces of the chemicals produced by the company, it would be premature to conclude that the water *was* safe to drink. It could be polluted in some other way.

We can see the errors in these two examples if we make the reasoning explicit. This sequence represents the reasoning in the first example:

1. If the chemicals are in the water, then it is not safe to drink.
2. The water is not safe to drink.

Therefore

3. The chemicals are in the water.

This conclusion clearly does not follow.

The following sequence makes the reasoning in the second example explicit:

1. If the chemicals are in the water, then it is not safe to drink.
2. The chemicals are not in the water.

Therefore

3. The water is safe to drink.

This conclusion doesn't follow either. Traditional deductive logic—the study of the forms of valid and invalid deductive reasoning—catego-

rizes these fallacious types of deductive reasoning. They are called "affirming the consequent" and "denying the antecedent," respectively. (The antecedent is the part of the conditional that comes after the "if," and the consequent is the part of the conditional that comes after the "then.") It is not as important, however, to learn the *names* for these fallacies as it is to understand *why* they are fallacies—the patterns of faulty reasoning they represent.

COMMON PROBLEMS IN THE USE OF CONDITIONAL REASONING

1. We do not think to draw conclusions by combining conditional statements with other relevant information.

2. We draw a conclusion to the effect that the antecedent is true because we have information that the consequent is true (the fallacy of Affirming the Consequent).

3. We draw a conclusion to the effect that the consequent is not true because we have information that the antecedent is not true (the fallacy of Denying the Antecedent).

Figure 18.1

Figure 18.1 contains a summary of the common problems people experience in conditional reasoning.

What Does Skillful Conditional Reasoning Involve?

Much of the time, our conditional reasoning is sound. Sometimes, however, our conditional reasoning misleads us into accepting conclusions that are not justified by the information we start with. Skillfulness in our reasoning can allow us to avoid these mistakes.

In skillful conditional reasoning, we combine information with a conditional statement and accept only conclusions that follow from this combination. Skill in conditional reasoning involves discriminating between conclusions that can and cannot be drawn from such given information. When we make such discriminations, we are determining the *validity* of inferences from the given information to the conclusion that is drawn.

Contexts for using skillful conditional reasoning. In practice, there are a number of different situations in which we use the skills involved in conditional reasoning. One such situation involves *someone else* presenting a conditional argument to you. Others involve *your* use of the skills of conditional reasoning to try to find information needed to draw a specific conclusion that follows from that information.

The first type of situation occurs, typically, when another person tries to convince you of something using a conditional argument. You should then reflect on whether their reasoning is correct (valid) and whether the information it is based on is accurate. This involves a *reactive* assessment of the arguments of others. It is a special case of the type of argument evaluation we discussed in Chapter 9: Finding Reasons and Conclusions.

For example, a Realtor might be trying to convince you to buy a particular house. You like the house and are interested in buying it but find that you cannot because the interest rates are too high and the mortgage will take too much of your monthly income. Suppose your Realtor shows you that if interest rates fall to below 7 percent, then you will be able to keep up with your monthly payments. A few weeks later, she calls you with good news. She tells you that she's found out that the interest rates have gone down to 6.75% percent. "You should buy now," she says. Should you change your mind? You are now in a position to assess whether her reasoning is sound. The three lines below make her reasoning explicit:

1. If the interest rates go down to below 7% percent, you can afford the house.
2. The rates have gone down to 6.75% percent.
Therefore,
3. You can now afford the house.

You agree with the Realtor about statement (1) and verify statement (2). Since the amount of the monthly payments was all that was deterring you from buying the house, her advice is good advice. The conditional reasoning above is quite correct.

Political speeches, letters to the editor, and advertising usually involve conditional reasoning. Once an argument is extracted from one of these contexts, it can be assessed to determine whether the conclusion follows from the given information.

In the second type of situation, we use conditional reasoning in our own thinking. Instead of reacting to the arguments of others, we actively draw and affirm conclusions that extend our knowledge. Although this dynamic is quite different from the more reactive assessment of arguments offered to us by others, at its roots we use the skills of conditional reasoning in the same way; we have to determine the validity of the conditional argument on which the conclusion is based.

One typical situation of this sort is when we select various conditional statements from our own ideas and gather relevant information to enable us to affirm conclusions that we previously only tentatively advanced. I might know that if I have driven more than 3,000 miles since my last oil change, my car is due for another oil change. So I seek information about when I had the oil changed last. When I find that it was more than 3,000 miles ago, I know I should make an appointment for an oil change. In this case, I did not have some of the information that I needed in order to draw a conclusion and had to seek the needed information. When I get that information, I can draw the conclusion. The three lines below represent this example:

1. If I have driven more than 3,000 miles since my last oil change, my car is due for another oil change.
2. My last oil change was 3,500 miles ago.
Therefore,
3. My car is due for another oil change.

My recognition that the reasoning in the above lines is valid conditional reasoning helps me realize that if I find out statement (2), I can determine whether I ought to take my car in for another oil change. This leads me to investigate when I had my last oil change.

Assessing conditional reasoning: can you legitimately draw the conclusion? As we have seen, for conditional reasoning to lead to an acceptable conclusion, two conditions must be satisfied:

- The conclusion follows from the information we combine to generate it; and
- The information we start with—the "premises" of the argument we develop—is acceptable.

Conditional reasoning can lead us to new insights; if one or both of these conditions fail, however, it can also lead us to accept conclusions that are quite false.

The first of these conditions relates to the validity of the conditional reasoning. How can we decide whether this condition is satisfied? In general, an informal strategy can be used to determine whether the conclusion follows from the premises in any deductive argument. If the conclusion of a deductive argument follows, then *the conclusion cannot be false if the premises are true*.

If it is true that my friend won last night's game and it is also true that if she won the game she became the champion, then it *has to be true* that she became the champion. It could not be otherwise. As we saw earlier, on the other hand, if I find out that my friend *did not* win last night's game and I conclude that she lost the championship, this would be invalid reasoning. Maybe she has two chances to win—if she wins tonight she could also win the championship. So there *is* a circumstance in which the conclusion would be false and yet the original premises true. It is, therefore, not legitimate to conclude that she did not win the championship because she did not win last night's game.

To summarize: When we engage in conditional reasoning we can ask, "Could there be circumstances in which the conclusion is false and yet the information on which it is based true?" If the answer to this question is "Yes," the reasoning is not valid; if the answer is "No," the reasoning is valid.

This strategy (trying to imagine a situation in which the information given could be true and the conclusion be false) is helpful in assessing many straightforward conditional arguments. If you can imagine such a circumstance, then the conclusion does not follow. Not finding an exception, however, does not guarantee that the conclusion does follow. Maybe you simply cannot think of an exception now. But if you spend some time trying to imagine an exceptional situation and cannot, chances are that none exists. So at least you've got a good reason for thinking that the conclusion follows.

Assessing conditional reasoning: are the premises acceptable? When we assess conditional reasoning in the way described, we are assessing the reasoning only—whether we can legitimately put together the two ideas and draw the conclusion in question. In many cases, however, we have to ask whether the information we use or the information given in the premises of someone else's argument are, in fact, accurate and acceptable. If it is not, then even though the reasoning may be valid, the conclusion won't be acceptable. You can start with the most outlandish conditional statements and add false information that the antecedent is true and then use valid conditional reasoning to draw false conclusions. Only if the information you start with is acceptable will the conclusion be acceptable.

It is, therefore, important to ask ourselves whether or not we have good reasons for accepting both the conditional statement and the additional information provided in the premises of a conditional argument, even if it is a valid argument. If either of the premises is problematic, we should note what information we need to be able to accept it. Perhaps the person who gave you the information about the winner of last night's game is a well-known trickster. Then you may have to ask someone else or find a news story about the game. Or maybe you aren't sure you have correctly calculated the number of games needed to win the championship. Then you can recalculate or consult an official to corroborate your own calculations.

To determine whether the information we start with is acceptable, we often have to make use of other critical thinking skills, such as determining the reliability of sources. If, when you do, you are satisfied that the premises are acceptable and that the argument is valid, you can accept the conclusion.

The diagram in figure 18.2 can help us record our thinking about the soundness of conditional arguments.

CONDITIONAL ARGUMENT EVALUATION CHECKLIST

1. Do circumstances allow the premises to be true but the conclusion false?   YES ☐ NO ☐

2. If so, describe that situation.

_____

_____

_____

A deductive argument is valid only if no circumstances allow the premises to be true and the conclusion false.   YES ☐ NO ☐

3. Is there any additional information needed to determine whether the conditional and/or the additional information is acceptable?   YES ☐ NO ☐

4. If so, what information is needed?

_____

_____

_____

The conclusion of a deductive argument is acceptable only if the argument is valid and the premises are acceptable.   ACCEPTABLE? YES ☐ NO ☐

Figure 18.2

This graphic organizer, like the corresponding one used in Chapter 9 on reasons and conclusions in arguments, serves two purposes. It provides us with a record of why we judge that a particular conditional argument is valid, invalid, sound, or unsound. It can also remind us what additional information to seek to make the conclusion more acceptable.

Alternative methods for assessing conditional reasoning. We can also assess validity in conditional reasoning by identifying the pattern of reasoning and comparing it to patterns of valid and invalid conditional reasoning. Many logic programs teach this method. Generally, conditional reasoning has two valid methods and two invalid methods. We have explored examples of some of these already. The valid types include:

- Additional information affirms what is stated in the antecedent of a conditional statement, and we conclude that what is stated in the consequent of the conditional is true. (The traditional name for this type of reasoning is "modus ponens.")

- Additional information denies the consequent, and we deny that what is stated in

the antecedent is true. (The traditional name for this type of reasoning is "modus tollens.")

The example about the tennis championship represents the first of these. You know that if your friend wins tonight's game, she will win the championship, and you find out that she won the game. Therefore, you know that she has won the championship.

The second is exemplified by a situation in which you find out that your friend did not win the championship. Then, you can conclude that she did not win the game last night.

The invalid types are the fallacies of affirming the consequent and denying the antecedent. Concluding that she lost the championship because she lost the game last night is an example of denying the antecedent. The example in which a person concludes that she won the game last night because she won the championship is an example of affirming the consequent.

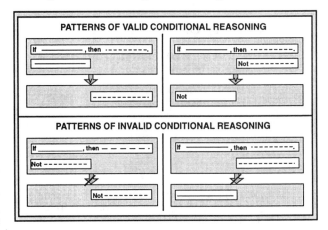

Figure 18.3

The chart in figure 18.3 provides a checklist of these patterns of valid and invalid conditional reasoning.

A third type of valid conditional reasoning is worth noting, though it is not as common as the previous types:

- We combine two conditional statements in which the consequent of the first conditional statement appears in the antecedent of another conditional statement and conclude that the antecedent in the first conditional statement is a condition for the consequent in the second one. (The traditional name for this type of reasoning is "hypothetical syllogism.")

This type of conditional reasoning is exemplified by the following reasoning: Suppose we have information that the person who wins the championship in this tennis match will be given a free trip to Paris. We can combine the information we have so far (if my friend wins tonight's game, then she will win the championship) with this new information (if my friend wins the championship, then she will receive a free trip to Paris) to conclude that if she wins tonight's game, she will win a free trip to Paris.

Tools for engaging in skillful conditional reasoning. We can learn to engage in sound conditional reasoning and avoid the fallacies by adopting some specific thinking strategies. The thinking map in figure 18.4 can guide us in making our conditional reasoning explicit and in assessing it.

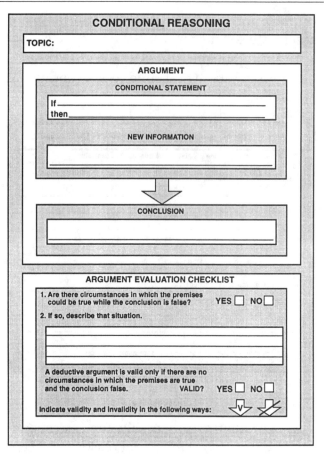

Figure 18.5

After the conditional reasoning is made explicit on the graphic organizer, we can use the strategy suggested to assess the reasoning. Then we can either indicate that the reasoning is valid or invalid using the markers on the diagram. If it is valid, a "V" should be put in the arrow; if it is invalid, a stroke across the arrow will mark this.

How to Teach Students to Engage in Skillful Conditional Reasoning

In teaching conditional reasoning, it is important to provide students with activities in which they consider examples of each of the main types of conditional arguments (modus ponens, modus tollens, affirming the consequent, and denying the antecedent). They should be asked to both analyze and assess arguments given by others as well as engage in their own conditional reasoning.

Many conditional-reasoning exercises in logic textbooks involve exercises in which students are given examples only of conditional arguments that others have offered and are asked to determine whether they are valid. Because the

---

CONDITIONAL REASONING

1. **What topic are you trying to get information about?**

2. **Formulate a conditional statement that you know about that topic.**

3. **What information do you have about the components of the conditional statement?**

4. **What conclusion are you considering?**

5. **Is it valid to reason from the conditional statement and the given information to the conclusion?**

   a. **If the conditional statement and the given information could be true but the conclusion false, then the reasoning is invalid.**

   b. **If the conditional statement and the given information couldn't be true but the conclusion false, the reasoning is valid.**

Figure 18.4

The graphic organizer in figure 18.5 can be used along with the thinking map.

Note that this graphic organizer is useful for recording someone else's conditional argument or our own conditional reasoning as we are guided by the questions on the thinking map.

active use of conditional reasoning plays such an important role in our lives, it is important not to overlook generating such examples as well. The lesson in this handbook does both.

Contexts in the Curriculum for Lessons on Conditional Reasoning

It is easy to construct examples of conditional reasoning in everyday situations, as we have done in this commentary. These often have little direct connection with the curriculum in school. However, similar examples can be constructed around content material that often appear in the secondary school curriculum. For example, characters in stories often employ conditional reasoning, either explicitly or implicitly, as they engage in dialogue or think about hard choices. The Declaration of Independence begins with a conditional statement and suggests an important conclusion—that the authors explain why they have decided to become independent.

More important for this volume of *Infusing Critical and Creative Thinking into Secondary Science*, however, is the role of conditional reasoning in the sciences. Conditional reasoning is extremely important in this field, and the secondary school curriculum in science is rich in contexts in which such reasoning is called for. Many of the basic laws and principles in science are formulated in " If..., then...." statements, as are a number of the principles in both algebra and geometry. As students learn and apply these in specific cases, they develop a repertoire of conditional knowledge that can be used—indeed is often used—as the basis for conditional reasoning.

The following is the most prevalent type of situation in which conditional reasoning plays a natural role in science and provides a multiplicity of good contexts for infusion lessons:

- *Students learn conditional statements and engage in inquiry to determine whether any information combined with the conditional will enable them to draw legitimate conclusions.* For example, students learn a variety of principles in science in the form of "if … then" statements. ("If the volume of water an object displaces is the same weight as the object, then the object will float.") Then they often gather data about what is in the antecedent (e.g., how much the water that an object displaces weighs). When they get this data, they can draw a conclusion using conditional reasoning (e.g., about whether the object will float).

One technique for constructing infusion lessons about conditional reasoning is to construct a vignette in which characters reason conditionally about the topic that students are studying, such as in the middle school science lesson *Growth and Digestion*. There, we include a dialogue between two students who use various types of conditional reasoning as they think about the scientific principles that they are learning (these are formulated as "If..., then..." statements) and discover additional information. The students who are taught this lesson are asked to determine what this reasoning involves and whether it is good conditional reasoning. Guiding students through such activities by using the thinking map and graphic organizer for conditional reasoning will begin to fix this strategy in their minds.

A second technique for such lessons is to put students in the position of searching for or researching specific information that they are then asked to combine with scientific principles formulated as conditional statements in order to draw conclusions by conditional reasoning. When they draw such conclusions, the students are then asked to determine what their reasoning was and whether it was valid. This, in effect, creates a real-life scenario analogous to the fictitious scenarios developed for lessons using the first technique. Hence, in effect, both types of lessons are structured in the same way.

To confirm students' understanding of conditional reasoning, you can ask students to try to figure out the rules for the valid and invalid types of conditional reasoning and then develop a plan for using these rules when they are thinking about conditionals. As they think about their thinking in this way, you can give them other examples in which they can use their thinking maps to monitor their own use of conditional reasoning. In doing this, they are, of course, starting to develop and guide themselves to follow some of the basic the principles of deductive reasoning.

Continued practice in which students reflect on, identify, and certify the conditional reasoning of others, together with examples in which the students themselves have to engage in conditional reasoning, will reinforce their skillful use of this type of reasoning.

As you construct lessons on conditional reasoning, be especially careful not to make these lessons center on issues about the accuracy or truth of the conditional statement itself that plays a role in conditional reasoning. Remember, conditional reasoning involves combining a given conditional statement with the right kind of additional information so that an appropriate conclusion follows.

If the information used to combine with the conditional statement is not the right kind of information and the conclusion is still drawn, an error of reasoning is being made.

The focus in lessons on conditional reasoning should be on these connections between given information and conclusions drawn and not on the accuracy or truth of the information you start with.

Critically evaluating this information, though necessary before you accept a conclusion that results from conditional reasoning, usually does not involve the use of skills for conditional reasoning itself but rather skills at accurate observation, determining the reliability of the sources of the information, and/or inferential skills like causal explanation.

A menu of suggested contexts for infusion lessons on conditional reasoning can be found at the end of the chapter.

Sample Lessons on Conditional Reasoning

We include one lesson on conditional reasoning in this chapter. It is a middle school science lesson on growth and digestion. This lesson provides students with samples of conditional reasoning and asks them to determine whether these are valid or invalid forms of this type of reasoning. It is a model that can easily be extrapolated into high school science.

As you read these lessons, consider the following:

- For the activities in these lessons, can you construct alternative examples in which students react to the arguments of others, on the one hand, and develop their own conditional reasoning, on the other?

- What additional examples of conditional reasoning can be used to reinforce this skill?

Tools for Designing Lessons on Conditional Reasoning

The thinking map for conditional reasoning is provided on the next page. The first graphic organizers can be used to record conditional arguments that others offer as well as conditional arguments that we develop ourselves. The chart of patterns of valid and invalid conditional arguments can serve as a checklist for specific conditional arguments.

The thinking maps and graphic organizers can guide you in designing the critical thinking activity in the lesson and can also serve as photocopy masters, transparency masters, or as models that can be enlarged and used as posters in the classroom. Reproduction rights are granted for use in single classrooms only.

CONDITIONAL REASONING

1. **What topic are you trying to get information about?**

2. **Formulate a conditional statement that you know about that topic.**

3. **What information do you have about the components of the conditional statement?**

4. **What conclusion are you considering?**

5. **Is it valid to reason from the conditional statement and the given information to the conclusion?**

   a. **If the conditional statement and the given information could be true but the conclusion false, then the reasoning is invalid.**

   b. **If the conditional statement and the given information couldn't be true but the conclusion false, the reasoning is valid.**

CONDITIONAL REASONING

TOPIC:

ARGUMENT

CONDITIONAL STATEMENT

If _____

then _____

NEW INFORMATION

CONCLUSION

ARGUMENT EVALUATION CHECKLIST

1. Are there circumstances in which the premises could be true while the conclusion is false?   YES ☐ NO ☐

2. If so, describe that situation.

A deductive argument is valid only if there are no circumstances in which the premises are true and the conclusion false.   VALID? YES ☐ NO ☐

Indicate validity and invalidity in the following ways:

CONDITIONAL ARGUMENT EVALUATION CHECKLIST

1. Do circumstances allow the premises to be true but the conclusion false?

YES ☐   NO ☐

2. If so, describe that situation.

_____

_____

_____

A deductive argument is valid only if no circumstances allow the premises to be true and the conclusion false.

YES ☐   NO ☐

3. Is there any additional information needed to determine whether the conditional and/or the additional information is acceptable?

YES ☐   NO ☐

4. If so, what information is needed?

_____

_____

_____

The conclusion of a deductive argument is acceptable only if the argument is valid and the premises are acceptable.

ACCEPTABLE?

YES ☐   NO ☐

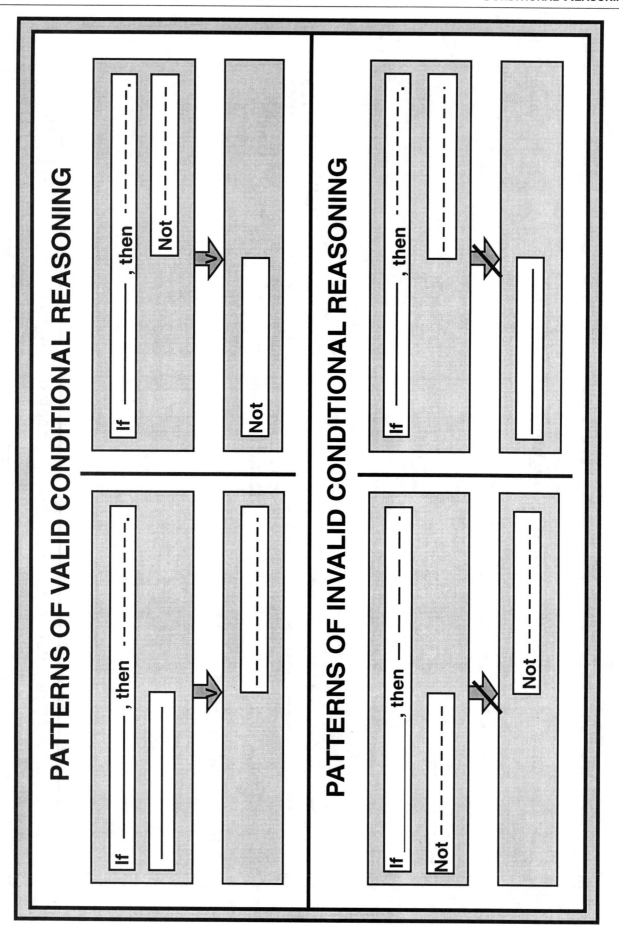

DIGESTION AND GROWTH

General Science, Biology, Anatomy, Physiology

Grades 6–12

OBJECTIVES

CONTENT

Students will learn about the role of the pancreas and pituitary glands in regulating growth and digestion. They will learn about the effects that various diseases of these glands have on growth and digestion and how these diseases can be treated.

THINKING SKILL/PROCESS

Students will learn to combine information that is conditional in character (information expressed by "if … then" statements) with other relevant information to draw conclusions.

METHODS AND MATERIALS

CONTENT

Students will gather information about the glands in the human body from their textbooks and outside reading. They will organize this information on a matrix indicating where each gland is, what it does, its disorders, and treatments of these disorders.

THINKING SKILL/PROCESS

Students will be guided to engage in conditional reasoning by structured questions and a graphic organizer. They will also work together in collaborative learning groups to develop rules for conditional reasoning.

LESSON

INTRODUCTION TO CONTENT AND THINKING SKILL/ PROCESS

- If you follow sports, you know that at a certain point in the season, if one of the teams wins just one more game, that team will win the championship. If you're a fan of that team and you find out that they have won that game, you've got reason to celebrate. You know your team has just won the championship!

- The kind of thinking you've done is called "conditional reasoning." You put two ideas of certain sorts together and drew a conclusion from them.

    One of these ideas was the conditional statement

    <u>If your team wins the next game</u>, <u>it will win the championship</u>.

    You found out that

    <u>Your team has won the game</u>.

    The conclusion you drew was

    <u>Your team has won the championship</u>.

    If someone did not know the first of these ideas but found out that your team won the game, that person wouldn't know what you know—that your team has won the championship. If someone knows that if your team won the next game, it would win the championship, but didn't know what happened in that game, that person also would not know what you know— that your team has won the championship. You know this because you combined these two ideas and drew a conclusion. That's one important way we can learn about things—by drawing good conclusions from things we already know. Can you think of another example in which you came to know something by conditional reasoning? Ask students to discuss the example with a partner and to write down the examples. Have two or three students tell the class about their examples. Write them on the board in the same form as the example about winning the championship. STUDENT ANSWERS VARY.

- These examples of conditional reasoning all have something in common. Notice the first statement. It is an "if ...then" statement. That sort of statement is called a "conditional" statement because it tells us that something will happen on the condition that something else happens. That is why this kind of thinking is called "conditional reasoning." Also notice that the second statement tells us that the condition has occurred. When you put these two ideas together you can draw the conclusion that the result has occurred. When you engage in conditional reasoning and you draw a conclusion that does follow from the information you start with, the reasoning is called "valid." A specific pattern describes this type of valid conditional reasoning. It looks like this chart. Show a larger copy of the conditional reasoning chart at the right.

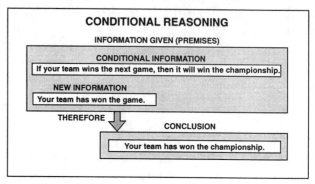

CONDITIONAL REASONING

INFORMATION GIVEN (PREMISES)

CONDITIONAL INFORMATION
If your team wins the next game, then it will win the championship.

NEW INFORMATION
Your team has won the game.

THEREFORE

CONCLUSION
Your team has won the championship.

- Work with your partner and write down some more conditional statements. Also, under "New Information," write a second sentence that describes the condition that you'd have to check out before concluding that the result will happen. Then, write the statement of what you would be able to conclude under the other part of the conditional statement with an arrow pointing to it from the first two statements. Finally, discuss with your partner how you might find out whether the condition actually existed. Write "If ___ then ___" on the board and ask the students to fill in the blanks to produce a group of conditional statements that they know are true. Get one example of a conditional statement from the students and write it on the board. Then, write the rest of the argument on the board in the same form as the sports example. Under the argument write, "If I can find out ___ I will be able to conclude ___," filled in appropriately. Ask the students to write similar sentences after they have constructed their arguments.

- Conditional reasoning is an example of <u>deductive reasoning</u>. If the reasoning is valid and the information you put together to draw the conclusion is true, this doesn't just make the conclusion likely or probable. The conclusion has to be true. That's what's special about deductive reasoning. Now you can see why this kind of thinking is so important. If you know a conditional statement and you find out that what comes after "if" is true (the condition is satisfied), then what comes after "then" will also be true.

- Conditional reasoning can be a great help to us in putting ideas together and finding out new things, but we can also make mistakes in conditional reasoning. Suppose you don't know how your team did in the last game they played but someone tells you that they won the championship. Can you conclude that they won the last game? Write this reasoning on the board starting with "If your team wins the game, they'll win the championship," and "Your team won the championship," and put a question mark after "Your team won the last game" as the conclusion. This conclusion does not necessarily follow from the information you start with. Your team might have won the championship because the second place team lost the game it was playing and was eliminated, not because your team won its game.

- When the conclusion doesn't follow from the given information, called the premise, the reasoning is "invalid." Here is how to find out if conditional reasoning is invalid. Imagine a situation in which the information you start with could be true and the conclusion false. That's what we just did when we imagined that the second-place team lost its game. Then, even though your team also lost, it could have won the championship because the second-place team was eliminated. Draw a diagram that represents the pattern of this invalid type of conditional reasoning. Put a slash through the arrow to indicate that the reasoning is invalid. See the diagram at the right.

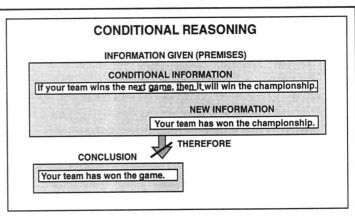

- We've been learning some things about our glands—the endocrine system—in science that can be stated conditionally. I'm going to tell you a story about two students—Bob and Sandy—who were studying the endocrine system and describe some of the things they were learning about conditional statements.

THINKING ACTIVELY

- This is a story about two students who had just worked together on a science project in which they had made a chart of the endocrine system. They were fascinated with the idea that tiny glands in various parts of the body can control so many things, such as the way we grow. The chart helped them organize the information they were learning so that they could see it all clearly. They made a column down the side of the chart and listed the glands in the endocrine system in each box. Across the top, they listed different things they were learning that they thought were important

ENDROCINE SYSTEM MATRIX

GLAND	IMPORTANT INFORMATION			
	LOCATION Where it is in the body	FUNCTION What it does for people	DISORDERS Possible problems with it	TREATMENT Cure s for disorders?

about these glands. In the cells under each of the headings, they wrote specific information they had gathered about the glands. Use your textbook or library resources to fill in the cells as Bob and Sandy might have. You can also gather supplementary material about the glands and make it available in the classroom.

- Bob became very concerned when he started writing the disorders of the various glands in the disorder column.

  "I didn't realize how important glands were," he said. "If your pancreas doesn't function properly, for example, you can't digest food properly. If your pituitary gland doesn't function properly when you're young, there's a danger that you won't grow."

  Sandy thought about this for a minute. "You know," she said, "I remember my mother telling me that one of the children in the new family that moved in down the street had a problem with his pituitary gland and that's why they were going to the hospital all the time. I didn't realize what that meant then, but I do now. That's scary!"

- Let's think about this to see what would be so scary about what Sandy remembered. What does Bob say that reminds Sandy about the child down the street? *If your pituitary gland doesn't function properly when you're young, there's a danger that you won't grow.*

- **What kind of statement is that?** *A conditional statement.*

- **Let's fill in the steps in Sandy's thinking so that we can determine what she thought about the child down the street that was scary to her. This time, we're going to use a graphic organizer that will allow us to state her reasoning. Write the condition you just identified in the top box marked "Conditional Statement."** Students should write "your pituitary gland doesn't function properly when you're young" after the "If" and "there's a danger that you won't grow" after the "then." Ask them what condition they wrote in the diagram and remind them of the conditional statement they just identified if they are having trouble. **What new information did Sandy recall when she heard Bob's conditional statement? Write it in the box for "New Information."** Students should write "A child down the street has a pituitary gland that doesn't function properly" in the box for new information. If they have a problem with this, ask the students what Sandy remembered about the child who just moved in down the street that has to do with the pituitary gland. **What conclusion do you think she drew based on these two pieces of information? Write the conclusion in the box for "Conclusion."** The conclusion is that the child who moved in down the street is in danger of not growing. If they write statements like "Something scary will happen," ask them what was scary. Tell them that sometimes people don't state their conclusions, but they make them all the same.

CONDITIONAL REASONING

TOPIC:

ARGUMENT

CONDITIONAL STATEMENT

If
then

NEW INFORMATION

CONCLUSION

ARGUMENT EVALUATION CHECKLIST

1. Are there circumstances in which the premises could be true while the conclusion is false?   YES ☐ NO ☐

2. If so, describe that situation.

A deductive argument is valid only if there are no circumstances in which the premises are true and the conclusion false.   VALID? YES ☐ NO ☐

Indicate validity and invalidity in the following ways:

- **Is this good conditional reasoning—was Sandy right to be concerned? Does the conclusion *follow from* the information she started with? Fill in the argument evaluation checklist on the graphic organizer and explain to your partner why you think the conclusion does or doesn't follow.** Ask for explanations from the students. POSSIBLE ANSWERS: *Yes, it follows, because if the first two statements are true, the conclusion must be true. Yes, because the new information tells us that the condition in the conditional statement is satisfied. The condition is that he has a problem with his pituitary gland, so what it is a condition for (that his growth may be stunted) must be true.* Students may also write out the argument in the form that was used in the team championship example and show that it falls into the same pattern.

- **Now, let's continue the story.**

    While Bob and Sandy were talking, Bob suddenly remembered something. The night before, his mother said she was having a bad case of indigestion. What she ate for dinner was just lying in her stomach like a lump, she said, and it didn't feel good. Bob got worried. "There must be something wrong with my mother's pancreas gland," he said. "And that's serious." He explained to Sandy what his mother had said. Sandy got concerned for a moment, too. But what Bob had said didn't seem quite right.

- **Can you help Sandy figure this out? Use the conditional reasoning graphic organizer to state Bob's reasoning. Then discuss with your partner whether this was valid reasoning. Use the form for checking the validity of conditional arguments to explain why you think it is or isn't**

valid. After a few minutes, ask students to report on how the graphic organizer should be filled in to represent Bob's reasoning.

In this case, the conditional statement is

<u>If your pancreas doesn't function properly, you can't digest food properly</u>.

The new information obtained from Bob's story about his mother is

<u>Bob's mother couldn't digest her food properly last night</u>.

The conclusion is

<u>Bob's mother's pancreas is not functioning properly</u>.

Students usually indicate that the above isn't valid conditional reasoning. POSSIBLE REASONS: *Maybe her indigestion had nothing to do with her pancreas; maybe it was because she ate too much. Maybe she had indigestion because the food was too spicy, not because of her pancreas. This isn't a good conclusion to draw because in the conditional statement we are not told that poor digestion is a condition of the pancreas not functioning properly. Rather, it is the other way around.*

- **As you can see with Bob and Sandy, it is sometimes very easy to get misled: You might think that a conclusion follows from a conditional statement and a new piece of information that appears in the wrong place in the conditional. When we make a mistake like this, it is called a "fallacy of reasoning." A fallacy is invalid reasoning in which you think a conclusion follows, but it really doesn't. Who committed the fallacy here, Bob or Sandy?** *Bob did.*

- **Let's think about the child who lives down the street from Sandy. What sorts of things can be done to help someone who has problems with their pituitary gland?** Remind the students that they have just drawn up a matrix that may have this information on it. If it doesn't, ask them to do some research in their textbooks or in the school library to try to see if there is anything that can be done for pituitary disorders and what that might be. Ask them to write their findings in their matrices and to report what they found to the class.

THINKING ABOUT THINKING

- **Let's think about your thinking now. What did you think about to be sure that valid conditional reasoning was used to draw conclusions? List what you thought about and draw up a thinking map to reflect these ideas.** POSSIBLE ANSWERS: *I thought about the conditional statement used to draw the conclusion about the boy who had a problem with his pituitary gland. I identified which part stated a condition and what the result would be if the condition occurred. I then thought about whether or not the condition had occurred. When I found out that the boy had the pituitary problem, I knew the conclusion that he wouldn't grow had to be true. If the statement, "If you have a problem with your pituitary gland, then you don't grow properly," is true and you find out that you have a problem with your pituitary gland, then it must be true that you won't grow properly. I thought about Bob's conclusion that his mother had a problem with her pancreas and wondered if you could really get that conclusion from the information given in the*

> **CONDITIONAL REASONING**
> 1. What topic are you trying to get information about?
> 2. Formulate a conditional statement that you know about that topic.
> 3. What information do you have about the components of the conditional statement?
> 4. What conclusion are you considering?
> 5. Is it valid to reason from the conditional statement and the given information to the conclusion?
>   a. If the conditional statement and the given information could be true but the conclusion false, then the reasoning is invalid.
>   b. If the conditional statement and the given information couldn't be true but the conclusion false, the reasoning is valid.

story. I then identified the conditional statement on the topic of glands and the new information Bob added to it to get his conclusion. I tried to imagine a situation in which the information he started with could still be true and yet the conclusion false. When I realized Bob's mother could have had indigestion because she ate too fast, I realized that Bob's reasoning was not valid. The thinking map should resemble the conditional reasoning map on the preceding page.

- In this lesson, we drew two diagrams of patterns of valid and invalid conditional reasoning. Can you formulate rules about conditional reasoning based on these patterns that will help you to draw valid conclusions from information and avoid committing fallacies? POSSIBLE ANSWERS: *If the stated condition is true, then you can draw the conclusion that what it is a condition for is true. If what comes after the "then" in a conditional statement is true, you still can't say that the condition is satisfied.*

- What advice would you give to someone who is having trouble with conditional reasoning to help them avoid making mistakes in their reasoning? POSSIBLE ANSWERS: *Be clear about the conditional statement that you start with and any other information corresponding to the components of the conditional. Before accepting the conclusion, check your reasoning for validity and make sure that the information you start with is acceptable.*

APPLYING THINKING

Immediate Transfer

- Suppose a music store were selling CDs at 50 percent off their regular price. Let's say the sale started last Saturday and was going to last a week. It is now Wednesday and you're in school. You know that on Thursday and Friday you have practice after school and can't get to the music store before it closes. You really want to get a particular CD at the sale price. When is the only time you can go to the store to get the record at the sale price? *Today after school (Wednesday).* Explain how you used conditional reasoning to conclude this. POSSIBLE AN-SWERS: *Two instances of conditional reasoning led to this decision. The first conditional statement can be formulated in the following way: "If I am to buy the CD at the sale price, I have to go to the store before Saturday." Since I do want the CD at the sale price, I can conclude that I have to go to the music store before Saturday. The second conditional statement is "If I am to go before Saturday, then I have to go on Wednesday, Thursday, or Friday." I can't go on Thursday or Friday, and today is Wednesday, so if I am to go before Saturday, I can go only today (Wednesday). The conditional statement combines with the fact that I do have to go before Saturday to yield my conclusion that I have to go today.* Ask students to fill in the graphic organizer for conditional reasoning as they explain these instances of reasoning. Then, ask them if these are good pieces of reasoning. They should refer to the rules they have formulated for conditional reasoning.

- Go back to your matrix about the endocrine system. Formulate any conditional statements you can based on the matrix. For example, you could formulate a different conditional statement about the pituitary gland: If a child grows normally, then that child's pituitary is functioning properly. After you've formulated some conditional statements, pick one and construct a story like the one I just told you. It should involve some conditional reasoning that is good and some conditional reasoning that contains a fallacy.

Reinforcement Later

Later in the school year, when your students are studying the following topics, prompt their use of conditional reasoning with the given dialog.

- We've been studying diet and nutrition. Formulate some conditional statements about the results of certain eating habits based on what you've learned. Based on these statements and what you know about your own eating habits, can you draw some conclusions about the results of how you eat?

- In *Alice in Wonderland*, after Alice went down the rabbit hole, she found a little bottle with a paper label tied around its neck saying "DRINK ME." Here's what happened next:

"It was all very well to say 'Drink Me,' but the wise little Alice was not going to do that in a hurry. 'No, I'll look first,' she said, 'and see whether it's marked *poison* or not.' She had read several nice little stories about children who had got burned, and eaten up by wild beasts, and other unpleasant things, all because they would not remember the simple rules their friends had taught them. But Alice did remember the rules she had learned. She knew that a red-hot poker will burn you if you hold it too long; and that, if you cut your finger very deeply with a knife, it usually bleeds. And she also never would forget that, if you drink from a bottle marked *poison*, it will almost certainly be harmful for you sooner or later. However, this bottle was not marked *poison*. Alice decided to taste it. …She very soon finished the whole bottle off."

What kind of reasoning leads Alice to drink what is in the bottle? Is this good reasoning? If you were there with her, what would you advise her to do, and why?

THINKING SKILL EXTENSION

You can build on what is taught in this lesson to teach your students other aspects of conditional reasoning. There are two important additional principles that deal with denying the consequent and denying the antecedent. You can use negative information about the consequent of a conditional (what comes after the "then") to draw a negative conclusion indicating that what is stated in the antecedent (what comes after the "if") is false. Suppose that if your team is to win the championship, then it has to win the last two games, and suppose your team didn't win the last two games. It is valid to conclude that your team didn't win the championship. On the other hand, suppose that your team doesn't win the championship. Then you can't conclude that it didn't win the last two games. It could have won the last two games and yet come in second to a team that won its last three games. All the conditional statement tells us is that if the team doesn't win the last two games, it will be eliminated. It doesn't tell us that if it wins those games that guarantees that it will win the championship. You can teach students these forms of conditional reasoning by using these examples, the example from *Alice in Wonderland*, or by constructing examples similar to the ones used in this lesson but representing these types of reasoning. A good way to bring this extension to a close is to ask your students to categorize the four patterns of reasoning repre-

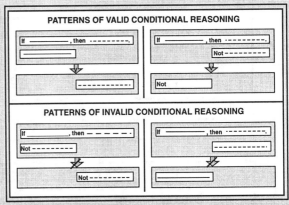

sented by the two in the lesson and these additional types. The diagram to the right represents one way to do this.

ASSESSING STUDENT CONDITIONAL REASONING

You can assess how well students engage in conditional reasoning by using any of the transfer or reinforcement examples and judging whether the students have followed the strategy outlined in the map for conditional reasoning. Make sure you ask students to explain their reasoning when you use these examples. They should use the correct terms (like "conditional") when they explain their reasoning.

Sample Student Responses • Growth and Digestion

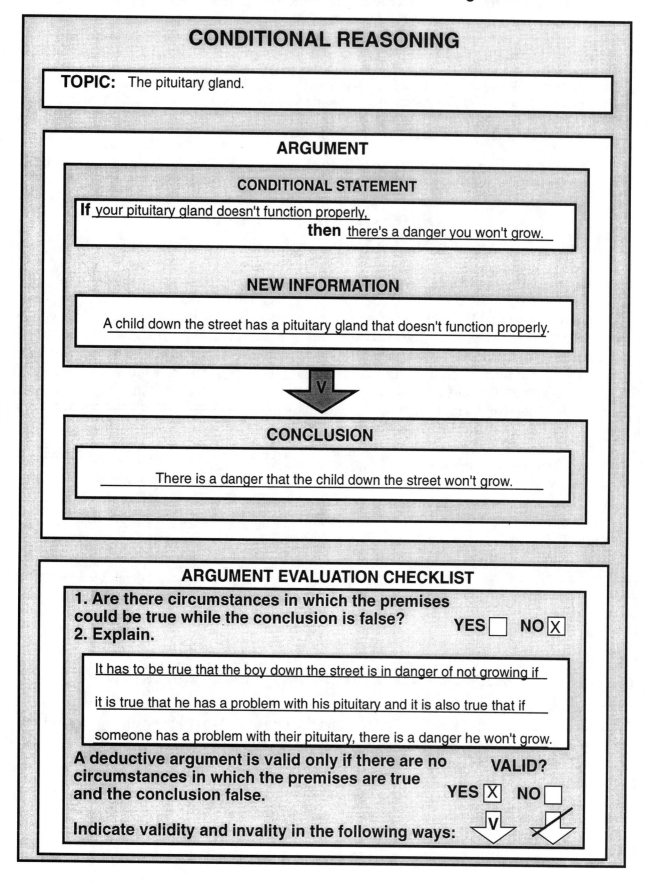

CONDITIONAL REASONING

TOPIC: The pituitary gland.

ARGUMENT

CONDITIONAL STATEMENT

If your pituitary gland doesn't function properly,

              then there's a danger you won't grow.

NEW INFORMATION

A child down the street has a pituitary gland that doesn't function properly.

V

CONCLUSION

There is a danger that the child down the street won't grow.

ARGUMENT EVALUATION CHECKLIST

1. Are there circumstances in which the premises could be true while the conclusion is false?

2. Explain.   YES ☐   NO ☒

It has to be true that the boy down the street is in danger of not growing if

it is true that he has a problem with his pituitary and it is also true that if

someone has a problem with their pituitary, there is a danger he won't grow.

A deductive argument is valid only if there are no circumstances in which the premises are true and the conclusion false.

VALID?

YES ☒   NO ☐

Indicate validity and invality in the following ways: V   ⤵

Sample Student Responses • Growth and Digestion

CONDITIONAL REASONING

TOPIC: The pancreas gland.

ARGUMENT

CONDITIONAL STATEMENT

If your pancreas gland doesn't function properly,

then you can't digest food properly.

NEW INFORMATION

Bob's mother couldn't digest her food properly last night.

CONCLUSION

Bob's mother's pancreas is not functioning properly.

ARGUMENT EVALUATION CHECKLIST

1. Are there circumstances in which the premises could be true while the conclusion is false? YES [X] NO [ ]
2. Explain.

It could be that Bob's mother had indigestion because she ate too much. In that case it might not be true that she has a problem with her pancreas even though the conditional and the information about her indigestion are correct.

A deductive argument is valid only if there are no circumstances in which the premises are true and the conclusion false.

VALID?
YES [ ] NO [X]

Indicate validity and invality in the following ways:

CONDITIONAL REASONING LESSON CONTEXTS

The following examples have been suggested by classroom teachers as contexts to develop infused lessons. If a skill or process has been introduced in a previous infused lesson, these contexts may be used to reinforce it.

GRADE	SUBJECT	TOPIC	THINKING ISSUE
6–8	Science	Plants	You know that if you fertilize your plants, they will grow more rapidly. What information would you need to be sure that they will grow more rapidly? What could you conclude if you didn't fertilize your plants?
9–12	Biology	Pollution	Write these facts on the board and ask students to construct a series of conditional arguments based on them, supporting the Clean Air Acts of the 70s and 80s: 1) Some coal is high in sulfur content. 2) Sulfur dioxide is produced during the combustion of fossil fuels. 3) Sulfur dioxide combines with oxygen in the air to produce sulfur trioxide. 4) Sulfur trioxide combines with atmospheric water to form sulfuric acid. Airborne sulfuric acid enters plants along with airborne oxygen. 5) Sulfur dioxide irritates the lungs. 5) Airborne sulfur gases can form acid rain in the lungs of animals. 6) Plants are damaged by acid rain. 7) Animals get respiratory diseases like chronic bronchitis from acid rain.
9–12	Biology	Carcinogens	Hundreds of studies have shown that there is a strong correlation between cigarette smoking and lung cancer. Tar, a major byproduct of burning tobacco, causes cancer in laboratory animals and in humans. What conclusion can you draw about the risk of cancer if you smoke a pack of cigarettes a day? What conditional allows you to draw this conclusion? Is the conditional acceptable? Suppose you find out that a person on your street has lung cancer. What conclusion can you draw about his or her smoking?
9–12	Biology	Cells	Imagine that you've been looking at various items under a microscope. You notice that one of them is spherical and has a wall that contains a complex of other substances. You remember that if a cell is a plant cell, then it will have a cell wall. You wonder if you have enough information to conclude that what you are seeing is a plant cell. Do you? If so, explain. If not, when else would you need to be able to conclude this?
9–12	Biology	Cells and chromosomes	Which of the following conditionals is correct: If a cell has 46 chromosomes, then it is a human cell; if a cell is a human cell, then it has 46 chromosomes? Suppose you discovered a cell with 23 chromosomes when you examined it under a microscope. What conclusion would you be justified in accepting using conditional reasoning?

CONDITIONAL REASONING LESSON CONTEXTS

GRADE	SUBJECT	TOPIC	THINKING ISSUE
9–12	Chemistry	Boyle's law	If the temperature of steam in a cylinder of a reciprocating steam engine remains constant, then you can predict the steam's pressure when it exhausts using Boyle's Law. Let's assume that the temperature remains constant. Let's also suppose that when the steam Ñ is preheated to 425 degrees, C Ñ enters the cylinder it occupies a volume of 30 cm cubed. At the end of the power stroke, the steam has expanded to occupy 140 cm cubed. The steam enters at a pressure of 2000 kPa. Because of the assumption about the temperature, we apply Boyle's law, yielding the result that the exit pressure should be 429kPa. This is what we expect will result. When we check the pressure, however, it is 270 kPa. Did we use valid conditional reasoning? What mistake in thinking accounts for this disparity?
9–12	Chemistry	Toxic waste	Your company has started using a new chemical process requiring that waste materials flow into a nearby river. You have ascertained that no toxic waste will be contained in these materials but that the temperature will be 6 degrees C higher than any waste has been in the past. At higher temperatures, less oxygen can dissolve in water, and some organisms cannot survive in oxygen-poor water. Should you be concerned? Set up a series of conditional arguments to determine what, if any, the effect of this new process will be on aquatic organisms.
9–12	Chemistry	Bonding	What conclusion can be drawn about the nature of the chemical bond between sodium and chlorine in sodium chloride? What conditional statements about valence and electronegativity allow you to draw this conclusion? Formulate a set of conditional statements that would allow you to draw conclusions about bonds formed in a variety of simple molecules (e.g., water and carbon dioxide) and compounds (e.g., magnesium chloride).
9–12	Geology	Earthquakes	Scientists know that if sufficient stress builds up between the two plates moving past each other along the coast of California, an earthquake will occur. Discuss the difficulties in getting the information necessary to be able to predict an earthquake in California based on this conditional.

PART 6

DESIGNING AND TEACHING INFUSION LESSONS

Chapter 19: Instructional Methods

Chapter 20: The Role of Metacognition

Chapter 21: Selecting Contexts for Infusion Lessons

INFUSION LESSON PLAN EXPLANATION

Subject: Grades:

OBJECTIVES

CONTENT	THINKING SKILL OR PROCESS
Statement of content objectives from curriculum guide or text outline	Description of the thinking skill/process the students will learn

METHODS AND MATERIALS

CONTENT	THINKING SKILL OR PROCESS
Use of instructional methods to teach the content effectively	Use of instructional methods to teach the thinking process effectively

Expository methods	Using manipulatives	Structured questioning strategies
Inquiry methods	Discourse/Socratic dialog	Specialized graphic organizers
Co-operative learning	Integrated arts	Collaborative learning, including Think/Pair/Share
Graphic organizers	Directed observation	Direct or inductive explanation of thinking processes
Advance organizers	Specialized software	
Higher order questions		Learner-generated cognitive maps (diagrams and pictures)

LESSON

INTRODUCTION TO CONTENT AND THINKING SKILL/PROCESS

Teacher's comments to introduce the content objectives
The lesson introduction should activate students' prior knowledge of the content and establish its relevance and importance.

Teacher's comments to introduce the thinking process and its significance
The lesson introduction should activate students' prior experience with the thinking skill/process, preview the thinking skill/process, and demonstrate the value and usefulness of performing the thinking skillfully. The introduction serves as an anticipatory set for the thinking process and should confirm the benefits of its skillful use.

THINKING ACTIVELY

Active thinking involving verbal prompts and graphic maps
The main activity of the lesson interweaves the explicit thinking skill/process with the content. This is what makes the content lesson an infused lesson. Students are guided through the thinking activity by verbal prompts (e.g., questions) in the language of the thinking skill/process and by graphic organizers.

THINKING ABOUT THINKING

Distancing activities that help students think about the thinking process
Students are asked direct questions about their thinking. The metacognition map guides the composition of these questions. Students reflect about what kind of thinking they did, how they did it, and how effective it was.

APPLYING THINKING

Transfer activities that involve student-prompted use of the skill in other examples
The two broad categories of transfer activities are (1) near or far activities that immediately follow the substance of the lesson and (2) reinforcement later in the school year. Both types of transfer involve less teacher prompting of the thinking process than in the Thinking Actively component of the lesson.

Immediate Transfer

Near transfer: Application of the process within the same class session or soon afterward to content similar to that of the initial infusion lesson. Decrease teacher prompting of the thinking.

Far transfer: Application of the process within the same class session or soon afterward to content different from that of the initial infusion lesson. Decrease teacher prompting of the thinking.

Reinforcement Later

Application of the process later in the school year to content different from that of the infusion lesson. Decrease teacher prompting of the thinking.

OPTIONAL EXTENSION ACTIVITIES
(Can occur at any time during the lesson)

REINFORCING OTHER THINKING SKILLS AND PROCESSES: Working on additional thinking skills/processes from previous infusion lessons that can play a role in this lesson.
RESEARCH EXTENSION: Gathering additional information that may be useful in reaching a conclusion or an interpretation in this lesson.
USE OF SPECIALIZED ASSIGNMENTS TO REINFORCE THE THINKING: Assigning written or oral tasks or projects that may further illustrate students' thinking about the content in this lesson.

ASSESSING STUDENT THINKING

Extended written or oral assignments or performance assessments of the effective use of the thinking skill or process.

CHAPTER 19
INSTRUCTIONAL METHODS IN INFUSION LESSONS ——

Instructional Methods That Prompt Student Thinking

Infusion lessons blend explicit instruction in thinking skills and processes with content instruction, using methods that enhance students' thinking and comprehension of the content. Understanding how these methods affect students' thinking enables you to select, combine, and vary instructional techniques in order to create rich and effective infusion lessons.

Various instructional methods are used in infusion lessons for different purposes: (1) to teach the thinking skills and processes, (2) to foster thinking collaboratively, (3) to prompt students to learn content. Instructional methods to achieve the first and second purposes as well as techniques to promote thoughtfulness and a thinking disposition are featured in all infusion lessons. Methods that promote thoughtful content learning will vary from discipline to discipline based on effectiveness in teaching objectives within a field.

Direct instruction strategies to teach thinking skills and processes. All infusion lessons feature an array of instructional strategies to promote clarity and reflection about the thinking skill or process involved. Structured questioning and the use of specialized graphic organizers make explicit what is involved in skillful thinking. These methods distinguish infusion lessons from other approaches. Modify these methods so lessons are varied and appropriate.

Thoughtful discussion of thinking and content through guided collaborative tasks. In infusion lessons, students carry out their thinking in a social context. A variety of collaborative and cooperative learning strategies are used in these lessons as students are given thinking challenges. These strategies encourage student thinking by prompting interaction with other students. In addition, each lesson varies small-group work with thinking tasks for individual students and class discussion. Blending individual thinking and group interaction shows students how the interplay of ideas enhances the ideas of each individual.

Instructional methods that prompt thoughtful content learning. The selection of instructional methods to teach content thoughtfully in infusion lessons varies based on appropriateness to the content and the level of cognitive development of students. Instructional methods include guided reading, asking higher order questions, prompting writing for reflection, directed essay writing, using standard graphic organizers or student-generated cognitive and concept maps, employing Socratic dialogue, engaging students in "hands-on" investigation, using mathematics manipulatives, prompting student questioning, and using whole-language techniques. Selecting such methods carefully and using them in developmentally appropriate and thoughtful ways promotes students' responsiveness to the thinking instruction that occurs in these lessons and enhances their deep understanding of the content.

General teacher behaviors that model good thinking dispositions. One of the goals of thinking instruction is to demonstrate the habits of mind that good thinkers are disposed to use—their thinking dispositions. Certain behaviors that teachers engage in model the thinking dispositions that we seek to foster in students. These habits of thoughtful interaction with students in any instructional context should occur regularly in infusion lessons as well. Practices that encourage the thinking dispositions we seek to foster in students include allowing wait time, prompting reconsideration, asking clarifying questions, responding with acceptance, listening with empathy, and (in order to clarify understanding) using precision in language and promoting precision of expression and requesting evidence.

Selecting instructional methods. Instructional methods from each category can blend to design lessons that are varied in form, robust in fostering students' thinking, rich in deep understanding of the content, and supportive of a classroom culture of thoughtfulness.

The outline in figure 19.1 (next page) summarizes how information on instructional methods is organized in this chapter.

Instructional Methods for Teaching Thinking Skills and Processing Infusion Lessons

Using thinking maps. The key element of an infusion lesson is the use of a thinking map to clarify and organize skillful thinking. A thinking map (also called a verbal map) is a structured list of key questions phrased in the language of the thinking skill or process that leads to an informed judgment. These are questions that thoughtful people ask when they engage in skillful thinking. They are sequenced in a natural order to represent effective ways of organizing one's thinking but may be reconsidered and raised again at any time during the process in order to give more depth and thoroughness to the thinking.

Thinking maps provide visual reminders in making the thinking skillful. They may be written on the blackboard, displayed on a poster, projected on a screen using a transparency, or photocopied for each student.

Whenever possible, derive the thinking map from students' understanding of the thinking skill or process. In the decision-making lesson "The Mystery of Silent Spring," students' comments are organized and transcribed to produce the thinking map of skillful causal explanation. In "Canals of Mars," students' questions about the reliability and accuracy of the observer's report are categorized, producing a thinking map to

---

INSTRUCTIONAL METHODS FEATURED IN INFUSION LESSONS

INSTRUCTIONAL METHODS TO TEACH THINKING DIRECTLY (FEATURED IN ALL INFUSION LESSONS)

    Using thinking maps
    Using specialized graphic organizers to guide thinking
    Using structured oral questioning

INSTRUCTIONAL METHODS TO FOSTER THOUGHTFULNESS AS STUDENTS ENGAGE WITH THE CONTENT (VARY IN INFUSION LESSONS)

    Using graphic organizers to depict information
    Using higher-order questioning
    Using whole-language techniques
    Using guided reading
    Using reflective writing
    Using essay writing
    Using manipulatives

INSTRUCTIONAL METHODS TO FOSTER THOUGHTFULNESS AS A SOCIAL ENDEAVOR (FEATURED IN ALL INFUSION LESSONS)

    Pooling background information
    Engaging in collaborative thinking tasks
    Using cooperative learning strategies
    Using think/Pair/Share
    Varying individual, small group, and class discussion

TEACHER BEHAVIORS THAT PROMOTE THOUGHTFULNESS AND THINKING DISPOSITIONS (FEATURED IN ALL INFUSION LESSONS)

    Allowing wait time
    Prompting reconsideration
    Using precision in language and promoting precision in students' responses
    Asking clarifying questions about students' responses
    Asking for reasons for students' judgments

Figure 19.1

---

guide making well-considered judgments about the reliability of any observation.

The use of thinking maps in infusion lessons prompts students to raise and answer a series of questions in order to make an organized and careful judgment. Students' own wording of the thinking map should be used in your initial lesson and in transfer activities.

Thinking maps, not necessarily expressed in the terms students will use, have a second role in the methodology of direct instruction. Teachers may follow the thinking maps to keep their own thinking focused as they guide discussion.

By posting the thinking maps in the classroom, students are reminded of the guiding questions for skillful thinking when called upon to use that kind of thinking in the future. The steps on the displayed thinking map serve as a checklist for key points in the thinking activity.

Using thinking maps promotes skillful thinking directly in the following ways:

- provides key questions that students answer as they engage in a thinking activity
- demonstrates to students that they have insight and are capable of critical thinking
- promotes recall of the thinking skill or process
- models reflection about one's thinking
- promotes students' confidence and competence in skillful thinking

Structured oral questioning. Questions prompt thinking. The kind of thinking prompted, however, is primarily a function of the type of question asked, the context in which it is asked, and students' background knowledge. Using structured questioning in a classroom is the oral equivalent of using a thinking map to guide students through a process of thinking. In structured questioning, the teacher or the students identify the type of thinking being done (e.g., causal explanation). The questions focus students on the key attention points in doing this type of thinking skillfully. Thus, as they are thinking about the mystery of the silent spring, the teacher asks students to put themselves in the position of someone who has to figure out what has caused the disaster and to suggest what some possible explanations might be. When

they have had a chance to explore this, she asks them to select a possible explanation and to answer the question, "What evidence might you find showing that this possible cause is likely to be the real cause?"

Structured questioning is an important ingredient in the direct instruction of thinking incorporated into infusion lessons. The repetition of these questions not only prompts students' thinking but also gets students used to these questions as they engage in the thinking activity. Thus, students will get used to hearing "What are the options?" as they go through a process of decision making or "What possible evidence would show that this was the cause?" as they go through a process of causal explanation.

This type of questioning goes hand-in-hand with the thinking maps. Together, they form the first step in internalizing these questions as students engage in these thinking activities.

Using structured questioning promotes skillful thinking in the following ways:

- identifies a sequence of important questions to ask in skillful thinking
- provides auditory cues that prompt students' organized thinking
- helps students to recall important questions to ask as they engage in a specific type of thinking

Use of specialized graphic organizers to guide the thinking process. This handbook provides graphic organizers that are specially designed for the thinking skills and processes that are taught in the lessons. These graphic organizers serve as visual guides to involve students in active thinking about the content.

The graphic organizers in this book have a number of purposes. Their most important feature is that they visually guide the process of thinking about a variety of factors. They are "graphic maps" of the thinking process, guiding students through the thinking by using diagrams, symbols, and the language of the thinking skill or process. Students "picture" the questions raised by the thinking maps. Using the graphic organizers also allows students to "download" information so that they can use it meaningfully, not relying on remembering large quantities of data.

In addition, the graphic organizers depict and organize complex information so that relationships among pieces of information are made clear and can be easily managed.

Graphic organizers that guide us through a process of skillful thinking are different from graphic organizers that merely depict information and its relationship. The data matrix used in the "Growth and Digestion" lesson helps us to organize information and to understand relationships but does not guide our thinking.

On the other hand, in the lesson *Desperado*, the graphic organizer, incorporating terms like "prediction," "possible evidence," "actual evidence," and "likelihood," prompts students to think systematically about evidence of the likelihood of the consequences of using various sorts of loops in the tracks to redesign and enhance the roller coaster ride. The graphic organizer holds details about the roller coaster and also visually represents how the likelihood of the consequences is weighed.

Graphic organizers can be used in two different ways in infusion lessons: (1) individual students or collaborative learning groups use them to *guide* their thinking and (2) individual students, a collaborative group, or an entire class can use them to *record* their thoughts.

Both in individual and small-group work, the primary purpose of using a graphic organizer is to guide students' thinking. Graphic organizers lose their effectiveness if individual students use them only to record the results of someone else's thought, such as copying comments. Teachers may, however, record students' responses on a transparency or drawing of the graphic organizer *after* the students have worked on their own diagrams.

When collaborative learning groups work on different issues (for example, when each group is working on a different solution using problem-solving diagrams), students can write their deliberations on a transparency of the graphic organizer or a large poster of the diagram and then display the group report. These completed graphics can then serve as the basis for the class to form its judgment on the issue.

After an infusion lesson, the graphic organizers can be posted as a bulletin board display to remind students of the thinking activity. The completed diagram of the initial thinking activity becomes a metacognitive tool to prompt students' recall of the process and to transfer the thinking skill to other contexts.

The goal in using graphic organizers is to demonstrate the usefulness of "drawing out one's thought" and not to make students' thinking conform to any given design. Ask students to critique graphic organizers used in lessons and to design other diagrams to represent their skillful thinking. Some diagrams, such as the cognitive maps used in the classification lesson on animals, are completely user-generated and follow no given form.

Using graphic organizers promotes skillful thinking in the following ways:

- guides students to organize their thinking
- utilizes figural/visual learning style to organize verbal information
- serves as a visual outline of relationships between pieces of information (particularly useful as a pre-writing tool)
- promotes retention of the thinking process, as well as a deep understanding of the content

Summary of instructional strategies used for direct instruction in skillful thinking. In infusion lessons, two instructional strategies (structured questioning using thinking maps and using specialized graphic organizers) blend to incorporate direct instruction in thinking into content lessons. These strategies may be used in a variety of ways to guide students' thinking, to prompt them to use the language of the thinking skill, and to help them recall and guide their own thinking in other situations.

Instructional Methods Used to Foster Students' Collaboration with Others

All infusion lessons involve collaborative learning experiences in discussion with one or more partners, working independently, and whole-class consideration of thinking issues. When accompanied by thinking prompts (e.g., questioning strategies), cooperative learning techniques help students reason together as a social endeavor and foster a variety of dispositions of good thinking, such as the willingness to listen to and respect the ideas of others.

Collaborative group thinking: open, small-group work on a thinking task. When students in small groups brainstorm possible causes of the extinction of the dinosaurs in the lesson "The Extinction of the Dinosaurs," they discuss ideas and generate possible explanations. Similarly, when they work together in small groups to make a judgment about the function of a specific part of the kestrel in the lesson "The Kestrel," they think together in a more convergent task. In both of these cases, students are prompted by the teacher to engage in a specific thinking task, but their group work is not structured for any other instructional or social purpose.

In such collaborative work, students' discussion stimulates ideas that they might not otherwise have developed. Prompted by the ideas of other students, they may modify their ideas in ways that they might not have considered if working on their own. Engaging students in metacognitive reflection on using the thinking strategy helps students to clarify their own thinking and learning processes.

Cooperative learning: organized, small-group work for thinking. Cooperative learning refers to specific procedures to achieve a variety of goals related to group interaction.

One popular method of cooperative learning involves using techniques to teach group functioning (face-to-face interaction, individual accountability, the development of social skills, positive interdependence, and group processing regarding how they work as a team). Other methods are more content focused, including teacher-assisted individualization involving mixed-ability groups working in teams on a mathematics task, cooperative reading, group demonstration of mastery, games and tournaments, and student teams' analysis of given answers. A third approach involves content-free "structures" that can be applied to content: answering in a prescribed order like "round robin," using a "jigsaw" in which each member teaches another a part, or "match ups" that make students partners. Other models involve group investigation, which stresses problem solving and reports.

You can use different cooperative learning systems in the same infusion lesson. Gathering information on energy sources in "Alternative Energy Sources" can serve as a context for teaching team functioning and social skills. Mixing ability groups and helping students demonstrate their understanding in a game-like presentation can enhance content learning, resulting in a lesson that is especially motivating and beneficial for low-ability students. A jigsaw activity allows students to teach each other the information they have gathered. Using group investigation procedures that incorporate problem-solving strategies can enrich students' research about energy sources.

Think/Pair/Share as a cooperative learning strategy. "Think/Pair/Share" commonly means any activity in which two students discuss what they are learning and reflect on how they think through or comprehend some new concept or process. A technique called "paired problem solving" involves one student listening and asking reflective questions as his/her partner talks about how he or she is working a mathematics problem. The listener prompts clarification of the partner's strategy for solving the problem.

In "RNA and DNA," Think/Pair/Share takes the form of a directed-listening activity in which students working in pairs take turns assisting each other to express their ideas as clearly as possible. The listener raises questions to assist the partner in reconsidering, refining, modifying, or extending his or her answer.

In this highly structured, collaborative activity, one partner reads and clarifies a statement, interpretation, or conclusion that he or she has thought about and written out. In order not to impose his or her own ideas on the speaker's, the listener may only ask:

• Questions of clarification: If the listener does not understand what a term means or does not follow the meaning of the statement, he or she may ask questions to clarify what is being said.

• Questions that extend the idea: If the listener thinks the partner is suggesting something important that needs elaboration, the listener may ask the partner for more details.

- Questions to challenge what is said: If the listener thinks the partner is misled or confused, the listener may ask questions to prompt the speaker to reconsider aspects of the statement.

After two minutes of reflection, the teacher signals students to change roles. After both partners have served as speaker and listener, students have the opportunity to reread, reconsider, or restate their statements.

Think/Pair/Share demonstrates the value of wait time, clarification, and reconsideration. It allows deliberate or uncertain students to review their responses and to express their ideas in their best choice of words.

Using collaborative and/or cooperative learning promotes thinking in the following ways:

- prompts ideas through student interaction
- engages students in reflective collaboration
- promotes self-correction and confidence in expressing ideas
- confirms the value of students' own words and understanding
- engages students in reflective understanding of information
- demonstrates active listening for another's ideas

Pooling information. Model infusion lessons usually involve a class discussion in which information about the content is pooled to provide a common knowledge base for the thinking activity. As students share what they know individually or after working in small collaborative learning groups, the teacher writes all responses on the board or on a transparency. In the lesson on creating metaphors, student responses create an array of attributes and details that other students may use to develop their own metaphors.

A more structured version of this strategy involves asking students what they know, think they know, and need to know about a subject and then recording what they say on a special information-organizing diagram. In the lesson "The Bottom of the Ocean," students' responses are listed on a Know/Think You Know/Need to Know organizer. This technique creates a class composite of significant information and research in order to conduct subsequent examination of an issue or concept.

Using the Know/Think You Know/Need to Know organizer promotes thinking in the following ways:

- identifies and utilizing students' prior and current knowledge to establish a common knowledge base for examining an issue or concept
- promotes retention
- validates students' understanding of concepts and information

Varying individual, small group, and class discussion of the thinking skill or process. Infusion lessons involve individual reflection (often as writing or drawing activities), small group discussion, and engagement in a thinking activity through discussion in the whole class. This progression from individual to group consideration affirms the value of individual perspective, small-group work for creativity and refining ideas, and whole-class participation to consider important information and processes.

Varying the levels of interaction fosters responses from students who are reluctant to volunteer responses and is often more time-efficient than waiting for individual students' responses in whole-group discussion.

In many infusion lessons, after a skill or process is demonstrated the first time, it is repeated with a smaller group, which reports its findings. In "The Mystery of Silent Spring," students work in collaborative learning groups on the likelihood of different possible causes and then pool their results with the rest of the class. They may even repeat the process with another possible cause, transferring the process to another context immediately.

After students have practiced a thinking process with sufficient guidance to do it skillfully, small groups may repeat the process by considering additional options, consequences, or evidence. After the whole class reviews the small group reports, the collected analysis helps students make an informed judgment in a time-effective manner. This practice provides immediate transfer of the thinking skill or process.

Providing many opportunities for individual, small-group, and large-group consideration promotes thinking in the following ways:

- demonstrates the role and value of individual reflection in the larger context of group discussion
- demonstrates the value of small-group discussion in an individual's thinking and learning
- demonstrates to students that they have background about the content and insight about the process

Other Instructional Methods Used in Infusion Lessons to Foster Thoughtfulness

Higher order questioning. Asking higher order questions commonly refers to posing questions that prompt students to be analytical, creative, or evaluative about the content that they are learning. This type of questioning contrasts with asking questions that require only recall.

Questions which ask why, how, and what-if are sometimes identified as higher order questions, but context also determines the type of thinking a student does. A "Why" question asked after students have read a passage that explains why something happens requires only recall.

Higher order questioning typically seeks to extend or clarify students' understanding of information by asking questions about relationships, by seeking new ideas about a subject, or by requesting support for judgment about the content. With these goals in mind, a broad range of questions beyond simple recall can be formulated to prompt higher order thinking. For example, expanding a why question into a "Why do you think..." question, or asking more directly, "What reasons do you have for thinking..." makes it clearly a request for reasons.

In "RNA and DNA," higher order questions clarify and extend students' responses in comparing and contrasting the two topics. The teacher asks questions about the cause, effect, significance, or implications of information that students cite. Since the compare/contrast process leads to an interpretation or conclusion about the two leaders based on their similarities and differences, such questioning brings out deeper understanding of the implications of the information that students have identified.

When you ask higher order questions, you model how thoughtful people interact with information that they are presented. While asking higher order questions prompts thoughtfulness about the subject, students do not learn a strategy for skillful thinking unless the thinking process is made explicit. Because including higher order questions in infusion lessons enhances understanding of the content, when students learn the thinking strategy, they gain more insight about the issue.

Using higher order questions in infusion lessons promotes thinking in the following ways:

- requests student responses about relationships, new ideas, and reasons for judgments
- models how thoughtful people interact with information

Using manipulatives. Manipulatives are concrete objects that the learner examines, moves, or operates as he or she is learning concepts or processes. Manipulatives usually prompt curiosity and can lead to discoveries in which students make use of various forms of thinking.

Manipulatives are commonly used to illustrate concepts in mathematics and science, but it is students' own manipulation of specimens, artifacts, or models that brings out their thinking. Where this is not possible, detailed pictures or sentence strips may also be used as semi-concrete manipulative tools.

Using manipulatives appeals to visual and tactile learning styles. Manipulatives are helpful in any instruction in which observing characteristics is required. Presenting the content in this way prompts the child to engage in active thinking about it.

"All the Mice in the World?" is conducted, for example, through hands-on activities by having them perform a series of experiments regarding the feeding habits of mice. This experiment often provides students with more authentic contexts for learning. Similarly, instead of just reading about waves, and perhaps looking at pictures of waves, students use a ripple tank in "Sound Waves" to identify properties of water waves.

In the "Living Things" lesson, detailed photographs are arranged on a cognitive map that learners design as a group. The pictures illustrate details about the animals that elementary students would have to read long passages to discover. Matching details visually prompts asking what those similar characteristics mean. Moving the picture around the cognitive map prompts consideration of the most appropriate classification of the animal.

Although manipulatives are important in teaching small children, they are helpful in any instruction in which observing characteristics is required.

Using manipulatives brings out student thinking in the following ways:

- prompts students' curiosity about the concrete objects they are working with

- allows students to perceive characteristics of objects in detail

- enriches interpersonal interaction in small-group discussion

Whole-language techniques. A whole-language approach to reading involves four basic ingredients. Students read *whole works* of children's literature, as compared to reading passages in basal readers designed to teach word attack skills. Instruction focuses on the *whole range of language experiences* (writing, speaking, drama, puppetry, choral reading, etc.) to increase students' comprehension and appreciation of a work. Students are encouraged to bring their *whole life experiences* (background, feelings, values, etc.) to interact with what they are reading. Finally, students also *examine a story holistically*, considering how all elements of the work (illustrations, background, format, style, etc.) affect its meaning. This is the approach used when infusion lessons involve reading in the content areas. In "The Bottom of The Ocean," for example, students read an article in which they must attend to illustrations, photographs, and charts as well as the text, and they must utilize a range of background knowledge to determine the likelihood that the information in the text is reliable and accurate.

Using whole-language techniques promotes thinking in the following ways:

- prompts a search for meaning in what is read

- appeals to a variety of learning styles

- examines children's literature and nonfiction in ways that are developmentally and socially appropriate

- confirms the value of students' own understanding and appreciation of a text

Guided reading. Guided reading involves prompting students to think about what they are reading in certain ways. For example, students may be asked to read for specific patterns or topics within a passage. In "RNA and DNA," they are asked to pick out and record similarities and differences between these two organic molecules as they read and compare two texts. Often, guided reading prompts students to interact with the text by "downloading" information onto a graphic organizer as they read the passage.

Guided reading promotes thinking in the following ways:

- identifies a common knowledge base for examining an issue or concept

- prompts students to search a text for specific items

Engaging students in writing for reflection. Writing for reflection includes a variety of writing experiences in which students "think on paper." They are asked to reflect on their analyses, conclusions, or interpretations by writing both their thoughts and the thinking involved in developing that understanding. The writing may be in journals, double-entry records, or short essays. Writing for clarification before discussion is especially helpful for students who are uncertain about their ideas. As the "hard copy" of a student's thinking, reflective writing is a valuable addition to thinking or writing portfolios. Such writing helps to confirm to students that they have sound ideas before they respond in class, allowing deliberate learners to gather their thoughts before expressing them in class discussion.

At any stage of an infusion lesson, students may write to reflect about their thinking, as well as about the content they are learning. They can write to recall prior experiences with the thinking skill or process in the lesson introduction, to clarify the thinking activity in the lesson as it happens, to record their metacognition retrospectively, or to preview when the thinking process might be helpful again.

An example of how writing helps students to reflect on the content occurs in the "RNA and DNA" lesson. Students write out their conclusions or interpretations before discussing it with a partner. Writing out their thoughts clarifies meaning, promotes students' ownership and confidence in their conclusions, and provides a record for reconsideration and restatement. Then, after discussing their statement with a partner, they rewrite their statement to make their ideas clearer.

In "The Mystery of the Silent Spring," students are also asked to write about how to improve their thinking when they engage in causal explanation. Then, after trying some of these ideas out, they write a plan for causal explanation that they can use in the future based on their retrospective metacognitive reflection about the activity of diagnosing the cause(s) of the silent spring.

Writing for reflection promotes thinking in the following ways:

- provides a vehicle for clear expression of one's thinking
- promotes self-correction and confidence in expressing ideas
- confirms the value of students' own words and understanding
- models the reflective understanding of information

Essay writing. Essay writing, as a follow-up to an infusion lesson, can extend student thinking about the content studied in the lesson. In "To Dissect or Not To Dissect?," students propose a recommendation for action, drawing on the thinking they've done in the lesson. They should always be asked to support the thesis of the essay with reasons or details, as appropriate. In "The Mystery of the Silent Spring," students write about the causal diagnoses they have made. They support these with information recorded on the graphic organizers they use, and the writing they do is guided by a writing template

that helps them translate their findings to a well-organized piece of expository prose. Essays offer an opportunity for students to engage in an extended consideration of the reasons for decisions, causal diagnoses, generalizations, etc.

Such follow-up essays should be of appropriate length to address the thinking issue in the lesson. The essay should confirm that the thinking process featured in the lesson is reflected in the student's writing.

Essays may also be used as a tool for personal reflection about a type of thinking students engage in. An especially important place for this to occur is in the introduction to an infusion lesson. For example, students write about their decisions and what they might do to improve them in "Alternative Energy Sources" as a prelude to developing a thinking map for skillful decision making. Any of the thinking skills and processes featured in this handbook can be addressed in essays.

Essay writing promotes thinking in the following ways:

- applies thinking processes in writing
- promotes self-correction and confidence in expressing ideas
- confirms the value of students' own words and understanding
- allows extended response and fully developed expression of students' thought

Using content-oriented graphic organizers. A variety of graphics (matrices, Venn diagrams, flow charts, branching hierarchical diagrams, story maps, web or fishbone diagrams) depict how information is related. These organizers complement the specialized graphics used to guide thinking. Information-organizing graphic organizers prompt students to think about and to organize the content they are learning, and many stimulate analytical forms of thinking.

In "Alternative Energy," a data matrix is used to record students' findings and to guide additional research. These matrices promote systematic investigation of the content and summarize information to assist students in forming generalizations about the material. By examining the energy information in rows, one summarizes the pros and cons of a particular source. By examining the columns, one compares various types of energy in light of a particular variable, such as ease of production or effect on the environment.

In "The Bottom of the Ocean," a Know/Think You Know/Need to Know graphic organizer is used in the introduction to help students make explicit the state of their background knowledge about the bottom of the ocean. This organizes information about the bottom of the ocean into categories that reveal how much students understand about the subject. A similar tool is used in "Growth and Digestion."

For some information-oriented graphic organizers, a standard design outlines information visually. Venn diagrams, flow charts, and matrices are common forms for showing relationships among data.

Some graphic organizers, on the other hand, like concept maps or cognitive maps, are user-generated to reflect the perceptions of its creator. Concept maps are weblike diagrams with many branches and "bubbles." They are used to record metaphors in creative thinking or to depict attributes in analytical thinking. In "Living Things," the shape of a cognitive map depends on how the student classifies an array of organisms.

Information-oriented graphic organizers promote student thinking in the following ways:

- provide a structure that organizes information in relation to other information;
- depict types of relationships among pieces of information;
- prompt the drawing of conclusions or interpretations by comparing pieces of information depicted on the graphic

Additional instructional methods for teaching the content in infusion lessons. Selecting the best fit of instructional methods for content improves the quality of instruction in infusion lessons, as it does in any classroom context. The outline shown in figure 18.1 summarizes some, but not all, instructional methods to improve thoughtfulness and understanding of content. We recommend that when teams of teachers work together to plan infusion lessons, careful attention be given to using a variety of well-researched instructional strategies that can promote student thinking.

Teacher Behaviors That Model Thinking Dispositions in the Classroom

One goal of thinking instruction is to help students develop the habits of mind of effective thinkers and to become disposed to demonstrate these thinking habits when needed. Some teaching behaviors, whatever the activity, promote a classroom climate of thoughtfulness. These behaviors model important thinking dispositions and should be employed in infusion lessons as well.

Using sufficient wait time. Research indicates that teachers commonly give students too little time to formulate thoughtful answers. Often, teachers ask students questions and then, when no one responds, answer the question themselves. In addition, many teachers acknowledge student responses too quickly and don't give other students time to think about their answers.

Both practices suggest that the teacher values facility over depth, short responses over rich consideration, and quick reply over careful reflection. Students who are deliberate thinkers and who take time to reflect about the richness or accuracy of their responses are sometimes characterized as "slow," suggesting lack of ability rather than speed of response. We as a society tend to misconstrue speed as capability.

Good thinkers take time to think things through carefully when time is available. Deliberate consideration is a valued disposition in sound thinking. Varying the pace of instruction demonstrates that some kinds of tasks warrant quick responses and others require careful thinking. Students begin to recognize that thinking tasks requiring judgment also require more time. Practice "wait time," time for thinking which may involve several minutes of silence. Cooperative group work on a thinking issue can also extend the time in which students arrive at a reflective response. Writing for reflection, if only jotting down preliminary ideas, improves the quality of students' responses.

Using wait time promotes thinking in the following ways:

- models reflective thinking
- promotes students' confidence in expressing ideas

- confirms the value of students' own words, understanding, and ideas
- values differences in thought processing and learning style

Asking several questions to clarify and extend a student's responses. Students often expect that they will only have one chance to explain their ideas before the teacher must give others a chance to respond. Teachers' habit of accepting one-sentence answers limits students' processing time to the few seconds before they reply, as well as limiting the amount of time that students expect to take to express their ideas. Asking additional questions to clarify and extend a student's response models other important dispositions of thinking: (1) continuing to develop and elaborate one's thought and (2) resisting impulsivity.

Often, higher order thinking skills are involved in follow-up questions: requesting reasons, possible causes, or results; inquiring about why certain information seemed important; asking what inferences can be drawn from what the student has reported, etc. As students realize that the teacher's additional questions about their ideas allow them to give richer answers, they curb their impulsivity, refraining from giving quick "sound bite" answers.

Asking several clarifying questions improves thinking in the following ways:

- models taking time for the elaboration and development of one's ideas
- models the resisting of impulsivity
- models a reflective understanding of information and active listening
- promotes students' confidence in expressing ideas by demonstrating that questions are not just "tests" for right answers but the means of understanding
- confirms the value of students' own words, understanding, and ideas

Prompt reconsideration. An extension of the previous technique is to ask the student directly to take time to reconsider and restate any conclusions or previous understanding of an issue. Reconsideration could be based on acquiring new information, using the thinking process in class, or arriving at new insight through collaborative dis-

cussion. Reconsideration of previously held views demonstrates open-mindedness and personal responsibility for one's views and actions.

Teachers may explicitly add time in writing activities for students to reread and reconsider what they have written and restate their ideas if necessary. Students must believe that there is more respect in reconsideration than in persisting to hold mistaken ideas. Acknowledge such intellectual maturity and be willing to share your own reconsideration of ideas.

Prompting reconsideration promotes thinking in the following ways:

- models open-mindedness
- promotes self-correction and confidence in expressing ideas
- confirms the value of students' own words and understanding
- models reflective understanding of information

Using precision in language and promoting precision of expression. Critical thinking both requires and promotes precision in the use of language. Being disposed to seek clarity is an important disposition of critical thinkers. Often, when students are unclear or ambiguous in oral and written responses, they are also unclear and ambiguous about the language of thinking. For example, adults and students often substitute "I feel" for "I believe." Whether the speaker is confused about whether an idea is well-supported or is intuitive, or he/she is unwilling to take the stronger stand of stating a belief rather than a hunch, the speaker is expressing an idea in a misleading way.

The habit of defining terms, concepts, and issues clearly is the mark of a critical thinker. Acknowledge and value responses in which a student restates for clarity or raises respectful questions about the clarity of others. Rethink and restate your own descriptions or explanations to demonstrate that you believe that clarity is important.

Demonstrating and valuing precision in language does the following:

- models the desirability of clarity
- promotes self-correction and confidence in expressing ideas

- confirms the value of students' own words and understanding
- models reflective understanding of information

Requesting evidence. Seeking reasons for ideas is an important thinking disposition. Not accepting an idea as true until we are convinced that it is well-supported is a mark of a critical thinker. Often, students develop beliefs without support for those views. Continuing to ask students for their reasons when they advance ideas, especially controversial ones, models this search for good reasons to support our judgments.

Often, when students request evidence to support or reject an idea, it is in a spirit of challenge or disagreement. When students realize that critical thinkers base their understandings, judgments, and actions on evidence, they recognize that requesting evidence is natural, commonly practiced, and essential in order to understand and evaluate what we learn. Model, acknowledge, and value students' open, curious, and respectful requests for evidence.

In addition to asking for reasons, you can model this disposition by citing information as evidence when you explain or describe content or offer reasons for your ideas. Requesting evidence:

- models the importance of having reasons for our judgments and beliefs
- confirms value of inquiry
- models reflective understanding of information
- demonstrates respectful, responsible discourse

Using instructional methods that promote thoughtfulness models the thinking dispositions and learning-to-learn skills that we want to promote in our students. These methods create a classroom climate that supports thinking, understanding, and growth. Often, the media and students' surroundings offer misleading views of how people come to believe what they do. Instructional methods to promote careful thinking and learning are more than devices to improve student performance. They set a climate and tone of careful thinking and demonstrate the value of thought. They are essential if schools are to become "a home for the mind." Our classrooms are workshops of thought, and the lessons we teach vehicles to provide students with a variety of tools and techniques for thinking and learning.

CHAPTER 20
THE ROLE OF METACOGNITION

Why is Metacognition Necessary?

Skillful thinkers reflect on and manage their own thinking. For example, when a decision is needed, skillful thinkers stop themselves from making hasty decisions and guide themselves through more careful decision making. Even after they've made a decision, they remain open to new information which may change their minds.

In effect, skillful thinkers have developed strategies for good thinking and manage their own thinking by using these strategies when needed.

The problems in thinking described in this handbook result when people do not manage their thinking very well. Impulse, association, and emotional appeal often determine their thoughts and judgments. Many of these thoughts and judgments do not serve them well.

Everyone, however, can learn to manage their thinking in more productive ways. One of the great insights growing out of recent work to improve thinking is that our thinking is as much within our own control as our actions are. Metacognition plays a key role in this. We have to be able to think about our thinking if we are to take charge of it.

To take charge of our actions, we have to know what we are doing, how we can act differently to do things in better ways, and what we can do to modify our behavior so that we do what we think is best.

To take charge of our thinking, we similarly have to understand what kind of thinking we are doing, how we do it, and how it can be done differently to improve it. We are then in a better position to change it according to our conception of what more skillful thinking is like.

Metacognition is the internal managing process that we use to take charge of and direct our own thinking so that it is no longer determined by impulse and association but by what we should do to be skillful thinkers.

What is Involved in Thinking about Our Thinking Skillfully?

What kind of thinking is metacognition? Some writers describe metacognition as a special kind of thinking skill or process different from the types of thinking we use to think about things, people, and events around us. On the contrary, metacognition is no different from certain types of thinking we use regularly.

We can, of course, think about our thinking in any number of ways. What *is* special about metacognition is that it involves certain ways of thinking about our thinking that are involved in our ability to manage things well. In metacognition we do the following things:

- We have to be able to *identify what kind* of thinking we do or plan to do. We should be able to tell that we are engaged in decision making, comparing and contrasting, determining the reliability of a source, etc.

- We have to be able to *analyze* how we presently do this kind of thinking in order to determine whether it needs improvement. For example, I may realize that I usually decide to do the first thing that comes into my mind and sometimes miss opportunities for better choices. Or I may realize that I often notice differences between things but rarely think about what the differences mean and, hence, do not gain much understanding of what I am comparing. Or I may realize that I usually accept as reliable anything in print and sometimes end up accepting things that are false or one-sided.

- We should be prepared to *distinguish component subtasks* that we usually do not engage in but that we think might be included in the thinking we have been performing. For example, I may wonder whether it is important to consider evidence for consequences when I make a decision, to draw conclusions from the similarities and differences I note, or to consider whether the publication in which an article appears is usually reliable.

- We should be ready to *evaluate* how we carry out our thinking and any new thinking strategy that we may be considering. How effective are these likely to be? For example, in making a decision, will my decisions be better if I consider consequences than if I do not consider them? If I consider the consequences, perhaps I can avoid being surprised by results that will affect me adversely and reject options that lead to problems. If I draw a conclusion from similarities and differences in the nutritional value of two foods, perhaps I can understand much better how my body benefits from eating them. If I consider whether a publication is usually reliable, perhaps I can avoid being misled into believing things that may not be true.

The thinking involved in these tasks is similar to the kind of thinking we do when we try to manage and improve any kind of performance. To improve my tennis performance, for example, I should be able to tell how I am handling the racket and moving my arm as I swing. I can then try to determine whether I am using my racket effectively. Do my shots go where I want them to? Do they hook to the right more than I would like? If my playing is not as effective as I would like, I can develop a plan to play differently. Perhaps I should hold the racket a little higher or lift my elbow. Then, I can execute this plan and see if it makes a difference.

Such management tasks are similar to metacognition, except that in metacognition, what we manage is our own thinking, not a physical performance. What is different about metacognition is *what we think about*—our thinking—and not *the kind of thinking we do* when we think about it.

The thinking map in figure 20.1 describes metacognitive thinking that aims at improving our performance in situations in which our thinking is what concerns us.

Notice the kinds of component thinking skills that are involved in skillful metacognition. When you are asking yourself about the type of think-

---

THINKING ABOUT THINKING SKILLFULLY

1. What type of thinking did you engage in?

2. How did you do the thinking?

3. Was that an effective way to do this thinking? Why or why not? If not, what can you do to improve this way of thinking?

4. How will you do this kind of thinking next time it is needed?

Figure 20.1

---

ing you are doing, you *classify* it. When you ask how you do it, you are *analyzing* it into its component parts. Both are ways of *describing* your thinking. Often in thinking skills programs, when students are asked to think about their thinking, they are asked only to engage in one or the other of these types of thinking.

If students are to manage their own thinking, however, the way they think about their thinking must involve more than just these types of thinking. Students must also be able to *evaluate* and *plan* their thinking. When you ask if your thinking strategy was effective or how you might improve it, you are *evaluating* your thinking. When you ask yourself how you will do it next time, your goal is a *decision* about how you will think in the future. You are *planning* your thinking.

Each of the four types of metacognitive thinking is essential for managing and planning our thinking effectively. When we blend them together to think about our thinking, they become a powerful strategy for self-directed thinking.

The graphic organizer in figure 20.2 (next page) provides spaces to jot down answers to the questions that we should ask ourselves to engage in skillful metacognition.

When do we engage in skillful metacognition? The wording of the thinking map of skillful metacognition (figure 20.1) indicates that metacognition sometimes occurs *after* we engage in some type of thinking. For example, after I have made a decision, I may realize that it was not a good one. I may have decided to go to a mountain resort for my vacation. After getting there, however, I discover that certain "hidden" costs that I

METACOGNITION LOG

What was the thinking skill you used in this lesson?	What questions or directions prompted you to engage in this thinking in the lesson?
How did you carry out this thinking? (What steps did you go through in your thinking as you did the lesson?)	
Describe how this way of thinking compares with other ways you might have thought about the issues in the lesson. Which do you prefer and why?	If you use this thinking in another situation, how would you plan to do it? Pick a specific example and describe what you would think about in some detail.

Figure 20.2

did not know about made the daily cost of the vacation almost double. I never would have gone to that resort if I had known in advance about these additional costs. I note that when I made my decision about where to go, I did not gather all of the relevant information I could have. I can't change my vacation now, but I must remember next time to gather all relevant information before I decide where to go. To do this carefully, I also decide that I should take some time to think about what information I will need before I try to get it.

In this case, I am reflecting on how I engaged in a certain kind of thinking that has already occurred, but I use these reflections to develop a plan that would modify the way I expect to engage in this type of thinking in the future.

I can also *monitor and correct* my thinking while I am doing it. This means shifting my attention back and forth from what I am thinking about to the thinking itself. For example, while trying to make a decision about where to go on my vacation, I may be tempted to go to the mountain resort, but I may step back from thinking about the vacation and ask myself, "Am I considering all

that I should to make a really well-informed decision?" I may then realize that I am considering just a few important factors and that I might need more information. So, I switch my strategy and decide to take time to gather more relevant information before I decide.

Being metacognitive before, during, and after we think maximizes the degree to which we can manage our own thinking.

Prerequisites for metacognition. What makes metacognition seem like a new and mysterious type of thinking is that, in order to do it well, we have to be able to distinguish different types of thinking as well as different elements in our thinking. To do this, we have to know and use the language of thinking and apply it to ourselves by attending to episodes of our own thinking. When I remark that I considered a number of options and then examined the consequences, searching for reasons why I should or should not choose one of my options, I am using the ordinary language of thinking that we all use, not a new technical language. I am using this language to describe my own thinking.

We are familiar with and use the language of thinking, but we do not describe our thought processes as frequently or insightfully as we might. We attend to events and things around us; but although our own thinking is constantly occurring, we rarely attend to it. Shifting our attention to our own thinking and applying the ordinary concepts and language of thinking to it is necessary for skillful metacognition.

Learning to use precise language to describe our thinking is no different from other situations in which we learn and apply a new conceptual framework and language to differentiate things around us. When we teach students to distinguish the parts of an automobile and to understand how they work, we teach them to apply and use automotive concepts and language. ("The carburetor is skipping and needs repair, but the spark plugs are in good working order.") Practice in using these concepts and words helps students identify the parts of automobiles with some facility. Discriminating and describing a variety of aspects of our thinking is, likewise, the result of getting used to attending to our thinking and applying the concepts and language of thinking to it.

This is why teaching thinking explicitly, as we do in infusion lessons, is so important in helping students develop and use their metacognitive abilities well. Explicit instruction uses the language of the thinking skills being taught to guide students' thinking and, hence, prompts application of this language to their own thinking. This is the first step in helping them to develop the habit of thinking about their thinking.

Teaching Students to Engage in Skillful Metacognition

Because metacognition is an essential ingredient in all forms of skillful thinking, this handbook includes no separate lessons on metacognition. Rather, metacognition is introduced in infusion lessons as a key component in learning to manage *any* type of thinking. An organized process of metacognitive questioning based on the thinking map in figure 19.1 appears in every infusion lesson.

The following summary of the dynamics of an infusion lesson makes it clear why it is important to include such organized metacognitive activities in each lesson:

- In infusion lessons, the teacher makes thinking strategies explicit and guides students in their use as they think about what they are learning in the curriculum (e.g., what energy sources are the best ones for our country to rely upon, the similarities and differences between pyramids and cones or RNA and DNA and their significance, the reliability of observation reports about the surface of the planet Mars).

- Then, students are asked to extract the strategy they used from the specific context of the lesson as they reflect on whether this strategy is an effective one.

- The students then create a plan in their own words for the thinking they have just done.

- Finally, they will use this plan the next time they must do the same kind of thinking.

With repeated applications, guided by their thinking plans, students develop the habits of thought that make them skillful thinkers. Meta-cognition plays a key role in the transition from teacher-guided thinking to students' using thinking strategies on their own.

Without metacognition, students' transfer and use of skillful thinking in other appropriate contexts will be minimal.

Becoming more metacognitive is not difficult. When you are learning to play tennis, it is initially difficult to monitor and evaluate your playing yourself. Your instructor gives you feedback, suggestions, and support. As you develop more expertise, you increasingly guide yourself. The same is true of skillful thinking. Your instructor guides you in doing it *and* in thinking about it so that you learn to guide yourself.

Four basic practices are used by teachers to teach metacognition in infusion lessons:

- distancing activities in which students are prompted to shift their attention to their thinking and away from what they have been thinking about

- structured questioning to prompt students to think skillfully about their thinking

- prompted practice in which students use specific thinking strategies to guide their own thinkings

- reflective writing to express students' thinking about their thinking

Distancing students from what they have been thinking about and shifting their attention to their own thinking. In the critical thinking activity in infusion lessons, students often get deeply involved in thinking about the content-related topic of the lesson. This could be what mathematical method to use, what caused the extinction of the dinosaurs, etc. A common strategy in distancing students from what they have been thinking about and shifting their attention to their thinking process is to ask them directly to do so. You might say, "Let's stop thinking about the dinosaurs now and focus our attention on how we thought about what caused their extinction" or "We're going to shift our attention away from the dinosaurs now and think about how you tried to figure out what caused their extinction." If students have become accustomed to doing this, your comments may be all that is needed to shift their attention.

If, on the other hand, students are not used to doing this, you may have to explain what you mean. You might say, "When you think about the dinosaurs, you remember many things about them: what they ate, etc. When you shift your attention to your thinking, you will be thinking about something different: What went through your mind as you were thinking about the dinosaurs? What questions did you ask? Did you remember things? That's thinking about thinking."

Structured questioning to prompt students to think skillfully about their thinking: kinds of questions to ask. When a teacher asks a student how he or she conducted a certain type of thinking, the teacher's question prompts the student to *describe the strategy used* for the thinking. After a decision-making activity, for example, the teacher's goal is to help the student become aware of and identify the degree to which options were considered, information about the consequences was considered, etc. If the student has difficulty doing this, the teacher can ask more specific questions about the process. She might even ask, "Did you think about options at all?" "What did you think about to weigh the consequences?" Guidance using prompting questions helps students to become familiar with their thinking, to learn how to describe it using the language of thinking, and to learn the kinds of questions they should ask themselves when they recall their own thinking.

Similarly, teachers use different types of questions to prompt students to *assess the effectiveness of their thinking.* Recurring questions like "Was that an effective way to make a decision?" and "How well did that strategy help you to assess the reliability of sources of information?" prompt students to evaluate their thinking and teach them questions they should ask themselves when they independently seek to evaluate how effectively they are engaging in a thinking task.

More specific evaluative questions can also be asked. For example, this is a question sequence from the *Mystery of the Silent Spring* lesson: "Is this a better way to find out what caused something than the way you ordinarily do? Why?" Such comparative questions are usually easier to answer than simply asking whether the kind of thinking students did was effective.

Guiding students even more specifically in evaluative metacognition may be necessary. A teacher might ask, "For example, do you think that developing a list of possible causes increases your chances of finding out what the cause really was? How?" or "Is thinking about options and consequences a way to avoid some of the defaults in decision making? How does using this way of making decisions help to avoid these defaults?"

Planning how we will engage in skillful thinking again is the culminating metacognitive activity in infusion lessons. Asking students to write out a plan or to create a flow chart that can guide them in doing the same kind of thinking again is a common practice. This creates ownership of these thinking plans, since students can put their plan in their own words. Soon, students become accustomed to planning how they will carry out their thinking in a variety of thinking tasks.

Students may also work together to develop a class plan for a specific thinking skill, then post it on the wall of the classroom. This, of course, blends cooperative work with metacognition.

Structured questioning to prompt students to think skillfully about their thinking: when to ask metacognitive questions. The main organized metacognitive activities in infusion lessons occur after the students have engaged in a well-developed thinking activity like deciding what sources of energy are the best for us to rely upon. The diagram in figure 20.3 shows the order in which metacognitive activities typically occur in infusion lessons.

INFUSION LESSON
Introduction
Thinking Actively
Thinking about Thinking
Applying your Thinking

Figure 20.3

At the same time, the more that metacognitive prompts are peppered throughout infusion lessons, the more practice students will get with the varied kinds of metacognition that are important for them to engage in as skillful thinkers. For example, the lessons provide numerous opportunities to prompt students to engage in metacognitive monitoring and reflection *while* the thinking activity is still going on. After they have generated a list of options, asking students a question like "What is important to think about next?" can distance them enough from thinking about alternative energy sources to prompt their thinking about the importance of thinking about consequences. The question allows them to focus on the consequences of the options they have generated.

A more sustained and subtle practice is to move from group to group as students are engaged in cooperative thinking tasks and ask the students if they are having any difficulties with the thinking they are doing. Are they finding anything hard? If so, what? When students describe the problems they are having, you can then discuss ways of overcoming these difficulties. You can also ask the group to help.

For example, as they move from group to group during a causal explanation activity, teachers commonly find that some students have problems generating many possible causes. Sometimes, teachers suggest ways of overcoming these problems, like piggybacking on the ideas of other students. Or, they may ask other students who are generating many options how they generate so many ideas. For example, in the lesson "The Mystery of Silent Spring," a student may respond that she asks herself, "What kind of possible cause is this?" about a possible cause that another student has mentioned. The possible cause of radiation contamination, which another student has raised, involves something that has happened elsewhere and then spreads contamination to the town, she notes. This prompts her to raise the question, "What are some other things that might have happened elsewhere that could have this effect on the town?" She then thinks of other possibilities: a disease that broke out in a nearby city, chemical contamination from a chemical accident at a chemical plant outside the town, like in Bhopal, India, etc.. Hearing this account, the teacher may suggest that the student who is having difficulty may also try this technique.

Helping students to monitor their thinking and overcome problems they are having while the thinking activity is going on is an excellent way to give students practice at thinking about their thinking. The more we couple monitoring thinking while we are doing it with retrospective metacognition and advanced planning, the more we help students become disposed to managing their own thinking.

The diagram in figure 20.4 contains important question stems that can be used to prompt skillful metacognition in students whether it occurs before, during, or after the main thinking activity. It is arranged to move from identifying a type of thinking to evaluating the strategy used in the thinking and planning how to do it again.

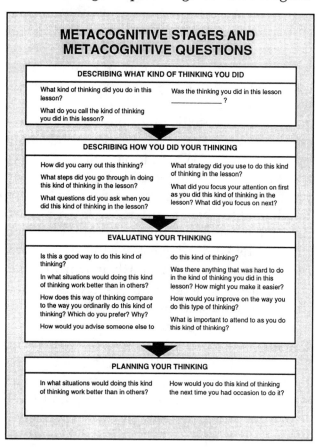

Figure 20.4

Prompting students to manage their own thinking. Describing, evaluating, and planning one's thinking are important stages in thinking about thinking skillfully. The questions teachers ask to guide students through these stages help students to learn how to do this kind of metacognition, and their continued use by teachers gets students accustomed to them. When the teacher does this, she is providing a supporting structure for students' metacognition. This "scaffolding" should be gradually removed, however, to allow students to take over this task themselves. Specific questioning strategies that you can use to accomplish this are featured in the application section of infusion lessons, part of the general strategy in these lessons to teach for the transfer and reinforcement of the thinking skills.

For example, after students have thought about RNA and DNA, and have engaged in metacognitive reflection about ways of comparing and contrasting effectively, the teacher may introduce a transfer activity by saying, "Go back to your plan for skillful comparison and use it to compare and contrast simple and compound interest."

With less scaffolding, the teacher might say, "Pick the thinking plan and graphic organizer that you think should be used here. Use it to be sure you understand simple and compound interest." Students will begin to guide their own thinking as they respond to these more general metacognitive prompts.

Thinking about thinking through writing. Reflective writing is an excellent vehicle for stimulating sustained thinking. Its use in infusion lessons provides an opportunity for students to do sustained thinking about the content they study in these lessons. It can also be used to support students' ongoing reflections about their thinking.

You can use one of two strategies for reflective writing about thinking. The first is to simply ask students to write about their thinking. For example, you can ask students to write out how they will try to solve a problem before they do it.

After working on decision making, for example, students can be asked to write a more extensive essay about the differences between the way they used to make decisions and the way they do now. They can write a recommendation for an effective decision-making strategy for people who make impulsive decisions. Many other possible situations are excellent examples for students to write about their thinking.

The second strategy relates to situations in which students keep journals as a regular writing activity. You can ask your students to use a second column in their journals to enter their thoughts or notes about their thinking as they do content-oriented writing in the first column.

You can suggest that they ask themselves questions like "Was this idea easy to understand? Was it hard to solve that problem? Why? Could I have improved on the way I tried to solve it?" Using double-entry journals can help students become accustomed to thinking about their thinking as they do it.

Summary: Teaching to Internalize Skillful Thinking

Figure 20.5 illustrates the role of metacognition as we move from highly scaffolded lessons with a significant amount of teacher guidance to students' management of their own thinking.

The diagram in figure 20.5 summarizes the way infusion, as an instructional strategy, makes good thinking stick with our students.

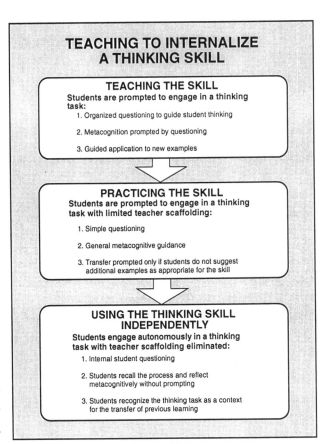

Figure 20.5

Tools for Designing Metacognition Sections of Lessons

The thinking map of sequenced questions for skillful metacognition can be used to guide you in designing the section of infusion lessons in which students think about their thinking. The thinking map and the graphic organizer can be used to supplement the Thinking About Thinking section of the lessons included in this handbook.

Both can be made into transparencies, photocopied, or enlarged as posters for display in the classroom. Reproduction rights are granted for single classroom use only.

THINKING ABOUT
THINKING SKILLFULLY

1. What type of thinking did you engage in?

2. How did you do the thinking?

3. Was that an effective way to do this thinking? Why or why not? If not, what can you do to improve this way of thinking?

4. How will you do this kind of thinking next time it is needed?

METACOGNITION LOG

What was the thinking skill you used in this lesson?

What questions or directions prompted you to engage in this thinking in the lesson?

How did you carry out this thinking? (What steps did you go through in your thinking as you did the lesson?)

Describe how this way of thinking compares with other ways you might have thought about the issues in the lesson. Which do you prefer and why?

If you use this thinking in another situation, how would you plan to do it? Pick a specific example and describe what you would think about in some detail.

METACOGNITIVE STAGES AND METACOGNITIVE QUESTIONS

DESCRIBING WHAT KIND OF THINKING YOU DID

What kind of thinking did you do in this lesson?

What do you call the kind of thinking you did in this lesson?

Was the thinking you did in this lesson _____ ?

DESCRIBING HOW YOU DID YOUR THINKING

How did you carry out this thinking?

What steps did you go through in doing this kind of thinking in the lesson?

What questions did you ask when you did this kind of thinking in the lesson?

What strategy did you use to do this kind of thinking in the lesson?

What did you focus your attention on first as you did this kind of thinking in the lesson? What did you focus on next?

EVALUATING YOUR THINKING

Is this a good way to do this kind of thinking?

In what situations would doing this kind of thinking work better than in others?

How does this way of thinking compare to the way you ordinarily do this kind of thinking? Which do you prefer? Why?

How would you advise someone else to do this kind of thinking?

Was there anything that was hard to do in the kind of thinking you did in this lesson? How might you make it easier?

How would you improve on the way you do this type of thinking?

What is important to attend to as you do this kind of thinking?

PLANNING YOUR THINKING

In what situations would doing this kind of thinking work better than in others?

How would you do this kind of thinking the next time you had occasion to do it?

TEACHING TO INTERNALIZE A THINKING SKILL

TEACHING THE SKILL
Students are prompted to engage in a thinking task:

1. Organized questioning to guide student thinking

2. Metacognition prompted by questioning

3. Guided application to new examples

PRACTICING THE SKILL
Students are prompted to engage in a thinking task with limited teacher scaffolding:

1. Simple questioning

2. General metacognitive guidance

3. Transfer prompted only if students do not suggest additional examples as appropriate for the skill

USING THE THINKING SKILL INDEPENDENTLY
Students engage autonomously in a thinking task with teacher scaffolding eliminated:

1. Internal student questioning

2. Students recall the process and reflect metacognitively without prompting

3. Students recognize the thinking task as a context for the transfer of previous learning

CHAPTER 21
SELECTING CONTEXTS FOR INFUSION LESSONS

Guidelines for Selecting Contexts

Selecting good contexts for infusion lessons is crucial in designing effective instruction. To accomplish the dual objectives of infusion lessons (teaching skillful thinking and improving students' understanding of the content), one should select contexts in which the thinking process relates naturally to the content being taught. For example, considering the viability of relying on alternative energy sources is a natural situation for skillful decision making. If students often confuse two important scientific or mathematical concepts such as RNA and DNA, it is natural to compare and contrast them.

The key to choosing appropriate contexts is finding examples that offer rich development of the thinking process and rich instruction of the content objective. For example, in "Sulfuric Acid," students' deep understanding of the chemical composition of this important acid and of the process of chemical equilibrium is enhanced as they learn and use a strategy for skillful causal explanation in trying to diagnose why the production of sulfuric acid is not reaching its anticipated levels in a specific processing plant.

If the content is too complex, however, the subject may overshadow the thinking activity to the extent that students remember the content but not the thinking process. For example, if you use an initial lesson on global warming as a context to introduce skillful decision making, the complexity of the topic and length of the unit may overshadow the decision-making process. Unless students are reminded of the decision-making strategy repeatedly, they may learn about global warming but remember little about decision making.

At the other extreme, students may understand the thinking clearly but learn little information about the subject. The content should be well-developed in the thinking lesson, not just a topic that is superficially touched on to provide a vehicle for teaching thinking. For example, if you use a real school incident as a context for a lesson on the reliability of observation reports,

students may learn how to evaluate an observation report but do not, unless prompted, apply the thinking skill to curriculum examples.

The curriculum abounds in opportunities for teaching all the thinking skills and processes featured in this handbook, but some thinking skills are taught more effectively in some disciplines than in others. For example, in mathematics, teaching compare/contrast or classification may be helpful in teaching basic mathematical concepts, but fewer contexts for lessons may be available on reliable sources or causal explanation. On the other hand, science instruction abounds in contexts for reliable sources and causal explanation lessons. As a rule of thumb, when contexts for using a specific type of skillful thinking (e.g., causal explanation) are plentiful in a field of study, this provides a *prima facie* reason for introducing students to that skill in that field of study. After the skill is introduced, it can be reinforced by more transfer examples in the same field. And once introduced in one field (e.g., causal explanation in science), its use in other fields in which the type of thinking is less prevalent or important (e.g., causal explanation in mathematics) can be undertaken not by another highly structured introductory lesson on the thinking skill, but by a more restricted application to examples that reinforce its transfer into the new field from the context of its original introduction.

Designing effective infusion lessons often depends on the clarity with which the content objective is stated. For example, "compare two substances" does not focus on the purpose of the comparison. Stating general similarities and differences adds interest but prompts little understanding. As you state the content objective, try to be as clear as possible about the kind of insight that you expect as the outcome of the comparison and contrast. For example, in "RNA and DNA," the content objective is stated in terms of the function of the two substances in the basic processes of the replication of living cells. This content emphasis yields a more robust lesson than would result if the lesson were a more general comparison and contrast.

The commentary that introduces each thinking skill or process includes suggestions for contexts that are rich examples for improving students' understanding of both the content and the thinking. The lesson menus at the end of each chapter list examples of contexts that teachers have suggested or used to design infusion lessons.

Sometimes, planning teams designate specific thinking skills and processes for particular grade levels and subject areas. This practice conveys the idea that students are not ready for or capable of certain types of thinking in the early grades. If instructional methods are developmentally appropriate, students from kindergarten through the elementary grades can skillfully demonstrate the thinking skills and processes featured in this handbook. This will take the burden off secondary school teachers for the introduction of a wide variety of new thinking skills. At the same time, secondary school teachers should be ready to introduce students to new and more sophisticated ways of engaging in the skillful forms of thinking covered in this handbook. You should also be ready to introduce many, if not all, of the skills in this handbook through the type of lessons included when you know that students have not been exposed to such thinking skills taught through infusion lessons in the elementary grades. Although it may take more than a year to design and gather lessons in all the skills and processes, our goal should be a full range of instruction and reinforcement of skillful thinking at any grade level.

Finding contexts by reviewing curriculum guides and texts is particularly important if instruction is assessed by procedures designed to indicate higher order thinking. If the school, district, or state assessment tests include extended writing or performance tasks that evaluate the quality of students' thinking, students should have sufficient practice to be competent and confident in carrying out the thinking required in such activities. Offering teachers a broad selection of contexts increases the likelihood of sufficient instruction for the transfer of the thinking skills and processes to become evident in students' performance.

Three Procedures for Finding Contexts for Infusion Lessons

We recommend three procedures for locating contexts for infusion lessons: creating menus of lesson ideas as you review demonstration lessons, scanning the curriculum for good opportunities to teach an infusion lesson, and reviewing curriculum guides and texts to identify many significant contexts. Using one or more of these procedures will depend on the following:

- the number of teachers working together to design infusion lessons
- how much experience you have in designing infusion lessons
- how much thinking instruction you want to infuse
- whether you are designing infusion lessons for your own classes or as part of a district team on thinking skills instruction
- the amount of available planning time for review and curriculum planning
- whether procedures for assessing students' thinking beyond those provided in the model lessons are used in the school or district

Menus

We use the term "menu" to describe a list of suggested contexts for infusion lessons. The menus in this handbook contain the grade level, subject, topic, and thinking issue for contexts that offer a rich opportunity to teach or reinforce a thinking skill or process. These topics have been suggested by teachers and have been reviewed by the authors. In this handbook, these menus are restricted to secondary school science.

Creating your own menus of lesson ideas as you become familiar with infusion lessons is a good way to apply the principles of designing infusion lessons. As you first read a chapter about a thinking skill or process, identify contexts in your curriculum in which to develop your first lessons. Take some time after reading each model lesson to suggest thinking instruction contexts. Use the menus at the end of the chapter as idea starters for developing your own lessons.

Creating menus as you first review the model lessons offers several lesson-planning advantages, including the following:

- **Immediate application**. Immediately applying the thinking skill or process to your own teaching fosters confidence and confirms the instructional relevance of the thinking instruction.

- **Versatility**. Menus demonstrate the usefulness and versatility of the thinking skills and processes in a variety of instructional uses across the curriculum.

- **Increased likelihood of use**. Creating menus while the thinking skill or process is fresh in your mind allows you to apply it to content objectives right away.

- **Significant contexts come to mind first**. Teachers' first ideas about good contexts for infusion lessons have often proven to be effective ones. Weeding out less effective ideas or restating them for clarity may be necessary.

The disadvantages of selecting contexts by developing menus include the following:

- **First responses**. These contexts are initial ideas that may or may not prove to be effective ones as you develop a lesson. Sometimes, a lesson suggestion may not be suitable because it does not work well for explaining the thinking skill or process clearly.

- **Not comprehensive**. Other significant contexts to which the thinking skill or process would be better suited may be overlooked.

- **Not sequential**. Although thinking skills and processes may usually be taught in any order, it is desirable but not necessary sometimes to teach some skills or processes before others. For example, compare and contrast involves classification and analogy. Identifying reasons and conclusions plays a role in uncovering unstated assumptions.

Suggesting menu ideas does not take into account when in the school year the lesson will be offered. The initial lesson on a particular thinking skill should occur early enough in the school year to allow time for transfer examples.

Looking for contexts and predicting how the thinking process may work in the lesson allows you to "replay" the thinking skill or process almost immediately and apply it to the content you teach. During the first year that you use this handbook, using the menus provided in the handbook or developing lessons from your own menus is sufficient to begin teaching thinking.

Scan the Curriculum for a Few Examples that Illustrate the Thinking Process Well

If you are the only teacher in your school or grade who is designing infusion lessons and have limited planning time, you can search your texts and curriculum guides for a few examples that "showcase" a particular thinking skill. Then, locate a few topics for transfer lessons to provide sufficient classroom practice. The goal in scanning the curriculum for rich contexts for thinking instruction is to identify key topics in your curriculum in which to implement the thinking instruction, followed by a few good transfer lessons.

Scanning the curriculum often produces a list of lesson ideas that addresses more significant curriculum objectives than generating menus from the first topics we identify. Even a quick survey of texts often prompts teachers to recognize key concepts that can be effective infusion lesson topics, examples that may not occur to us when we read these lessons for the first time.

The advantages of scanning the curriculum for contexts that best fit the thinking activity include the following:

- *Creating powerful examples* of using the thinking skill or process

- Designing *a few good lessons* that demonstrate the thinking skill well

- Selecting contexts that teach *significant curriculum objectives*.

Disadvantages of this approach include the following:

- Lesson suggestions *are not comprehensive*

- The thinking skills and processes may appear *less versatile or relevant*. Since you are looking for only a few good examples, you may not realize how often using the thinking skill or process might be beneficial or how important it can be in helping students understand issues.

If you want to assure that you are offering adequate thinking instruction in the fewest lessons, you may find the "best fit" approach helpful. Teacher educators or staff development personnel who must demonstrate infusion lessons for other teachers may want to pick only the most dramatic uses of the thinking processes for clarity and to add interest in their lessons.

As you begin to teach infusion lessons, you may recognize in the middle of a lesson that this would have been a rich opportunity for teaching an infusion lesson. While the idea is fresh, highlighting with markers or using tabs to mark these contexts in the teacher's manual provides a reminder for planning next year's lessons. You are "scanning the curriculum for contexts" by being alert to good opportunities.

A Comprehensive Review of Curriculum Guides and Texts

After you have practiced writing infusion lessons and have some experience locating rich contexts, you may serve on a planning team to review local or state curriculum guides and textbooks to locate a variety of other topics well suited for infusion lessons. Given sufficient planning time, reviewing the curriculum for the best contexts for teaching the thinking skills and processes results in well-articulated thinking instruction and effective content instruction.

The matrix for integrating content objectives by thinking skills (figure 2011) serves as a worksheet for each planning-team participant to identify contexts for one grade. Teachers write district or state codes for curriculum objectives or topics from texts in each square. Suggested contexts are then reviewed with other teachers in the same field or at the same grade level. Then teachers from several grade levels may combine their contexts grade-by-grade, producing a collection of lesson contexts for thinking instruction across grade levels and subjects. Contexts from all grades and subjects can be combined using an extended form of the same diagram. Indicate which contexts seem best suited for initial lessons and transfer examples, taking into account the order in which content objectives are usually taught in the school year.

Advantages of this systematic review process include the following:

- **Comprehensive**. The review process identifies an array of contexts to which the thinking skill or process is well-suited.

- **Serves as a basis for integrating the curriculum**. Ways of integrating the curriculum abound, among them cross-indexing topics around common themes, using district goals, or featuring a thinking skill or process as a principle for unifying the curriculum. Developing interdisciplinary units around a thinking skill or process correlates concepts and emphasizes transfer of the thinking process.

- **Demonstrates the relevance** of the thinking skill or process to a wide range of content objectives. Teachers recognize the versatility of the thinking skills and processes and the wide range of content objectives in which improved student thinking will result in more effective learning of content.

- **Creates a larger pool of contexts**. Selecting the best examples from many topics promotes a well-designed thinking program. Having many contexts allows teachers to schedule initial lessons and transfer examples systematically in the school year.

- **Identifies concepts and thinking processes featured on tests and other assessment procedures**. The review process allows teachers to select topics about which students' understanding and thinking will be evaluated. As extended response writing and alternative assessment procedures are commonly used, demonstrating organized thinking—as well as thorough knowledge—becomes increasingly important.

- **Articulation across grade levels**. Teachers recognize topics that other teachers at earlier and later grade levels teach, confirming the significance and continuity of their efforts to design infusion lessons.

- **Lasting**. While curriculum reforms come and go, the core concepts of the curriculum are constant. The curriculum guides for teaching thinking that grow out of this review process will be useful, regardless of other curriculum initiatives. The school's or district's commitment of time for cur-

INTEGRATING CONTENT OBJECTIVES BY THINKING SKILL OR PROCESS

SKILL	LANGUAGE ARTS	MATHEMATICS	SOCIAL STUDIES	SCIENCE/ HEALTH	ART/MUSIC/GUIDANCE PHYSICAL EDUCATION
Comparing/ contrasting					
Classifying					
Parts/whole					
Sequencing					
Uncovering assumptions					
Reliable sources/ accurate observations					
Reasons/ conclusions					
Causal explanation					

Figure 21.1

INTEGRATING CONTENT OBJECTIVES BY THINKING SKILL OR PROCESS

SKILL	LANGUAGE ARTS	MATHEMATICS	SOCIAL STUDIES	SCIENCE/ HEALTH	ART/MUSIC/GUIDANCE PHYSICAL EDUCATION
Prediction					
Reasoning by analogy					
Generalization					
Conditional reasoning					
Generating possibilities					
Generating metaphors					
Decision making					
Problem Solving					

Figure 21.1 continued

riculum review and planning demonstrates its recognition that improving students' thinking is an ongoing priority.

- **Reduces planning time for individual teachers**. A grade, school, or district compilation of content objectives allows each teacher to select identified contexts, rather than spending time looking for good topics.

- **Available to new teachers coming into the school or district**. As new teachers become familiar with designing infusion lessons, an array of curriculum contexts is already available to them.

- **Addresses state and local instructional priorities and testing practices**. Teaching thinking addresses the school improvement strategic plans of many individual schools and school districts. Relating thinking contexts to specific content objectives demonstrates that the thinking program is being addressed systematically.

Disadvantages of reviewing curriculum guides and texts include the following:

- **Lack of planning/curriculum development time**. Preparing a systematic guide for infusing thinking skills and processes requires significant planning time for thoughtful review of curriculum guides and texts.

- **Training and/or experience of planning team staff**. The curriculum review process should begin after teachers have had considerable experience designing and teaching infusion lessons. After designing infusion lessons, teachers understand more fully what is entailed in skillful thinking and how thinking instruction adds to students' understanding of content.

- **Less ownership and involvement by individual teachers**. Teachers who did not participate in the review process may not utilize a curriculum guide that is "handed to them." Many curriculum guides are not used because teachers either do not know they exist or are unfamiliar with them.

As the effectiveness of teaching thinking prompts schoolwide implementation, groups of teachers may help each other write lessons and edit the curriculum for contexts. However, school or district commitment to this instruction merits sufficient review time to allow teachers to plan systematic implementation.

Using any of these techniques for locating contexts, teachers recognize the versatility of the core thinking processes and the heightened understanding that teaching thinking adds to content learning. These approaches represent the stages that individual teachers experience as they design infusion lessons.

Integrating Instruction by Infusing the Teaching of Critical and Creative Thinking

The significance of infusing thinking skills and processes into instruction becomes more apparent as one recognizes how the curriculum can be unified by common threads of thought. Using the thinking maps and graphic organizers demonstrates to students the relatedness of thinking in one subject to the same kind of thinking in another. As teachers become more aware of the interconnectedness within the curriculum, they may integrate learning by using either of two organizing principles:

1. The thinking skill or process itself may be a unifying theme for a period of instruction.

2. The kinds of thinking involved in an interdisciplinary theme add to students' understanding of that theme.

Thinking skills or processes to unify curriculum for a given period of instruction. In this approach to unifying curriculum, the thinking skill or process is stated and displayed as an instructional focus for a given period. This can be done at the individual classroom level, for all the classes at a given grade level, or for the whole school.

For example, for a period of about four weeks, teachers in a middle school taught infusion lessons on decision making. The thinking map and graphic organizers for decision making were displayed as posters in the halls, on bulletin boards, and in parent communications. In each class, teachers issued reminders that good decision making is important and that students in that school were learning to do it well.

Students recognized that decision making in different subjects involves the same strategy. They also recognized the variety of opportunities for skillful decision making and the significance of doing it well. The graphic organizers and student products, including their writing about decisions, were displayed. The Parents Night featured skits showing parents the decision-making strategy their children were learning. Teacher-prepared webbing displays showed parents the many contexts in which their children were learning decision making.

This approach can be enhanced by studying the strategies used by important decision makers and by considering specialized kinds of decisions. Case studies may include decisions made by presidents, by scientists, by artists, and by athletes. Students recognize both the common strategy and the important factors that can make decision making somewhat different in different disciplines or contexts.

The goal in using this approach was to demonstrate transfer, relevance, and versatility of the thinking skill or process; thematic connection between disciplines was of little concern. The summary of the goals, curriculum modification, and planning process on the next page explains this way of unifying instruction. The graphic organizer in figure 21.1 serves as a guide for this kind of planning. The outline in figure 21.2 summarizes the goals, curriculum modifications, and planning process to integrate curriculum by a selected thinking skill or process. The sample webbing organizer in figure 21.3 illustrates this curriculum integration for teaching decision making in a fifth-grade class.

Teaching the thinking skills or processes involved in a theme that has been selected for instruction across disciplines. In this case, the goal of curriculum integration is for students to develop a deep understanding of the theme selected for interdisciplinary instruction. For example, if the theme is "change," the teacher planning the unit would ask, "What kinds of thinking will help students understand change meaningfully?" The student may need to know what causes change in a variety of situations, to predict what may happen as the result of change, to understand how decisions affect change, to understand common patterns of change, to compare and contrast different forms or indicators of change, and to suggest ways to bring about change. Then one selects topics from various subjects that offer good examples of both the theme and the thinking process.

The outline in figure 21.4 summarizes the goals, curriculum modifications, and planning process to integrate curriculum by teaching selected thinking skills or processes involved in an interdisciplinary theme.

The webbing diagram in figure 21.5 shows the topics of lessons by which a fifth-grade class may study change. They think skillfully about causal explanation, prediction, decision making, sequencing, comparing and contrasting, and generating possibilities as they gain insight into the thinking involved in different forms of change.

Themes that are as versatile as change can be meaningfully implemented across the curriculum. However, all academic subjects should not necessarily be included in a single integrated unit. Some themes, such as relationships, responsibility, rules, families, or cities, involve general principles of individual or social behavior and can be developed in social studies, science, and literature. Some themes, such as functions, systems, planning, or proportions, relate to science, mathematics, or information processing. Some themes, such as seasons or animals, are more limited topics that involve only a few subjects but still apply thinking skills and processes effectively.

INTEGRATING CURRICULUM:
THINKING SKILL ➨ SUBJECT AREA ➨ TOPIC

Goals of Integrating Content by Thinking Skill or Process

- Emphasize thinking skills and processes as a curriculum goal.
- Transfer teaching thinking skills and processes systematically and with frequent use.
- Demonstrate curricular and noncurricular examples of thinking skills and processes schoolwide, including displays of the thinking maps and graphic organizers.
- Convey the goals of thinking instruction to parents and the community.
- Demonstrate students' interpersonal and individual behaviors based on skillful thinking.
- Demonstrate the versatility, usefulness, and variations in the skillful use of the thinking skill or process in different contexts.

Instructional Modifications to Integrate the Curriculum by Thinking Skills or Processes

- The school/district desires curriculum integration, but teaching themes is not the primary goal. Teaching thinking is an organizing principle of curriculum integration.
- Schoolwide commitment to infusing thinking instruction into the total curriculum.
- School climate that values thinking and makes it a priority.

Planning Process

- Identify versatile thinking skills and processes.
- Identify significant contexts for teaching thinking skills or processes meaningfully.
- Identify periods in the school year in which most teachers can easily implement their lessons on this skill and plan to offer all lessons on the selected thinking skill at that time.
- Check that all curriculum objectives have been addressed in units based on thinking skills and processes.
- Prepare classroom displays of webbing diagrams showing how the thinking skill or process was implemented in various subjects.
- Rotate the total list of thinking skills and processes on a two-to-three-year cycle.

Figure 21.2

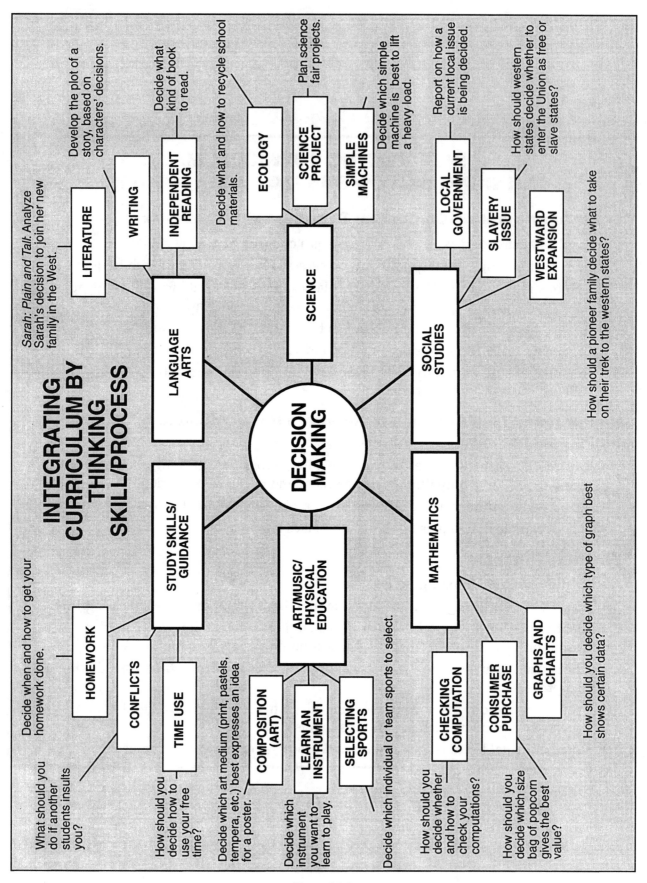

Figure 21.3

INTEGRATING CURRICULUM:
THEME ➡ THINKING SKILL ➡ SUBJECT AREA ➡ TOPIC

Goals of Integrating Content around the Significant Thinking Skills and Processes Involved in Selected Themes

- Enhance students' insight about the theme selected for curriculum integration.
- Implement thinking instruction within an interdisciplinary unit plan.
- Demonstrate and frequently transfer thinking skills and processes.

Instructional Modifications to Integrate Curriculum

- School- and/or districtwide commitment to integrated curriculum, expressed in connecting some, but not necessarily all, curriculum to each interdisciplinary theme.
- Common themes have already been identified in curriculum objectives that serve as organizing principles for scheduling thinking instruction.
- Schoolwide commitment to thinking instruction.

Planning Process

- Identify significant themes and key ideas that students should understand.
- Identify thinking skills and processes that offer insights about these key ideas.
- Examine curriculum objectives for significant contexts in which selected thinking skills or processes can be meaningfully taught.
- Schedule each theme in the school year when it most naturally falls.
- Check that the introductory lesson on each thinking skill or process occurs early enough in the school year to allow sufficient transfer lessons.

Figure 21.4

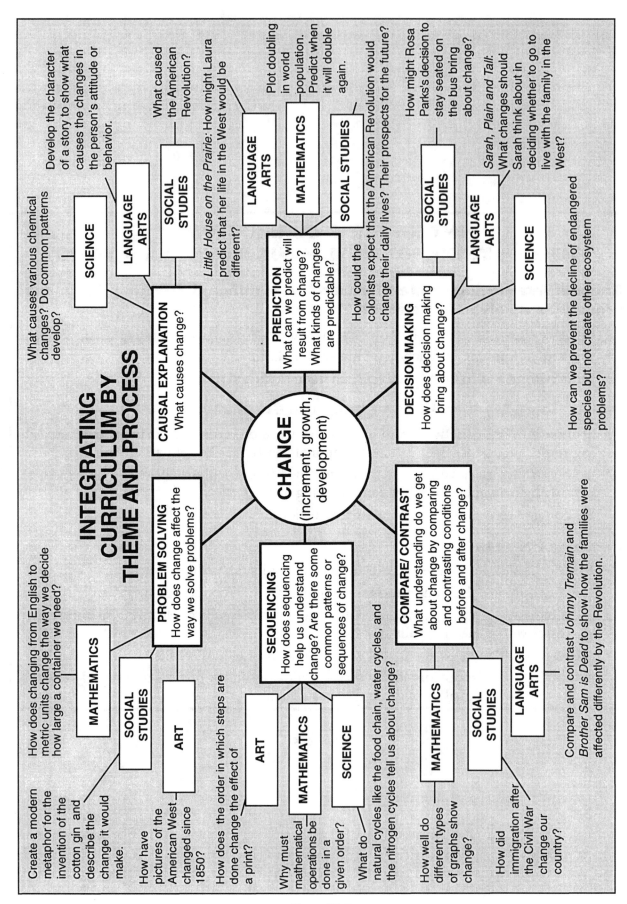

Figure 21.5